The ideas of particle physics

1082

This is the second edition of a book that has already been well received as a clear and readable introduction to the world of particle physics. It bridges the gap between traditional textbooks on the subject and the popular accounts that assume little or no background in the physical sciences on the part of the reader. The first edition has been carefully revised throughout to provide a completely up-to-date and comprehensive overview of our understanding of elementary particle physics. There are also four completely new chapters covering quantum gravity, super-unification, the relationship between particle physics and cosmology, and superstrings.

All the important concepts in our modern understanding of particle physics are covered. Historical developments are also discussed, together with the most important recent experiments, and the theoretical development of the subject is traced from its foundations in relativity and quantum mechanics through to the very latest theories. Although entirely self-contained, some familiarity with simple physics concepts, such as energy, mass, momentum etc., is assumed, allowing a fuller discussion of recent progress in the field.

The book is intended for anyone with a background in the physical sciences who wishes to go deeper into the subject than popular science books allow. It should also be of value to students of physics wishing to gain an introductory overview of the subject before getting down to the details of the formalism.

The Ideas of Particle Physics

AN INTRODUCTION FOR
SCIENTISTS

G. D. Coughlan and J. E. Dodd

SECOND EDITION

CHESTER COLLEGE

ACC. No.
00920952

DEPT.

CLASS No.
539.72 COU

LIBRARY

The right of the
University of Cambridge
to print and sell
all manner of books
was granted by
Henry VIII in 1534.
The University has printed
and published continuously
since 1584.

Cambridge University Press

CAMBRIDGE

NEW YORK PORT CHESTER

MELBOURNE SYDNEY

Published by the Press Syndicate of the University of Cambridge
The Pitt Building, Trumpington Street, Cambridge CB2 1RP
40 West 20th Street, New York, NY 10011-4211, USA
10 Stamford Road, Oakleigh, Melbourne 3166, Australia

© Cambridge University Press, 1984, 1991

First published 1984
Reprinted 1986, 1988
Second edition 1991
Printed in Great Britain at the University Press, Cambridge

British Library cataloguing in publication data

Coughlan, G. D. (Guy D.)
The ideas of particle physics.—2nd. ed.
1. Elementary particles
I. Title   II. Dodd, J. E. (James E.)
539.721

Library of Congress cataloguing in publication data

Coughlan, G. D.
The ideas of particle physics: an introduction for scientists /
G. D. Coughlan and J. E. Dodd.—2nd ed.
    p.   cm.
Rev. ed. of: The ideas of particle physics / J. E. Dodd.
Includes bibliographical references.
ISBN 0–521–38506–7. —ISBN 0–521–38677–2 (pbk)
1. Particles (Nuclear physics)   I. Dodd, J. E. (James Edmund),
1952–   . II. Title.
QC793.2.D6   1991
539.7′2—dc20   90–47884   CIP

ISBN 0 521 38506 7 hardback
ISBN 0 521 38677 2 paperback

Contents

To Orla, Mary and our families

Preface

The last twenty years have seen an enormous advance in our understanding of the microworld. We now have a convincing picture of the fundamental structure of observable matter in terms of certain point-like elementary particles. We also have a comprehensive theory describing the behaviour of and the forces between these elementary particles, which we believe provides a complete and correct description of all non-gravitational physics.

Matter, so it seems, consists of just two types of elementary particles: quarks and leptons. These are the fundamental building blocks of the material world, out of which we ourselves are made. The theory describing the microscopic behaviour of these particles has, over the past decade or so, become known as the 'standard model', providing as it does an accurate account of the force of electromagnetism, the weak nuclear force (responsible for radioactive decay), and the strong nuclear force (which holds atomic nuclei together). The standard model has been remarkably successful; all experimental tests have verified the detailed predictions of the theory.

The standard model is based on the principle of 'gauge symmetry', which asserts that the properties and interactions of elementary particles are governed by certain fundamental symmetries related to familiar conservation laws. Thus, the strong, weak and electromagnetic forces are all 'gauge' forces. They are mediated by the exchange of certain particles, called gauge bosons, which are, for example, responsible for the interaction between two electric charges, and for the nuclear processes taking place within the sun. Unsuccessful attempts have been made to fit the only other known force – gravity – into this gauge framework. However, despite our clear understanding of certain macroscopic aspects of gravity, a microscopic theory of gravity has so far proved elusive.

The above picture of the microworld has emerged slowly since the late 1960s, at which time only the electromagnetic force was well understood. It is the story of the discoveries which have been made since that time to which this book is devoted. The telling of the story is broadly in chronological order, but where appropriate this gives way to a more logical exposition in which complete topics are presented in largely self-contained units. The advances described in Parts 5–8, for example, were made more or less simultaneously, but no attempt is made here to relate an accurate history. Instead, we focus on the logical development of the individual topics and give only the main historical interconnections.

Our main concern in writing this book has been to communicate the central ideas and concepts of elementary particle physics. We have attempted to present a comprehensive overview of the subject at a level which carries the reader beyond the simplifications and generalisations necessary in popular science books. It is aimed principally at graduates in the physical sciences, mathematics, engineering, or other numerate subjects. But we must stress that this is *not* a textbook. It makes no claim whatsoever to the precision and rigour required of a textbook. It contains no mathematical derivations of any kind, and no complicated formulae are written down (other than for the purpose of illustration). Nevertheless, simple mathematical equations are frequently employed to aid in the explanation of a particular idea, and the book does assume a familiarity with basic physical concepts (such as mass, momentum, energy, etc.).

This book is organised in nine parts each consisting of four or five short chapters. However, Part 9 is more substantial. Dealing with the most exciting of current research topics, it consists of seven chapters which are rather longer than average and which will require more time and concentration on the part of the reader. We draw the reader's attention to the Glossary (Appendix 2), which gives concise definitions of the most important of particle physics nomenclature. It should prove useful as a memory prompt, as well as a source of supplementary information.

The story begins in Part 0 at the turn of the century when physicists were first beginning to glimpse the remarkable nature of ordinary matter. Out of this period came the two elements essential for the understanding of the microworld: the theories of special relativity and quantum mechanics. These are the unshakeable foundations upon which the rest of the story is based.

Part 1 introduces the four known fundamental forces, and is followed by a more detailed discussion of the physics of the strong and weak (nuclear) forces in Parts 2–4. It was the desire to understand the weak force, in particular, which led eventually to recognition of the role of gauge symmetry as a vital ingredient in theories of the microworld. Gauge theory is the subject of Part 5, which introduces the Glashow–Weinberg–Salam theory of the electromagnetic and weak forces. This theory, often called the 'electroweak model', has been spectacularly verified in many experiments over the past two decades. The most impressive of these was the discovery at CERN in 1983 of the massive $W^{\pm}$ and Z^0 gauge bosons which mediate the weak force.

At about the same time as the electroweak model was being developed, physicists were using 'deep inelastic scattering' experiments to probe the interior of the proton. These experiments, which are described in Part 6, provided the first indication that the proton was not truly elementary, but composed of point-like objects (called quarks). As the physical reality of quarks gained wider acceptance, a new gauge theory was formulated in an attempt to explain the strong forces between them. This theory is called 'quantum chromodynamics' and attributes the strong force to the exchange of certain gauge bosons called gluons. It is described in Part 7. Together, quantum chromodynamics and the Glashow–Weinberg–Salam electroweak theory constitute the 'standard model' of elementary particle physics.

Part 8 completes our discussion of the standard model with a description of an important class of experiments involving collisions between electrons and positrons. These experiments have, over two decades, been instrumental in confirming the physical reality of quarks and in testing many of the predictions of quantum chromodynamics and the electroweak theory.

No sooner had the standard model been proposed and received modest experimental support, than particle theorists began to explore new ideas. Grand unified theories were put forward as an attempt to unite the elements of the standard model in a single formulation. Superunified theories made the first attempts to include gravity. More recently, superstrings have been proposed as a 'theory of everything'. These are some of the current research topics discussed in Part 9.

During the preparation of this second edition we have benefited greatly from discussions with Drs Graham Ross, Tim Hollowood, Jonathan Evans, Tien Kieu and Paul Tod, who join all those similarly acknowledged in the first edition. Their many comments, suggestions and penetrating questions are greatly appreciated. In addition, we are indebted to Dr Robert Taylor for patiently proof-reading the entire manuscript and suggesting numerous improvements and clarifications to the text.

Guy Coughlan
James Dodd
Oxford

Part 0
Introduction

1

Matter and light

1.1 Introduction

The physical world we see around us has two main components, matter and light, and it is the modern explanation of these things which is the purpose of this book. During the course of the story, these concerns will be restated in terms of material particles and the forces which act between them, and we will most assuredly encounter new and exotic forms of both particles and forces. But in case we become distracted and confused by the elaborate and almost wholly alien contents of the microworld, let us remember that the origin of the story, and the motivation for all that follows, is the explanation of everyday matter and visible light.

Beginning as it does, with a laudable sense of history, at the turn of the century, the story is one of twentieth-century achievement. For the background, we have only to appreciate the level of understanding of matter and light around 1900, and some of the problems in this understanding, to prepare ourselves for the story of progress which follows.

1.2 The nature of matter

By 1900 most scientists were convinced that all matter is made up of a number of different sorts of atoms, as had been conjectured by the ancient Greeks millennia before and as had been indicated by chemistry experiments over the preceding two centuries. In the atomic picture, the different types of substance can be seen as arising from different arrangements of the atoms. In solids, the atoms are relatively immobile and in the case of crystals are arranged in set patterns of impressive precision. In liquids they roll loosely over one another and in gases they are widely separated and fly about at a velocity depending on the temperature of the gas; see Figure 1.1. The application of heat to a substance can cause phase transitions in which the atoms change their mode of behaviour as the heat energy is transferred into the kinetic energy of the atoms' motions.

Many familiar substances consist not of single atoms, but of definite combinations of certain atoms called molecules. In such cases it is these molecules which behave in the manner appropriate to the type of substance concerned. For instance, water consists of molecules, each made up of two hydrogen atoms and one oxygen atom. It is the molecules which are subject to a specific static arrangement in solid ice, the molecules which roll over each other in water and the molecules which fly about in steam.

Fig. 1.1. (*a*) Static atoms arranged in a crystal. (*b*) Atoms rolling around in a liquid. (*c*) Atoms flying about in a gas.

(*a*)

(*b*)

(*c*)

The laws of chemistry, most of which were discovered empirically between 1700 and 1900, contain many deductions concerning the behaviour of atoms and molecules. At the risk of brutal over-simplification the most important of these can be summarised as follows:

(1) Atoms can combine to form molecules, as indicated by chemical elements combining only in certain proportions (Richter and Dalton).

(2) At a given temperature and pressure, equal volumes of gas contain equal numbers of molecules (Avogadro).

(3) The relative weights of the atoms are approximately multiples of the weight of the hydrogen atom (Prout).

(4) The mass of each atom is associated with a specific quantity of electrical charge (Faraday and Webber).

(5) The elements can be arranged in families having common chemical properties but different atomic weights (Mendeleeff's periodic table).

(6) An atom is approximately 10^{-10} m across, as implied by the internal friction of a gas (Loschmidt).

One of the philosophical motivations behind the atomic theory (a motivation we shall see repeated later) was the desire to explain the diversity of matter by assuming the existence of just a few fundamental and indivisible atoms. But by 1900 over 90 varieties of atoms were known, an uncomfortably large number for a supposedly fundamental entity. Also, there was evidence for the disintegration (divisibility) of atoms. At this breakdown of the 'ancient' atomic theory, modern physics begins.

1.3　Atomic radiations

1.3.1　*Electrons*

In the late 1890s, J. J. Thomson of the Cavendish Laboratory at Cambridge was conducting experiments to examine the behaviour of gas in a glass tube when an electric field was applied across it. He came to the conclusion that the tube contained a cloud of minute particles with negative electrical charge – the electrons. As the tube had been filled only with ordinary gas atoms, Thomson was forced to conclude that the electrons had originated within the supposedly indivisible atoms. As the atom as a whole is electrically neutral, on the release of a negatively charged electron the remaining part, the ion, must carry the equal and opposite positive charge. This was entirely in accord with the long-known results of Faraday's electrolysis experiments, which required a specific electrical charge to be associated with the atomic mass.

By 1897, Thomson had measured the ratio of the charge to the mass of the electron (denoted *e/m*) by observing its behaviour in magnetic fields. By comparing this number with that of the ion, he was able to conclude that the electron is thousands of times less massive than the atom (and some 1837 times lighter than the lightest atom, hydrogen). This led Thomson to propose his 'plum-pudding' picture of the atom, in which the small negatively charged electrons were thought to be dotted in the massive, positively charged body of the atom (see Figure 1.2).

1.3.2 *X-rays*

Two years earlier in 1895, the German Wilhelm Röntgen had discovered a new form of penetrating radiation, which he called X-rays. This radiation was emitted when a stream of fast electrons (which had not yet been identified as such) struck solid matter and were thus rapidly decelerated. This was achieved by boiling the electrons out of a metallic electrode in a vacuum tube and accelerating them into another electrode by applying an electric field across the two, as in Figure 1.3. Very soon the X-rays were identified as another form of electromagnetic radiation, i.e. radiation that is basically the same as visible light, but with a much higher frequency and shorter wavelength. An impressive demonstration of the wave nature of X-rays was provided in 1912 when the German physicist Max von Laue shone them through a crystal structure. In doing so, he noticed the regular geometrical patterns characteristic of the diffraction which occurs when a wave passes through a regular structure whose characteristic size is comparable to the wavelength of the wave. In this case, the regular spacing of atoms within the crystal is about the same as the wavelength of the X-rays. Although these X-rays do not originate from within the structure of matter, we shall see next how they are the close relatives of radiations which do.

1.3.3 *Radioactivity*

At about the same time as the work taking place on electrons and X-rays, the French physicist Becquerel was conducting experiments on the heavy elements. During his study of uranium salts in 1896, Becquerel noticed the emission of radiation rather like that which Röntgen had discovered. But Becquerel was doing nothing to his

Fig. 1.3. The production of X-rays by colliding fast electrons with matter.

Fig. 1.2. Thomson's 'plum-pudding' picture of the atom.

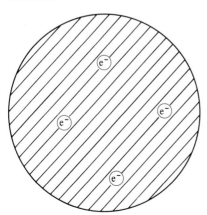

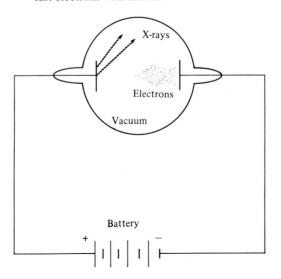

uranium: the radiation was emerging spontaneously. Inspired by this discovery, Pierre and
Marie Curie began investigating the new radiation. By 1898, the Curies had discovered that the
element radium also emits copious amounts of
radiation.

These early experimenters first discovered
the radiation through its darkening effect on photographic plates. However, other methods for
detecting radiation were soon developed, including scintillation techniques, electroscopes and a
primitive version of the Geiger counter. Then a
great breakthrough came in 1912 when C. T. R.
Wilson of the Cavendish Laboratory invented the
cloud chamber. This device encourages easily visible water droplets to form around the atoms,
which have been ionised (i.e. have had an electron
removed) by the passage of the radiation through
air. This provides a plan view of the path of the
radiation and so gives us a clear picture of what is
happening. (Sophisticated variants of the device
are still in use today as the massive bubble
chambers in high-energy accelerator experiments.)

If a radioactive source such as radium is
brought close to the cloud chamber, the emitted
radiation will trace paths in the chamber. When a

Fig. 1.4. Three components of radioactivity
displayed in a cloud chamber. ⊙ signifies that
the direction of the applied magnetic field is
perpendicular to, and out of the plane of, the
paper.

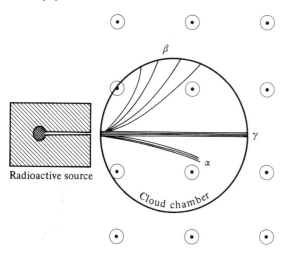

magnetic field is placed across the chamber, then
the radiation paths will separate into three components which are characteristic of the type of
radiation (see Figure 1.4). The first component of
radiation (denoted α) is bent slightly by the magnetic field, which indicates that the radiation carries electric charge. Measuring the radius of curvature of the path in a given magnetic field can tell us
that it is made up of massive particles with two
positive electric charges. These particles can be
identified as the nuclei of helium atoms, often
referred to as α particles. Furthermore, these α
particles always seem to travel a fixed distance
before being stopped by collisions with the air
molecules. This suggests that they are liberated
from the source with a constant amount of energy
and that the same internal reactions within the
source atoms are responsible for all α particles.

The second component of the radiation
(denoted γ) is not at all affected by the magnetic
field, showing that it carries no electric charge, and
it is not stopped by collisions with the air molecules. These γ-rays were soon identified as the
close relatives of Röntgen's X-rays but with even
higher frequencies and even shorter wavelengths.
The γ-rays can penetrate many centimetres of lead
before being absorbed. They are the products of
reactions occurring spontaneously within the
source atoms, which liberate large amounts of
electromagnetic energy but no material particles,
indicating a different sort of reaction to that
responsible for α-rays.

The third component (denoted β radiation)
is bent significantly in the magnetic field in the
opposite direction to the α-rays. This is interpreted
as single, negative electrical charges with much
lesser mass than the α-rays. They were soon identified as the same electrons as those discovered by J.
J. Thomson, being emitted from the source atoms
with a range of different energies. The reactions
responsible form a third class distinct from the
origins of α- or γ-rays.

The three varieties of radioactivity have a
double importance in our story. Firstly, they result
from the three main fundamental forces of nature
effective within atoms. Thus the phenomenon of
radioactivity may be seen as the cradle for all of
what follows. Secondly, and more practically, it
was the products of radioactivity which first

allowed physicists to explore the interior of atoms and which later indicated totally novel forms of matter, as we shall see in due course.

1.4 Rutherford's atom

In the first decade of the twentieth century, Rutherford had pioneered the use of naturally occurring atomic radiations as probes of the internal structure of atoms. In 1909, at Manchester University, he suggested to his colleagues, Geiger and Marsden, that they allow the α particles emitted from a radioactive element to pass through a thin gold foil and observe the deflection of the outgoing α particles from their original paths (see Figure 1.5). On the basis of Thomson's 'plum-pudding' model of the atom, they should experience only slight deflections, as nowhere in the uniformly occupied body of the atom would the electric field be enormously high. But the experimenters were surprised to find that the heavy α particles were sometimes drastically deflected, occasionally bouncing right back towards the source. In a dramatic analogy attributed (somewhat dubiously) to Rutherford: 'It was almost as

incredible as if you fired a 15-inch shell at a piece of tissue paper and it came back and hit you!'

The implication of this observation is that a very strong repulsive force must be at work within the atom. This force cannot be due to the electrons as they are over 7000 times lighter than the α particles and so can exert only minute effects on the α-particle trajectories. The only satisfactory explanation of the experiment is that all the positive electric charge in the atom is concentrated in a small nucleus at the middle, with the electrons orbiting the nucleus at some distance. By assuming that the entire positive charge of the atom is concentrated with the atomic mass in a small central nucleus, Rutherford was able to derive his famous scattering formula which describes the relative numbers of α particles scattered through given angles on colliding with an atom (see Figure 1.5).

Rutherford's picture of the orbital atom is in contrast with our perception of apparently 'solid' matter. From the experiments he was able to deduce that the atomic nucleus, which contains 99.9% of the mass of the atom, has a diameter of about 10^{-15} m compared to an atomic diameter of about 10^{-10} m. For illustration, if we took a cricket ball to act as the nucleus, the atomic electrons would be ping-pong balls 5 km distant! Such an analogy brings home forcibly just how sparse apparently solid matter is and just how dense is the nucleus itself. But despite this clear picture of the atom, indicated from the experiment, explaining how it works is fraught with difficulties, as we shall see in Chapter 3.

1.5 Two problems

Just as these early atomic experiments revealed an unexpected richness in the structure of matter, so too, theoretical problems forced upon physicists more-sophisticated descriptions of the natural world. The theories of special relatively and quantum mechanics arose as physicists realised that the classical physics of mechanics, thermodynamics and electromagnetism were inadequate to account for apparent mysteries in the behaviour of matter and light. Historically, the mysteries were contained in two problems, both under active investigation at the turn of the century.

Fig. 1.5. The Geiger and Marsden experiment. According to Rutherford's scattering formula, the number of α particles scattered through a given angle decreases as the angle increases away from the forward direction.

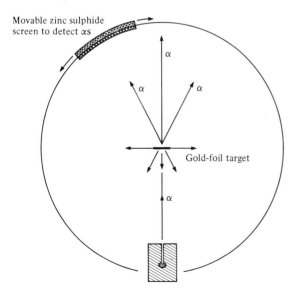

1.5.1 *The constancy of the speed of light*

Despite many attempts to detect an effect, no variation was discovered in the speed of light. Light emerging from a torch at rest seems to travel forward at the same speed as light from a torch travelling at arbitrarily high speeds. This is very different from the way we perceive the behaviour of velocities in the everyday world. But, of course, we humans never perceive the velocity of light, it is just too fast! This unexpected behaviour is not contrary to common experience, it is beyond it! Explanation for the behaviour forms the starting point for the theory of special relativity, which is the necessary description of anything moving very fast (i.e. nearly all elementary particles); see Chapter 2.

1.5.2 *The interaction of light with matter*

All light, for instance sunlight, is a form of heat and so the description of the emission and absorption of radiation by matter was approached as a thermodynamical problem. In 1900 the German physicist Max Planck concluded that the classical thermodynamical theory was inadequate to describe the process correctly. The classical theory seemed to imply that if light of any one colour (any one wavelength) could be emitted from matter in a continuous range of energy down to zero, then the total amount of energy radiated by the matter would be infinite. Much against his inclination, Planck was forced to conclude that light of any given colour cannot be emitted in a continuous band of energy down to zero, but only in multiples of a fundamental quantum of energy, representing the minimum negotiable bundle of energy at any particular wavelength. This is the starting point of quantum mechanics, which is the necessary description of anything very small (i.e. all atoms and elementary particles); see Chapter 3.

As the elementary particles are both fast moving and small, it follows that their description must incorporate the rules of both special relativity and quantum mechanics. The synthesis of the two is known as relativistic quantum theory and this is described briefly in Chapter 4.

2

Special relativity

2.1 Introduction

A principle of relativity is simply a statement reconciling the points of view of observers who may be in different physical situations. Classical physics relies on the Galilean principle of relativity, which is perfectly adequate to reconcile the points of view of human observers in everyday situations. But modern physics requires the adoption of Einstein's special theory of relativity, as it is this theory which is known to account for the behaviour of physical laws when very high velocities are involved (typically those at or near the speed of light, denoted by c).

It is an astonishing tribute to Einstein's genius that he was able to infer the special theory of relativity in the almost total absence of the experimental evidence which is now commonplace. He was able to construct the theory from the most tenuous scraps of evidence.

To us lesser mortals, it is challenge enough to force ourselves to think in terms of special relativity when envisaging the behaviour of the elementary particles, especially as all our direct experience is of 'normal' Galilean relativity. What follows is of course only a thumbnail sketch of relativity. Many excellent accounts have been written on the subject, not least of which is that written by Einstein himself.

2.2 Galilean relativity

Any theory of 'relativity' is about the relationships between different sets of coordinates against which physical events can be measured. Coordinates are numbers which specify the position of a point in space (and in time). However, for these numbers to have any meaning, we must also specify the particular coordinate system (or frame of reference) they refer to. For example, we might choose the origin of our coordinates to be the Royal Greenwich Observatory, and choose to specify coordinates in terms of the distance east of the observatory, the distance north, and the height. Hence, the choice of a coordinate system involves specifying (1) an origin from which to measure coordinates (e.g. the observatory), and (2) three independent directions (e.g. east, north and up). So, relative to any chosen coordinate system, the position of a point in space is specified in terms of three independent coordinates, which we may write as (x, y, z). These three coordinates can be denoted collectively as a vector, $\mathbf{x} = (x, y, z)$. A further coordinate, t, is required to specify time.

Galileo's simple example is still one of the clearest descriptions of what relativity is all about. If a man drops a stone from the mast of a ship, he will see it fall in a straight line and hit the deck below, having experienced a constant acceleration

Fig. 2.1. The transformations of Galilean relativity.

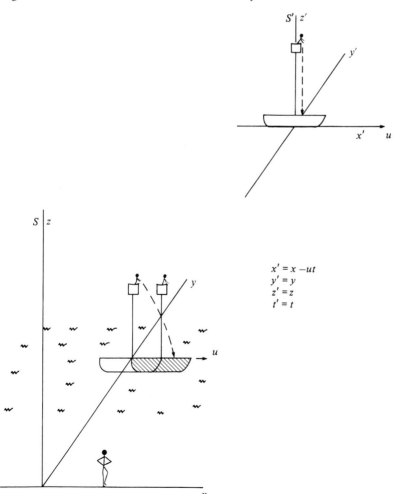

$$x' = x - ut$$
$$y' = y$$
$$z' = z$$
$$t' = t$$

due to the force of gravity. Another man standing on the shore and watching the ship sail past will see the stone trace out a parabolic path, because, at the moment of release, it is already moving with the horizontal velocity of the ship. Both the sailor and the shoreman can write down their views of the stone's motion using the mathematical equations for a straight line and a parabola respectively. As both sets of equations are describing the same event (the same force acting on the same stone), they are related by transformations between the two observers. These transformations relate the measurements of position ($\mathbf{x}'$), time (t'), and velocity ($\mathbf{v}'$) in the sailor's coordinate system S', with the corresponding measurements ($\mathbf{x}$, t, $\mathbf{v}$) made by the shoreman in his coordinate system S. This situation, assuming that the ship is sailing along the x-axis with velocity u, is shown in Figure 2.1.

Important features of the Galilean transformations are that velocity transformations are additive and that time is invariant between the two coordinate frames. Thus if a sailor throws the stone forward at 10 m per second in a ship travelling forward at 10 m per second, the speed of the stone to a stationary observer on shore will be 20 m per second. And if the sailor on a round trip measures the voyage as one hour long, this will be the same duration as observed by the stationary shoreman.

Lest the reader be surprised by the triviality of such remarks, let him or her be warned that this is not the case in special relativity. At the high velocities, such as are common in the microworld, velocities do not simply add to give the relative velocity, and time is not an invariant quantity. But before we address these sophistications, let us see how the idea came about.

2.3   The origins of special relativity

The fact that Galilean transformations allow us to relate observations made in different coordinate frames implies that any one inertial frame (a frame at rest or moving at constant velocity) is as good as another for describing the laws of physics. Nineteenth-century physicists were happy that this should apply to mechanical phenomena, but were less happy to allow the same freedom to apply to electromagnetic phenomena, and especially to the propagation of light.

The manifestation of light as a wave phenomenon (as demonstrated in the diffraction and interference experiments of optics) encouraged physicists to believe in the existence of a medium called the ether through which the waves might propagate (believing that any wave was necessarily due to the perturbation of some medium from its equilibrium state). The existence of such an ether would imply a preferred inertial frame, namely, the one at rest relative to the ether. In all other inertial frames moving with constant velocity relative to the ether, measurement and formulation of physical laws (say the force of gravitation) would mix both the effect under study and the effect of motion relative to the ether (say some sort of viscous drag). The laws of physics would appear different in different inertial frames, due to the different effects of the interaction with the ether. Only the preferred frame would reveal the true nature of the physical law.

The existence of the ether and the law of the addition of velocities suggested that it should be

Fig. 2.2. Anticipated variation in the propagation of light reflected to and fro along a distance L due to the earth's motion through space v_E.

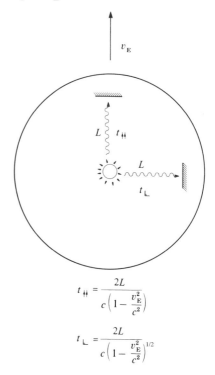

$$t_{\mathbin{\|}} = \frac{2L}{c\left(1 - \dfrac{v_E^2}{c^2}\right)}$$

$$t_L = \frac{2L}{c\left(1 - \dfrac{v_E^2}{c^2}\right)^{1/2}}$$

possible to detect some variation of the speed of light as emitted by some terrestrial source. As the earth travels through space at 30 km per second in an approximately circular orbit, it is bound to have some relative velocity with respect to the ether. Consequently, if this relative velocity is simply added to that of the light emitted from the source (as in the Galilean transformations), then light emitted simultaneously in two perpendicular directions should be travelling at different speeds, corresponding to the two relative velocities of the light with respect to the ether (see Figure 2.2).

In one of the most famous experiments in physics, the American physicists Michelson and Morley set out in 1887 to detect this variation in the velocity of propagation of light. The anticipated variation was well within the sensitivity of their measuring apparatus, but absolutely none was found. This experiment provided clear proof that no such ether exists and that the speed of light is a constant regardless of the motion of the source.

2.4 The Lorentz–Fitzgerald contraction

Around the turn of the century, many physicists were attempting to explain the null result of the Michelson and Morley experiment. The Dutch physicist Lorentz and the Irish physicist Fitzgerald realised that it could be explained by assuming that intervals of length and time, when measured in a given frame, appear contracted when compared with the same measurements taken in another frame by a factor dependent on the relative velocity between the two. Their arguments were simply that the anticipated variations in the speed of light were cancelled by compensating changes in the distance and time which the light travelled, thus giving rise to the apparent constancy observed. It is possible to calculate geometrically that an interval of length x measured in one frame is found to be x' when measured in a second frame travelling at velocity v relative to the first where:

$$x = \frac{x'}{\left(1 - \frac{v^2}{c^2}\right)^{1/2}}. \tag{2.1}$$

Here, c is the speed of light, which is approximately equal to 2.998×10^8 metres per second. And, similarly, the intervals of time observed in the two frames are related by:

$$t = \frac{t'}{\left(1 - \frac{v^2}{c^2}\right)^{1/2}}. \tag{2.2}$$

These empirical relationships, proposed on an *ad hoc* basis by Lorentz and Fitzgerald, suggest that because the 'common-sense' Galilean law of velocity addition fails at speeds at or near that of light, our common-sense perceptions of the behaviour of space and time must also fail in that regime. It was Einstein who, quite independently, raised these conclusions and relationships to the status of a theory.

2.5 The special theory of relativity

The special theory of relativity is founded on Einstein's perception of two fundamental physical truths which he put forward as the basis of his theory:

(1) All inertial frames (i.e. those moving at a constant velocity relative to one another) are equivalent for the observation and formulation of physical laws.
(2) The speed of light in a vacuum is constant.

The first of these is simply the extension of the ideas of Galilean relativity to include the propagation of light, and the denial of the existence of the speculated ether. With our privileged hindsight, the amazing fact of history must be that the nineteenth-century physicist preferred to cling to the idea of relativity for mechanical phenomena whilst rejecting it in favour of the concept of a preferred frame (the ether) for the propagation of light. Einstein's contribution here was to extend the idea of relativity to include electromagnetic phenomena, given that all attempts to detect the ether had failed.

The second principle is the statement of the far-from-obvious physical reality that the speed of light is truly independent of the motion of the source and so is totally alien to our everyday conceptions. Einstein's achievement here was to embrace this apparently ludicrous result with no qualms. Thus the theory of relativity, which has had such a revolutionary effect on modern thought is, in fact, based on the most conservative assumptions compatible with experimental results.

Given the equivalence of all inertial frames for the formulation of physical laws and this bewildering constancy of the speed of light in all frames, it is understandable intuitively that measurements of space and time must vary between frames to maintain this absolute value for the speed of light. The relationships between measurements of space, time and velocity in different frames are related by mathematical transformations, just as were measurements in Galilean relativity, but the transformations of special relativity also contain the Lorentz–Fitzgerald contraction factors to account for the constancy of the speed of light (see Figure 2.3).

The first feature of the transformations to note is that when the relative velocity between frames is small compared with that of light (i.e. all velocities commonly experienced by humans), then $u/c \approx 0$, and the transformations reduce to the common-sense relations of Galilean relativity.

The unfamiliar effects of special relativity contained in the transformations can be illustrated by a futuristic example of Galileo's mariner: an astronaut in a starship travelling close to the speed of light (c).

Because of the transformations, velocities no

Fig. 2.3. The Lorentz transformations of special relativity.

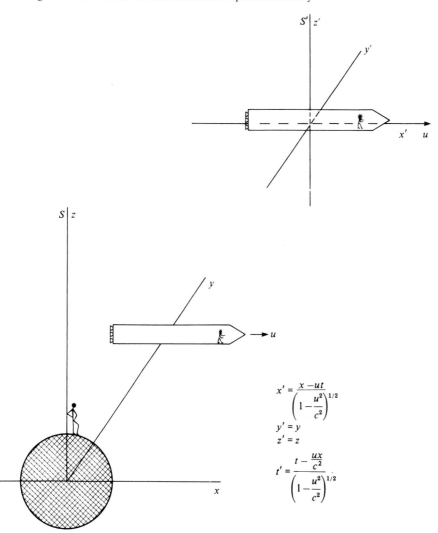

$$x' = \frac{x - ut}{\left(1 - \dfrac{u^2}{c^2}\right)^{1/2}}$$

$$y' = y$$

$$z' = z$$

$$t' = \frac{t - \dfrac{ux}{c^2}}{\left(1 - \dfrac{u^2}{c^2}\right)^{1/2}}$$

longer simply add. If, say, the astronaut fires photon torpedoes forward at speed $1c$ from the starship, which itself may be travelling at $0.95c$, the total velocities of the photon torpedoes as observed by a stationary planetary observer is not the sum, $1.95c$, but is still c, the constant speed of light. Also, time is dilated. So a voyage which to the stationary observer is measured as a given length of time will appear less to the kinetic astronaut.

Another intriguing feature of the transformations is that continued combinations of arbitrary velocities less than c can never be made to exceed c. Thus the transformations imply that continued attempts to add to a particle's velocity (by successive accelerations) can never break the light barrier. Indeed, the transformations themselves do not cater for velocities greater than c, as when $u > c$ the equations become imaginary, indicating a departure from the physical world. Special relativity therefore implies the existence of an ultimate limiting velocity beyond which nothing can be accelerated.

2.6 Mass momentum and energy

If the transformation laws of special relativity show diminishing returns on any attempts to accelerate a particle (by application of some force), it is reasonable to expect some compensating factor to bring returns in some other way, and so maintain energy conservation. This compensating factor is the famous increase in the mass of a particle as it is accelerated to speeds approaching c.

By requiring the laws of conservation of mass and conservation of momentum to be invariant under the Lorentz transformations, it is possible to derive the relationship between the mass of a body m and its speed v,

$$m = \frac{m_0}{\left(1 - \frac{v^2}{c^2}\right)^{1/2}}, \qquad (2.3)$$

where m_0 is the mass of the body in a frame in which it is at rest. Multiplying the equation by c^2 and expanding the bracket we obtain:

$$mc^2 = m_0 c^2 + \frac{m_0 v^2}{2} + \cdots. \qquad (2.4)$$

We can identify the second term on the right-hand side of the equation as the classical kinetic energy of the particle. The subsequent terms are the relativistic corrections to the energy whilst the first is describing a quantity of energy arising only from the mass itself.

This is the origin of the mass–energy equivalence of special relativity expressed in the most famous formula of all time:

$$E = mc^2. \qquad (2.5)$$

From this formula several others follow immediately. One can be obtained by substituting an expression for the momentum ($\mathbf{p}$) into the expansion for m in the above:

$$\mathbf{p} = m\mathbf{v},$$

so

$$E^2 = m_0^2 c^4 + p^2 c^2. \qquad (2.6)$$

For a particle with no rest mass, such as the photon, this gives:

$$\frac{E}{p} = c. \qquad (2.7)$$

2.7 The physical effects of special relativity

The effects which we have just introduced are all wholly unfamiliar to human experience and this is perhaps one reason why, even today, the reality of special relativity is repeatedly challenged by sceptical disbelievers (see Figure 2.4). But all the effects are real and they can all be measured.

A roll call of the effects of special relativity provides a useful checklist which we should remember when envisaging the behaviour of elementary particles.

2.7.1 *The ultimate speed c*

It is possible to measure directly the velocity of electrons travelling between two electrodes by measuring the time of flight taken. It is observed that the speed does not increase with the energy which the electrons have been given as it would under classical Newtonian theory, but instead tends to a constant value given by c.

2.7.2 *Addition of velocities*

Under special relativity, only when individual velocities are much smaller than c can they be simply added to give the relative velocities. At speeds approaching c, velocities do not add, but combine in a more complicated way so that the total of any combination is always less than c. This can be tested directly by an elementary particle reaction. One kind of elementary particle we shall encounter is the neutral pion π^0 which often decays into a pair of photons. If the pion is travelling say at $0.99c$ when it emits a photon, we would expect the photon to have a total velocity of $1.99c$ under the laws of Galilean relativity. This is not observed. The photon velocity is measured to be c, showing that very high velocities do not add, but combine according to the formula:

$$v_{\text{total}} = \frac{v_1 + v_2}{\left(1 + \dfrac{v_1 v_2}{c^2}\right)}.$$

Fig. 2.4. Special relativity in trouble? An advertisement from *New Scientist* magazine, 27 May 1982.

DELTA PUBLICATIONS

7305, Aram Nagar, New Delhi-110055, INDIA

A BIG HOWLER

EINSTEIN'S THEORY OF SPECIAL RELATIVITY

Dr S. P. Gulati & Dr (Mrs) S. Gulati, Associate Professors, Cuttington University College, Liberia; January 1982; 106 pages; Price US$12·50, STG·6·25 (Air Parcel Postage free).

This book is an open challenging invitation to 'Einsteinians'—particularly so to persons like Professor A. I. Miller of USA who in his recent book 'Albert Einstein's Theory of Special Relativity' (Addison-Wesley, 1981) has undertaken to apotheosize Einstein whose work if not an act of straight plagiarism is definitely 'A BIG HOWLER': infested with infidelities. The 'Transformation Maze' is another interesting feature of the book. Besides, it also contains outlines of the authors' 'SIMILARITY THEORY', perhaps, the only valid alternative.

The book is obtainable either directly from the publisher or through your book-seller or the authors.

2.7.3 *Time dilation*

This is the effect which causes moving clocks to run slow and it has been measured directly in an experiment involving another type of elementary particle. The experiment looks at a species of elementary particle called the muon, which is produced in the upper atmosphere by the interactions of cosmic rays from outer space. The muon decays into other particles with a distribution of lifetimes around the mean value of 2.2×10^{-6} s when measured at rest in the laboratory. By measuring the number of muons incident on a mountain top, it is possible to predict the number which should penetrate to sea level before decaying. In fact, many times the naive prediction are found at sea level, indicating that the moving particles have experienced less time than if they were stationary. Muons moving at, say, $0.99c$ seem to keep time at only one-quarter the rate when stationary with respect to us.

2.7.4 *Relativistic mass increase*

The last effect we shall illustrate is the well-known increase in the apparent mass of a body as its velocity increases. This has been measured directly by observing the electric and magnetic deflections of electrons of varying energies (see Figure 2.5).

Fig. 2.5. Relativistic mass increase as a function of velocity.

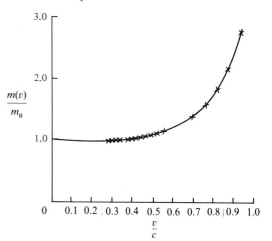

2.8 Using relativity

As we have seen, relativity tells us how to relate the formulations of physical laws in different frames of reference, but it does not tell us how to formulate them in the first place. This is the rest of physics! In this pursuit, special relativity is introduced by adopting kinematical prescriptions which the dynamical variables must obey.

2.8.1 *Space–time diagrams*

In classical relativity, space and time are entirely separate, but in special relativity they are treated on an identical basis and are mixed together in the Lorentz transformations. So specifying only spatial dimensions in one frame will lead to specification in both space and time dimensions in another. Thus it makes little sense to visualise events as occurring only in space. A better context in which to visualise them is space–time. Space–time diagrams can be used to display events at the expense, for the purposes of visualisation, of making do with only one, or possibly two, spatial dimensions (see Figure 2.6). A point in space–time is frequently called an *event*.

2.8.2 *Four-vectors*

Just as ordinary vectors $(\mathbf{x} = (x, y, z))$ (three-vectors) define the components of a position of velocity in ordinary space, we can define four-vectors $(\mathbf{x}, ct) = (x, y, z, ct)$ to define an event in space–time. The fourth coordinate is the time coordinate multiplied by c to give an equivalent

distance, so matching the other three distance components.

The benefit of writing equations in three-vector form is to ensure their covariance under spatial rotations. (Covariance is not quite the same as invariance, which means that absolutely nothing changes. Covariance means that both sides of an equation change in the same way, preserving the validity of the equation.) This permits freedom in the orientation of the coordinate system employed and also ensures the conservation of angular momentum (see Chapter 6). If we can write the laws of physics in four-vector form, then the benefit is that the laws will be covariant under rotations in space–time (which are equivalent to the Lorentz transformations of special relativity).

In addition to the position three-vector $(\mathbf{x})$, the momentum of a particle is also a vector quantity $(\mathbf{p})$. By examining the effects of the Lorentz transformations on the momentum and the energy of a particle, it is possible to form a four-vector from the quantities $(\mathbf{p}, E/c)$. This four-vector is used to specify the dynamic state of a particle. It does not specify an event in space–time.

2.8.3 *Relativistic invariants*

Although special relativity illustrates how perceptions of space and time may vary according to the observer's frame, it also accommodates absolutely invariant quantities, which we might expect to vary under Galilean relativity. The speed of light in a vacuum is the obvious invariant upon which the theory is founded. Another quantity is the square of the space–time interval between an event and the origin of the coordinate system,

$$s^2 = x^2 - (ct)^2 = x'^2 - (ct')^2. \tag{2.8}$$

This is just a special case of the square of the space–time interval between two events, which is the difference between their four-vectors,

$$\Delta s^2 = (\Delta x)^2 - (c\Delta t)^2.$$

Another invariant quantity is the rest mass of a given material particle. All observers will agree on the mass of the same particle at rest in their respective frames:

$$m_0^2 = \frac{E^2}{c^4} - \frac{p^2}{c^2}. \tag{2.9}$$

Fig. 2.6. A space–time diagram particle collision (*b*) shown sequentially in (*a*).

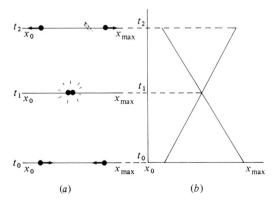

(*a*)   (*b*)

Relativistic invariants are useful in high-energy physics because, once measured, their value will be known in all other circumstances. Here it is worth appreciating that high-energy experiments regularly exercise the idea of Lorentz transformations. One experiment may arrange for two protons travelling with equal energies in opposite directions to collide head-on, whilst another experiment may collide moving protons and a stationary target. The centres of mass of the two experiments will be moving relative to one another with some velocity which is likely to be an appreciable fraction of the speed of light. This will require the Lorentz transformations to relate measurements in the two experiments.

This concludes our brief sketch of special relativity and now we pass on to the second of the two great pillars of twentieth-century physics: quantum mechanics.

3

Quantum mechanics

3.1 Introduction

It is fascinating to reflect on the fact that both quantum mechanics and special relativity were conjured into being in the first five years of this century, and interesting to compare the development of the two. Whereas special relativity sprang as a complete theory (1905) from Einstein's genius, quantum mechanics emerged in a series of steps over a quarter of a century (1900–25). One explanation of this is that, whereas in special relativity the behaviour of space and time follow uniquely from the two principles, in quantum theory there were no such simple principles which, known at the beginning, allowed the derivation of all quantum phenomena. Rather, each of the steps was a fresh hypothesis based on, or predicting, some new experimental facts and these do not necessarily follow logically one from another, still less from just one or two fundamental principles. So quantum mechanics emerged, hypothesis hand-in-hand with experiment, over twenty-five years or so. As indicated by the sub-headings of this chapter, most of the steps in the progression can be associated closely with just one man, and we will use the examination of each of these in turn as our introduction to quantum mechanics.

3.2 Planck's hypothesis

As we have mentioned in Section 1.5.2, quantum theory came into being when Max Planck attempted to explain the interaction of light with matter. That is, for instance, how hot metal emits light and how light is absorbed by matter.

Using the well-known and highly trusted classical theories of thermodynamics and electromagnetism, Planck derived a formula describing the power emitted by a body, in the form of radiation, when the body is heated. To find the total power radiated, it is necessary to integrate over all the possible frequencies of the emitted radiation. But when Planck tried to do this using his classical formula, he found that the total radiated power was predicted to be infinite – an obviously nonsensical prediction!

Planck was able to avoid this conclusion only by introducing the concept of a minimum amount of energy which can exist for any one frequency of the radiation – a *quantum*. By assuming that light can be emitted or absorbed by matter only in multiples of the quantum, Planck derived a formula which gives the correct prediction for the total amount of power emitted by a hot body. A convenient analogy here may be the economic wealth of an individual, which is normally thought of as a continuously variable quantity. Yet when the individual is in economic interaction (i.e. goes shopping), his or her wealth is quantised in multiples of the smallest denomination coin available. The minimum quantum of energy E, allowed at a given frequency v, is given by Planck's formula

$$E = hv, \tag{3.1}$$

where h is Planck's quantum constant with dimensions of energy per frequency and the minute value of 6.625×10^{-34} joule seconds. The appearance of Planck's constant in the equations of physics is a valuable diagnostic device. When we set $h = 0$, then we are ignoring the existence of the quantum and so should recover the results of classical physics. However, when we examine formulae (or parts of formulae) which are proportional to h, then we are looking at wholly quantum effects which would not be predicted by classical physics.

3.3 Einstein's explanation of the photoelectric effect

The next major step in quantum theory was taken by Einstein in the same year as his formulation of special relativity. This was his explanation of the photoelectric effect, or how metal can be

made to emit electrons by shining a light on it. Planck had suggested that only light in interaction with matter would reveal its quantum behaviour at low energies. Again it was left to Einstein to generalise the idea (as he had generalised the idea of relativity to include electromagnetism). He proposed that all light exists in quanta and set out to show how this might explain the photoelectric effect.

He assumed that the electrons need a definite amount of energy to escape from the metal. If the light of a given colour which is shone on the metal consists of a large number of quanta, each of energy hv, then quanta which collide with the electrons provide them with the energy they need to escape. The electrons will pop out of the metal with an energy which is the difference between that of the quanta and that needed to escape the surface. If the light is below a certain frequency, then no matter how much of it is used, no single quantum will be able to give an electron enough energy to escape. Ignoring multiple quanta–electron collisions, no electrons will emerge. But if the frequency of the light is increased, scanning up the spectrum from red to blue, the electrons will suddenly appear when the quanta have just enough energy to liberate them. As the frequency is increased further still, the electrons will be ejected with higher and higher energies.

This picture exactly fits the experimental facts of the photoelectric effect discovered in 1902 by Lenard. These are that the energy of the electrons emitted depends only on the frequency of the light and not on the intensity (the number of quanta), and that the number of electrons emitted depends only on the intensity but not the frequency.

Einstein's explanation of the photoelectric effect confirmed the quantum theory of light (and won him the Nobel prize). This resurrection of a corpuscular theory of light causes immediate conceptual problems because light is quite demonstrably also a continuous wave phenomenon (as demonstrated by diffraction and other interference experiments). It appears to be both a discrete particle (a *photon*) and an extended wave! How can this be?

Resolution of this apparent paradox requires the introduction of a new entity which reduces to both particle and wave in different circumstances. This entity turns out to be a *field* which we shall discuss further in Chapter 4. But before going on to this we will come to appreciate that not only light is subject to such schizophrenic behaviour.

3.4   Bohr's atom

We saw in Chapter 1 how Rutherford's scattering experiments led to a picture of the atom in which the light, negatively charged electrons orbit the small, massive, positively charged nucleus located in the centre, the vast majority of the volume of the atom being empty space. This appealing picture has fundamental difficulties. Firstly, in the classical theory of electrodynamics, all electric charges which experience an acceleration should emit electromagnetic radiation. Any body constrained to an orbit is subject to an acceleration by the force which gives rise to the orbit in the first place. Thus the electrons in Rutherford's atom should be emitting radiation constantly. This represents a loss of energy from the electrons which, as a result, should spiral down into lower orbits and eventually into the nucleus itself. This 'radiation collapse' of atoms is an inescapable consequence of classical physics and represents the failure of the theory in the atomic domain. Another problem of the Rutherford atom is to explain why all the atoms of any one element are identical. In classical physics, no particular configuration of electronic orbits is predicted other than on the grounds of minimising the total energy of the system. This does not explain the identity of the atoms of any element. A fundamentally new approach is needed to describe the Rutherford atom.

It was the Danish physicist Niels Bohr who in 1913 suggested a new quantum theory of the atom, which, at a stroke, dismissed the problem of the radiation collapse of atoms, explained the way in which light is emitted from atoms and incorporated the new quantum ideas of Planck and Einstein.

Bohr's basic hypothesis was the simplest possible application of the quantum idea to the atom. Just as Planck had hypothesised that light exists only in discrete quanta, so Bohr proposed that atoms can exist only in discrete quantum states, separated from each other by finite energy differences, and that *when in these quantum states the atoms do not radiate*. A simple way to think of these quantum states is as a set of allowed orbits for the electron around the nucleus, the space between the orbits being forbidden to the electrons.

The allowed orbits are specified as those in which the orbital angular momentum of the electron is quantised in integral units of Planck's quantum constant divided by 2π and denoted $\hbar$. It may seem odd that angular momentum should be one of the few quantities to be quantised (like energy and electric charge but unlike mass, linear momentum and time). But we may have suspected as much on first meeting Planck's constant. Its rather unusual units of energy per frequency are in fact identical to the dimensions of angular momentum.

Although the atom is assumed not to radiate light when all its electrons are safely tucked into their quantum orbits, it will do so when an electron makes a transition from one of the allowed orbits to another. This process of emission should explain the behaviour of light observed in the real world. The light from a gas discharge lamp, say a neon or mercury vapour tube, has a distinctive appearance. The atoms in a gas or vapour are widely separated and interact with each other relatively seldom. This means that the light they emit will be characteristic of the particular atoms involved. It is a mixture of just a few separate frequencies which can be split up by a prism. The resulting spectrum of frequency lines is a unique property of the element which is emitting the light (see Figure 3.1). Late in the nineteenth century, researchers such as Balmer, Lyman and Paschen looked at the spectra of many different elements and noted that they all fall into simple mathematical patterns – with several discrete patterns per element. These patterns had long defied explanation, essentially because they defy the smooth way in which quantities vary in classical physics. But with the quantum theory, Bohr was able to forward a convincing explanation of the origin of these lines. Each pattern of frequency lines represents the energy difference between a particular quantum state and all the others in the atom from which the electron can reach that state by emitting light (see Figure 3.2).

Fig. 3.1. The characteristic spectrum of a gas discharge lamp.

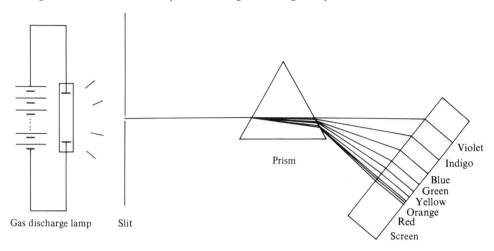

Gas discharge lamp   Slit   Prism

Violet
Indigo
Blue
Green
Yellow
Orange
Red
Screen

With Bohr's model of the atom, physicists were able to calculate, in great detail, many of the spectroscopic results obtained by the experimenters of previous decades. On the basis of this understanding of atoms, Bohr himself was able to propose a tentative explanation of Mendeleeff's periodic table of elements. The periodic table, which classifies the elements into groups reflecting

Fig. 3.2. The discrete patterns of frequency lines in a given element (such as hydrogen) arise from transitions into each available state from all the others. Each state is labelled by its Bohr orbital quantum number n.

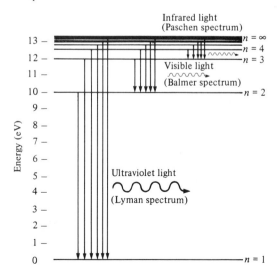

Infrared light
(Paschen spectrum)

$n = \infty$
$n = 4$
$n = 3$

Visible light
(Balmer spectrum)

$n = 2$

Ultraviolet light
(Lyman spectrum)

$n = 1$

Energy (eV)

their chemical behaviour, is explained by the way the electronic orbits are filled in the different elements. The chemical properties of an element are determined predominantly by the number of electrons in its outermost orbit, and so by proposing the electronic orbital structures of the elements it is possible to reproduce the pattern of the table (see Figure 3.3).

Whilst the Bohr atom was an enormous step forward, and the concept of electronic orbits is a mental crutch for our imaginations operating so far beyond their normal domain, it is important to realise that it is only the simplest quantum model of the atom and that more-sophisticated portrayals of electronic behaviour are necessary, as we are about to see.

3.5 De Broglie's electron waves

The next major conceptual advance in quantum theory came much later, in 1924. The young French physicist Louis de Broglie suggested in his doctoral thesis that just as light waves could act like particles in certain circumstances, so too could particles manifest a wavelike behaviour. In particular, he proposed that the electrons, which had previously been regarded as hard, impenetrable, charged spheres could in fact behave like extended waves undergoing diffraction and interference phenomena just like light or water waves.

According to de Broglie, the wavelength of a particle wave is inversely proportional to its

Fig. 3.3. A fragment of the periodic table and the associated electronic orbital structure.

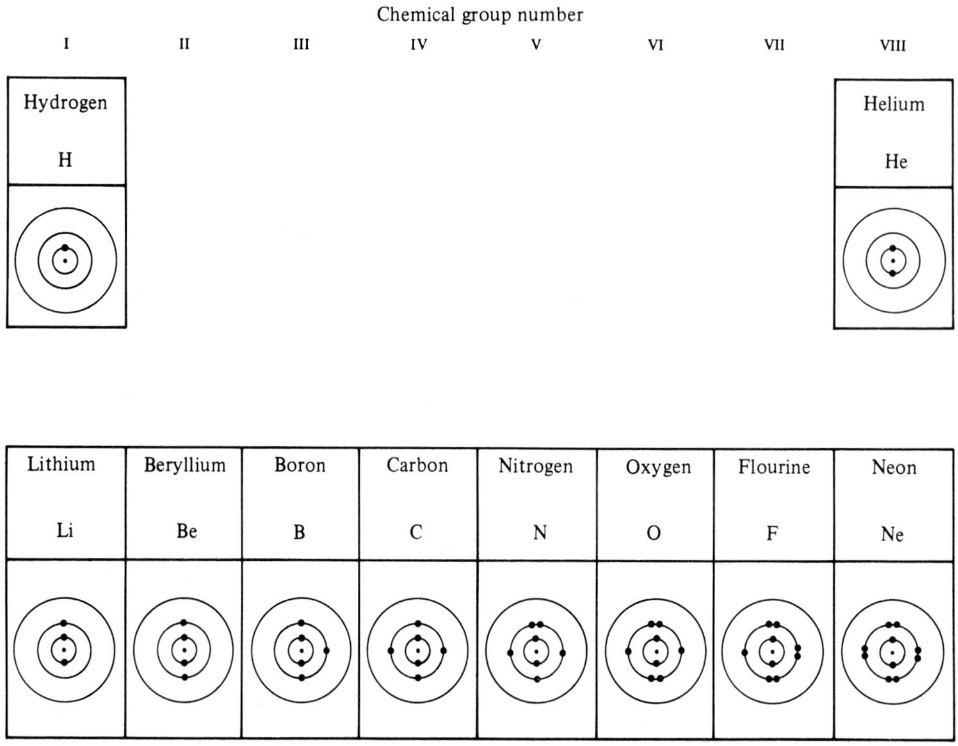

Chemical group number

I	II	III	IV	V	VI	VII	VIII

Lithium	Beryllium	Boron	Carbon	Nitrogen	Oxygen	Flourine	Neon
Li	Be	B	C	N	O	F	Ne

momentum, the constant of proportionality being Planck's quantum constant:

$$\lambda = \frac{h}{p}. \tag{3.2}$$

So the higher the momentum of a particle, the smaller its wavelength. It is worth appreciating that de Broglie's hypothesis applies to all particles, not just to electrons and the other elementary particles. For instance, a billiard ball rolling across the table top will have a wavelength, but because Planck's constant is so minute and the ball's momentum is so comparatively large, the billiard ball's wavelength is about 10^{-34} m. This, of course, is many orders of magnitude different from the typical dimensions of billiards, and so the wave character of the ball never reveals itself. But for electrons, their typical momenta can give rise to wavelengths of 10^{-10} m, which are typical of atomic distance scales. So electrons may be expected to exhibit a wavelike character during interaction with atomic structures.

This wavelike character was observed in 1927 by the US physicists Clinton Davisson and Lester Germer, and independently by G. P. Thomson (J. J.'s son) who was at the time Professor of Natural Philosophy at the University of Aberdeen in Scotland. They demonstrated that electrons undergo diffraction through the lattice structure of a crystal in a fashion similar to the diffraction of light through a grating. Davisson and Thomson were jointly awarded the 1937 Nobel Prize for Physics.

De Broglie's hypothesis also provided the first rationale for Bohr's model of the atom. The existence of only certain specific electronic orbits can be explained by allowing only those orbits which contain an integral number of de Broglie wavelengths. This reflects the momentum of the electron involved (and so the energy of the orbit); see Figure 3.4.

Adoption of de Broglie's idea requires the comprehensive assimilation of particle–wave duality. For any entity in the microworld, there will be situations in which it is best thought of as a wave

Fig. 3.4. Allowed orbits are explained as containing an integral number of de Broglie wavelengths.

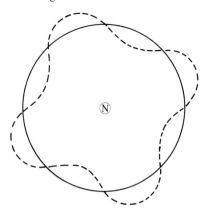

Fig. 3.5. Schrödinger's wave equation.

$$-\frac{\hbar^2}{2m}\frac{\partial^2\psi}{\partial x^2} + V\psi = i\hbar\frac{\partial\psi}{\partial t}$$

and situations in which it is best thought of as a particle. Neither is a truer representation of reality than the other, as both are the coarse product of our human macroscopic imaginings.

The advent of de Broglie's ideas marks the spark which started the intellectual bush fire of quantum theory proper. Up to the early 1920s, quantum theory was a series of prescriptions (albeit revolutionary ones) but not a dynamical theory of mechanics to transcend that of Newton. The second wave of the quantum revolution (1924–27) was to provide just such a theory.

3.6 Schrödinger's wavefunction

Following on directly from de Broglie's ideas, the Austrian physicist Erwin Schrödinger developed the idea of particle waves into a wave mechanics proper. Schrödinger's starting point was essentially the wave equation describing the behaviour of light waves in space and time. Just as this is the accurate representation of optical phenomena (which can be described approximately by the light *rays* of geometrical optics), Schrödinger formulated a matter wave equation which he put forward as the accurate representation of the behaviour of matter (which is described approximately by the particle dynamics). Schrödinger's equation (Figure 3.5) describes a particle by its wavefunction, denoted ψ, and goes on to show how the particle wavefunction evolves in space and time under a specific set of circumstances.

One such circumstance of very great interest is that of a single electron moving in the electric field of a proton. Using his wave equation, Schrödinger was able to show that the electron wavefunction can assume only certain discrete energy levels, and that those energy levels are precisely the same as the energies of the electronic orbits of the hydrogen atom, postulated earlier by Bohr.

The particle wavefunction is an extremely significant concept which we shall use frequently in the coming chapters. It is a mathematical expression describing all the observable features of a particle. Collisions between particles are no longer necessarily viewed as some variant of billiard-ball behaviour but, instead, as the interference of wavefunctions giving rise to effects rather like interference phenomena in optics.

But now that we have introduced the particle wavefunction, and claimed that an equation governing it can predict the behaviour of particles, what exactly is its significance? Should we think of an electron as a localised ball of stuff, or as some extended wave? And if a wave, what is doing the waving? After all, there is no such thing as a light wave; it is a handy paraphrase for time- and space-varying electric and magnetic fields. What, then, is a matter wave?

Before we answer these intriguing questions, we need one more principle of quantum theory. This is the 'uncertainty principle' which the German physicist Werner Heisenberg derived from his alternative formulation of a quantum mechanics, developed simultaneously with Schrödinger's wave formulation, but from a very different starting point.

3.7 Heisenberg's mechanics and the uncertainty principle

Heisenberg took as his starting point the quantum state of the system under consideration (e.g. a single electron, an atom, a molecule etc.), and argued that the only sensible way to formulate a mechanics of the system was by modelling the act of observation on it. Here, by the word 'observation' we mean any interaction experienced by the system, such as the scattering off it of light or of an electron. In the absence of any interaction, the system would be totally isolated from the outside world and so totally irrelevant. Only by some form of interaction or observation does the system exist in a definite state.

Heisenberg's approach is the literal manifestation of Wittgenstein's parting philosophical rejoinder, 'concerning that of which we cannot speak, we must pass over in silence'. We can speak (or write equations) only of what we observe, and so observation is to have pride of place in quantum theory.

Heisenberg represented observations on a system as mathematical operations on its quantum state. This allowed him to write equations governing the behaviour of a quantum system and so led to results which were identical to the somewhat more accessible wave mechanics of Schrödinger (say in predicting the energy levels of the hydrogen atom). The equivalence of the two approaches can be appreciated by realising that the expressions Heisenberg used to represent the observations are differential operators and that they act on the quantum state, which is represented by the wavefunction of the system. So this approach will result in a differential equation in the wavefunction ψ, identical to the wave equation which Schrödinger obtained by analogy with the wave equation for light.

Heisenberg's uncertainty principle results from the realisation that any act of observation on the quantum system will disturb it, thus denying perfect knowledge of the system to the observer. This is best illustrated by analysis of what would happen if we were to attempt to observe the position of an electron in an atomic orbit by scattering a photon off it (Figure 3.6). The photon's wavelength is related to its momentum by the same equation as for any other particle:

$$\lambda = \frac{h}{p}.$$

So the greater the photon's momentum, the shorter its wavelength and vice versa. If then we wish to determine the position of the electron as accurately as possible, we should use the photon with the highest possible momentum (shortest wavelength), as it is not possible to resolve distances shorter than the wavelength of the light used. However, by using a high-momentum photon, although we will gain a good estimate of the electron's position at the instant of measurement, the electron will have been violently disturbed by the high momentum of the photon and so its momentum will be very uncertain. This is the essence of Heisenberg's uncertainty principle. Knowledge of any one parameter implies uncertainty of some other so-called 'conjugate' parameter. This is expressed mathematically by requiring that the product of the uncertainties in the two conjugate parameters must always be greater than or equal to some small measure of the effect of measurement. Not surprisingly, this measure turns out to be none other than Planck's ubiquitous constant:

$$\Delta p \, \Delta x \geqslant \hbar \quad \text{with} \quad \hbar = \frac{h}{2\pi}.$$

Fig. 3.6. A long-wavelength (low-momentum) photon can give only a rough estimate of the position of the electron, but does not disturb the atom very much. A short-wavelength (high-momentum) photon localises the electron more accurately, but causes great disturbance.

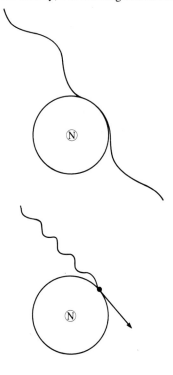

A similar trade-off occurs when attempting to measure the energy of a quantum system at a given time. An instantaneous measurement implies a high-frequency probe (one wavelength over in a short time), but this means a high-energy probe which will mask the energy of the quantum state itself. Conversely, a very low-energy probe, which will not unduly affect the energy of the quantum state, implies a low frequency probe, which means the time to be associated with the measurement is uncertain, thus:

$$\Delta E \, \Delta t \geqslant \hbar.$$

Heisenberg's uncertainty principle is an enormously powerful result when we realise that the uncertainty in a quantity provides a good guide to its minimum value. For instance, if we know that the uncertainty in a particle's lifetime is 1 s, then the lifetime is unlikely to be less than $\frac{1}{2}$ s as the uncertainty could not otherwise be accommodated. Similarly, if we know that a particle is confined to a small volume (say the nucleus $\Delta x \approx 10^{-15}$ m), then we can conclude that the momentum of the particle must be greater than

$$p_{min} \approx \frac{\Delta p}{2} \approx \frac{\hbar}{2\Delta x} \approx 100 \text{ MeV}/c.$$

If the particle is confined to the nucleus, then this is a reasonable guide to the strength (energy) of the force which is keeping it there.

Armed with these ideas, we can turn to the thorny problem of just what a matter wave is.

3.8 The interpretation of the wavefunction ψ

Firstly, let us address the question of whether an electron is to be regarded as a localised ball or an extended wave. Which of these two descriptions applies is very much a matter of the circumstances the electron finds itself in (see Figure 3.7).

For an electron which is travelling through space with a definite momentum ($\Delta p = 0$) and so isolated from all interactions, the uncertainty in its position is infinite. Thus its wavefunction is a sine wave of definite wavelength extended throughout space. The electron is in no sense a localised particle. If an electron is vaguely localised, say we know it has disturbed an atom, then with Δx as the dimension of the atom, we know that there will be an uncertainty Δp in the electron momentum (due

to its interaction with the atom) and so a spread in the wavelength of the wavefunction, $\Delta \lambda = h/(\Delta p)$. This spread in wavelengths (frequencies) causes the formation of a localised wavepacket in the wavefunction reflecting the rough localisation of the electron.

When the electron is very specifically localised, say in a quasi-point-like, high-energy collision with another particle, then the uncertainty in its momentum (and so the spread in the wavelength components of the wavefunction) is large, and the wavepacket becomes very localised, in which case it is sensible to regard the electron as a particle.

This picture of the electron wave makes rather a nonsense of the simple Bohr picture of the orbiting electrons. The dimensions of the electronic wavefunction are comparable to that of the atom itself. Until some act of measurement localises the electron more closely, there is no meaning to ascribing any more detailed a position for the electron. However, this explanation is not altogether satisfactory as it stands, as we have left the electron with a rather poorly defined role in the atom. Progress in understanding this aspect is related to our other outstanding question about the wavefunction: what is it?

Fig. 3.7. A particle's wavefunction reflects its localisation (see text).

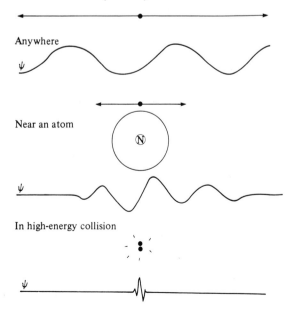

In 1926 the German physicist Max Born ventured the suggestion that the square of the amplitude of the wavefunction at any point is related to the probability of finding the particle at that point. The wavefunction itself is proposed to have no direct physical interpretation other than that of a 'probability wave'. When squared, it gives the chance of finding the particle at a particular point on the act of measurement. Hence, the probability density for finding the particle at the position **x** at time *t* is

$$\text{probability density} = |\psi(\mathbf{x}, t)|^2.$$

So the location of the electron in the atom is not wholly indeterminate. The solution to Schrödinger's equation for an electron in the electrical field of the proton will give an amplitude for the wavefunction as a function of distance from the proton (as well as the energy levels mentioned earlier). When squared, the amplitude gives the probability of finding the electron at any particular point. Thus we can give only a probability for finding an electron in its Bohr orbit, a probability for determining its position within the orbit and probabilities for finding it in the space between orbits. There is even a small probability of this so-called orbital electron existing actually inside the nucleus!

Schrödinger's wavefunction associates with every point in space (and time) two numbers: the *magnitude* (or size) of the wavefunction, and its *phase*. In general, the phase of a wave corresponds to the position in its cycle, with respect to an arbitrary reference point. In other words, it is a measure of how far away one is from a wave crest or trough. The phase is usually expressed as an angle. In contrast to the wavefunction's magnitude (which is related to the probability), its phase can never be directly observed – it is unobservable. Only differences in phase are observable (e.g. as interference patterns in optics).

3.9　Electron spin

Having just developed a rather sophisticated picture of the electronic wavefunction, we shall immediately retreat to the comfortingly familiar picture of Bohr's orbital atom to explain the next important development in quantum theory!

By 1925, physicists attempting to explain the nature of atomic spectra had realised that not all was correct. Where, according to Bohr's model, just one spectral line should have existed, two were sometimes found very close together. To explain this and other similar puzzles, the Dutch physicists Sam Goudsmit and George Uhlenbeck proposed that the electron spins on its axis as it orbits around the nucleus (just as the earth spins around the north–south axis as it orbits around the sun; see Figure 3.8).

The splitting of the spectral lines is explained by the existence of magnetic effects inside the atom. The electron orbit around the nucleus forms a small loop of electric current and so sets up a magnetic field; the atom behaves like a small magnet. The spin of the electron is an even smaller loop of electric current and this sets up another magnetic field which is referred to as the 'magnetic moment of the electron'. This can either add to, or subtract from, the main magnetic field of the atom, depending on the way in which the electron is spinning. This will lead to a slight difference in the energy of the electronic orbit for the different spins of the electron, and will result in the splitting of the spectral line associated with the Bohr orbit.

The above is a nice classical picture, but it has its limitations. The fact that the spectral line splits into just two components indicates that the electron cannot be spinning around at any arbitrary

Fig. 3.8. In the orbital picture of a particle electron, the electron spins on its own axis.

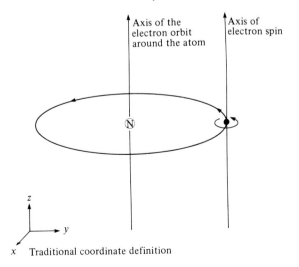

angular momentum but must be such that it has just two values along the line of the atom's magnetic field (or, in the case of a free electron, along the line of any applied magnetic field). The components of the spin in this direction are referred to as the 'z components' (see Figure 3.8) or the 'third components' of spin and are measured to be quantised in half-integral units of Planck's quantum constant (divided by 2π),

$$s_z = \pm \frac{\hbar}{2}.$$

Although the picture of the electron as a spinning ball is attractive, it is important to remember that it is simply a model. In fact, electron spin is purely a quantum concept (it is directly proportional to h). We must be prepared to think also of the electron as an extended wave which carries a quantum of intrinsic angular momentum, just like its quantum of electric charge.

Other particles also carry spin. The proton and the neutron carry spin quanta which are half-integral multiples of Planck's constant, just like the electron. The photon also has spin, but the quantum is of one complete unit $\hbar$. As the photon is simply a packet of electric and magnetic fields this shows that intrinsic angular momentum can be a feature of purely non-material fields. As we shall soon see, the difference in particle spins is very important. On a fundamental level it gives a method of categorising the behaviour of the wavefunctions of particles under the Lorentz transformations of special relativity (a connection we shall discuss further in Chapter 6). On a practical level, it signifies very different behaviours of ensembles of particles (see next section).

3.10 The Pauli exclusion principle

A straightforward look at the Bohr model of the atom tells us that some fundamental principle must be missing. For there is seemingly nothing to prevent all the electrons of any atom from performing the same orbit. Yet we know that a typical atom will have its electrons spread over several different orbits. Otherwise, transitions between them would be rare, in contradiction to the observations of atomic spectra. So some rule must keep the electrons spread out across the orbits of the atom.

In 1925 the Austrian physicist Wolfgang Pauli derived the principle that no two electrons can simultaneously occupy precisely the same quantum state (i.e. have identical values of momentum, charge and spin in the same region of space). He reached this conclusion after examining carefully the atomic spectra of helium. He found that transitions to certain states were always missing, implying that the quantum states themselves were forbidden. For instance, the lowest orbit (or ground state) of helium in which the two electrons have the same value of spin is not present. But the state in which the two electron spins are opposite is observed.

The power of this principle in atomic physics can hardly be overstated. Because no two electrons can exist in the same state, the addition of extra orbital electrons will successively fill up the outer-lying electron orbits and will avoid overcrowding in the lowest one. Just two electrons are allowed in the ground state because the only difference can be the two values of spin available. More electrons are allowed in the higher orbits because their quantum states can differ by a wide range of orbital angular momenta around the nucleus (which also turns out to be quantised). It is the Pauli exclusion principle which is responsible for the chemical identities of all atoms of the same element, as it is this principle which determines the allowed arrangements of the atomic electrons.

Although we have focused on the atom, the exclusion principle applies to any quantum system, the extent of which is defined principally by the wavefunctions of the component particles. In the case of totally isolated electrons of definite momentum whose wavefunctions extend over all space, the exclusion principle means that only two electrons with opposite spins can have the same momentum. In the case of electrons confined to a crystal (i.e. electrons whose wavefunctions extend over the dimensions of the crystal), the rule will apply to all electrons in the crystal.

Pauli's exclusion principle can be expressed alternatively in terms of the behaviour of the wavefunction of a quantum system. Although we have talked so far only of the wavefunctions of individual particles, these can be aggregated for any quantum system to give a wavefunction describing the whole system. For example, the total wavefunction of the helium atom can describe the behaviour of two electrons at the same time. Just as the wavefunction of a single electron is a wavepacket reflecting the localisation of the electron, a double-electron wavefunction will contain two wavepacket humps reflecting the localisations of the two electrons. The consequence of the exclusion principle for any multiple-electron wavefunction is that it must change sign under the interchange of any two electrons. Wherever the amplitude is positive it must become negative and vice versa. The wavefunction is said to be antisymmetric under the interchange of two electrons. This effect can be understood by considering the two-electron helium atom. If the two electrons cannot be in the same quantum state (differing spins), they must be distinguishable under interchange and the wavefunction must signal this. On the other hand, all we are doing is relabelling the electrons and this should make no difference to the physical results (e.g. energy levels and probability densities). The antisymmetry of the wavefunction allows just this. As all physical quantities are proportional to its square, changing only its sign will make no difference.

Particles such as the electron and the proton with spin $\frac{1}{2}\hbar$ (and other more exotic particles that we shall meet with other half-integral spins $\frac{3}{2}\hbar, \frac{5}{2}\hbar, \ldots$) obey the exclusion principle, give rise to antisymmetric wavefunction under the interchange of two identical such particles and are referred to as *fermions*. This is because an ensemble of fermions obey statistics governing their dynamics, which were first formulated by the Italian physicist Enrico Fermi, and the Englishman Paul Dirac. Fermi–Dirac statistics show how momentum is distributed amongst the particles of the ensemble. Because of the exclusion principle in any quantum system, there is a limit to the number of particles which can adopt any particular value of momentum and so this leads to a wide range of momentum carried by the particles. Particles such as the photon with spin $\hbar$ (and other particles we shall meet with integral spins $0, \hbar, 2\hbar, 3\hbar, \ldots$) do not obey the exclusion principle and are called *bosons*. Their wavefunction does not alter under the interchange of two particles. An assembly of bosons obeys dynamical statistics first formulated by the Indian physicist Satiendranath

Bose and Albert Einstein. In Bose–Einstein statistics there is no limit to the number of particles which can have the same value of momentum, and this allows the assembly of bosons to act coherently, as in the case of laser light.

This last principle concludes our whistle-stop tour of quantum mechanics. Although brief, the tour has included most of the new concepts introduced by the theory. For the purposes of the rest of the book, the most important of these is the wavefunction interpretation of a particle, although we will use the uncertainty and exclusion principles from time to time. As in the case of relativity, it is a constant challenge to shrug off our everyday imaginings in the microworld and learn to think in terms of these unfamiliar ideas. But before we are quite ready to approach the subject we must look at what happens when relativity and quantum mechanics are put together.

4

Relativistic quantum theory

4.1 Introduction

Quantum mechanics, just like ordinary mechanics and electrodynamics, must be made to obey the principles of special relativity. Because the entities (particles, atoms etc.) described by quantum theory quite often travel at speeds at or near c, this becomes an essential requirement. Special relativity will not just give corrections to conventional Newtonian mechanics, but will dictate dominant, unconventional relativistic effects.

We will see that the synthesis of relativity with quantum theory predicts wholly new and unfamiliar physical consequences (e.g. antimatter). This requires us to develop a new way of looking at matter via quantum fields. If we can then go on to develop the mechanics of interacting quantum fields, this will provide us with the most satisfactory description of the behaviour of matter (both the conventional matter we have discussed so far, and the unconventional antimatter we have had to introduce along the way).

4.2 The Dirac equation

At the same time as Schrödinger and Heisenberg were formulating their respective versions of the quantum theory, Paul Dirac was attempting the same task. But, in addition, he was concerned that the quantum theory should manifestly respect Einstein's special relativity. This implies two distinct requirements: firstly, that the theory must predict the correct energy–momentum relation for relativistic particles,

$$E^2 = m_0^2 c^4 + p^2 c^2$$

and, secondly, that the theory must incorporate the phenomenon of electron spin in a Lorentz covariant fashion.

In one of the most celebrated brainstorming sessions of theoretical physics, Dirac simply wrote down the correct equation! He was guided in this task by realising that Schrödinger's equation for the electronic wavefunction cannot possibly satisfy the requirements of special relativity because time and space enter the equation in different ways (as first- and second-order derivatives respectively). Schrödinger's equation is perfectly adequate for particles moving with velocities much less than c, and it predicts the correct Newtonian energy–momentum relationship for particles,

$$E = \frac{p^2}{2m} = \frac{mv^2}{2}.$$

But because space and time are not treated correctly, it does not predict the correct relativistic relationships or incorporate energy–mass equivalence.

In the spirit of special relativity, Dirac sought an equation treating space and time on an equal basis. In this he succeeded, but found that in doing so the electron wavefunction ψ could no longer be a simple number. Incorporating time and space on an equal basis requires the electron wavefunction ψ to contain two separate numbers which may be interpreted as reflecting the probabilities that the electron is spin up (with spin quantum $+\hbar/2$) or spin down (with spin quantum $-\hbar/2$). Thus ψ is written as a two-component *spinor*, $\psi = \begin{pmatrix} a \\ b \end{pmatrix}$. In fact, in the full theory it is a four-component object, for reasons which will become clear in the next section.

So in attempting to incorporate special relativity into quantum mechanics it was necessary to invent electron spin! It is fascinating to wonder whether, if electron spin had not been proposed and discovered experimentally, it would have been proposed theoretically on this basis. And if so, would it still have become known as electron spin or would it have been introduced as some sort of 'Lorentz charge'?

Dirac's equation can be used for exactly the same purposes as Schrödinger's, but with much greater effect. In Section 3.9 we saw that the spin of the electron gives rise to a splitting in the energy levels of the hydrogen atom. This is because the magnetic moment of the electron may either add to, or subtract from, the magnetic field set up by the electron's orbital angular momentum. It was noticed in experiments that the half-integral unit of spin angular momentum $\hbar/2$ produced as big a magnetic moment as a whole integral unit of orbital angular momentum (i.e. spin is twice as effective in producing a magnetic moment as is orbital angular momentum). This is quantified by ascribing the value of 2 to the gyromagnetic ratio (the g-factor) of the electron. This is effectively the constant of proportionality between the electron spin and the magnetic moment resulting. In non-relativistic quantum mechanics, $g = 2$ is an empirical fact. With the Dirac equation, it is an exact prediction.

The Dirac equation can also explain the fine splitting and hyperfine splitting of energy levels within the hydrogen atom. These result from the magnetic interactions between the electron's orbital angular momentum, the electron spin and the proton spin.

4.3 Antiparticles

One immediate consequence of predicting the relativistic relationship between energy and momentum for the electron wavefunction is that the Dirac equation seems to allow the existence of both positive- and negative-energy particles:

$$E = \pm(m_0^2 c^4 + p^2 c^2)^{1/2}.$$

In an amazing feat of intellectual bravado, Dirac suggested that this prediction of negative-energy particles was not rubbish but, instead, the first glimpse of a hidden universe of antimatter.

The concept of negative-energy entities is wholly alien to our knowledge of the universe. All things of physical significance are associated with varying amounts of positive energy. So Dirac did not ascribe a straightforward physical existence to these negative-energy electrons. Instead, he proposed an energy spectrum containing all electrons in the universe (see Figure 4.1). This spectrum consists of all positive-energy electrons which inhabit a band of energies stretching from $m_0 c^2$, the rest mass, up to arbitrarily high energies. These are the normal electrons which we observe in the laboratory and whose distribution over the energy spectrum is determined by the Pauli exclusion principle. Dirac then went on to suggest that the spectrum also contains the negative-energy electrons which span the spectrum from $-m_0 c^2$ down to arbitrarily large negative energies. He proposed that these negative-energy electrons are unobservable in the real world. To prevent the real, positive-energy electrons simply collapsing down into negative-energy states, it is necessary to assume that the entire negative-energy spectrum is full and that double occupancy of any energy state in the continuum is prevented by the Pauli exclusion principle. No electrons inhabit the energy gap between $-m_0 c^2$ and $m_0 c^2$. (There are no fractions of electrons in evidence.)

Viewed picturesquely, it is as if the world of physical reality conducts itself whilst hovering over an unseen sea of negative-energy electrons.

But if this sea of negative-energy electrons is to remain unseen, what is its effect on the everyday world? The answer to this is that elementary particle interactions of various sorts can occasionally transfer enough energy to a negative-energy electron to boost it across the energy gap into the real world. For instance, a photon with energy $E \geqslant 2m_0 c^2$ may collide with the negative-energy electron and so promote it to reality. But this cannot be the end of the story, as we seem to have created a unit of electrical charge, whereas we are convinced that this is a quantity which is conserved absolutely. Also, we started out with a photon of energy $E \geqslant 2m_0 c^2$ and have created an electron with an energy just over $m_0 c^2$. Where has the energy difference of $m_0 c^2$ gone? We believe that positive energy is also conserved absolutely; it does not disappear into some negative-energy sea.

These problems of interpretation are resolved by proposing that the hole left in the negative-energy sea represents a perceptible, positive-energy particle with an electrical charge opposite to that on the electron. (The absence of a negative-energy particle represents the presence of a positive-energy particle.) This particle is

Fig. 4.1. Dirac's energy spectrum of electronic states (*a*) and its interpretation (*b*).

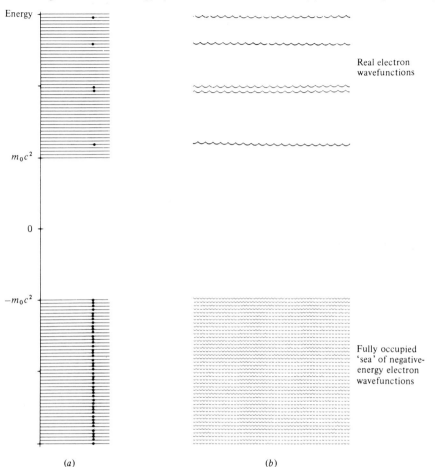

referred to as the *antiparticle* of the electron, is called the *positron*, and is denoted by e$^+$.

The positron was first discovered in 1931 by the American physicist Carl Anderson in a cloud chamber photograph of cosmic rays.

Although the arguments given here have concentrated specifically on the electron and the positron, it is important to appreciate that the Dirac equation applies to any relativistic spin-$\frac{1}{2}$ particle, and so too do the ideas of a negative-energy sea and antiparticles. Both the proton p and neutron n can be described by the Dirac equation and seas of negative-energy protons and neutrons may be proposed as coexisting with those of the electrons. The holes in those seas, the antiprotons denoted p̄, and antineutrons denoted n̄, took somewhat longer to discover than the

positron as, in their case, $2m_0c^2$ is large. It requires high-energy accelerators to provide probes which are energetic enough to boost the antiprotons into existence. These were not available until the mid-1950s.

The electron wavefunction which is described in the Dirac equation can now be appreciated in its full four-component form. These components describe, respectively, the spin-up and spin-down states of both the electron and the positron.

The development of the next concept in the microworld is contained in the behaviour of particles and antiparticles. We suggested that an energetic photon can promote a negative-energy electron from the sea, thus leaving a hole. So the photon can create an electron–positron pair from

Fig. 4.2. Pair creation by a photon γ in the Dirac picture in (a), and in a space–time diagram in (b). Energy and momentum conservation require the subsequent involvement of a nearby nucleus.

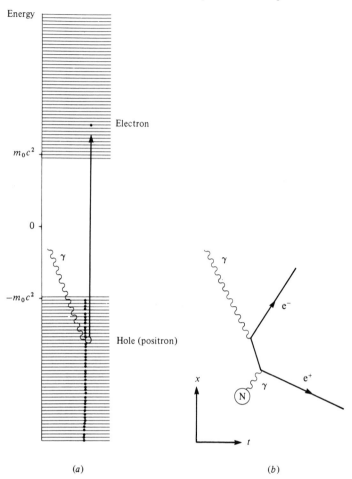

(a)   (b)

the vacuum. (In fact, this must take place in the presence of another particle to ensure conservation of energy and momentum; see Figure 4.2.) Similarly, an electron and a positron can annihilate each other and give rise to energetic photons. The upshot of this is that particles such as the electron can no longer be regarded as immutable, fundamental entities. They can be created and destroyed just like photons, the quanta of the electromagnetic field.

4.4 Quantum field theory

In the most sophisticated form of quantum theory, all entities are described by fields. Just as the photon is most obviously a manifestation of the

electromagnetic field, so too is an electron taken to be a manifestation of an electron field and a proton of a proton field. Once we have learned to accept the idea of an electron wavefunction extending throughout space (by virtue of Heisenberg's uncertainty principle for a particle of definite momentum), it is not too great a leap to the idea of an electron field extending throughout space. Any one individual electron wavefunction may be thought of as a particular frequency excitation of the field and may be localised to a greater or lesser extent dependent on its interactions.

The electron field variable is, then, the (Fourier) sum over the individual wavefunctions, where coefficients multiplying each of the individual wavefunctions represent the probability of the creation or destruction of a quantum of that particular wavelength (momentum) at any given point. The representation of a field as the summation over its quanta, with coefficients specifying the probabilities of the creation and destruction of those quanta, is referred to as *second quantisation*.

First quantisation is the recognition of the particle nature of a wave or of the wave nature of a particle (the Planck–Einstein and de Broglie hypotheses respectively). Second quantisation is the incorporation of the ability to create and destroy the quanta in various reactions.

There is a relatively simple picture which should help us to appreciate the nature of a quantum field and its connection with the notion of a particle. A quantum field is equivalent, at least mathematically, to an infinite collection of harmonic oscillators. These oscillators can be thought of as a series of springs with masses attached. When some of the oscillators become excited, they oscillate (or vibrate) at particular frequencies. These oscillations correspond to a particular excitation of the quantum field and hence to the presence of particles, i.e. field quanta.

We are familiar with the electromagnetic and gravitational fields because, their quanta being bosons, there are no restrictions on the number of quanta in any one energy state and so large assemblies of quanta may act together coherently to produce macroscopic effects. Electron and proton fields are not at all evident because, being fermions, the quanta must obey Pauli's exclusion principle and this prevents them from acting together in a macroscopically observable fashion. So although we can have concentrated beams of coherent photons (laser beams), we cannot produce similar beams of electrons. These instead must resemble ordinary incoherent lights (e.g. torchlights) with a wide spread of energies in the beam.

4.5 Interacting fields

Having introduced this new, rather nebulous, concept of a field representation of matter, we must now set about using it. Our ultimate objective must be to predict the values of physical quantities which can be measured in the laboratory such as particle reaction cross-sections, particle lifetimes, energy levels in bound systems, etc. We hope to achieve this by using the idea of quantum fields to tell us the probabilities of the creation and destruction of their quanta in various reactions, and to provide us with descriptions of the behaviour of the quanta between creation and destruction (the wavefunctions). This will then allow us to calculate the probabilities associated with physical processes.

To formulate this, we must be guided by some principle from which all mechanics can be derived. In fact, the same principle, Hamilton's variational principle, can be used to derive Newtonian mechanics, quantum mechanics and quantum field theory. The *Lagrangian L* for any system is the difference between its kinetic energy (KE) and its potential energy (PE),

$$L = KE - PE.$$

For a classical particle, say a cricket ball, moving through the gravitational field of the earth, the potential energy is due to its height x above the earth ($PE = mgx$), and its kinetic energy is due to its velocity ($KE = \frac{1}{2}mv^2$).

Hamilton's principle states that the evolution of any system is such as to minimise L. So the path of the cricket ball is dictated by those values of position x and velocity v for which L remains at its minimum value. Put another way around, by minimising L with respect to x and v we can obtain the equations of motion of the cricket ball:

$$\delta L(x, v) = 0 \quad \text{gives} \quad F = ma.$$

This entirely general principle can also be used in quantum mechanics. (In the quantum version, however, as we are dealing with wavefunctions (or, more properly, fields) which extend throughout space, we do not deal with the total Lagrangian L directly, but with the Lagrangian density $\mathscr{L}$. The total Lagrangian can then be found by integrating the Lagrangian density over all space. Although in future discussions we shall be talking about the properties of the Lagrangian, the comments will properly apply to the Lagrangian density, a fact which we will acknowledge by using the symbol $\mathscr{L}$.)

It is possible to write down the expression for the Lagrangian density of a free electron as a function of the free-electron wavefunction. By minimising $\mathscr{L}$ with respect to the wavefunction and its time and space derivatives, it is possible to derive Dirac's equation of motion of the electron, denoted $D\psi_e = 0$:

$$\delta\mathscr{L}(\psi_e) = 0 \quad \text{gives} \quad D\psi_e = 0.$$

In both classical and quantum mechanics it was, of course, the equations of motion which were discovered first.

We regress to the Lagrangian and work forward from there for the sake of generality. But in the case of elementary particles *in interaction* we do not know in general the equations of motion and, where we do, we cannot solve them. We cannot therefore proceed to calculate the quantities of physical interest resulting from the motions of particles.

4.6 Perturbation theory

To describe elementary particle reactions in which quanta can be created and destroyed, it is necessary to propose an expression for the Lagrangian of the interacting quantum fields. Let us concentrate on interacting electron and photon fields only. The Lagrangian will contain parts which represent free electrons $\mathscr{L}_0(\psi_e)$ and free photons $\mathscr{L}_0(A)$, where A denotes a four-vector representing the electromagnetic field. It will also contain parts which represent the interactions between electrons and photons, $\mathscr{L}_{INT}(\psi_e, A)$, whose form will be dictated by general principles. These will include, for instance, Lorentz invariance and various conservation laws which the interactions are observed to respect (such as the conservation of electrical charge). In Chapter 21 we shall see how these principles can be expressed in terms of the symmetry of the Lagrangian under various groups of transformations.

The total Lagrangian is then the sum of all these parts:

$$\mathscr{L} = \mathscr{L}_0(\psi_e) + \mathscr{L}_0(A) + \mathscr{L}_{INT}(\psi_e, A).$$

This is the top-level specification of the fields being described and the way in which they interact. We can proceed to predict the values of physical quantities by carrying out a modern version of Hamilton's variational principle. Instead of being the equations of motion for free electrons and photons, the equations are now modified by the existence of interactions. The variational principle now describes the propagation of the fields in terms of (1) probabilities of the creation and the destruction of the quanta of the fields and (2) wavefunctions of the quanta (referred to as propagators in this context).

In the late 1940s the American physicist Richard Feynman derived a set of rules which specifies the propagation of the fields as the sum of a set of increasingly complicated sub-processes involving the quanta of the interacting fields. Each sub-process in the sum can be represented in a convenient diagram referred to as a Feynman diagram. The rules associate with each diagram a mathematical expression describing the wavefunctions of the particles involved. To calculate the probability of occurrence P of any physical event involving the quanta of the fields, it is first necessary to specify the initial and final states being observed, denoted $|i\rangle$ and $\langle f|$ respectively, and then to select all the Feynman diagrams which can connect the two. The mathematical expression for each diagram is then worked out: essentially, the wavefunctions of the quanta are multiplied together to give the quantum-mechanical *amplitude m* for the sub-process. The amplitude for a number of the individual sub-processes may then be added to give the total amplitude M which is then squared to give the required probability of occurrence:

$$P = |\langle f|M|i\rangle|^2$$
$$M = m_1^{(1)} + m_2^{(1)}$$
$$m_1^{(2)} + m_2^{(2)} + m_3^{(2)} + \cdots.$$

In this notation $m_i^{(1)}$ denotes the 'first-order' diagrams with just two photon–electron vertices involved, $m_i^{(2)}$ denotes 'second-order' diagrams with four photon–electron vertices, $m_i^{(3)}$ denotes 'third-order' diagrams and so on.

For example, in the case of electron–positron elastic scattering, the initial and final states are $|e^+e^-\rangle$ and $\langle e^+e^-|$ respectively. A few of the simplest Feynman diagrams connecting the two are shown in Figure 4.3. The first sub-process, amplitude $m_1^{(1)}$, is the exchange of a photon between the electron and the positron; the second, $m_2^{(1)}$, is the annihilation of the electron and the positron into a photon and its subsequent reconversion; the third $m_1^{(2)}$, is the exchange of two photons, and so on.

Fig. 4.3. The perturbation series containing the various sub-processes possible in electron–positron scattering.

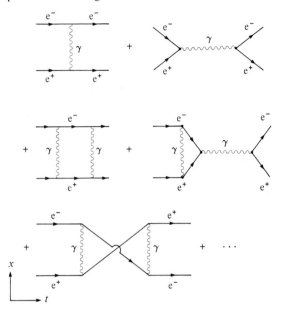

orders is reduced by $e^2/(\hbar c) = \frac{1}{137}$. So only the first few sub-processes need be calculated to achieve an acceptable approximation to the exact answer.

Summary

The Lagrangian (L)	specifies the form of the interaction between the fields.
The variational principle	provides the equations of motion from L.
The perturbation principle	approximates the equations of motion by a series of . . .
Feynman diagrams	which show sub-processes between initial and final states involving quanta which may be calculated to give . . .
Probabilities of physical events	which may be stated as cross-sections, lifetimes, etc.

The probability of occurrence (i.e. of the transformation between initial and final states) may then be restated as the cross-sectional area of two colliding particles, as the mean lifetime for a particle to decay, or as some other appropriate measurable parameter. This is achieved by adopting the kinematical prescriptions which take into account factors like the initial flux of colliding particles, the density of targets available in a stationary target and so on.

The reason why this approach can be adopted is that only the first few of the simplest Feynman diagrams from the infinite series need be considered. This is because the energy of the interaction between electrons and photons (the strength of the electromagnetic force) is small compared with that of the free particles. It can be regarded as a perturbation of free-particle-type behaviour. Another way of stating this is that the probability of the electron or positron interacting with a photon is small. In fact, each photon–electron vertex multiplies the probability of occurrence of the diagram by $e/\sqrt{(\hbar c)}$. As each new order of diagram contains a new photon line with two vertices, the relative magnitude of successive

4.7 Virtual processes

It is important to understand that the dynamics of the individual field quanta within any sub-process of the perturbation expansion are *not* constrained by energy or momentum conservation, provided that the sub-process as a whole does conserve both. This microscopic anarchy is permitted by Heisenberg's uncertainty principle which states that energy can be uncertain to within ΔE for a time Δt, such that

$$\Delta E \, \Delta t \geqslant \hbar.$$

So an electron may emit an energetic photon, or a photon may convert into an electron–positron pair over microscopic timescales, provided that energy conservation is preserved in the long run.

These illicit processes are known as 'virtual processes'. They form the intermediate states of elementary particle reactions. So although we do not see them, we must calculate the probabilities of their occurrence and add them all up to find the number of different ways for a particle reaction to get from its initial to its final state. A good example of a virtual process is the annihilation of an e^+e^- pair into a photon. The energy of the e^+e^- pair must be

$$E_{e^+e^-} = (m_{e^+}^2 c^4 + p_{e^+}^2 c^2)^{1/2} \\ + (m_{e^-}^2 c^4 + p_{e^-}^2 c^2)^{1/2},$$

whereas the energy momentum relation of the photon is

$$E_\gamma = p_\gamma c.$$

So it is not possible to have both

$$E_{e^+e^-} = E_\gamma \quad \text{and} \quad p_\gamma = p_{e^+} + p_{e^-}$$

because of the rest mass of the e^+e^- pair. This means that the virtual photon can exist only as an unobservable intermediate state before dissolving into a collection of material particles which do conserve energy and momentum. Virtual particles are said to be 'off mass-shell', because they do not satisfy the relationship $E^2 = p^2 c^2 + m^2 c^4$. Massless particles, such as photons, are 'off mass-shell' if $E \neq pc$.

4.8 Renormalisation

In writing down all the Feynman diagrams of the sub-processes we find some whose amplitude (product of wavefunctions) appears to be infinite. These diagrams are generally those with bubbles on either electron or photon wavefunctions or surrounding electron–photon vertices; see Figure 4.4. These diagrams give infinite contributions owing to ambiguities in defining the electron and the photon.

An ordinary electron propagating through space is constantly emitting and absorbing virtual photons. It is enjoying self-interaction with its own electromagnetic field (of which its own charge is the source). So the wavefunction of the electron is already dressed up with these virtual photons; see Figure 4.5(*a*). Similarly, a photon propagating through space is free to exist as a virtual e^+e^- pair, and the full photon wavefunction already contains the probabilities of this occurring (Figure 4.5(*b*)). Also, the electric charge, which we denote e, already contains the quantum corrections implied by the diagram of Figure 4.4(*c*).

In 1949, Feynman, Schwinger, Dyson and

Tomonaga showed how the infinite contributions to the perturbation series can be removed by redefining the electron, photon and electric charge to include the quantum corrections. When the real electrons, photons and charges appear, the infinite diagrams are included implicitly and should not be recounted. The mathematical proof of this demonstration is known as 'renormalisation'.

Renormalisation is a necessary formal process which shows that the particles in the theory and their interactions are consistent with the principles of quantum theory. These may seem like hollow words for the familiar interactions of electrons with photons. But in the more esoteric quantum field theories we are going to encounter, both the particle content of the theories and the form of their interactions are largely unknown. In these cases, the ability to renormalise the perturbation expansion of the Lagrangian is a good guide to the acceptability of the theory.

Fig. 4.4. (*a*), (*b*) and (*c*). Diagrams with 'bubbles' which give infinite contributions to the perturbation series.

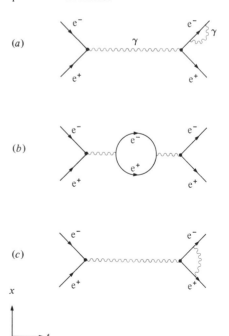

Fig. 4.5. (*a*) The completed ('dressed') electron wavefunction already contains its quantum corrections (interactions with virtual photons). (*b*) The photon propagator likewise.

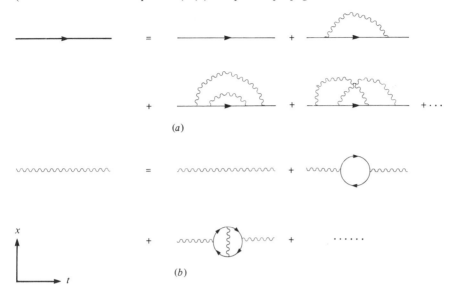

4.9 The quantum vacuum

In classical (non-quantum) physics, empty space–time is called the *vacuum*. The classical vacuum is utterly featureless. However, in quantum mechanics, the vacuum is a much more complex entity: it is far from featureless and far from empty. Actually, the quantum vacuum is just one particular state of a quantum field. It is the quantum-mechanical state in which no field quanta are excited, that is, no particles are present. Hence, it is the 'ground state' of the quantum field, the state of minimum energy.

Let us recall the analogy, introduced above in Section 4.4, between a quantum field and an infinite collection of harmonic oscillators (masses connected to springs). In the vacuum, every oscillator is in its ground state. For a classical oscillator, this means that it is motionless: the spring holds the mass in a fixed position. However, for a quantum oscillator, the uncertainty principle means that neither position nor momentum is precisely fixed, and both are subject to random quantum fluctuations. These fluctuations are called zero-point oscillations, or zero-point vibrations. So, the quantum vacuum is full of fluctuating quantum fields. There are no real particles involved, only virtual ones. Virtual particle–antiparticle pairs continually materialise out of the vacuum, propagate for a short time (allowed by the uncertainty principle) and then annihilate.

These zero-point vibrations mean that, in the vacuum – the state of minimum energy – there is a zero-point energy associated with any quantum field. Since there is an infinite number of harmonic oscillators per unit volume, the total zero-point energy density is, in fact, infinite. We have already seen that some sense can be made of infinite quantities through the process of renormalisation. As it is usually implemented, this yields a zero energy density for the standard quantum vacuum.

It is extremely difficult to observe these vacuum fluctuations, since there is no state of lower energy with which the vacuum can be compared. However, there is one situation in which its effects can be seen indirectly. In 1948, Hendrik Casimir predicted that two clean, neutral, parallel, microscopically flat metal plates attract each other by a very weak force that varies inversely as the fourth power of the distance between them. The 'Casimir effect' was experimentally verified in 1958. It can be understood in the following way. The zero-point energy filling the vacuum exerts pressure on everything. In most circumstances, this pressure is not noticeable, since it acts in all directions and the effect cancels. However, the quantum vacuum has different properties between the two metal plates. Some of the zero-point vibrations of the electromagnetic field are suppressed, namely, those with wavelengths too long to fit between the plates. So, the zero-point energy density between the plates is *less* than that of the standard vacuum, i.e. it is negative. From this it follows that the pressure outside is greater and hence the plates feel an attractive force.

4.10 Quantum electrodynamics

This is the name (often abbreviated to QED) given to the relativistic quantum field theory describing the interactions of electrically charged particles via photons. The discovery of the perturbation expansion revealed the existence of an infinite number of ever-decreasing quantum corrections to any electromagnetic process. The renormalisability of QED means that we can avoid apparently infinite contributions to the perturbation expansion by careful definition of the electron and photon. Therefore we can calculate the value of observable parameters of electromagnetic processes to any desired degree of accuracy, being limited only by the computational effort required to evaluate the many hundreds of Feynman diagrams which are generated within the first few orders (first few powers of $e^2/(\hbar c)$) of the perturbation expansion. This has led to some spectacular agreements between theoretical calculations and very accurate experimental measurements.

The g-factor of the electron is not, in fact, exactly equal to 2 (as predicted by the Dirac equation). Its value is affected by the quantum corrections to the electron propagator illustrated in Figure 4.5(*a*). Essentially, the virtual photons of the quantum corrections carry off some of the mass of the electron while leaving its charge unaltered. This can then affect the magnetic moment generated by the electron during interactions. The measure of agreement between QED and experi-

mental measurement is given by the figure for the modified g-factor:

$$\frac{g}{2} = \pm \frac{1.001\,159\,652\,41}{0.000\,000\,000\,20} \quad \text{experimental measurement}$$

$$\frac{g}{2} = \pm \frac{1.001\,159\,652\,38}{0.000\,000\,000\,26} \quad \text{theoretical prediction.}$$

There are several other such amazing testaments to the success of QED, including numbers similar to the above for the g factor of the muon (a heavy brother of the electron which we will meet soon), and yet more subtle shifting of the exact values of the energy levels within the hydrogen atom, the so-called Lamb shift.

This success makes QED the most precise picture we have of the physical world (or at least the electromagnetic phenomena in it). For this reason we shall look at QED again in Part 6 in an attempt to discover the fundamental principles behind it (i.e. behind the form of the interaction between the fields). This is so that we can attempt to repeat the theory's success for the other forces in nature.

4.11 Postscript

We have now looked at the frontiers of physics as they appeared at the turn of the century and have seen that relativity and quantum mechanics emerged in turn from the vacuum of knowledge beyond those frontiers. The realisation that relativity and quantum mechanics must be made mutually consistent led to the discovery of antiparticles, which led in turn to the concept of quantum fields. The theory of interacting quantum fields is the most satisfactory description of elementary particle behaviour. All calculations in quantum field theory follow from the specification of the correct interaction Lagrangian, which is determined by the conservation laws obeyed by the force under study.

We have developed this picture of the world almost exclusively in terms of the particles interacting by the electromagnetic force. It is now time to turn our attention to the other particles and forces in nature to see if they are amenable to a similar treatment.

In what follows, we shall often use the language of particle wavefunctions rather than that of quantum fields. This is perfectly acceptable as far as we are concerned, because once the Feynman rules for any theory have been derived, the origin of the wavefunction in the underlying quantum field is of largely historical interest. A particle wavefunction is a slightly more convenient and intuitive concept in most situations. However, there will be occasions in later chapters in which a proper understanding of certain phenomena demands that we consider the quantum fields themselves rather than wavefunctions.

Part 1
Basic particle physics

5

The fundamental forces

5.1 Introduction

It is an impressive demonstration of the unifying power of physics to realise that all the phenomena observed in the natural world can be attributed to the effects of just four fundamental forces. These are the familiar forces of gravity and electromagnetism, and the not-so-familiar weak and strong nuclear forces (generally referred to as the 'weak' and 'strong' forces). Still more impressive is the fact that the phenomena occurring in the everyday world can be attributed to just two: gravity and electromagnetism. This is because only these forces have significant effects at observable ranges. The effects of the weak and strong nuclear forces are confined to within, at most, 10^{-15} m of their sources.

With this in mind, it is worthwhile summarising a few key facts about each of the four forces before going on to look at the variety of phenomena they display in our laboratories. In each case we are interested in the sources of the force and the intrinsic strength of the interactions to which they give rise. We are interested also in the space–time properties of the force: how it propagates through space and how it affects the motions of particles under its influence. Finally, we must consider both the macroscopic (or classical) description of the forces (where appropriate) and the microscopic (or quantum-mechanical) picture (where possible).

5.2 Gravity

Gravity is by far the most familiar of the forces in human experience, governing phenomena as diverse as falling apples and collapsing galaxies. The source of the gravitational force is mass and, because there is no such thing as negative mass, this force is always attractive. Furthermore, it is independent of all other attributes of the bodies upon which it acts, such as electric charge, spin, direction of motion, etc.

The gravitational force is described classically by Newton's famous inverse square law, which states that the magnitude of the force between two particles is proportional to the product of their masses and inversely proportional to the square of the distance between them:

$$F = G\frac{m_1 m_2}{r^2}.$$

The strength of the force is governed by Newton's constant, G, and is extremely feeble compared with the other forces (see Table 5.1). We notice the effects of gravity only because it is the *only* long-range force acting between electrically neutral matter. In the microworld, the effects of gravity are mainly negligible. Only in exotic situations, such as on the boundary of a black hole and at the beginning of the universe, do the effects of gravity on the elementary particles become important.

The mechanism which gives rise to this force in the classical picture is that of the gravitational field, which spreads out from each mass-source to infinity. A test mass will interact not with the mass source directly, but with the gravitational field. At each point in space, this carries the knowledge of its parent's mass and of what potential to offer the test mass (i.e. what force to exert on it). However, according to Newton's theory, when a mass-source moves, its gravitational field changes instantaneously to accommodate its new position. This instantaneous change is fundamentally incompatible with the theory of special relativity, which holds that disturbances cannot propagate faster than light. This motivated Einstein to formulate his theory of general relativity, which he completed in 1915.

A further feature of Newton's formula is that the quantity characterising a source of gravity – its *gravitational* mass – is identical to the quantity – its *inertial* mass – which characterises its response to an applied force, as given by another of Newton's famous equations:

$$F = ma.$$

This equivalence between gravitational and inertial mass, which was known to many generations of physicists before him, led Einstein to speculate on the connection between gravity and acceleration. The principle of equivalence is the apotheosis of this connection, and formed the basis of his conceptual leap from the theory of special relativity to the theory of general relativity.

5.2.1 *General relativity*

We saw how, in special relativity, observers' perceptions of time and space are modified by factors depending on their relative velocities. From this it follows that during acceleration

Table 5.1. *Relative strengths of forces as expressed in natural units*

Force	Range	Strength	Acts on
Gravity	∞	$G_{\text{Newton}} \approx 6 \times 10^{-39}$	All particles
Weak nuclear force	$<10^{-18}$ m	$G_{\text{Fermi}} \approx 1 \times 10^{-5}$	Leptons, hadrons
Electromagnetism	∞	$\alpha = \frac{1}{137}$	All charged particles
Strong nuclear force	$\approx 10^{-15}$ m	$g^2 \approx 1$	Hadrons

The dimensionality of the forces is removed by dividing out the appropriate powers of $\hbar$ and c, to leave a dimensionless measure of the forces' intrinsic strengths. Note that for gravity and the weak force, a mass must also be introduced to give a dimensionless quantity. In the table we have used the proton mass.

(changing velocity) an observer's scales of time and space must become distorted. By the principle of equivalence, an acceleration is identical to the effects of a gravitational field, and so this too must give rise to a distortion of space–time. Einstein's general relativity goes on to explain the somewhat tenuous reality of the gravitational field as the warping of space–time around a mass-source. Thus a mass will distort space–time rather like a bowling ball laid on a rubber sheet. And the effect of gravity on the trajectory of a passing particle will be analogous to rolling a marble across the curved rubber sheet (see Figure 5.1).

So, general relativity suggests that instead of thinking of bodies as moving under the influence of a gravitational force, we should think of them as moving freely through a warped, or curved, space–time. Hence, the force of gravity is reduced to the curved geometry of space–time. Geometry has, as we know, different rules on a curved surface. For example, on the curved surface of the earth, two

north-pointing lines which are parallel at the equator (lines of longitude) actually meet at the north pole; whereas on a flat surface, two parallel lines never meet. In fact, in curved space–time, straight lines must be replaced by *geodesics* as the shortest path between two points; free particles move along geodesics. (On the surface of the earth, geodesics correspond to great circles.) Einstein embodied this interpretation of gravity as geometry in his field equations of general relativity:

$$G_{\mu\nu} = 8\pi G T_{\mu\nu}$$

which loosely translates as

$$\begin{pmatrix} \text{geometry of} \\ \text{space–time} \end{pmatrix} = 8\pi G \times \begin{pmatrix} \text{mass and} \\ \text{energy} \end{pmatrix}.$$

So, mass and energy determine the curvature and geometry of space–time; and the curvature and geometry of space–time determine the motion of matter. In other words, 'matter tells space–time

Fig. 5.1. According to general relativity, mass bends space–time, which gives rise to the trajectories associated with gravity.

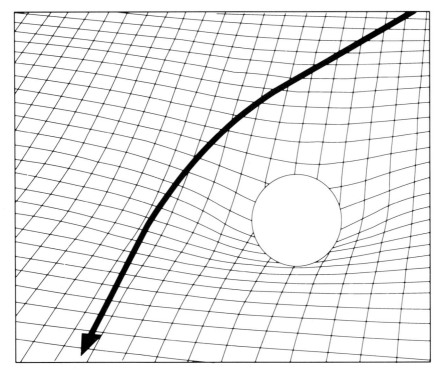

how to bend', and 'space–time tells matter how to move'.

The theory goes on to predict the existence of *gravitational waves* propagating through space as the result of some changes in a mass-source (such as the collapse of a star into a neutron star or black hole). In this event, distortions in space–time will spread out spherically in space at the speed of light, rather like ripples spread out circularly across the surface of a still pond into which a stone is dropped. There has been considerable experimental effort devoted to attempts to detect such waves resulting from cosmic events out in space, but none has so far succeeded, most probably because the disturbances are too small to affect noticeably even our most sensitive detectors.

5.2.2 *Quantum gravity*

It is important to remember that Einstein's general relativity is still a classical theory; it does not account for gravity in the quantum-mechanical regime. A successful quantum theory of gravity has not yet been formulated, and the reconciliation of general relativity with quantum mechanics is one of the major outstanding problems in theoretical physics. It is straightforward enough to take the first few steps towards such a theory, following an analogy with quantum electrodynamics. We can propose that the gravitational field consists of microscopic quanta called *gravitons* which must be massless (to accommodate the infinite range of gravity) and of spin 2 (for consistency with general relativity). The gravitational force between any two masses can then be described as an exchange of gravitons between them. Problems arise, however, because, unlike quantum electrodynamics, certain graviton sub-processes always seem to occur with an infinite probability – quantum gravity is not renormalisable. We shall return to quantum gravity in Chapter 40.

5.3 Electromagnetism

This is the force of which we have the fullest understanding despite its rather complicated nature. This is possibly a reflection of its physical characteristics: it is of infinite range, allowing macroscopic phenomena to guide our understanding of classical electromagnetism, and it is a reasonably strong force, allowing its microscopic phenomena to be observed and to guide our formulation of QED. The strength of the electromagnetic force is characterised by the *fine structure constant*: $\alpha = e^2/\hbar c = \frac{1}{137}$.

The source of this force is, of course, electric charge which can be either positive or negative, leading to an attractive force between unlike charges and a repulsive force between like charges. When two charges are at rest, the electrostatic force between them is given by Coulomb's law, which is very similar to Newton's law of gravity, namely that the magnitude of the force is proportional to the product of the magnitudes of the charges involved (empirically observed to exist only as multiples of the charge on an electron) and inversely proportional to the square of the separation between them:

$$F = K \frac{N_1 e \cdot N_2 e}{r^2}$$

where N_1 and N_2 are integral multiples of the charge on the electron e and K is a constant depending on the electrical permittivity of free space. New mysteries are introduced by the concept of electric charge. What is it, other than a label for the source of a force we observe to act? Why does it exist only in quanta? Why is the charge quantum on the electron exactly opposite to that on the proton? These are largely taken for granted in classical electromagnetism and are only now being addressed in the modern theories described in Part 9.

Unlike gravity, when electric charges start to move, qualitatively new phenomena are introduced. A moving charge has associated with it not only an electric field, but also a magnetic field. A test charge will always be attracted (or repelled) along the direction of the electrical field, i.e. along a line joining the centres of the two charges. But the effect of the magnetic field is that a test charge will be subjected to an additional force along a direction which is mutually perpendicular to the relative motion of the source charge and to the direction of the magnetic field (Figure 5.2). These properties imply that the combined electromagnetic force on a particle cannot be described simply by a number representing the magnitude of the force but, instead, by a vector quantity describing

Fig. 5.2. The motion of a charged particle in a magnetic field directed out of the plane of the paper.

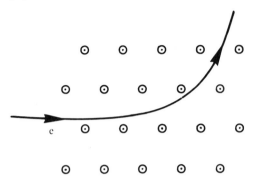

the magnitude of the forces acting in each of the three directions.

When a charge is subject to an acceleration, then a variation in electric and magnetic fields is propagated out through space to signal the event. If it is subject to regular accelerations, as may occur when an alternating voltage is applied to a radio aerial, then the charge emits an electromagnetic wave which consists of variations in the electric and magnetic fields perpendicular to the direction of propagation of the wave – see Figure 5.3. Such an electromagnetic wave is part of the electromagnetic spectrum which contains, according to the frequency of oscillation of the fields, radio waves, infra-red waves, visible light, ultraviolet light, X-rays and gamma rays. See Figure 5.4.

Electromagnetic phenomena are all described in the classical regime by Maxwell's equations, which allow us to calculate, say, the electric field resulting from a particular configuration of charges, or the wave equation describing the propagation of electric and magnetic fields through space.

One interesting feature of these equations is that they are asymmetric owing to the absence of a fundamental quantum of magnetic charge. It is possible to conceive of a source of a magnetic field which would give rise to an elementary magnetostatic force. Such a magnetic charge would appear as a single magnetic pole, in contrast to all examples of terrestrial magnets which consist invariably of north-pole–south-pole pairs. These conventional magnets are magnetic dipoles which are the result of the motions of the atomic electrons. The possibility of the existence of truly fundamental magnetic monopoles has been revived recently following their emergence from the most modern theories and reported sightings (see Part 9).

We have already seen, in Part 0, how we can formulate the quantum theory of electrodynamics by describing the interactions of charged particles via the electromagnetic field as the exchange of the quanta of the field, the photons, between the particles involved. QED is the paradigm quantum theory towards which our descriptions of the other forces all aspire.

5.4 The strong nuclear force

When the neutron was discovered by James Chadwick in 1932, it became obvious that a new force of nature must exist to bind together the neutrons and protons (referred to generically as *nucleons*) within the nucleus. (Prior to this discovery physicists seriously entertained the idea that the nucleus might have consisted of protons and electrons bound together by the electromagnetic force.) Several features of the new force are readily apparent.

Fig. 5.3. The propagation of an electromagnetic wave resulting from the regular accelerations of a charge.

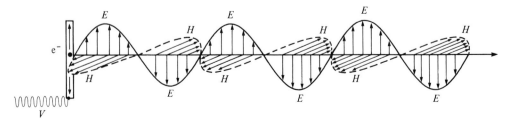

Fig. 5.4. The electromagnetic spectrum.

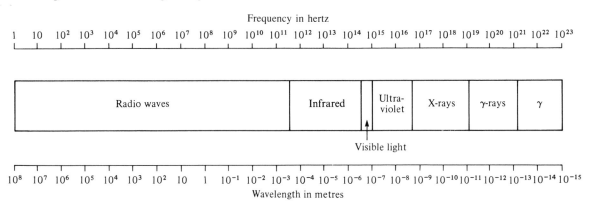

Firstly, as the nucleus was realised to consist only of positively charged protons and neutral neutrons confined within a very small volume (typically of diameter 10^{-15} m), the strong force must be very strongly attractive to overcome the intense mutual repulsions felt by the protons. The binding energy of the strong force between two protons is measured in millions of electronvolts, MeV, as opposed to typical atomic binding energies which are measured in electronvolts (see Appendix 1 for definitions of these units of energy).

The second fact concerning the strong nuclear force is that it is of extremely short range. We know that the electromagnetic force alone can explain the arrangement of electronic orbits within atoms (corresponding typically to radii of 10^{-10} m). Moreover, Rutherford's early scattering experiments of α particles through atoms could be described by the electromagnetic force alone. Only at higher energies, when the α particles are able to approach the nuclei more closely, are any effects of the strong force found. In fact, the force may be thought of as acting between two protons only when they are actually touching, implying a range of the strong force similar to that of a nuclear diameter of about 10^{-15} m.

Finally, the last fact we shall mention is that the strong nuclear force is independent of electric charge in that it binds both protons and neutrons in a similar fashion within the nucleus.

One consequence of the solely microscopic nature of the strong force is that we should expect it to be a uniquely quantum phenomenon. We can expect no accurate interpretation in terms of classical physics but only in the probabilistic laws of quantum theory.

One of the prime sources of early information on the strong force was the phenomenon of radioactivity and the question of nuclear stability. This involves the explanation of the neutron/proton ratios of the stable or nearly stable nuclei. These can be displayed as a band of stability on the plane defined by the neutron number, N, and the proton number, Z, of the nucleus, as in Figure 5.5.

The fact that it is predominantly the heavier nuclei which decay confirms our picture of the short-range nature of the strong force. If we naively think of the nucleus as a bag full of touching spheres, then if the force due to any one nucleon source were able to act on all other nucleons present, we would expect nuclei with more nucleons to enjoy proportionately stronger binding and thus greater stability. (Adding the nth nucleon to a nucleus would give rise to an extra $(n-1)$ nuclear bonds and so a binding energy which increases with n.) This is observed not to be the case. It is the heavier nuclei which suffer radioactive decay, indicating an insufficient binding together of the nucleons. This is because the nuclear force acts only between touching, or 'nearest-neighbour' nucleons. The addition of any extra nucleon will then give rise only to a constant extra binding energy whereas the electric repulsion of the protons is a long-range force and does grow with the number of protons present.

Fig. 5.5. Nuclear stability against radioactive decay is governed by the ratio of protons to neutrons.

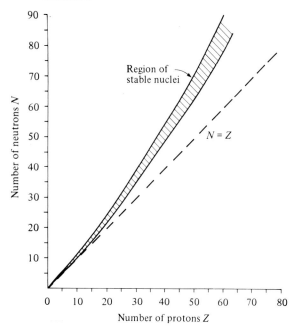

Fig. 5.6. Nuclear stability can be expressed in terms of the binding energy available per nucleon in the nucleus.

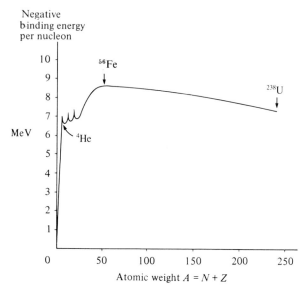

Thus the question of nuclear stability may be described in part by the balance of the repulsive electrical forces and the attractive strong forces affecting the nucleons. It is possible to calculate the sum of these two forces for each nucleus and so to calculate the average binding energy per nucleon in each case, the more negative the binding energy indicating that the more strongly bound are the nucleons within the nuclei. This can be shown graphically as in Figure 5.6. The relatively small negative binding energy of the light atoms results from them not having enough nucleons to saturate fully the nearest-neighbour strong interactions available. The most strongly bound nuclei are those in the mid-mass range, like iron, which more efficiently use the strong force without incurring undue electric repulsion. The heavier nuclei suffer because the electric repulsion grows by an amount proportional to the number of protons present.

As nature attempts to accommodate heavier and heavier nuclei, a point is reached where it becomes energetically more favourable for the large nuclei to split into two more-tightly bound, mid-mass-range nuclei. This gives rise to an upper limit on the weights of atoms found in nature,

occupied by uranium-238 with 92 protons and 146 neutrons. By bombarding uranium with neutrons, it is possible to exceed nature's stability limit causing the uranium + neutron nucleus to split into two. This is nuclear fission.

Radioactive α decay occurs when an element is not big enough to split into two, but would still like to shed some weight to move up the binding energy curve to a region of greater stability. The α particle (which is a helium nucleus consisting of two protons and two neutrons) will have existed as a 'nucleus within a nucleus' prior to the decay. By borrowing energy for a short time according to Heisenberg's uncertainty principle, it will be able to travel beyond the range of the strong attractive forces of the remaining nucleons to a region where it is subject only to the electrical repulsion due to the protons. Thus the nucleus is seen to expel an α particle; see Figure 5.7. Because the energy is borrowed according to the probabilistic laws of quantum theory, it is not possible to specify a particular time for α decay, but only to specify the time by which there will be a, say, 50% probability of a given nucleus having undergone the decay (corresponding to the average time needed for 50% of a sample to decay). This is called the *half-life* of the element, denoted by $\tau_{1/2}$.

Fig. 5.7. Radioactive α decay. An α particle
within the nucleus borrows enough energy via
Heisenberg's uncertainty principle to overcome
the potential binding energy of the strong force.

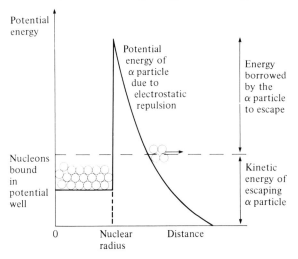

5.5 The weak nuclear force

One of the most obvious features of the
neutron is that it decays spontaneously into a
proton and an electron with a half-life of about 10
minutes. This period is much longer than any of
the phenomena associated with the strong force,
and it is difficult to imagine how the electromagnetic force could be responsible for this comparative
stability. So, we are led to the conclusion that
neutron decay is due to some qualitatively new
force of nature.

It is this 'weak' force causing neutron decay
which lies behind the phenomenon of the radioactive β decay of nuclei described in Chapter 1. The
decay of the neutron into a proton allows a nucleus
to relieve a crucial neutron surplus which, because
of the action of the Pauli exclusion principle, may
be incurring a substantial energy penalty and eroding the binding energy of the nucleus.

The same interaction may also allow the
reverse reaction to occur in which a nuclear proton
transfers into a neutron by absorbing an electron.
(This may occur because of the very small but
finite chance that the electron may find itself actually inside the nucleus, according to the positional
uncertainty represented by the electron wavefunction.) This reaction will allow a proton-rich nucleus suffering from undue electric repulsion to
dilute its proton content slightly, thereby strengthening its binding.

One problem soon encountered in attempts
to explain radioactive β decay is that the electrons
which are emitted from decaying nuclei are seen to
emerge with a range of energies up to some maximum which is equal to the difference in the masses
of the initial and final nuclei involved. When the
electrons emerge with less than this maximum
figure we seem to have lost some energy. To avoid
this apparent violation of energy conservation
(and also an accompanying apparent violation of
angular momentum conservation), Pauli postulated in 1930 that another, invisible, particle was
also emitted during the decay, which carries off the
missing energy and angular momentum. As the
original reactions do conserve electric charge, then
this new particle must be neutral. On the strength
of this, Fermi called it the *neutrino*.

Several properties of the neutrino are apparent from the facts of β decay. Because of conser-

Another feature of nuclear stability can be
explained by the action of another quantum principle. Although we have explained why too many
protons in a massive nucleus may cause it to break
up, we have not explained why this cannot be
countered by simply adding an arbitrary number of
neutrons to gain extra attractive strong forces. The
reason is due to Pauli's exclusion principle. Because both protons and neutrons are fermions, no
two protons and no two neutrons can occupy the
same quantum state. We cannot simply add an
arbitrary number of neutrons to dilute the repulsive effects of the electric charges on the protons,
as the exclusion principle forces the neutrons to
stack up in increasingly energetic configurations
leading to a reduction in the negative binding
energy per nucleon and so to decreased stability.

Although we have now reviewed some facts
about the strong force (its short range, charge
independence and spin dependence via the exclusion principle, etc.), we have done nothing to
explain the mechanism of its action apart from
noting that only a quantum picture will be suitable
for such a microscopic phenomenon. Yukawa's
formulation of his meson theory of the strong force
is the point of departure into particle physics
proper from the inferences of nuclear physics just
discussed, and is described in Chapter 7.

vation of energy, it is necessary that the neutrino be very light or indeed massless (because some electrons do emerge with the maximum energy allowed by the mass difference). Similarly, because of angular momentum conservation the neutrino must be spin $\frac{1}{2}$. Another interesting feature is that the neutrino interacts with other particles only by the weak force and gravity (because the strong interaction is obviously not present in neutron decay, and because the uncharged neutrino experiences no electromagnetic effects).

The apparent invisibility of the neutrino is due to the very feebleness of the weak force, as indicated in Table 5.1. This reluctance to interact allows it to pass through the entire mass of the earth with only a minimal chance of interaction en route. Because of this, the neutrino was not observed (i.e. collisions attributable to its path were not identified) until large neutrino fluxes emerging from nuclear reactors became available. This was achieved in 1956 by Reines, some 26 years after Pauli's proposal.

The weak force, like its strong counterpart, acts over microscopic distances only. In fact, to all intents and purposes it makes itself felt only when particles come together at a point (i.e. below any resolving power available to physics, say less than 10^{-18} m). This allows its description in terms of quantum physics only, which we shall discuss further in Parts 3, 4 and 5.

6

Symmetry in the microworld

6.1 Introduction

In the everyday world, symmetries in both space and time have a universal fascination for the human observer. In nature, the symmetry exhibited in a snowflake crystal or on a butterfly's wings might be taken to indicate some divine guiding hand, whilst in art the pleasures of design or of a fugue may be seen as its imitation. Pleasing as symmetry may be, however, its significance generally remains unappreciated.

In the world of physics, and especially in the microworld, symmetries are linked closely to the actual dynamics of the systems under study. They are not just interesting patterns or an artistic disguise for science's passion for classification. Indeed, it is no exaggeration to say that symmetries are the most fundamental explanation for the way things behave (the laws of physics).

Historically this has not always been appreciated. It is, of course, the case that physicists notice natural phenomena and write down equations of motion to describe them (notably Newton, Einstein and Dirac, to name but an illustrious few). But in describing the microworld it is generally far too difficult to write down the equations of motion straight away; the forces are unfamiliar and our experiments provide only ground-floor windows into the skyscraper of the high-energy domain. So we are forced to consider first the symmetries governing the phenomena under study, generally indicated by the action of conservation rules of one sort or another (e.g. energy, momentum and electric charge). The symmetries may then guide our investigation of the nature of the forces to which they give rise.

Symmetry is described by a branch of mathematics called 'group theory'. A group is simply a collection of elements with specific interrelationships defined by group transformations. The notion is made non-trivial by the demand that repeated transformations between elements of the group are equivalent to another group transformation from the initial to the final elements. When a symmetry group governs a particular physical phenomenon (if the Lagrangian governing the phenomenon does not change under the group transformations), then this implies the existence of a conserved quantity. This connection is due to a mathematical proof called Noether's theorem which states that, for every continuous symmetry of a Lagrangian, there is a quantity which is conserved by its dynamics. This quantity is given by the generator of the group.

We can proceed to put some flesh on this theoretical skeleton by examples of four kinds of symmetry used extensively in particle physics: continuous space–time symmetries, discrete symmetries, dynamical symmetries, and internal symmetries. After a look at each of these, we will mention in closing how even broken symmetries can provide a useful guide to the formulation of physical laws.

6.2 Space–time symmetries

Foremost amongst these are the operations of translation through space, translation through time and rotation about an axis. The physical laws governing any process are formulated with a particular origin and a particular coordinate system in mind; for instance, laws of terrestrial gravity might use the centre of the earth as their origin, whilst the laws of planetary motion might use the centre of the sun.

However, the physical laws should remain the same whatever the choice and so any mathematical expression of the laws should remain the same under any of these transformations. Application of Noether's theorem, then, reveals the conserved quantity corresponding to each particular invariance. Invariance under a translation in

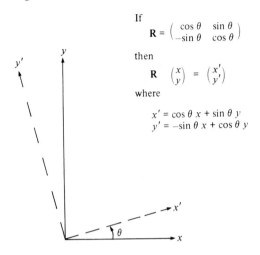

Fig. 6.1. Rotations redefine a coordinate system. Invariance of the laws of physics with respect to such a rotation implies the conservation of angular momentum.

If
$$R = \begin{pmatrix} \cos\theta & \sin\theta \\ -\sin\theta & \cos\theta \end{pmatrix}$$
then
$$R \begin{pmatrix} x \\ y \end{pmatrix} = \begin{pmatrix} x' \\ y' \end{pmatrix}$$
where
$$x' = \cos\theta\, x + \sin\theta\, y$$
$$y' = -\sin\theta\, x + \cos\theta\, y$$

time (i.e. that the laws of physics predict the same evolution of identical processes regardless of when they occur) implies that conservation of energy is built into the laws describing the process. Invariance under a translation in space (e.g. that physics is the same in London as in New York) implies conservation of momentum. And invariance under spatial rotations implies conservation of angular momentum; see Figure 6.1.

The laws of physics are also invariant under the Lorentz transformations of special relativity (Figure 2.3 in Section 2.5). More generally, physical laws are unchanged under any combination of a Lorentz transformation and a space–time translation. These are called Poincaré transformations after the French mathematician Henri Poincaré. Invariance under the complete group of Poincaré transformations incorporates all of the above space–time symmetries.

6.3 Discrete symmetries

The continuous space–time symmetries are called *proper* Lorentz transformations because they can be built up from a succession of infinitesimally small ones. However, there are also *improper* symmetries which cannot be so built up. These improper, or discrete, symmetries do not have corresponding conservation laws as important as those of the continuous symmetries. However, they have proved very useful in telling us which particle reactions are possible with a given force and which are not. We shall deal with the three most important improper symmetries in turn.

6.3.1 *Parity or space inversion*

In this operation, denoted **P**, the system under consideration (e.g. a particle wavefunction) is reflected through the origin of the coordinate system, as in Figure 6.2(*a*). An alternative way of thinking of this is as the reversal of a left-handed coordinate system into a right-handed one, as shown in Figure 6.2(*b*). The parity operation is equivalent to a mirror reflection followed by a rotation through 180°.

If a system (a particle, or collection of particles) is described by a wavefunction $\psi(\mathbf{x}, t)$, then

Fig. 6.2. A parity transformation reverses the spatial coordinates of an event E (*a*) or, equivalently, converts a left-handed coordinate system into a right-handed one (*b*).

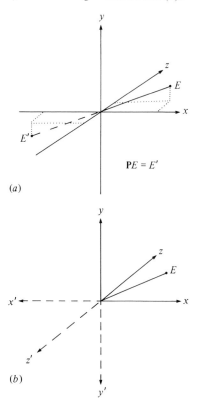

(*a*)

(*b*)

the parity operation will reverse the sign of the *spatial* coordinates:

$$\mathbf{P}\psi(\mathbf{x}, t) = \psi(-\mathbf{x}, t).$$

If the system is to remain invariant under the parity operation, then the observable quantity which must not change is the probability density, which is just the absolute square of the wavefunction:

$$|\psi(\mathbf{x}, t)|^2 = |\psi(-\mathbf{x}, t)|^2.$$

So, if ψ describes a state of definite parity, then we have

$$\psi(\mathbf{x}, t) = \pm\psi(-\mathbf{x}, t),$$

and hence

$$\mathbf{P}\psi(\mathbf{x}, t) = \pm\psi(\mathbf{x}, t).$$

So if the system is to remain invariant under the parity operation, the system's wavefunction may either remain unchanged, $\mathbf{P}\psi = +\psi$, in which case we say the system is in an *even* parity state; or the system's wavefunction may change sign, $\mathbf{P}\psi = -\psi$, in which case we say the system is in an *odd* parity state.

If the forces governing the system respect parity, an even parity state cannot change into an odd one or vice versa. This helps us define the ways in which a system may evolve.

An example of this is the way in which light is emitted from atoms. Each state an electron can occupy has a definitive parity assignment, even or odd, which is determined by the magnitude of the orbital angular momentum of the electron about the nucleus and by the orientation of electron spin. As the electromagnetic force respects parity, and the photon has odd intrinsic parity (see below), transitions can only occur between atomic states of opposite parity. This limits the transitions possible and so prescribes the energies of the photons emitted. By observing the spectral lines emitted from atoms we can thus check the conservation of parity.

It is also necessary to consider the *intrinsic parity* of a single particle for which the operation of space inversion is not so obvious. This is illustrated by the decay of one particle into two. The final state of two particles with some well-defined motion with respect to one another can be examined under parity transformations and either even or odd parity assigned. If then the interaction responsible for causing the decay conserves parity, the initial one-particle state must also be a state of well-defined parity. Thus a particle can be assigned some intrinsic parity, even $(+1)$ or odd (-1), which is multiplied together with the spatial parity to obtain the overall parity of the state.

Intrinsic parity has meaning only because particles can be created or destroyed. If the particles in a system were always the same, then the product of their intrinsic parities in any initial or final states would always be the same and so would be a meaningless quantity. In this hypothetical immutable world we should be free to assign any particle either even or odd parity with no reason. In the real world this arbitrariness allows us to define the intrinsic parity of certain particles (normally the nucleons are given even parity) and then the intrinsic parities of all other particles are established from experiment.

6.3.2 *Charge conjugation symmetry*

Another useful symmetry in particle physics is that of the interchange of particles with their antiparticles, denoted **C**. This symmetry means that if physical laws predict the behaviour of a set of particles, then they will predict exactly the same behaviour for the corresponding set of antiparticles. For example, a collision between an electron and a proton will look precisely the same as a collision between a positron and an antiproton (see Figure 6.3).

The symmetry applies also to the antiparticles of particles with no electric charge, such as the neutron. The interaction of a proton and a neutron is the same as that of an antiproton and an antineutron.

Fig. 6.3. Symmetry under the charge conjugation operation implies the equivalence of (*a*) particle and (*b*) antiparticle reactions.

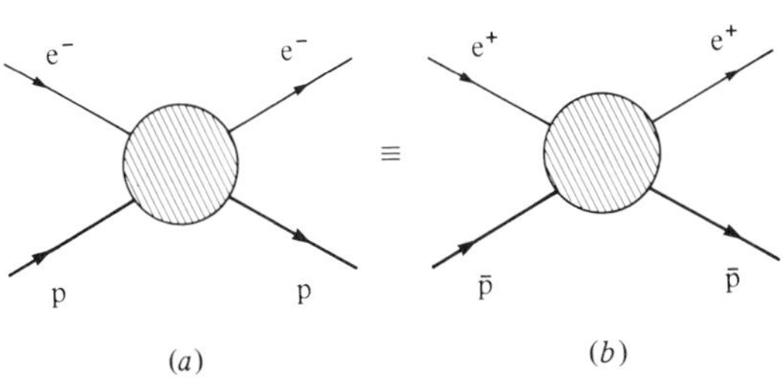

As with the parity operation, the wavefunction of a system may be even or odd under the action of the charge conjugation operation:

$$\mathbf{C}\psi = \pm\psi.$$

An example of the use of this symmetry is provided by particle decay into photons by the electromagnetic force. A single photon is odd under **C** symmetry. Observing the decay of a particle into two photons, then, determines the particle to be even under charge conjugation symmetry, as this is given simply by the product of the two photons' symmetries $(-1)^2$. We then know that the particle cannot decay into an odd-charge conjugate state, such as three photons, if **C** symmetry is to be preserved.

6.3.3 *Time reversal*

The last of the three discrete symmetries, denoted **T**, connects a process with that obtained by running backwards in time. Despite the rather intriguing name, the operation refers simply to that process obtained by reversing the directions of motion within the system. Symmetry under 'time reversal' implies that if any system can evolve from a given initial state to some final state, then it is possible to start from the final state and re-enter the initial state by reversing the directions of motion of all the components of the system.

6.4 The CPT theorem

It is possible also to define product symmetries which can be obtained by operating two or more of these discrete symmetries simultaneously. For instance, a system of particles in a coordinate system can be subject to the operations of parity and charge conjugation simultaneously to reveal a system of antiparticles in the reverse-handed coordinate system. If the laws governing the system are invariant under **CP** operation, then the two systems will behave in exactly the same way. Also, it is possible to assign an even or odd symmetry under the combined **CP** symmetry to any state and so require that the system evolves to a state of the same symmetry.

There are no utterly fundamental reasons for supposing that the individual symmetries should be preserved by the various forces of nature (that there should be symmetry between a process and its mirror image, between a process and its anti-process, and so on). But it seems a reasonable assumption and was taken for granted for many years. In fact, as we will soon see, the symmetries are *not* exact and there do exist phenomena which display slight asymmetries between process and mirror process, and process and antiprocess (see Chapter 12).

But there are very good reasons for supposing that the combined **CPT** symmetry is absolutely exact. So that, for any process, its mirror image, antiparticle and time-reversed process will look exactly as the original. This is the so-called **CPT** theorem, which can be derived from only the most fundamental of assumptions, such as the causality of physical events (cause must precede effect), the locality of interactions (instantaneous action at a distance is not possible) and the connection between the spin of particles and the statistics governing their collective behaviour.

The consequences of the **CPT** theorem are that particles and their antiparticles should have exactly the same masses and lifetimes, and this has always been observed to be the case. Another consequence is that if any one individual (or pair) of the symmetries is broken, as mentioned, there must be a compensating asymmetry in the remaining operation(s) to cancel it and so ensure exact symmetry under **CPT**.

6.5 Dynamical symmetries

The symmetries of space and time give rise to universal conservation laws such as those of energy, momentum and angular momentum. As these laws must be respected by all processes, the Lagrangian of any system must be invariant under the groups of transformations through time, space and angular rotations, respectively.

But other conservation laws are also known to exist, such as the conservation of electrical charge. This can be represented by requiring the Lagrangian to remain invariant under arbitrary shifts in the phases of the charged particle wavefunctions appearing in the Lagrangian (see Figure 6.4).

We will learn that there are many other quantities which are conserved in interactions arising from the various forces of nature. This implies that the Lagrangians describing these interactions

Fig. 6.4. A symbolic representation of the action of a dynamical symmetry.

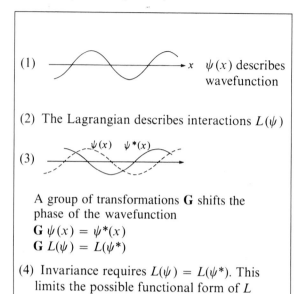

(1) $\psi(x)$ describes wavefunction

(2) The Lagrangian describes interactions $L(\psi)$

(3) $\psi(x)$ $\psi^*(x)$

A group of transformations **G** shifts the phase of the wavefunction
$$\mathbf{G}\,\psi(x) = \psi^*(x)$$
$$\mathbf{G}\,L(\psi) = L(\psi^*)$$

(4) Invariance requires $L(\psi) = L(\psi^*)$. This limits the possible functional form of L

must be invariant under appropriate symmetry operations. We will see that demanding such invariances gives rise to physically significant predictions such as the existence of new particles and values for their electric charges, spins and other quantum numbers yet to be introduced.

6.6 Internal symmetries

The symmetry operations we have introduced so far are the fundamental ways of describing the conservation laws we observe to obtain in particle interactions. But symmetry can also help us categorise particles according to their intrinsic properties.

In addition to the familiar particles carrying only electrical charge, we will soon meet particles with wholly new quantum numbers such as 'strangeness', 'charm' and so on. The values of these quantum numbers carried by the particles allow them to be classified into fixed patterns or multiplets. We shall see this at work in Part 2.

Suffice at this stage to say that, in the micro-world, symmetry does fulfil its traditional role of arranging disparate elements into regular patterns (just like the periodic table of elements).

6.7 Broken symmetries

Symmetries are sufficiently valuable that even broken ones can be useful. For many purposes a broken mirror is as good as a whole one! We have mentioned already how the individual reflection symmetries **P**, **C** and **T** might be broken in some classes of reaction (which turn out to be those governed by the weak nuclear force). But for other forces which do respect them, they are still a valuable guide for indicating which reactions are possible and which are forbidden.

Similarly, conservation laws and the internal symmetries on which they are based may not be exact. The first successful internal symmetry scheme for classifying the reactions of the strongly interacting particles was known from the start to be badly broken but, nevertheless, it provided a valuable ordering effect on the variety of reactions observed.

Of particular interest is the case when the Lagrangian governing the dynamics due to some force or forces is not quite invariant under some group of gauge transformations, but only under a restricted group or when additional particles have been introduced. This indicates that the relatively more complicated forces arising from the imperfect or restricted symmetries have their origin in a truly general symmetry (and its simpler forces) which may have obtained under different circumstances. This is the gist of the modern approach to the unified theory of the forces of nature in which approximate symmetries are a guide to the nature of forces in unfamiliar circumstances (e.g. just after the Big Bang).

7

Mesons

7.1 Introduction

Modern particle physics can be thought of as starting with the advent of mesons. For these are not constituents of everyday matter, as are the protons and the electrons, but were first proposed to provide a description of nuclear forces. The subsequent discoveries of a bewildering variety of mesons heralded an unexpected richness in the structure of matter, which took many decades to understand.

7.2 Yukawa's proposal

In attempting to describe the features of the strong nuclear force, physicists in the 1930s had to satisfy two basic requirements. Firstly, as the force acts in the same way on both protons and neutrons, it must be independent of electric charges and, secondly, as the force is felt only within the atomic nucleus, it must be of very short range. In 1935 the Japanese physicist H. Yukawa suggested that the nuclear force between protons is mediated by a massive particle, now called the pi-meson or pion, denoted by π, in contrast to the massless photon which mediates the infinite-range electromagnetic force. It is the mass of the mediating particle which ensures that the force it carries extends over only a finite range. This is indicated by Heisenberg's uncertainty principle which allows the violation of energy conservation for a brief period. If the proton emits a pion of finite mass, then energy conservation is violated by an amount equal to this mass energy. The time for which this situation can obtain places an upper limit on the distance which the pion can travel, and this distance is a guide to the maximum effective range of the force.

From the α-particle scattering experiments, we know that the effective range of the strong force is about 10^{-15} m, which gives a pion mass of about 300 times that of the electron, or about 150 MeV. To account for all the possible interactions between nucleons, the pions must come in three charge states. For instance, the proton may transform into a neutron by the emission of a positively charged pion or, equivalently, by the absorption of a negatively charged pion. But the proton may also remain unchanged during a nuclear reaction, which can be explained only by the existence of an uncharged pion. So the pion must exist in three charge states: positive, neutral and negative (π^+, π^0, π^-).

7.3 The muon

In 1937, five years after his discovery of the positron, Anderson observed in his cloud chamber yet another new particle originating from cosmic rays. The particle was found to exist in both positive- and negative-charge states with a mass some 200 times that of the electron, about 106 MeV. At first, the particle was thought to be Yukawa's pion and only gradually was this proved not to be the case. Most importantly, the new 'mesons' seemed very reluctant to interact with atomic nuclei, as indicated by the fact that they are able to penetrate the earth's atmosphere to reach the cloud chamber at ground level. For particles which were expected to be carrying the strong nuclear force such behaviour was unlikely. Also, there was no sign of the neutral meson. Theorists eventually accepted that this new particle was not the pion; instead it was named the *muon*, and is denoted by μ.

The muon was a baffling discovery as it seemed to have no purpose in the scheme of things. It behaves exactly like a heavy electron and it decays into an electron in 2×10^{-6} s; and so is not found in ordinary matter. Although we shall see later how the muon can fit into a second generation of heavy elementary particles, the reason for this repetition is still by no means obvious. So, the muon is not a meson at all, but a *lepton* like the electron.

7.4 The real pion

If Yukawa's pion is to interact strongly with atomic nuclei, it is unreasonable to expect it to penetrate the entire atmosphere without being absorbed. So experiments at ground level are unlikely to detect it. In 1947 C. Powell, C. Lattes and G. Occhialini from Bristol University took photographic plates to a mountain top to reduce the decay distance which pions created at the top of the atmosphere had to traverse before being detected. They found Yukawa's pion, which quickly decays into a muon, which itself then decays (Figure 7.1). The mass of these charged pions $\pi^\pm$ was determined to be 273 times the mass of the electron (140 MeV), very close to Yukawa's original estimate. Since the initial discovery of the charged pions, it has been established that decay into a muon and a neutrino is their main decay mode with a lifetime of about 2.6×10^{-8} s. Other decay modes do exist but are thousands of times less likely.

The uncharged pion π^0 was eventually discovered in accelerator experiments in 1950. The delay was due to the fact that uncharged particles

Fig. 7.1. The pion decays to a muon, which then decays to an electron. Neutrinos are emitted to ensure conservation of energy and angular momentum.

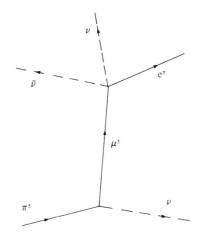

leave no obvious trace in most particle detectors and so cannot be observed directly. The most likely decay mode of the π^0 is into two photons which also leave no tracks. Only by observing the electron–positron pairs created by the photons can the existence of the π^0 be inferred (Figure 7.2). The mass of the π^0 was found to be slightly less than that of its charged counterparts at 264 times the mass of the electron, but its lifetime is much shorter at 0.8×10^{-16} s. The reason for this large difference in lifetimes is that the π^0 decays by the action of the electromagnetic force, as indicated by the presence of the two photons, whereas the charged pions decay by the weak force, as indicated by the presence of the neutrino.

In 1953 the pions were established spin 0 by comparing the relative magnitudes of the cross-sections of the reactions:

$$p + p \rightarrow d + \pi^+,$$
$$\pi^+ + d \rightarrow p + p.$$

The relative magnitude of the two can depend only on the spins of the particles in the collision, and knowing the spins of the protons (p) and the deuteron (d) determines that of the pion (π^+). Such reactions also establish the intrinsic parity of the pion (relative to the nucleons). This is found to be odd (-1).

7.5 Terminology

At this point it is worth both introducing some of the generic names which are used for these particles and defining their essential features. A few of the most often-used are:

nucleons: neutrons and protons;

hadrons: all particles affected by the strong nuclear force;

baryons: hadrons which are fermions (half-integral spin particles) such as the nucleons;

mesons: hadrons which are bosons (integral spin particles) such as the pion;

leptons: all particles not affected by the strong nuclear force, such as the electron and the muon.

Particles which are baryons are assigned a baryon number B which takes the value $B = 1$ for the nucleons, $B = -1$ for the antinucleons and $B = 0$ for all mesons and leptons. In all particle reactions, the total baryon number is found to be conserved (i.e. the total of the baryon numbers of all the ingoing particles must equal that of all outgoing particles). Similarly, particles which are leptons are assigned a lepton number which is also conserved in particle reactions. This is explained further in Chapter 14.

7.6 Isotopic spin

We have met, so far, two sorts of particles which differ only slightly in their masses but which have different electric charges: the nucleon (the proton and the neutron) and the pions. The strong nuclear force seems to ignore totally the effects of electric charge and influences all nucleons in the same way and all pions in the same way. As far as the strong force is concerned, there is only one nucleon and only one pion. In 1932 Heisenberg described this mathematically by introducing the concept of *isotopic spin* or *isospin*. This concept is the prototype of both the elementary particle classification schemes and the modern dynamical theories of the fundamental forces, so it merits some attention.

Recall that the two different orientations in (real) space of the 'third components' of the spin of the electron (see Chapter 3) provide two distinct

Fig. 7.2. The decay of the neutral pion.

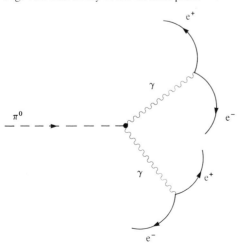

states in which the electron can exist (in the presence of a magnetic field). Analogously, Heisenberg proposed that different orientations in an abstract charge space of the third components of an imaginary isotopic spin would be a mathematically convenient way of representing the charge states within a family of particles (in the presence of electromagnetism) (see Figure 7.3). Similarly, just as the different components of electron spin are separated in energy by a magnetic field (causing the fine structure in spectral lines), so too the different components of isotopic spin in a particle family are separated in mass by the effects of the electromagnetic force (causing the slight

Fig. 7.3. The analogy between particle spin in real space and particle isospin in abstract charge space.

Spin assignment	Particle	Orientation of components of spin in space in the presence of a magnetic field
$s = \frac{1}{2}$	e^-	$s_z = +\frac{1}{2}$ $s_z = -\frac{1}{2}$
Isospin assignment	Particle	Orientation of components of isospin in charge space in the presence of electromagnetism
$I = \frac{1}{2}$	N	$I_3 = +\frac{1}{2}$ $I_3 = -\frac{1}{2}$
$I = 1$	π	$I_3 = +1$ $I_3 = 0$ $I_3 = -1$

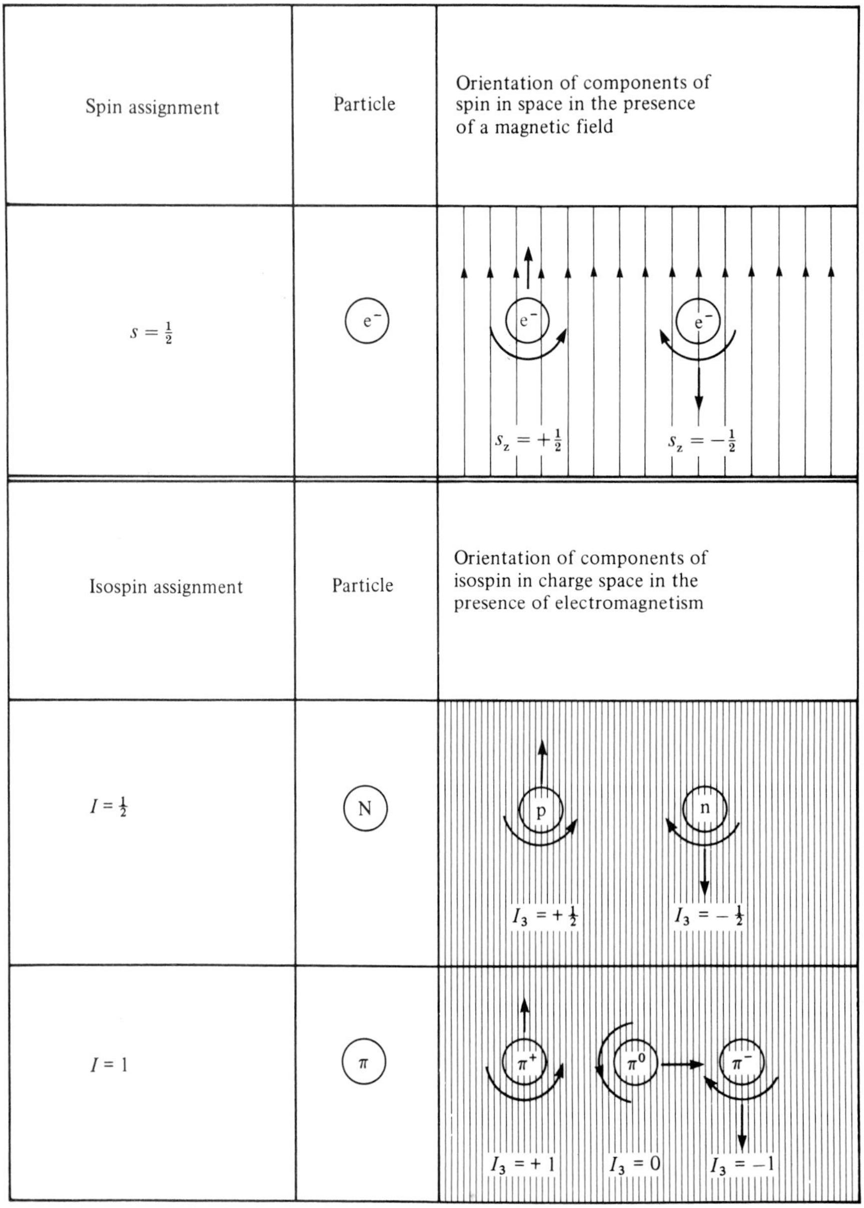

mass differences between the proton and the neutron, and between charged and neutral pions).

The electric charges of the hadrons, Q, are related to their isospin assignments by the simple formula,

$$Q = e\left(I_3 + \frac{B}{2}\right).$$

So for the pions which have zero baryon number ($B = 0$), the charges are simply the units of the electronic charge corresponding to the three 'third' components of spin I_3 $(1, 0, -1)$. For the nucleon which has unit baryon number ($B = 1$), the two isospin states with third component $+\frac{1}{2}$ and $-\frac{1}{2}$ become the positive and neutral charge states respectively.

8

Strange particles

8.1 Introduction

In 1947 the British physicists G. D. Rochester and C. C. Butler observed more new particles, about a thousand times more massive than the electron, in cloud chamber photographs of cosmic rays. As these particles were often associated with V-shaped tracks, they were at first called V particles (see Figure 8.1). Their origin and purpose were an entire mystery. If we remember that this same year saw the discovery of the real pion and

Fig. 8.1. (*a*) A neutral V^0 particle decays into pions. (*b*) A charged V^+ decays into a muon and a neutrino. These V particles are now called *kaons*, and denoted K^0 and K^+.

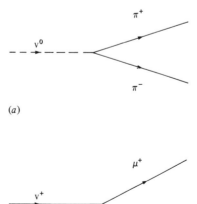

(*a*)

(*b*)

the subsequent redundancy of the muon, it is fair to think of it as the beginning of the baroque era of particle physics, in which an increasing number of particles were discovered, seemingly with no other purpose than to decorate cloud chambers. For the following six years, the V particles were observed in cosmic ray experiments and two kinds became apparent. There are those whose decay products always include a proton and are called *hyperons*, and there are those whose decay products consist only of mesons and are called *K mesons*, or *kaons*.

The hyperons and kaons soon became known as the *strange* particles because of their anomalous behaviour. They were observed frequently enough to indicate production by the strong nuclear force, say, between two protons, or a pion and a proton, and so we would expect a decay time typical of a strong nuclear process (i.e. about 10^{-23} s). But, from the length of their tracks in the photographs, it was possible to estimate their average lifetimes at about 10^{-10} s, the time-scale typical of weak interaction processes. This behaviour seemed to contradict the microscopic reversibility of reactions and required explanation.

8.2   Associated production

The first step in search of this explanation was provided in 1952 by the American physicist A. Pais. He suggested that the strange particles could not be produced singly by the strong interaction, but only in pairs. This was confirmed in experiments at the Brookhaven accelerator in 1953, when strange particles were man-made for the first time. The strange particles always emerged in pairs in reactions such as

$$\pi^- + p \rightarrow \Lambda^0 + K^0$$

where Λ^0 denotes the hyperon and K^0 denotes a neutral kaon.

In the same year Gell-Mann and Nishijima explained this mechanism of associated production by proposing the introduction of a new conservation law, that of *strangeness*, which applies only to the strong interaction. Each particle is assigned a quantum number of strangeness, in addition to its quantum numbers of spin, intrinsic parity and isospin. Then, in any strong inter-

action, the total strangeness of all the particles before and after the reaction must be the same.

Associated production can now be explained by assigning a positive strangeness to one of the strange particles produced and a negative strangeness to the other, so that the total strangeness of the final state is zero, the same as that of the non-strange initial state:

$$\pi^- + p \rightarrow \Lambda^0 + K^0 \qquad \text{Strangeness}$$
$$(0) + (0) \rightarrow (-1) + (+1) \qquad \text{assignments.}$$

The decay of strange particles into non-strange particles cannot proceed by the strong interaction, which must, by definition, conserve strangeness. Instead, such decays proceed by the weak interaction, which need not, and which allows the strange particles a comparatively long life:

$$\Lambda^0 \xrightarrow[\Delta S=1]{} \pi^- + p \quad (\tau \approx 10^{-10}\,\text{s}).$$

The strangeness of the strongly interacting hadrons is defined by

$$Q = e\left(I_3 + \frac{B+S}{2}\right).$$

When $S = 0$, we recover the equation of Section 7.6 which relates the charge to the third component of isospin for pions and nucleons and other non-strange hadrons.

8.3   The kaons

There are two charged strange mesons K^+ and K^- which each have a mass of 494 MeV, and a neutral one K^0 of mass 498 MeV. This makes the K mesons about three times more massive than the pions. But, like the pions, the kaons were found to be spin 0 and to have odd intrinsic parity. They are thus in some sense close relations of the pions. However, they have a very different multiplet structure. Let us recall that the three charge states of the pion (π^+, π^0, π^-) are the different I_3 states of the same $I = 1$ pion, and that the uncharged pion is its own antiparticle. This is not the case with the kaons because of complications due to the strangeness quantum number. If we assign to the neutral kaon K^0 a value of strangeness $S = 1$, then from the formula in Section 8.2, the value of total isospin $I = 1$ is ruled out and the kaons cannot form any isospin triplet like the pions. Instead, the kaons are grouped into isospin doublets as shown in Figure 8.2. From this we can see that the uncharged kaon must come in two versions with opposite strangeness if the scheme is to work. So although the K^- is the antiparticle of the K^+, the K^0 is not its own antiparticle, which must have different strangeness:

$$\overline{K^+} \equiv K^- \quad \text{but} \quad \overline{K^0} \neq K^0.$$

Because the K^0 is different from the $\overline{K^0}$ only by the value of its strangeness, it might somehow be able to exhibit some effects directly attributable to strangeness. After all, so far we have merely categorised observed particle decay patterns by awarding the particles different values of a hypothetical quantum number. If we could observe some experimental effect due to strangeness, then we might be more convinced of its physical reality. This thought occurred at the time to Fermi, who challenged Gell-Mann to prove the worth of his strangeness by demonstrating some difference between the K^0 and the $\overline{K^0}$. This led to some very important work, as we shall see in Chapter 15.

We can very neatly summarise our knowledge of the mesons discussed so far by plotting their assignments of isospin and strangeness (Figure 8.3). These graphs are known as multiplets for particles of the same spin and intrinsic parity and we shall see how they form the basis of the elementary particle classification scheme in Chapter 11.

8.4 The hyperons

The hyperons are the strange particles which eventually decay into a proton and which, like the proton, have spin $\frac{1}{2}$ and are baryons with baryon number 1. The lambda hyperon Λ^0 is the least massive at 1115 MeV and has isospin zero (it exists only as a neutral particle). The sigma hyperon Σ has a mass of 1190 MeV and has isospin 1 and so exists in three different charge states ($\Sigma^+, \Sigma^0, \Sigma^-$). Finally, the xi hyperon Ξ, known also as the cascade particle, has mass 1320 MeV and isospin $\frac{1}{2}$ and has strangeness -2. To decay into non-strange particles, it therefore needs to undergo two weak interactions, as the weak force can only change strangeness by one unit at a time:

$$\Xi^0 \to \Lambda^0 + \pi^0 \quad |\Delta S| = 1$$
$$\quad\quad \hookrightarrow \pi^- + p \quad |\Delta S| = 1$$

For the hyperons, we often prefer to use *hypercharge Y* as the distinguishing quantum number, which is the sum of baryon number and strangeness:

$$Y = B + S.$$

Those Λ, Σ and Ξ hyperons of spin $\frac{1}{2}$ which have been mentioned form only the basic set of those which exist. There are very many more massive hyperons which have spins $\frac{3}{2}, \frac{5}{2}$ or even $\frac{7}{2}$. These resonances are short-lived and generally decay quickly into one of the hyperons in the basic set by the strong interaction (conserving strangeness, or hypercharge) before these eventually decay by the weak interaction back into non-strange baryons.

Fig. 8.3. A multiplet of pions and K mesons arranged according to value of strangeness and third component of isospin.

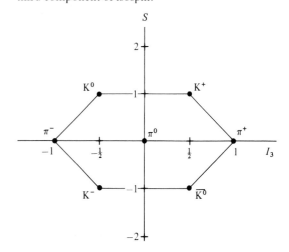

Fig. 8.2. Isospin doublets of the K mesons.

$$S = 1, \ I = \frac{1}{2} \quad \begin{cases} I_3 = +\frac{1}{2} & K^+ \\ I_3 = -\frac{1}{2} & K^0 \end{cases}$$

$$S = -1, \ I = \frac{1}{2} \quad \begin{cases} I_3 = +\frac{1}{2} & \overline{K^0} \\ I_3 = -\frac{1}{2} & K^- \end{cases}$$

8.5 Summary

In Figure 8.4 all the particles we have mentioned so far are plotted according to their masses and are categorised according to their generic names. The origin of the names is clear from the diagram: the leptons are the lightweights, the mesons are the middleweights and the baryons are the heavyweights. We also show the applicability of the fundamental forces to the various categories of particles. We may think it more than just coincidence that the strongly-interacting hadrons are the most massive category if we believe that the mass of the particles somehow arises from the interactions they experience.

We now know that the mass alone is not a reliable way to categorise the species. Recent experiments have found both leptons and mesons more massive than the baryons. Nowadays the classifications are taken to refer to the interactions experienced by the various classes, which is taken to be a more fundamental attribute than mass.

Fig. 8.4. The basic set of elementary particles known by the early 1950s.

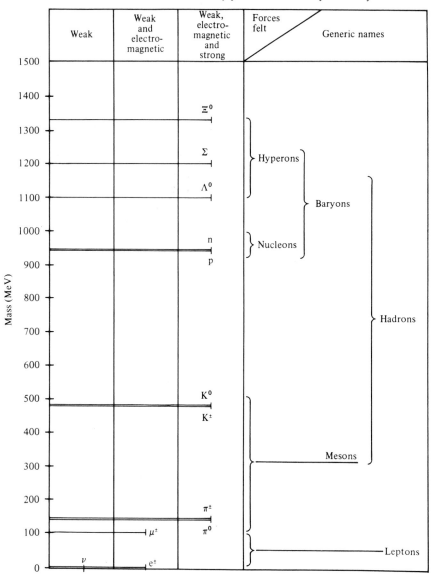

Part 2
Strong interaction physics

9
Resonance particles

9.1 Introduction

Most of the particles which we have discussed up to this point have lifetimes sufficient for them to leave observable tracks in bubble chambers or other detectors, say greater than about 10^{-12} s. But there is no reason for us to demand that anything we call a particle should necessarily have this property. It may be, for instance, that some particles exist only for a much shorter time before decaying into others. In this case we should expect not to detect them directly, but to have to infer their existence from the indirect evidence of their decay products. These transient particles are called *resonance particles* and many have been discovered with widely varying properties. It was the attempts to categorise the large number of resonances which first led to an appreciation of the need for a more fundamental pattern of order, which in turn led to the idea of quarks.

9.2 Resonance particle experiments

Resonance particles can be produced in two different types of experiment: resonance formation and resonance production experiments. In the formation experiments, two colliding particles come together to form a single resonance which acts as an intermediate state between the original colliding particles and the final outgoing products of the collision. The presence of the resonance is indicated when the cross-section for the collision (i.e. the effective target area of the colliding particles) varies dramatically over a small range of collision energy centred on the mass of the resonance (see Figure 9.1). The value of the energy range corresponding to one-half of the height of the resonance peak is referred to as the *width* of the resonance and this is a measure of the uncertainty in the mass of the particle.

Only if a particle is perfectly stable can it be thought of as having a uniquely defined mass, as time is needed for the act of measurement defining the mass. For an unstable particle there will always be uncertainty in the value of its mass, given by Heisenberg's uncertainty principle:

$$\Delta E \, \Delta t \geqslant \hbar.$$

Fig. 9.1. An example of resonance formation. A large increase in the pion–proton cross-section, σ, plotted against the pion beam momentum $p_{\pi\text{beam}}$ signals the formation of a resonance particle.

From this we can see that the narrower the width ΔE of the resonance, the larger will be the uncertainty in the lifetime, thereby implying a longer-lived particle. Conversely, if the resonance is broad, this implies a short lifetime. Typical widths for hadronic resonances, such as the N* resonances in pion–proton scattering, are a few hundred MeV, which correspond to lifetimes of about 10^{-23} s. This makes them the most transient phenomena studied in the natural world.

In resonance production experiments, the presence of a resonance is inferred when it is found that the outgoing particles prefer to emerge with a particular value of combined mass. Finding the resonances in this fashion is more difficult because it is first necessary to look at all the possible combinations of outgoing particles which might have arisen from the resonance, and then to plot the combined masses of the combinations to see if any preferred values exist (see Figure 9.2).

One advantage of the production method is that it does not require us to study only the reso-

Fig. 9.2. An example of resonance production.

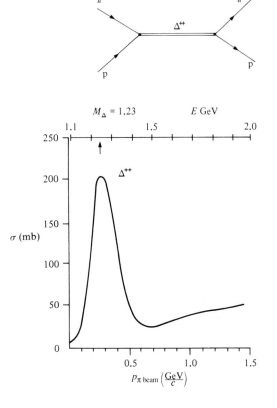

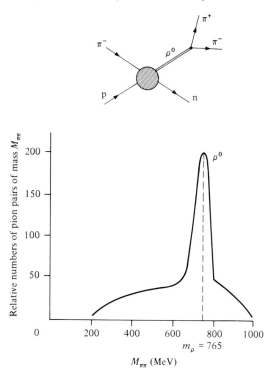

nances which can be formed by the ingoing particles. In high-energy collisions, any number of new and interesting particles may emerge and it is possible to see if they have originated from some previously unknown resonance. In this fashion we can study the resonances made from πs and Ks only, even though we cannot arrange collisions between only πs and Ks as the ingoing particles. These methods have allowed physicists to build up a picture of literally hundreds of resonances, all of which may be legitimately regarded as just as elementary as the neutron or the pion.

Over the years, interesting regularities in the resonance spectrum became apparent. Often a particular set of quantum numbers for isospin and strangeness, such as for the pion ($I = 1, S = 0$) and the kaons ($I = \frac{1}{2}, S = 1$) are duplicated by particles with higher masses and spins. These higher-mass versions of the quantum numbers generally decay very quickly back down to the least-massive particle with those particular numbers, by the strong interaction. This least-massive version can then itself decay more slowly by the weak force, violating quantum-number conservation as it does so (see Table 9.1).

Table 9.1. *Two mass series of meson resonances*

Meson symbol	I	S	Mass (MeV)	Spin	Decay	Force acting
π	1	0	140	0	$\mu\nu$	weak
ρ	1	0	768	1	$\pi\pi$	strong
a_2	1	0	1320	2	$\rho\pi$	strong
ρ_3	1	0	1690	3	4π	strong
K	$\frac{1}{2}$	1	494	0	$\mu\nu$	weak
K*	$\frac{1}{2}$	1	892	1	$K\pi$	strong
K_2^*	$\frac{1}{2}$	1	1425	2	$K\pi$	strong

10

SU(3) and quarks

10.1 Introduction

By the early 1960s it became clear that many hundreds of so-called 'elementary' resonance particles exist, each with well-defined values of the various quantum numbers such as spin, isospin, strangeness and baryon number and with widths which are generally seen to increase (or lifetimes which are generally seen to decrease) as their masses become larger. At that time, the most urgent task for physicists was to discover the correct classification scheme for the particles, which would do for the elementary particles what Mendeleeff's periodic table had done for the variety of elements known in the nineteenth century. A closely related problem was whether or not it was sensible to regard such a plethora of different particles as truly elementary. Most of the resonance particles are very massive compared with, say, the electron and occupy a finite region of space with a radius of about 10^{-15} m, while many have high values of spin and the internal quantum numbers. All these factors argue in favour of the existence of more-fundamental constituents combining in a variety of ways to make up the known hadrons, just as a few fundamental atomic constituents (electrons, protons and neutrons) can combine to make up the variety of elements. But, historically, it was not possible to pass directly to the analysis of these fundamental constituents, which at the time were extremely speculative.

Initially, it was necessary to classify the bewildering variety of hadrons according to some symmetry scheme from which clues to the nature of the constituents could be derived.

10.2 Internal symmetry

Such a classification scheme is provided by an internal symmetry group, as described in Chapter 6. The scheme was proposed independently by Murray Gell-Mann and Yuval Ne'eman in 1961. The starting point of this symmetry group is the charge independence of the strong nuclear forces, as expressed in Chapter 7 by the concept of isospin. By regarding the neutron and the proton as the isospin down and isospin up components of a single nucleon, the strong interaction's indifference to 'neutron-ness' and 'proton-ness' can be expressed as the invariance of strong interactions to rotations in the abstract isospin space. The group of transformations which achieves these rotations is the *Special Unitary* group of dimension 2 called $SU(2)$, which acts on the 2-dimensional space defined by the proton and the neutron, redefining the proton and neutron as a mixture of the original two:

$$\mathbf{G}^{SU(2)}\begin{pmatrix} p \\ n \end{pmatrix} \rightarrow \begin{pmatrix} p^* \\ n^* \end{pmatrix}.$$

Of course, the same must also be true of the pions, which form a 3-dimensional space (π^+, π^0, π^-), and the Δ baryons $(\Delta^{++}, \Delta^+, \Delta^0, \Delta^-)$, which form a 4-dimensional space. These are referred to as the 2-, 3- or 4-dimensional representations of $SU(2)$.

When conservation of strangeness is added to that of isospin as a property of the strong interaction, it is clear that the strongly interacting particles are governed by a bigger symmetry group. Although it seems obvious, it took a great deal of work to show that $SU(3)$ is the appropriate group. The transformations of the $SU(3)$ group generate many dimensional representations (multiplets), **1**, **3**, **6**, **8**, **10**, **27**, etc., each of which is a well-defined quantum-number pattern. It was a triumph for the originators of the scheme to find that some of these exactly fitted the quantum-number structure of the observed hadrons (see Figure 10.1). The identification of the correct symmetry group for the strong interactions, and

the assignment of hadrons to the multiplets, led to the prediction in 1962 of a new hadron necessary to complete the spin-$\frac{3}{2}$ baryon decuplet **10**. This is the famous Ω^- particle with strangeness assignment -3. Its spectacular discovery in 1964 in bubble chamber photographs at Brookhaven convinced a previously sceptical world of the validity of $SU(3)$.

Fig. 10.1. $SU(3)$ representations provide the quantum-number patterns for the elementary particles.

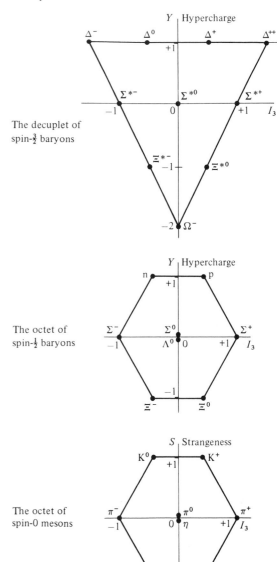

The decuplet of spin-$\frac{3}{2}$ baryons

The octet of spin-$\frac{1}{2}$ baryons

The octet of spin-0 mesons

The correct symmetry group having been found, a major problem remained. It was necessary to explain why the mesons filled some multiplets and the baryons fitted others, but other multiplets had no particles. In particular it seemed odd that the fundamental 3-dimensional representation should remain unfilled (i.e. the most basic representation of the $SU(3)$ group). In an unsuccessful symmetry scheme prior to that of Gell-Mann and Ne'eman, the proton, neutron and hyperon were assigned to this triplet, but the logical consequences of such an assignment were incompatible with the experimental evidence.

10.3 Quarks

In 1964, Gell-Mann and George Zweig pointed out that the representations of $SU(3)$ which were occupied by particles could be chosen from amongst all those mathematically possible by assuming them to be generated by just two combinations of the fundamental representation. Gell-Mann called the entities in the fundamental representation *quarks* (a word abstracted from the novel *Finnegan's Wake* by James Joyce). The three varieties of quark, or flavours as they are now called, have since come to be known as the up, down and strange quarks, the up and down labels referring to the orientation of the quarks' isospin. The combinations of quarks which give the occupied representations of $SU(3)$ are a quark–antiquark pair for the meson multiplets and three quarks for the baryon multiplets. This is expressed mathematically by combining the representations of the group:

$$\mathbf{q} \otimes \mathbf{q} \otimes \mathbf{q} \equiv \mathbf{3} \otimes \mathbf{3} \otimes \mathbf{3} \rightarrow \mathbf{1} \oplus \mathbf{8} \oplus \mathbf{8} \oplus \mathbf{10}$$
$$\mathbf{q} \otimes \bar{\mathbf{q}} \equiv \mathbf{3} \otimes \mathbf{3}^* \quad \rightarrow \mathbf{1} \oplus \mathbf{8}$$

The quark constituents of the baryon decuplet and of the meson octet are illustrated in Figure 10.2.

One significant consequence of this scheme is that if three quarks are to make up each baryon with a baryon number 1, then the quarks themselves must have baryon number $\frac{1}{3}$. From the formula relating charge to isospin and baryon number, this means that they must also have fractional electronic charge. Also, to ensure that the baryons generated are fermions and that the mesons are bosons, it is necessary to assign the quarks spin $\frac{1}{2}$. A summary of their properties is shown in Table 10.1.

Table 10.1. *The quantum number assignments of the early quarks*

Quark	q	Spin	Charge	I	I_3	S	B
Up	u	$\frac{1}{2}$	$+\frac{2}{3}$	$\frac{1}{2}$	$+\frac{1}{2}$	0	$\frac{1}{3}$
Down	d	$\frac{1}{2}$	$-\frac{1}{3}$	$\frac{1}{2}$	$-\frac{1}{2}$	0	$\frac{1}{3}$
Strange	s	$\frac{1}{2}$	$-\frac{1}{3}$	0	0	-1	$\frac{1}{3}$

Fig. 10.2. The quark content of the $SU(3)$ representations. $(qqq)'$ signifies the summation over the cyclic permutations of the quarks.

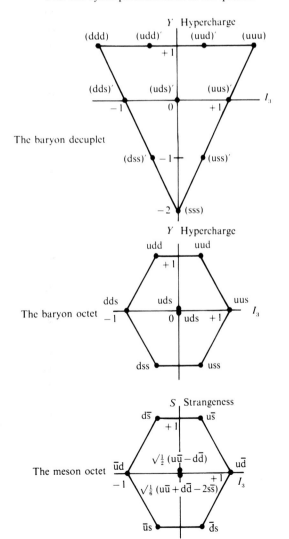

Quarks are also useful in providing a qualitative understanding of various hadronic processes through what are called *quark line diagrams*. As an example, consider the decay of the Δ^{++} (uuu) into a proton (uud) and a pion (u$\bar{\text{d}}$) shown in Figure 10.3. Forward-directed arrows indicate quarks whilst arrows directed backwards in time indicate antiquarks. Note that these are not the same as Feynman diagrams because the quarks are confined inside hadrons and the strong interactions between them are generally not shown. Because of conservation of baryon number, quark lines cannot be broken.

The quarks were referred to earlier as entities rather than particles for good reason. It is not necessary to assume their existence as observable particles to enjoy the successes of the $SU(3)$ flavour scheme. They may be thought of as the mathematical elements only for such a scheme, devoid of physical reality. This was a fortunate escape clause at the beginning of the quarks' career because their fractional electronic charges and the failure to observe them in experiments encouraged scepticism in the naturally conservative world of physics. As we shall see, indirect evidence for the physical reality of quarks is now very convincing – despite the fact that they have never been seen directly in isolation. But this evidence has mounted only rather slowly since 1968 with the beginning of the 'deep inelastic' experiments at the Stanford Linear Accelerator Center (SLAC) in California. Prior to this, most physicists preferred to reserve judgement on the reality of quarks, content to rely on the mathematics of $SU(3)$ only.

Because of this historical background of doubt, conclusions which rely on the mathematics of group theory are put together as the $SU(3)$ scheme of the elementary particles, and con-clusions which rely on the physical reality of quarks are referred to as the 'quark model'. The mathematics of $SU(3)$, in addition to generating the multiplet structure of the observed particles, can also provide simple predictions of relationships between the masses of the particles in an $SU(3)$ multiplet. If the $SU(3)$ symmetry were perfect, then all the particles in the same $SU(3)$ multiplet would have to have the same mass. This is obviously not true and so we know that $SU(3)$ cannot be a perfect symmetry: it is broken. But by making assumptions about just how the symmetry fails, it is possible to derive mass formulae which seem to hold good:

$$\tfrac{1}{2}(m_{\text{N}} + m_{\Xi}) = \tfrac{1}{4}(3m_{\Lambda^0} + m_{\Sigma}) \qquad \text{baryons,}$$
$$m_{\text{K}}^2 = \tfrac{1}{4}(3m_{\eta}^2 + m_{\pi}^2) \qquad \text{mesons.}$$

In the quark model, the effects of symmetry break down can be described by saying that the strange quark has a larger mass than either of the equal mass up and down quarks. Also, in the quark model, it is possible to assume the existence of forces holding the quarks together and then to generate the spectrum of elementary particle masses for particles of the same quantum number and different spins. The predictions are found to match the experimentally measured masses really rather well, better than the approximations of the model would seem to justify in fact. But the simple quark model is unable to explain the outstanding problems surrounding the quarks. Why are they not seen? Why do they seem to form only in certain combinations? What is the nature of the forces which they experience? These questions, as we will see, had to await the advent of a theory of quarks on a par with the QED theory of electrons.

Fig. 10.3. Quark line diagram for the decay $\Delta^{++} \rightarrow \text{p}\pi^+$. Forward-directed arrows indicate quarks and backward-directed arrows antiquarks.

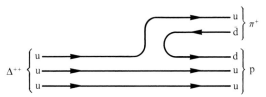

11

Hadron dynamics

11.1 Introduction

The unresolved problems of the quark model led the majority of physicists in the 1960s to restrict their attention to the dynamics of the observed hadrons only. Although it was generally realised that no fundamental theory of the strong interaction was likely to result, this work did lend some order to the bewildering variety of effects seen in hadron collisions, such as those between pions and protons (πp), kaons and protons (Kp) and protons and protons (pp) which constituted the majority of the experimental programmes for the 1960s. This investigation into hadron dynamics performs the complementary task to the $SU(3)$ internal symmetry scheme which orders the static properties of the hadronic resonance spectrum.

11.2 Regge trajectories

In simple two-particle scattering, Regge theory describes how the angular momentum of the two-particle system varies with the energy of the collision. So, in the resonance region, Regge theory tells us how the spin of the resonances varies with their mass. If we know the form of the forces acting between the particles, we can predict the spectrum of resonances on a graph of spin against mass. These are known as *Regge trajectories* (see Figure 11.1).

Regge theory was first used for the low-energy scattering of particles in a known potential, such as in the electric field in an atom. Strictly speaking, the theory can be justified only in this area. It came as a pleasant surprise to physicists to find that the resonances observed in higher-energy collisions lay on particularly simple Regge trajectories. They are all straight lines. This provides a classification scheme which relates the spins and the masses of all resonances with the same internal quantum numbers, in contrast to the $SU(3)$ scheme which relates particles of different quantum numbers with the same values for spin (and mass in perfect $SU(3)$). Some of the Regge trajectories for mesons and baryons are shown in Figure 11.2.

11.3 Hadron collisions

The usefulness of Regge theory does not stop simply at providing another classification scheme for the resonances, it can describe hadron collisions also. Looking back to Chapter 7, we remember that the earliest picture of the strong nuclear force was that of it being carried between two colliding hadrons (say protons) by a pion, in analogy with the photon exchange mechanism of an

Fig. 11.1. Particles approach with given energy and angular momentum to form a resonance of given mass m and spin J. The spectrum of resonances lies on a Regge trajectory relating mass and spin.

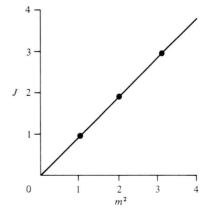

Fig. 11.2. Regge trajectories for mesons and baryons. The spin is denoted by J.

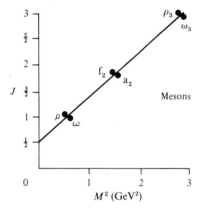

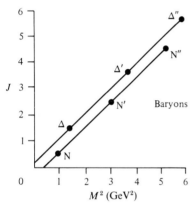

electromagnetic process. Unfortunately, this simple picture has severe limitations. Firstly, the strength of the nuclear forces means that there is no justification for ignoring two-, three- or more pion exchange processes; in principle they may be considerably more important than simple pion exchange. Secondly, the calculations which can be performed for single-pion exchange give only limited agreement with experimental results.

And finally, one pion exchange cannot alone suffice: many different mesons are known and it is unlikely that these have no part to play in hadron dynamics. Also, there are some types of collision (pion–proton scattering, for example), where pion exchange is actually forbidden by conservation of parity or some other quantum number. This indicates that some other mechanism must be at work.

An improved description of high-energy hadron collisions is obtained when it is not just one pion (or any other meson) which is taken as being exchanged by the colliding particles, but the entire contents of one or more Regge trajectories. This is often referred to as *reggeon* exchange. A single reggeon exchange will therefore represent the exchange of all those resonances with differing masses and spins with an otherwise identical set of internal quantum numbers (see Figure 11.3).

The calculations of the reggeon exchange process are far less certain than the Feynman rules of QED, but they do a similar job. In particular they predict that the total cross-sections for hadron collisions are nearly constant over a very wide range of collision energies, just as is observed in experiments (Figure 11.4).

Fig. 11.3. The exchange of a reggeon is equivalent to the exchange of many different spin resonances.

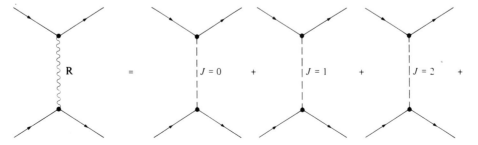

For hadron collisions at very high energies, say at about 50 times the rest mass energy of the proton, the dominant feature of most events is the sheer number of particles produced. This number is called the *multiplicity* of the reaction (see Figure 11.5). Regge theory can be extended to describe these reactions by introducing the concept of multiple reggeon exchange. Formulae describing these processes can then be used as the basis for a reggeon field theory of hadron collisions in which it is the reggeons that play the same role as do the photons in QED.

The picture of hadron collisions is made consistent with the $SU(3)$ classification scheme of the hadrons by using quark diagrams to show the flow of internal quantum numbers in a collision

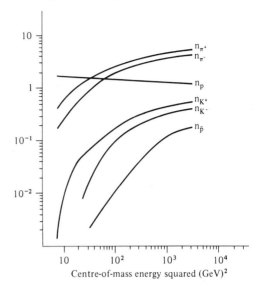

Fig. 11.5. Multiplicity data in hadron collisions.

process whose dynamics are described by Regge theory. Thus it is possible to represent electric charge or strangeness being carried by the reggeon from one external particle to another, as in Figure 11.6. As mentioned previously, the use of quarks and quark line diagrams in no way requires the physical existence of the quarks – they may be thought of purely as convenient mnemonics for the

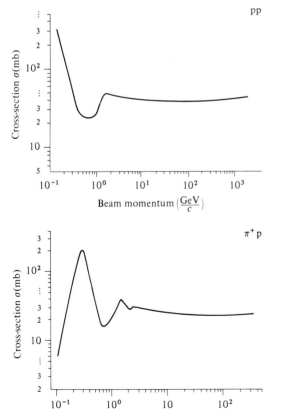

Fig. 11.4. Examples of the behaviour of cross-section data obtained in high-energy hadron collisions.

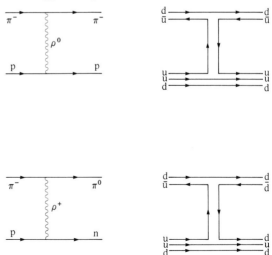

Fig. 11.6. Quark line diagrams corresponding to reggeon diagrams.

$SU(3)$ symmetry of the observed hadrons. On the other hand, the progression to quark line diagrams is highly suggestive of a more fundamental role.

11.4 Summary

It was the plethora of resonance particles discovered mainly in hadron collisions which first gave rise to the two classification schemes of $SU(3)$ flavour symmetry and Regge trajectories. The $SU(3)$ scheme successfully relates particles of different quantum numbers and of the same spin, and in so doing introduces elemental quantum-number entities which act as building blocks, certain combinations of which can account for all the observed hadrons. These entities were, as we shall see, later established to correspond to new fundamental particles called quarks. The Regge theory, on the other hand, relates hadrons of different spins and masses, but of the same quantum numbers. It also provides a mathematical description of high-energy hadron collisions, but with no fundamental justification. So, Regge theory is only a 'macroscopic' approximation to the real dynamics of the fundamental quarks.

Part 3
Weak interaction physics I

12

The violation of parity

12.1 Introduction

The decay of the strange kaons led to a great deal of confusion in the early 1950s. Two decay modes in particular seemed so different that they were at one time thought to originate from two different parent particles, called the τ and θ mesons:

$$\tau^+ \to \pi^+ + \pi^- + \pi^-,$$
$$\theta^+ \to \pi^+ + \pi^0.$$

However, detailed study of the two- and three-pion final states indicated that the τ and θ were indeed both manifestations of the same charged kaon, K^+. In both cases the mass was the same, and so too was the lifetime – about 10^{-8} s, a timescale which indicates it is the weak force that is responsible for the decays. The decays were thought to be incompatible because the parities of the two final states are different. If they originate from the same initial particle, they imply that parity is not conserved by the force responsible for the decays. This means that the force behaves differently in left-handed and right-handed coordinate systems: it can distinguish left from right, or image from mirror image.

Such a revolutionary conclusion was not seriously entertained until 1956, when T. D. Lee and C. N. Yang pointed out that, although evidence existed for the conservation of parity by the strong and electromagnetic forces, there was no evidence for its conservation by the weak force. Certainly, the τ–θ puzzle indicated that the weak force did not conserve parity, and Lee and Yang proposed that this was a general feature of all weak interactions.

12.2 β decay of cobalt

Within months of Lee and Yang's suggestion, experiments were performed to test for parity violation in other weak processes. The first and most famous was conducted on the β decay of cobalt by C. S. Wu and E. Ambler at the National Bureau of Standards in Washington. The point of the experiment was to observe some spatial asymmetry in the emission of β-decay electrons from

the cobalt, which could lead to a distinction between β decay and its mirror-image process. The process in question was the ordinary radioactive β decay of cobalt into nickel:

$$^{60}\text{Co} \rightarrow {}^{60}\text{Ni} + e^- + \bar{\nu}.$$

Firstly it was necessary to establish some direction in space of which the cobalt was aware, and with respect to which the emission of β-decay electrons could be measured. This was done by putting a magnetic field across a specimen of cobalt, cooled to a very low temperature. In this situation, the spin of the nuclei align predominantly along the direction of the magnetic field. By measuring the

Fig. 12.1. (a) If no asymmetry were detected in the emission of decay electrons, the real world and the mirror world would be indistinguishable. (b) If asymmetry were detected, this would result in a distinction between the two. This latter case is observed in experiments.

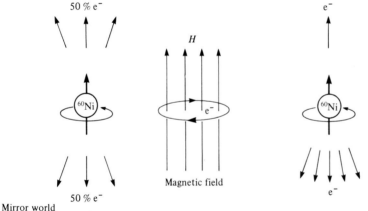

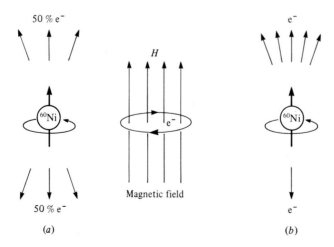

(a)　　　　　　　　　　　　　　　　　　　　(b)

emission of the β-decay electrons along or against the orientation of nuclear spin (the orientation of the magnetic field), any asymmetry can be detected. It is possible to show that the direction of spin will not change under mirror reflection, nor will the direction of the magnetic field. But the direction in which the β-decay electrons are emitted will change under mirror reflection and so any asymmetry of electron emission measured with respect to the magnetic field direction will appear to be reversed in the mirror-image process (see Figure 12.1). Hence the process and its mirror image are distinguishable, and the weak force responsible for nuclear β decay can tell its right hand from its left.

The world of physics could scarcely have been more surprised when Wu and her colleagues duly observed the asymmetry which Lee and Yang's work had implied. Not since the discovery of the quantum nature of light had Nature seemed so contrary to common perception. The shock is reputed to have led one eminent physicist to accuse God of being 'weakly left-handed'.

Other experiments soon confirmed this

Fig. 12.2. How the basic ^{60}Co experiment transforms under repeated **C** and **P** transformations.

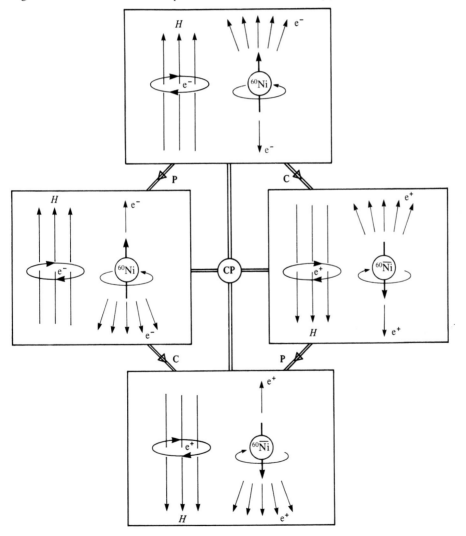

parity-violating effect. One example is provided by the decay of the hyperon in the process,

$$\pi^- + p \to \Lambda^0 + K^0$$
$$ \hookrightarrow \pi^- + p.$$

It is possible to define a plane formed by the tracks of the incoming π^- and the outgoing hyperon. If parity were conserved, equal numbers of outgoing π^- would emerge on either side of this plane. In 1957 an experiment yet again detected an asymmetry indicating parity violation.

12.3 Absolute-handedness and CP invariance

It is an interesting, if rather academic, question of philosophy to ask whether or not it is possible to distinguish absolutely between left and right, using the parity-violating effects of the weak force. The famous thought-test of such a distinction is to attempt to communicate our convention for left and right (or clockwise and anticlockwise) to an intelligent alien in a distant galaxy.

We might think of achieving this by instructing the alien to perform the ^{60}Co experiment and telling him that our definition of anticlockwise is the direction required of the electron loop that provides the magnetic field, when viewed from the direction towards which most of the β-decay electrons are emitted. This would certainly suffice for aliens in our galaxy, but would not necessarily for aliens in more distant parts of the universe. This is due to the possibility that distant aliens may be made of antimatter and may be conducting an anticobalt-60 experiment in which precisely the same procedures would lead to the opposite of our intended conclusions. The reason for this is that although the weak force violates parity, it also violates the symmetry of charge conjugation (matter–antimatter interchange) in such a way that the product symmetry of the two, denoted **CP**, is almost exactly conserved.

Starting from our original experiment in which the majority of β-decay electrons emerge in the direction of the field, we can see that the operation of space inversion will lead to an observable difference: the electrons are emitted against the direction of the field. However, if we then imagine the *additional* operation of charge conjugation we are led to an exact copy of the original process: the particles are emitted along the direction of the field (see Fig. 12.2). So a real-matter alien looking at our original experiment from the direction in which most decay products are emitted will see a clockwise current loop, but the antimatter alien will see an anticlockwise current loop. Thus an alien observing that the emitted particles come out preferentially in the direction of the field would know that *either* his conventions about left–right and particles–antiparticles were the same as ours *or* that they were both different.

Of course, we have not been able to test the validity of the **CP** conservation using anticobalt, although other experiments have been conducted to show that it is preserved to a high degree. But this is not the end of this particular story, as we will see in Chapter 15.

13

Fermi's theory of the weak interactions

13.1 Introduction

Prior to the early 1960s, just three different leptons were recognised: the electron, the muon and the neutrino (together with their antiparticles). The best place to study the weak interaction is in processes involving these leptons only. This ensures that there are no unwanted strong interaction effects spoiling the picture. Unfortunately, early opportunities to study purely leptonic reactions were limited, being restricted to the muon decay into an electron and neutrino. The most common weak interactions available for study are the radioactive β decay of nuclei and the decay of the pions and kaons (which are described generically as the weak decay of hadrons), and it was predominantly these reactions which formed the basis for the first description of the weak interactions formulated by Fermi in 1933.

13.2 Fermi's theory of β decay

The simplest manifestation of β decay is the decay of a free neutron into a proton, an electron and an antineutrino (see Figure 13.1):

$$n \rightarrow p + e^- + \bar{\nu}.$$

Fermi took this to be the prototype for the weak interactions, which he thus described as four fermions reacting at a single point. He expressed this mathematically by saying that, at a single point in space–time, the quantum-mechanical wavefunction of the neutron is transformed into that of the proton and that the wavefunction of the incoming neutrino (equivalent to the outgoing antineutrino we actually see) is transformed into that of the electron. So a description of this reaction is provided by multiplying the wavefunctions by unknown factors Γ which effect the transformations, and by another factor G_F called the Fermi coupling constant. This is the quantity which governs the intrinsic strength of the weak interactions, and so the rate of the decay. Thus the amplitude for β decay is given by

$$M = G_F(\bar{\psi}_p \Gamma \psi_n)(\bar{\psi}_e \Gamma \psi_\nu).$$

The factors Γ contain the essence of the weak interaction effects which give rise to the transformation of the particles. The challenge was to discover the nature of these quantities (whether they are just numbers (scalars) or vectors, tensors, etc.). By examining the angles of emission between the outgoing products of β decay and their various energies, it is possible to narrow down the choice. This took many years: their nature was not confirmed until the parity-violating effect of the weak force was known.

In 1956 Feynman and Gell-Mann proposed that the interaction factors Γ be a mixture of vector and axial-vector quantities, to account for the parity-violating effects of the interactions.

A vector quantity has well-defined properties under a Lorentz transformation. For instance, it will change sign if rotated through 180° and will appear identical after rotation through 360°. An axial-vector quantity will transform just like a vector under rotations, but will transform with the opposite sign to a vector under improper transformations such as parity. Thus, if the interaction

Fig. 13.1. The β decay of a free neutron.

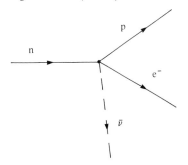

comprises vector and axial-vector components, it will look different after a parity transformation (the components might add together instead of cancelling), which is just what we need to describe the weak interactions. By inserting this form of interaction factor Γ into the amplitude M for β decay, it is possible to calculate the features of particle emission in free neutron decay.

13.3 The polarisation of β-decay electrons

We have already seen how parity violation manifests itself in β decay as an asymmetry in the direction of emission of electrons. But it also affects the spins of the emitted electrons. In the absence of parity violation, as many left-handed as right-handed electrons should be emitted (see Figure 13.2). But because of it, the electrons show a net preference to spin in one of the two possible ways. We define a polarisation P of the electrons to quantify this preference:

$$P = \frac{N_R - N_L}{N_R + N_L},$$

where N_L (N_R) is the number of left-spinning (right-spinning) electrons in a measured sample. When $P = 1$, all the electrons are right-spinning

Fig. 13.2. The emission of left-handed and right-handed electrons in β decay. Left-handed electrons are found to predominate experimentally.

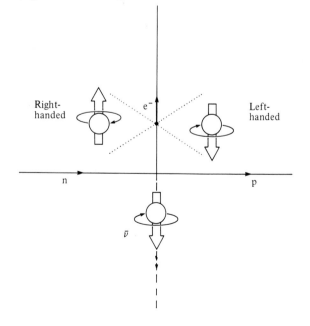

and when $P = -1$, all the electrons are left-spinning. Assuming that the final-state proton does not recoil, the polarisation can be calculated to be equal to minus the ratio of the electron's speed to that of light:

$$P = -\frac{v_e}{c}.$$

So when the electrons are emitted slowly, $v_e \approx 0$ and there is no net polarisation. But when the electrons are emitted relativistically, $v_e \approx c$, they are nearly all left-handed. In 1957, F. Frauenfelder and his colleagues observed the polarisation of the electrons from the β decay of ^{60}Co by scattering them through a foil of heavy atoms. They found a net polarisation of -0.4 for electrons travelling at $0.49c$, which is taken as a satisfactory agreement with the prediction.

13.4 Neutrino helicity

The four-fermion interaction constrains, very tightly, the spins which can couple through the weak interactions. In fact, when we look at the neutrino spins which are allowed to couple through the Γ we have specified, we find that only left-handed neutrinos and right-handed antineutrinos can take part in the weak interactions. As the neutrinos interact only by the weak interactions, only these two possible cases are ever seen (Figure 13.3). (Neutrinos and antineutrinos of the

Fig. 13.3. Only left-handed helicity neutrinos (and right-handed antineutrinos) have been observed. Helicity is the component of a particle's spin along its direction of motion. It is simply a convenient way of defining the spin of moving particles.

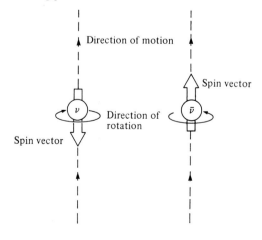

opposite helicities have never been observed, yet may still exist. If so, they would not take part in any of the known interactions, except for gravity.)

To measure the neutrino helicity we must look for a particularly simple example of β decay and deduce it from the helicities of the other decay products, using the principle of angular momentum conservation.

Such an experiment was performed by M. Goldhaber and his colleagues in 1958. The spin-0 nucleus ^{152}Eu is observed to undergo hybrid β decay in which an electron is captured and a neutrino emitted, leaving an excited state of the nucleus ^{152}Sm with spin 1. This then decays to its spin-0 ground state by the emission of a photon (Figure 13.4):

$$^{152}\text{Eu} + \text{e}^- \rightarrow {}^{152}\text{Sm}^* + \nu$$
$$\hookrightarrow {}^{152}\text{Sm} + \gamma.$$

The initial and final states of the nuclei are spin 0, so if there is no angular momentum between the neutrino and the photon, then their spins must be opposite. By observing the photon helicity, that of the neutrino can be inferred, and it is found to be negative (i.e. left-handed).

13.5 In conclusion

We can say that these experiments conducted in the late 1950s lend support to Fermi's original idea of a four-fermion interaction acting at a single point, and that the phenomena of parity violation can be incorporated by choosing the interaction factors Γ to accommodate the observed spin effects, e.g. left-handed neutrinos (right-handed antineutrinos) coupling predominantly to left-handed electrons.

Other reactions went on to confirm this picture. In particular, the purely leptonic decay of the muon is an obvious candidate for Fermi's description as a four-fermion interaction (see Figure 13.5):

$$\mu^\pm \rightarrow \text{e}^\pm + \nu + \bar{\nu}.$$

The amplitude which describes this reaction is the same as that for the β decay of the free neutron, with the appropriate wavefunctions substituted. Thanks to this, it is possible to establish that the value of G_F, which is necessary to account for the observed rate of muon decay, is equal to within 2% of the value needed to account for neutron β decay. In this way we are sure that it is the same force that is responsible for these two very different processes.

The weak decay of the pion (see Figure 13.6(a)) may at first be thought not to fit into the four-fermion description, viz.,

$$\pi^\pm \rightarrow \mu^\pm + \nu,$$

Fig. 13.4. The helicity of the photon emitted in the hybrid decay of ^{152}Eu shows the single-handedness of the neutrino.

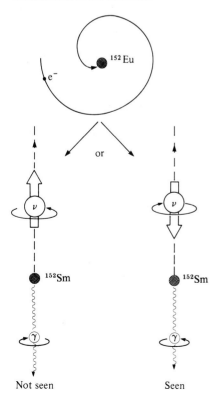

Fig. 13.5. The four-fermion picture of the decay of the muon.

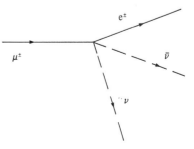

or

$$\pi^\pm \to e^\pm + \nu.$$

But these can be accommodated by imagining the pion to dissociate into a virtual nucleon–antinucleon pair by borrowing energy for a time allowed by Heisenberg's uncertainty principle, so the process becomes (Figure 13.6(*b*))

$$\pi^\pm \to N + \bar{N} \to \mu^\pm + \nu.$$

However, this rather contrived way of describing the decay is avoided in the more recent quark picture of the weak decays of hadrons, as we will soon see.

Fig. 13.6. The weak decay of the (*a*) charged pion and (*b*) its four-fermion interpretation.

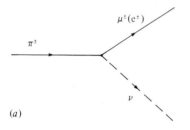

(*a*)

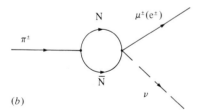

(*b*)

14

Two neutrinos

14.1 Introduction

Before nuclear β decay was fully understood, it was not known if the neutrinos emitted in neutron β decay were the same as those emitted in proton β decay or if they were different. (Remember that proton β decay can occur only within the nucleus, as a free proton is stable to all intents and purposes.) As the positron emitted in proton decay is the antiparticle of the electron emitted in neutron decay, it was suggested that a neutrino is emitted from one and an antineutrino from the other. This allows us to formulate a law of lepton-number conservation which was first put forward in 1953 by Konopinski and Mahmoud. If we assign a lepton number $+1$ to the electron, the negatively charged muon and the neutrino, and a lepton number -1 to the positron, the positively charged muon and the antineutrino, and a lepton number 0 to all other particles, then in any reaction the sum of lepton numbers is preserved. These assignments are summarised in Table 14.1. We can check this law in the weak interactions we have met so far:

In neutron β decay

$$
\begin{array}{cccc}
\mathrm{n} \to & \mathrm{p} + & \mathrm{e}^- + & \bar{\nu} \\
(0) \to & (0) + & (1) + & (-1)\checkmark
\end{array}
$$

In proton β decay

$$
\begin{array}{cccc}
\mathrm{p} \to & \mathrm{n} + & \mathrm{e}^+ + & \nu \\
(0) \to & (0) + & (-1) + & (1)\checkmark
\end{array}
$$

Table 14.1. *The assignment of simple lepton number*

Particle	$e^- \mu^- \nu$	$e^+ \mu^+ \bar{\nu}$	Others
Lepton number	1	−1	0

Table 14.2. *The assignment of lepton-type number*

Particle	$e^- \nu_e$	$e^+ \bar{\nu}_e$	$\mu^- \nu_\mu$	$\mu^+ \bar{\nu}_\mu$	Others
Electron number	1	−1	0	0	0
Muon number	0	0	1	−1	0

In pion decay

$$\pi^\pm \to \mu^\pm + \bar{\nu}(\nu)$$
$$(0) \to (\mp 1) + (\pm 1) \checkmark.$$

The assignment of these lepton numbers is shown to have a physical significance by the absence of reactions which do not conserve them, but which otherwise seem feasible. For instance

$$\bar{\nu} + p \to e^+ + n$$

is observed, whereas the similar reaction

$$\bar{\nu} + n \to e^- + p$$

is not observed.

14.2 A problem in the weak interactions

The Fermi theory of the weak interactions and the lepton-number conservation law with the experiments of β, pion and muon decays formed the content of weak interaction physics until 1960. And although the theory could adequately explain the experimental observations, it could not equally well explain what was not observed. In particular, the decay of a muon into an electron and a photon was not observed despite seeming to be a perfectly valid electromagnetic transition,

$$\mu^- \to e^- + \gamma.$$

The solution to this impasse is the proposal that any neutrino belongs to either of two distinct species: one associated with the electron and one associated with the muon, and that electron-type neutrinos ν_e can never transform into muons, nor muon-type neutrinos ν_μ into electrons. So the β decay of the neutron involves only electron-type antineutrinos:

$$n \to p + e^- + \bar{\nu}_e;$$

and the muon decay of the pion involves only muon-type antineutrinos:

$$\pi^- \to \mu^- + \bar{\nu}_\mu.$$

The muon can decay into an electron only if a muon-type neutrino carries off the 'muon-ness' and an electron-type antineutrino cancels out the 'electron-ness' of the electron:

$$\mu^- \to e^- + \bar{\nu}_e + \nu_\mu.$$

The decay of the muon into an electron and a photon is forbidden because the 'muon-ness' is apparently transformed into 'electron-ness'.

All this means that the law of lepton conservation is now extended to that of lepton-type conservation and works in a similar fashion. Both electron-type number and muon-type number must be conserved separately in each reaction. Using the revised lepton-number assignments in Table 14.2, we can see how this works for muon decay:

$$\mu^- \to e^- + \bar{\nu}_e + \nu_\mu$$

Muon number $\quad (1) \to (0) + (0) + (1) \checkmark$

Electron number $(0) \to (1) + (-1) + (0) \checkmark$

Strictly speaking, each time we have written down the symbol for the neutrino in the preceding chapters, we should have associated with it a suffix denoting electron- or muon-type, a procedure we shall follow from now on.

Once again, the adoption of a new conservation law has side-stepped our difficulties (the last occasion being the introduction of strangeness in Chapter 8). It will not be the last time we adopt such an approach. Our inescapable conclusion from this conservation law is that the electron-type and muon-type neutrinos should be physically different particles, and the first modern neutrino experiment was designed to illustrate just this.

14.3 The two-neutrino experiment

Neutrino experiments have peculiar difficulties due to the extremely feeble nature of the weak interactions. It is possible to use the low-energy neutrino flux from a nuclear reactor for some experiments, but others require higher-energy neutrinos which have the benefit of interacting more frequently with the targets presented. The first source of high-energy neutrinos became available in the early 1960s with the construction of the first of the big accelerators, the Alternating Gradient Synchrotron in Brookhaven in the USA. With this machine, protons could be collided into a solid target such as beryllium to produce a large flux of pions. As we have seen, these pions then decay predominantly into muons and muon-type neutrinos. It is possible to separate out just the neutrinos by passing the beam through vast quantities of iron (at thicknesses of about 20 m) to filter out the muons and any other extraneous particles (see Figure 14.1).

If there is no distinction between electron- and muon-neutrino types, then we would expect the two possible reactions to occur with equal likelihood:

$$\nu_\mu + n \rightarrow \mu^- + p, \tag{14.1}$$
$$\nu_e + n \rightarrow e^- + p. \tag{14.2}$$

However, if the two types really are distinct we would expect the first reaction to occur to the almost total exclusion of the second, as the neutrino beam consists almost entirely of muon-type neutrinos.

In the first really massive accelerator experiment of modern physics, Leon Lederman, Melvin Schwartz and Jack Steinberger showed that the first (muon) reaction does indeed predominate. In 25 days of accelerator time, some 10^{14} neutrinos traversed their spark chamber, which produced just 51 reactions resulting in a final-state muon. The ratio of the electrons to muons produced was later measured at CERN to be 0.017 ± 0.005, so conclusively demonstrating the existence of two separate neutrino types. This experiment demonstrated the validity of lepton-type conservation and explained the observed absence of decays which would otherwise be allowed. Lederman, Schwartz and Steinberger shared the 1988 Nobel Prize for their discovery.

Fig. 14.1. Schematic diagram of the two-neutrino experiment.

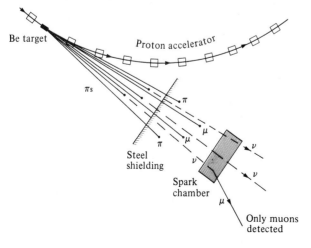

15

Neutral kaons and CP violation

15.1 Introduction

Soon after the observation of the weak interaction's violation of parity, it was discovered that it does not preserve charge conjugation symmetry **C** either. This was demonstrated by examining the spins of the electrons and positrons emitted in the decays of positively charged and negatively charged muons respectively. But physicists hoped that these two symmetry violations cancelled each other out exactly, so that the combined **CP** symmetry would be preserved by the weak interactions. To test this it is necessary to define an elementary particle state which is either even or odd under the **CP**-symmetry operation, allow the weak interaction to act, and then check that the final result has the same **CP** symmetry. It is possible to assign even or odd parity to an elementary particle state because the nature of the state remains unchanged under the parity operation, the only effect being the possible change in sign of the state wavefunction for a state of odd parity. Unfortunately, it is not possible to assign a well-defined **CP** symmetry to the K^0. This is because the operation always transforms it into its antiparticle and so changes the identity of the wavefunction. To compare the nature of the wavefunctions before and after an intended symmetry operation we must at least be sure they represent the same particle. We phrase this technically by saying that the K^0 is not an *eigenstate* of the **CP** operation.

Were **CP** to be a good symmetry (one preserved by the weak interaction), it would follow that to describe the interaction satisfactorily, we should consider it as acting on states which have a well-defined value of **CP**, i.e. on the eigenstates of **CP**. As K^0 and $\overline{K^0}$ are not these eigenstates it means that the weak interaction does not really 'see' these particles, but some others instead which are eigenstates. The simplest of these are simple mixtures of the original particles:

$$K_1^0 = \frac{1}{\sqrt{2}}(K^0 + \overline{K^0}),$$

$$K_2^0 = \frac{1}{\sqrt{2}}(K^0 - \overline{K^0}).$$

The wavefunctions of these two states keep their identity under the **CP** operation; the K_1^0 has even **CP** symmetry and the K_2^0 odd **CP** symmetry (i.e. $\mathbf{CP}\,K_1^0 = +K_1^0$ and $\mathbf{CP}\,K_2^0 = -K_2^0$). The weak interaction acts on these states, not on the neutral K mesons produced by the strong forces.

15.2 What is a neutral kaon?

The answer to this question depends on the interaction by which the particle is observed. The kaon which is produced in the strong interaction is either K^0 or $\overline{K^0}$, both of which are eigenstates of the parity operation with odd intrinsic parity and which have definite assignment of strangeness. The 'particle' which decays by the weak interaction is K_1^0 or K_2^0 which are eigenstates of the combined **CP** operation, but which do not have a well-defined strangeness quantum number.

Proof that the K_1^0 and K_2^0 are like real particles to the weak interaction as are K^0 and $\overline{K^0}$ to the strong can be found from their decays. The K_1^0 (which is even under **CP** symmetry) can decay only to states which are also even, such as a state of two pions. But K_2^0, which is odd under **CP**, can decay only to **CP** odd states, such as a state of three pions. This gives rise to very different mean life-times of the two particles:

$$K_1^0 \rightarrow 2\pi \qquad \tau = 0.9 \times 10^{-10}\,\text{s},$$
$$K_2^0 \rightarrow 3\pi \qquad \tau = 5.2 \times 10^{-8}\,\text{s}.$$

Another remarkable distinction between K_1^0 and K_2^0 is that they have different masses, although they are both equal mixtures of K^0 and $\overline{K^0}$ which have identical masses. This seemingly paradoxical conclusion was reached in 1961 when the mass difference was measured experimentally by a method which neatly shows up the identity crisis suffered by neutral K mesons.

When a neutral K meson is first produced by the strong interaction, it is definitely either K^0 or $\overline{K^0}$, depending on the strangeness of the reaction, e.g.

$$\pi^- + p \to \Lambda^0 + K^0. \tag{15.1}$$

Immediately, at the point of creation, the K^0 is an equal mixture of K_1^0 and K_2^0. However, we know that K_1^0 has a much shorter mean lifetime than K_2^0 and so the longer the time since its creation, the more likely it is to be a K_2^0. When the time elapsed is much greater than the mean lifetime of the K_1^0 we can say that the kaon is almost entirely K_2^0. This means that, according to the equation

$$K_2^0 = \frac{1}{\sqrt{2}}(K^0 - \overline{K^0}),$$

what originally started out as a particle (i.e. K^0), now has a 50% chance of being its antiparticle (i.e. $\overline{K^0}$). We can see this transformation explicitly by using the kaon produced from the reaction above to undergo the reaction which produces hyperons

$$\overline{K^0} + N \to \Lambda + \pi. \tag{15.2}$$

Because of strangeness conservation, this can occur only with $\overline{K^0}$ but not with K^0. So if we take the neutral kaon in (15.1) and wait for the K_1^0 content to drop, we should start to see reaction

(15.2) occur when a suitable target is placed in the beam. The frequency of reaction (15.2) will depend on the fraction of $\overline{K^0}$ which is generated in the beam by the K_1^0 component decaying. The intensity of the $\overline{K^0}$ component of the beam can be plotted as a function of time and it turns out that the nature of the variation depends on any mass difference which exists between K_1^0 and K_2^0. Experiments indicate the existence of a mass difference of about 3.5×10^{-6} eV. Bearing in mind the mass of K^0 of 498 MeV, the mass difference between K_1^0 and K_2^0 is about one part in 10^{14}!

15.3 Violation of CP symmetry

In 1964, Christenson, Cronin, Fitch and Turlay decided to check that the weak interaction did conserve **CP** symmetry exactly, and so to justify the use of states K_1^0 and K_2^0 as the particles appropriate to the weak force. They chose simply to observe a beam of K_2^0 and look for any decay into just two pions. If any were observed, then this would mean that the K_2^0 particle with **CP** $= -1$ had transformed into the two-pion state with **CP** $= +1$ and that the weak interaction does not conserve the symmetry. In the experiment the beam was allowed to travel about 18 m to ensure as few K_1^0 present as possible. The products of the particle decays of the K_2^0 beam were then observed as they left their tracks through the particle detectors, which also measured their energies (see Figure 15.1). They observed just a few of the forbidden decays of the K_2^0 beam into pairs of oppositely charged pions: about 50 out of a total of 23 000

Table 15.1. *The strong and weak forces 'see' different kaon eigenstates*

Interaction	Relevant eigenstates
Strong	K^0, $\bar{K}^0$
Weak:	
(1) were **CP** conserved	$K_1^0 = \dfrac{1}{\sqrt{2}}(K^0 + \bar{K}^0)$
	$K_2^0 = \dfrac{1}{\sqrt{2}}(K^0 - \bar{K}^0)$
(2) with **CP** violated	$K_S^0 = K_1^0 - \varepsilon K_2^0$
	$K_L^0 = K_2^0 + \varepsilon K_1^0$

Fig. 15.1. A schematic drawing of the **CP**-violation experiment of Christenson *et al.* (1964). Any decay of K_2^0 into just two pions represents the violation of **CP** symmetry.

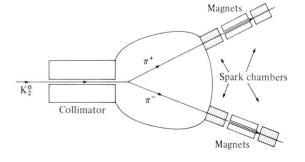

decays. This was far higher than any background event rate which may have resulted from the accidental presence of K_1^0 particles still in the beam, and the team concluded that the K_2^0 could decay into just two pions and that **CP** symmetry was not preserved exactly by the weak interaction after all.

Because of this **CP**-violating effect, it follows that K_1^0 and K_2^0 are not the exact eigenstates of **CP** and so are not quite the particles 'seen' by the weak interaction. Instead, the particles, as 'seen' by the weak interaction, are basically the **CP** eigenstates K_1^0 and K_2^0, but each with a small admixture of the other. These are called the long-lived K_L, and short-lived K_S, kaons, respectively, where,

$$K_L = K_2^0 + \varepsilon K_1^0, \quad K_S = K_1^0 - \varepsilon K_2^0,$$
$$\varepsilon \approx 2 \times 10^{-3}.$$

This small admixture of K_1^0 (i.e. the 'wrong' **CP** eigenstate) in K_L^0 is due to a transition between K^0 and $\bar{K}^0$. It is a higher-order weak-interaction process and hence very small.

The violation of **CP** has profound theoretical consequences which are not properly understood even now. In most weak interaction theories, **CP** is normally treated as being exact, with the violation appearing as some small perturbation to an otherwise **CP**-conserving force. One possible explanation is that the weak interaction does conserve **CP**, and that the violation is due to some 'superweak' force which is an entirely new interaction. If, however, the origin of **CP** violation is indeed the weak interaction, then it should also be manifest in kaon decays. A CERN experiment conducted in 1988 claims evidence for this 'direct' **CP** violation, which would seem to rule out the 'superweak' hypothesis. But this result has been disputed by other groups. One intellectually satisfying consequence of **CP**-symmetry violation is that we can at last convey to our intelligent alien the absolute distinction between left and right. Violation of **CP** symmetry gives rise to an observable difference in the probabilities of occurrence (or branching ratios) of the reactions:

$$K_2^0 \rightarrow \pi^+ + e^- + \bar{\nu}_e,$$
$$K_2^0 \rightarrow \pi^- + e^+ + \nu_e.$$

We can now communicate that we define the neutrino by specifying the branching ratio of the reaction in which it is present. This establishes a common matter–antimatter convention which allows our alien to identify uniquely our handedness convention.

Part 4

Weak interaction physics II

16

The current–current theory of the weak interactions

16.1 Introduction

In Part 3 we examined some of the processes of the weak interactions and learnt some of their physical attributes (relatively long lifetimes of weak decays, parity-violating effects, etc.). What we want to do now is to introduce a framework which relates the very disparate phenomena of the weak force, ranging from nuclear β decay and muon decay to high-energy neutrino collisions with matter. Because some of the processes involve hadrons, it will be necessary to ensure that our framework incorporates the consequences of the internal $SU(3)$ flavour symmetry of the hadrons, and furthermore, it is desirable that it be able to accommodate quarks as the origin of this symmetry.

This framework provides a description of the weak interactions in terms of the interaction of two 'currents', specifying the flow of particles. For example, in β decay, one current converts a neutron into a proton, and the other creates an electron and its antineutrino. We begin the construction of this framework by dividing the weak interactions into three classes reflecting the categories described above:

(1) *leptonic reactions* involving only leptons, such as muon decay, $\mu^- \rightarrow e^- + \nu_\mu + \bar{\nu}_e$;

(2) *semi-leptonic reactions* involving both leptons and hadrons, such as neutron β decay, $n \rightarrow p + e^- + \bar{\nu}_e$;

(3) *hadronic weak reactions* involving only had-

rons such as the pionic decay of the kaons, $K_1^0 \rightarrow \pi^+ + \pi^-$.

16.2 The lepton current

The ultimate aim is to achieve a common description of all three classes of weak interactions. But we start by concentrating only on one, the leptonic reaction. What we have seen of the weak force provides us with our description. In these reactions we saw that whenever an electron-neutrino is absorbed, an electron is created; and equivalently, whenever an electron-neutrino is created a positron has to be created also. This is as a result of the laws of conservation of lepton number and lepton-type number. This means that in our description, the lepton wavefunctions must always come in pairs. Also, from our knowledge of β decay, we know that these wavefunctions must be coupled together via an interaction factor Γ which combines the spins together in a correct, parity-violating way. We can now write down a 'lepton current' L^W which describes the flow of leptons during the weak interaction:

$$L^W = \bar{\psi}_e \Gamma \psi_{\nu_e} + \bar{\psi}_\mu \Gamma \psi_{\nu_\mu},$$
$$\bar{L}^W = \bar{\psi}_{\nu_e} \Gamma \psi_e + \bar{\psi}_{\nu_\mu} \Gamma \psi_\mu.$$

The second line is essentially the antiworld equivalent of the familiar process.

We can now generate the first-order amplitudes $m^{(1)}$ of all the leptonic processes by specifying interactions between leptonic currents. In fact, the simple product of the two lepton currents shown seems to generate all the reactions we see:

$$m^{(1)} \supset G_F \bar{L}^W L^W.$$

We can illustrate the currents and their couplings diagrammatically, as in Figure 16.1. Inspecting the interaction diagrams, we must bear in mind that the destruction of a particle is equivalent to the creation of its antiparticle, so the same diagram can describe, for instance, both electron–muon scattering and muon decay.

The weak leptonic current shown above flows between a charged lepton (e.g. e^-) and its neutrino (ν_e). Because these particles have different electric charge, it is called a *charged* current. It was widely believed for many years that all weak interactions were charged-current processes. However, in 1973 the discovery of a weak interaction *neutral* current (in which particles do not change identity) was made at CERN. We shall return to neutral currents shortly.

16.3 Higher-order interactions

We can use this description to show also the higher-order interactions which can occur between the leptons. These result when the weak force acts on the leptons more than once and are described essentially by the product of two amplitudes at successive instants. For the second-order ampli-

Fig. 16.1. The leptonic current L^W can be multiplied with its antiworld partner $\bar{L}^W$ to generate all the observed weak interactions of leptons.

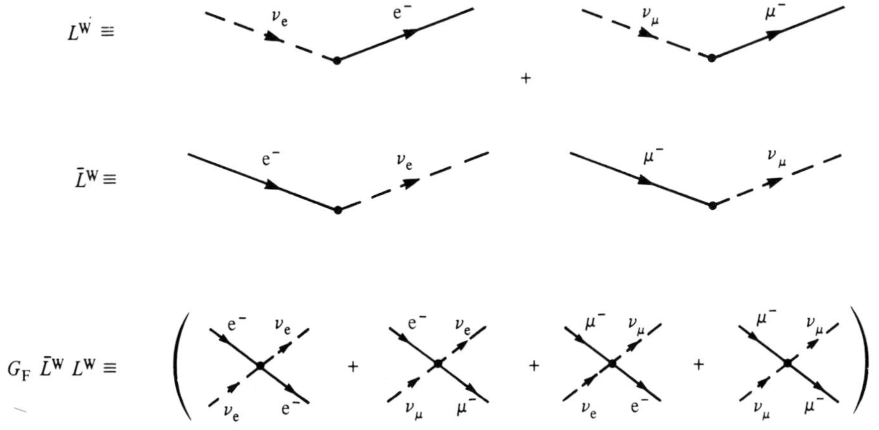

Fig. 16.2. Higher-order interactions (repeated weak interactions) can be generated by multiplying the current–current interaction with itself.

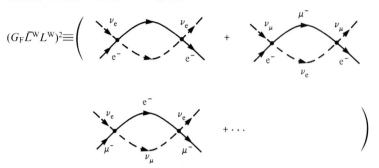

tudes $m^{(2)}$ this is given by the square of the simple current–current interaction:

$$m^{(2)} \supset (G_F \bar{L}^W L^W)^2.$$

These are shown diagrammatically in Figure 16.2. In fact, these higher-order interactions are not of practical importance for the weak interaction. This is because the second-order amplitudes are proportional to G_F^2 and the nth-order amplitudes are proportional to G_F^n.

Because the weak interaction is weak, the Fermi coupling constant G_F is small (compared with 1) and the higher powers of G_F are even smaller:

$$1 \gg G_F \gg G_F^2 \gg \cdots \gg G_F^n.$$

So to achieve the accurate description of a process, only the first-order term is significant. But the higher-order terms are of theoretical significance. It is desirable in principle that they should be calculable, and it was this motivation which has led us to the most recent theory of the weak force, as we shall soon see in Part 5.

17

An example leptonic process: electron–neutrino scattering

17.1 Introduction

Elastic electron-neutrino–electron scattering provides us with an example of the weak interaction at its simplest. The amplitude for the process is found in the current–current interaction:

$$m^{(1)}(\nu_e + e^- \to e^- + \nu_e) \\ = G_F(\bar{\psi}_e \Gamma \psi_{\nu_e})(\bar{\psi}_{\nu_e} \Gamma \psi_e).$$

By inserting the mathematical expression for the wavefunctions and interaction factors into the amplitude it is possible to calculate the cross-section for the process in the laboratory frame of reference: σ_{LAB}. The final answer is of a particularly simple form when the incoming neutrino energy E_{ν_e} is large:

$$\sigma_{LAB}(\nu_e + e^- \to e^- + \nu_e) = \sigma_0 \frac{E_{\nu_e}}{m_e}$$

where σ_0 is a constant factor arising from the calculation with a value of

$$\sigma_0 = 9 \times 10^{-45} \, \text{cm}^2.$$

This is the very tiny effective area of interaction between the neutrino and an electron. So it is easy to understand why neutrino interactions are so rare. We can perform a very similar calculation to work out the cross-section for antineutrino–electron scattering. The answer as we might expect, is very similar at high energies:

$$\sigma_{LAB}(\bar{\nu}_e + e^- \to e^- + \bar{\nu}_e) = \frac{\sigma_0}{3} \frac{E_{\nu_e}}{m_e}$$

where we shall later see how the difference of a factor of 3 between the two results from the different-handedness of neutrinos and anti-neutrinos.

The fact that these cross-sections rise linearly with the energy of the incoming neutrino presents us with an almost ironic situation. When the neutrino energy is low, say $E_\nu \approx 5$ MeV, which corresponds to E_ν/m_e of about 10, then the cross-section remains around 10^{-43} cm^2. This is a minute cross-section, even for the microworld. The neutrinos which emerge from nuclear reactors are of about this energy, and so experiments using them to observe these neutrino–electron collisions must use a very high flux of neutrinos and must be prepared to wait a very long time to gather enough observations. However, when the neutrino energy is higher, say around 5 GeV (corresponding to E_ν/m_e of about 10 000), then the cross-section rises to around 10^{-40} cm^2 and the reactions are, in principle, much more accessible. Unfortunately, the only such high-energy neutrinos so far produced are from the decay of pion beams in the high-energy accelerators. Not only does this mean that the neutrino flux is limited to a rather meagre level, but also that they will nearly all be muon-type neutrinos. We can see from Figure 16.1 that elastic muon-neutrino–electron scattering is not included in the basic current–current interaction, and so such a process, if in fact it does exist, could still not provide data to test our answers. (In fact, examples of this class of process, the so-called 'neutral current' reactions, have been discovered and they necessitate modifications to the current–current interaction; we shall discuss this further in Part 5.)

So we are forced back to waiting for rare events involving neutrinos from nuclear reactors to check our answers for the cross-sections. At present, the results of the experiments can be summarised by saying that they are not inconsistent with the cross-section having the predicted levels.

17.2 The role of the weak force in astrophysics

Electron-neutrino–electron scattering processes may play a role in astrophysics by allowing substantial numbers of $(\nu_e, \bar{\nu}_e)$ pairs to transfer energy from the interiors of stars to their outer layers. Normally, photons might be thought of as fulfilling this role, but in a stellar environment they are absorbed too quickly to perform an effective transfer. So it is left to the more weakly interacting neutrinos. When a heavy star has finished burning all its hydrogen (i.e. fusing hydrogen nuclei into helium and releasing energy), it moves on to a stage in which helium is burnt, after which it burns carbon and then successively heavier elements, with each stage being hotter than the previous one. At higher temperatures, the neutrinos transfer heat more efficiently throughout the star, leading to, for example, a shorter carbon-burning phase than would otherwise be the case. As a consequence, we expect a larger ratio of helium-burning stars to carbon-burning stars than we would were neutrino scattering processes absent.

Observation of stellar populations, when combined with various other astronomical and astrophysical observations, allows us to put crude limits on the strength of the weak interaction coupling constant governing the leptonic interactions involving only electron types, denoted $G_{\nu_e e^-}$. We find

$$\frac{G_F}{10} < G_{\nu_e e^-} < 10 G_F,$$

where G_F is the coupling measured in muon decay which governs the leptonic interaction involving both electron and muon types. So we must conclude that, owing to the experimental difficulties and observational uncertainties in measuring ν_e, e^- interactions, we are unable to be sure of the identity $G_F \equiv G_{\nu_e e}$. The best we can say is that the hypothesis is not inconsistent with data.

This is just one example of the close connection between particle physics and astrophysics/cosmology. We shall meet others in Chapter 42.

18

The weak interactions of hadrons

18.1 Introduction

Having written down a lepton current which provides a description of the purely leptonic reactions, we must now extend the concept to include the semi-leptonic processes, such as nuclear β decay, which are historically more important, and also the hadronic processes. Both these categories are divided up into strangeness-conserving reactions and strangeness-changing reactions, as a first step in categorising the effects of the weak interactions. Examples of the reactions in each category are shown in Table 18.1. The most obvious and serious difficulty in writing down a hadronic current (just as we wrote down the leptonic current) is that it is impracticable to write down wavefunctions for the observed hadrons; there are simply too many of them! If we were to use a separate wavefunction for each hadron, a current describing all the possible interactions would fill a book by itself. This is too complicated to be true.

18.2 The hadronic current

To proceed we can avoid the problem of wavefunctions for the hadrons and simply characterise the hadronic current in terms of its effect on the quantum numbers of the participating particles. So we can write the total weak interaction current as a sum of leptonic and hadron components:

$$J^W = L^W + H^W.$$

Table 18.1. *Categories of the weak interactions of hadrons*

Reaction	Class Strangeness-conserving	Strangeness-changing
Semi-leptonic	$\pi^{\pm} \to l + \nu(\bar{\nu})$ $n \to p + e^- + \bar{\nu}_e$ $\mu^- + p \to \nu_\mu + n$ $K^0 \to \pi^{\pm} + e^{\mp} + \nu(\bar{\nu})$ $\bar{\nu} + p \to e^+ + n$	$K^{\pm} \to l + \nu$ $K^{\pm} \to \pi^0 + l + \nu$ $\Lambda^0 \to p + l + \bar{\nu}$ $\Xi \to \Lambda + l + \bar{\nu}$ $\bar{\nu} + p \to e^+ + \Lambda$
Hadronic	Parity-violating effects in ordinary hadron physics, e.g. $p + p \to p + p$	$K^0 \to n\pi \ (n = 2, 3)$ $\Lambda^0 \to \pi^- p$ $\Sigma \to n\pi$

l denotes $e^{\pm}$ or $\mu^{\pm}$.

As before, the weak interaction amplitudes are generated by the product of the total current with its antimatter conjugate multiplied by the Fermi coupling:

$$G_{\mathrm{F}} \bar{J}^{\mathrm{W}} J^{\mathrm{W}} = G_{\mathrm{F}}(\bar{L}^{\mathrm{W}} L^{\mathrm{W}} + \bar{L}^{\mathrm{W}} H^{\mathrm{W}} + \bar{H}^{\mathrm{W}} L^{\mathrm{W}} + \bar{H}^{\mathrm{W}} H^{\mathrm{W}}).$$

This contains all the leptonic reactions ($\bar{L}^{\mathrm{W}} L^{\mathrm{W}}$), the semi-leptonic reactions ($\bar{L}^{\mathrm{W}} H^{\mathrm{W}} + \bar{H}^{\mathrm{W}} L^{\mathrm{W}}$), and the purely hadronic reactions ($\bar{H}^{\mathrm{W}} H^{\mathrm{W}}$). The form of the hadronic current (less the wavefunctions) consists of one part which conserves the strangeness of the participating hadrons, $h^{\pm}$, and of one part which changes it, $s^{\pm}$:

$$H^{\mathrm{W}} = h^{\pm} \cos \theta_{\mathrm{C}} + s^{\pm} \sin \theta_{\mathrm{C}}.$$

The relative strength of the two components is governed by the Cabbibo angle θ_{C} which is a parameter intrinsic to the weak interactions and which must be measured from experiments.

At this point we must remember that just as the leptonic current L^{W} contains the interaction factors Γ (which are a mixture of vector and axial-vector quantities), so too do the separate components of the hadronic current to ensure the correct parity-violating coupling of hadronic spins in the reactions. So the hadronic weak current has four separate components:

(1) a vector current which conserves strangeness;
(2) a vector current which changes strangeness;
(3) an axial-vector current which conserves strangeness;
(4) an axial-vector current which changes strangeness.

18.3 Current algebra

Current algebra is concerned with deriving relationships between the various components of the hadronic current on the basis of the internal symmetry scheme governing the hadrons. Because of this symmetry group $SU(3)$ there exist relationships between the generators of the symmetry transformations (the group algebra) which dictate the allowed quantum-number structure of the hadrons. The fact that the components of the hadronic weak current can change the quantum numbers of the hadrons suggests they have an intimate connection with the symmetry scheme and, in fact, both the vector and axial-vector components of the hadronic current form separate octet representations of $SU(3)$. This allows us to categorise the effects which the strangeness-conserving and the strangeness-changing components (each the sum of vector and axial-vector parts) have on the quantum numbers of the hadrons in an interaction:

$$h^{\pm} \Rightarrow (\Delta Y, \Delta I, \Delta I_3) = (0, 1, \pm 1),$$
$$s^{\pm} \Rightarrow (\Delta Y, \Delta I, \Delta I_3) = (\pm 1, \tfrac{1}{2}, \pm\tfrac{1}{2}).$$

Armed with this knowledge of the effects of the weak interaction on the quantum numbers of the hadrons, it is possible to write down expressions for the amplitudes of the various reactions (using some imagination when it comes to the hadron wavefunctions), and to proceed to a description of the reactions.

The rate for pion decay into leptons,

$$\pi^{\pm} \to l^{\pm} + \nu(\bar{\nu})$$

can be calculated with just one unknown remaining in the formula as a relic of the pion wavefunction. Pion β decay, which is a close relative to free-neutron β decay, explores different components of the hadronic current

$$\pi^{\pm} \to \pi^{0} + e^{\pm} + \nu_e(\bar{\nu}_e)$$

and, most importantly, allows the experimental determination of the Cabbibo angle. This turns out to be

$$\cos \theta_C = 0.97$$

Inserting this into the hadronic current gives

$$H^W = 0.97h^{\pm} + 0.24s^{\pm}$$

which shows the strangeness-conserving component of the current to be much more important than the strangeness-changing part. Building on this information, we can go on to establish relationships for reactions such as nuclear β decay, neutrino–nucleon scattering and the purely hadronic decays. These predictions enjoy varying degrees of success, but all suffer the deficiency of the inadequate treatment of the hadron wavefunctions.

18.4 The hadron current and quarks

The hadron current takes on a very simple form when written in terms of the quarks. Naively, we can think of the weak current as simply changing one flavour of quark (u, d or s) into another, and so changing the quantum numbers of the parent particle.

In fact, it is not quite as simple as this because the weak interaction does not specify a unique transformation say, from a u quark into a d quark. As we have seen, the weak current has both strangeness-conserving and -changing components, which implies that the u quark can have a certain probability of transforming into a d quark,

Fig. 18.2. The quark picture of quantum-number flow during the semi-leptonic decay of the kaon.

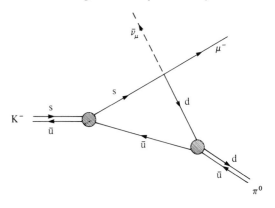

and another of transforming into an s quark. So the current is written

$$H^W = \bar{u}\Gamma(d \cos \theta_C + s \sin \theta_C).$$

This current is shown symbolically in Figure 18.1 The decay of a strange meson is illustrated in Figure 18.2.

This representation is very useful for envisaging the flow of quantum numbers during a reaction. Unfortunately, it is of limited use in calculating dynamical quantities because we have very little idea of how to represent the confinement of quarks mathematically. In the diagram the ignorance is contained within the shaded blobs.

Fig. 18.1. The weak interaction transforms quarks.

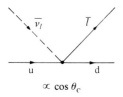 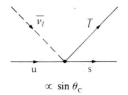

19

The W boson

19.1 Introduction

It is true to say that we can explain all the data from low-energy weak interaction processes with the Fermi theory (expressed in the framework of the current–current theory of lepton processes). Unfortunately, this theory of a point-like interaction makes unacceptable predictions for high-energy weak interactions. We have just seen how the theory predicts that cross-sections for neutrino–electron scattering will rise linearly with the energy of the incoming neutrino. But we must realise that this prediction cannot be true for arbitrarily high energies. For instance, if it were true, neutrinos with exceedingly high energies (say those in cosmic rays, originating in space) would have a very high cross-section for interacting with matter, so we would expect neutrino collisions to be commonplace events in cosmic ray photographs and more common in laboratory bubble chambers. This is just not true, and we must accept that our formula for the neutrino cross-section is valid only at small energies. Also, there are extremely well-founded theorems resting only on assumptions, such as causality, which constrain the rate at which cross-sections can rise with energy.

To solve this problem, and also to put the description of the weak interaction on a common footing with the theories of electromagnetism and the strong nuclear force, it is necessary to abandon the four-fermion point-like interaction and replace it with a particle exchange mechanism (just like, say, pion exchange between nucleons).

The particle which carries the weak interaction is called the intermediate vector boson, denoted W (see Figure 19.1). What we must do is to go back and describe all the weak interaction phenomena in terms of a particle exchange mechanism, which must approximate to the four-fermion point-like interaction at low energies to preserve its successful explanation of the data.

19.2 The W boson

The essence of the W-boson mechanism is that the two currents involved in a weak interaction process no longer couple directly to each other at a single point. Instead, each current couples to the W-boson wavefunction at different space–time points and the W boson mediates the interaction between the currents. The basic weak interaction amplitude $m^{(1)}$ is thus the coupling of the current with the W-boson wavefunction W (see Figure 19.2):

$$m^{(1)} \equiv g(L^W \bar{W} + W \bar{L}^W)$$
$$\equiv g(\bar{\psi}_e \Gamma \psi_{\nu_e} \bar{W} + W \bar{\psi}_{\nu_e} \Gamma \psi_e).$$

The first thing to note about the W boson is that it must come in both positively and negatively charged versions if it is to allow the transformations of positrons and electrons into antineutrinos and neutrinos respectively. Also, if we wish to describe neutral current phenomena, we must allow the existence of a neutral W boson as well. The role of the differing charge states is shown in Figure 19.2.

Fig. 19.1. Just as the pion carries the strong force between hadrons, so the W boson carries the weak force between leptons.

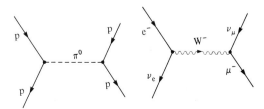

Fig. 19.2. Basic lepton processes defining the charge states required of the W boson.

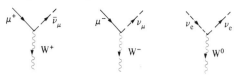

Another property of the W boson, which is easily established, is that it must be very massive. Recalling our simple argument of Chapter 7, using Heisenberg's uncertainty principle: the range of the force may be thought of as typified by the maximum distance which its carrier can travel in the time element allowed by the uncertainty principle. The more massive the carrier, the shorter the range of the force. As the Fermi theory managed quite satisfactorily with a point-like assumption, it follows that the W boson must be much more massive than the pion (which allows the strong force the measurable range of 10^{-15} m).

It is clear that the simple amplitudes of Figure 19.2 represent the creation (or destruction) of a W boson from the (into the) familiar leptons. The reactions involving only leptons as external particles will result from higher-order interactions, represented by products of these basic amplitudes:

$$m^{(2)} = g^2(L^W \bar{W} W \bar{L}^W)$$
$$= g^2(\bar{\psi}_e \Gamma \psi_{\nu_e} \langle \bar{W} W \rangle \bar{\psi}_{\nu_e} \Gamma \psi_e).$$

Figure 19.3 shows the W-exchange amplitudes for some of the basic weak-interaction processes we have met so far. The factor in the amplitude describing the propagation of the virtual W boson $\langle \bar{W} W \rangle$ is known as the W-boson propagator, which acts as its wavefunction between its creation and its destruction. It is this new factor which improves

the unacceptable high-energy behaviour of total cross-sections. The mathematical expression for the W-boson propagator describes its mass and its spin (spin 1 for a vector particle) and allows us to relate the old Fermi coupling G_F to the new lepton–W coupling g. At low energies we find:

$$G_F \propto \frac{g^2}{M_W^2}.$$

Knowing the expression for the W propagator as well as the wavefunction for the external leptons and the interaction factors allows us to recalculate the cross-sections for all processes of interest. At low energies, the answers are the same as for the Fermi theory, as desired.

19.3 Observing the W boson

We have suggested that the quantum of the weak force, the W boson, has spin 1 (like the photon), comes in three charge states (W^+, W^0, W^-), and decays into the familiar leptons. Its discovery was anticipated for many years, during which time it was generally believed that its mass was so large as to prevent it being seen in experiments. However, as we shall see in the next part, its mass can be predicted from the modern theory of the weak interactions (and the relevant experimental data). Furthermore, in experiments carried out at CERN in 1983, the W boson finally revealed itself, just as predicted with a mass around 80 GeV.

Fig. 19.3. Basic weak interaction processes mediated by W exchange.

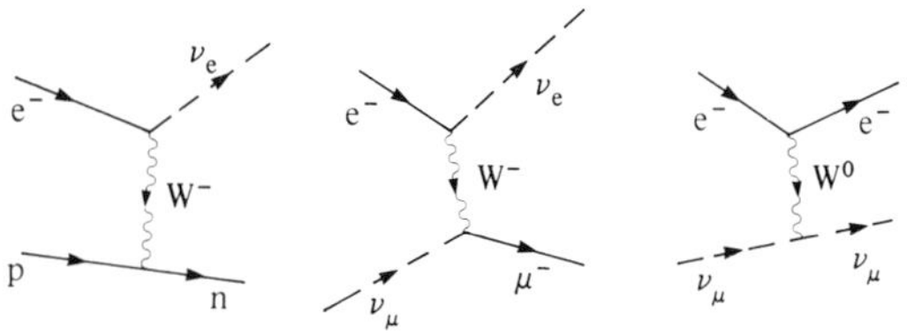

Part 5
Gauge theory of the weak interactions

20

Motivation for the theory

20.1 Introduction

The description of the weak interactions of leptons afforded by the current–current theory of Chapter 16 provides a good account of the experimental observations. This description includes the use of wavefunctions for the leptons (possible because there are just a few of them) and can also incorporate the use of W bosons in the role of 'the photons of the weak interactions'. We might then wonder why it is not possible to write down a fully fledged theory of the weak interactions of leptons mediated by the W bosons, similar to the QED theory of electrons and photons. Although there are no pressing practical reasons for doing so, there is considerable theoretical motivation. Firstly, it would be satisfying to be able to calculate answers to any desired degree of accuracy, in contrast to using only the simplest interactions to which our current understanding of the W bosons limits us. Secondly, we would like to have a predictive theory which could demonstrate its relevance by actually revealing new phenomena. And finally, if we can derive a theory similar to QED, this may allow us to formulate a unified theory of weak and electromagnetic forces.

20.2 Problems with the W bosons

Given the motivation and the basic building blocks of the theory, we immediately discover several problems arising in the use of the W bosons. These problems originate in the fact that these particles have spin 1 *and* non-zero mass. Let us investigate the consequences of this seemingly innocuous combination.

When we first talked about quantum-mechanical spin, we noted that although the naive picture of the spinning ball is helpful, it is a simplification of a more fundamental attribute. In fact, the spin of particles is the method for categorising their transformation properties under Lorentz transformations. A spin-1 particle is said to *transform* like a vector, which means that it has three components to define its orientation (i.e. its polarisation) at any point. Normally, we define the components such that two (i.e. the transverse) are perpendicular to and one (i.e. the longitudinal) is parallel to the direction of the momentum of the particle (see Figure 20.1). This is appropriate for a *massive* spin-1 particle, but for a *massless* spin-1 particle, like the photon, the transformations of special relativity show that the longitudinal degree of freedom has no physical significance. The photon must always be polarised in the plane transverse to its direction of motion. The difference becomes important when the propagators carry very high momentum. The transverse propagator for the massless photon behaves like $1/p^2$ and becomes very small at high momentum (i.e. large p^2). But for the massive vector particle, the presence of the extra longitudinal component in its propagator spoils this behaviour. At very high momenta, the massive propagator approaches a constant value of $1/M^2$.

We have seen that, in perturbation theory, the probability of an event occurring is given by the sum of contributions from a series of increasingly more complex Feynman diagrams. Each of these contributions should be, to all intents and purposes, a dimensionless number – probability carries no dimensions. Furthermore, as we have seen in QED, it is necessary to sum over all the possible values for the unobserved momenta of all the internal virtual particles in any diagram. However, at very high momenta a massive W-boson propagator contributes a factor of $1/M^2$. To compensate for this, each propagator (i.e. each wiggly line in the diagram) must be multiplied by corresponding factors of the momentum in order to give a dimensionless contribution of the form p^2/M^2c^2.

The presence of these momentum factors multiplying each propagator means that the mathematical expressions for the increasingly more complicated diagrams will be infinite when summed over all possible internal momenta. The diagrams are said to *diverge*.

In summary, it is the mass factor in the W-boson propagator which leads to an ever-increasing number of infinite contributions to the perturbation series. It is not possible for these to

Fig. 20.1. Vector propagators. A massive vector particle can have three components of polarisation (*a*), a massless vector particle only two (*b*).

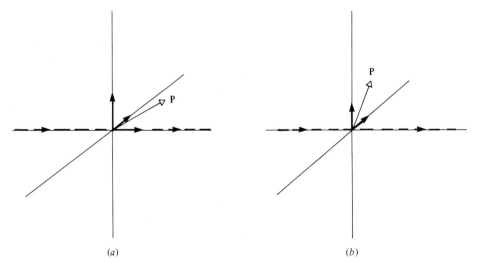

(*a*)   (*b*)

be reabsorbed into redefinitions of the masses and couplings. The theory is unrenormalisable and incapable of providing sensible answers to an arbitrary degree of accuracy.

Related to the problem of the theory's lack of renormalisability is the bad high-energy behaviour exhibited by some processes involving the W bosons. The presence of the mass factor in the W-boson propagator causes the cross-sections for these processes to rise with energy faster than is allowed by fundamental theorems (see Figure 20.2). This bad high-energy behaviour is precisely the problem which the W bosons were introduced to cure! As it is the mass of the W boson which seems to be the source of the trouble, the best thing for us to do is to study the origins of this particle and its mass further.

Fig. 20.2. A process with bad high-energy behaviour caused by the mass of the W bosons.

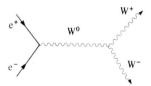

21

Gauge theory

21.1 Introduction

The principle of gauge invariance is perhaps the most significant of the concepts used in modern particle theories, as it is the origin of the fundamental forces themselves. It appears to apply to *all* of the four known forces described in Chapter 5 (in one guise or another), and so may eventually provide us with the basis for a comprehensive unified theory. The basic method of gauge theory is to ensure that the Lagrangian describing the interaction of particle wavefunctions remains invariant under certain symmetry transformations which reflect conservation laws observed in nature. As a first step, we can see how this works in QED.

21.2 The formulation of QED

QED seeks to explain the interaction of charged particles, say electrons, in such a way that total electric charge is always conserved. To represent this, the Lagrangian which describes the electron wavefunction must be invariant under a certain group of symmetry transformations **G**. (See Figure 6.4.) We write this symbolically:

$$\mathbf{G}\mathcal{L}(\psi_e) \rightarrow \mathcal{L}(\psi_e^*).$$

In fact, the group in question, denoted $U(1)$, corresponds to a simple shift in the phase of the electron wavefunction. This is called a *global* phase transformation because it represents an identical operation at all points in space–time.

From Section 3.8, we know that the actual value of the phase of the electron wavefunction is unobservable – but we can observe differences in phase. However, before so doing, we must first establish a convention as to the starting point from which such phase differences are measured. Clearly, if the results we obtain are to make any sense, they must be independent of whichever convention we choose. Furthermore, performing a global phase transformation corresponds to changing this convention. So, the global phase symmetry is just a statement of the fact that the laws of physics are independent of the choice of phase convention.

The global phase symmetry is a relatively simple symmetry and does not place a very strong constraint on the form of the Lagrangian. The exercise becomes more interesting if we demand that the theory be invariant under *local* phase transformations which vary according to position:

$$\mathbf{G}(x)\mathscr{L}(\psi_e) \to \mathscr{L}^*(\psi_e^*)$$

where $x = (\mathbf{x}, t)$ denotes a space–time four-vector. These are called *local gauge transformations*. A local gauge transformation corresponds to choosing a convention for defining the phase of the electron wavefunction, which is different at different space–time points. That is, the convention can be decided independently at every point in space and at every moment in time. But because of the space–time dependence, the Lagrangian representing the electron wavefunction is changed by the transformation ($\mathscr{L} \neq \mathscr{L}^*$), and the theory is not invariant under this more demanding symmetry. However, it comes as a pleasant surprise to find that by introducing another field, which compensates for the local change in the electron wavefunction, we can obtain a Lagrangian which indeed exhibits such symmetry.

The required field must have infinite range, since there is no limit to the distances over which the phase conventions must be reconciled. Hence the quantum of this new field must be massless. In fact, the field required for local gauge invariance is none other than the electromagnetic field whose quantum is, of course, the photon.

We already know that the interaction of two electrons should most correctly be described in terms of one electron interacting with a photon at one point, the propagation of the photon, and its subsequent interaction with the other electron at another space–time point. It so happens that the changes in the photon wavefunction cancel out the changes in the Lagrangian resulting from a local phase transformation. So, the introduction of the photon leads to local gauge invariance. We write this symbolically as

$$\mathbf{G}(x)\mathscr{L}(\psi_e, A) \to \mathscr{L}(\psi_e^*, A^*)$$

where the symbol A denotes a four-vector describing the electromagnetic field (i.e. the photon's wavefunctions).

So, invariance of the Lagrangian under local gauge transformations requires the existence of a massless gauge boson: the photon, which is the quantum of the long-range electromagnetic field. A physical explanation of its role is that the photon communicates the different space–time-dependent conventions, which define the phase of the electron wavefunction, between different points in space–time. Furthermore, the gauge symmetry implies a conservation law (see Section 6.1), namely, the conservation of electric charge. Hence, the gauge theory of QED successfully explains the interactions of electrons (and other charged particles).

An interesting fact is that the presence in the Lagrangian of a term attempting to describe a hypothetical mass for the photon would destroy the gauge invariance. As we have already seen, massive spin-1 particles generally give rise to non-renormalisable theories. So we may suspect local gauge invariance to be a good guide to the renormalisability of theories.

21.3 Generalised gauge invariance

Our next task is to generalise the principle of gauge invariance to other particles and other forces. The most convenient example is a historical one which, in fact, turned out to be untrue, but which follows on naturally from our previous discussions. It was originally proposed as a theory describing the strong interaction.

The charge independence of the strong nuclear force means that it acts identically on both protons and neutrons. It cannot distinguish between them but instead 'sees' only one basic nucleon N. This led us to categorise the proton and neutron as the two isospin components of the isospin-$\frac{1}{2}$ nucleon. The charge independence of the force may then be expressed as its invariance under rotations in isospin space, and associated with the invariance is the conservation of the total isospin of the system on which the forces act. (Of course, this isospin symmetry is broken by the electromagnetic interaction, which discriminates between protons and neutrons. But as far as the strong interaction is concerned, isospin is a good symmetry.) The Lagrangian which describes the interaction of nucleons should then be invariant under the group of global isospin rotations $SU(2)$:

$$\mathbf{G}^{SU(2)}\mathcal{L}(N) \to \mathcal{L}(N^*)$$

where the group $SU(2)$ effectively rotates a proton into a neutron and vice versa. As the strong force cannot distinguish between the two, we must establish a convention for what we call a proton and what we call a neutron: any possible mixture of the two can be used to define a nucleon. The global transformation essentially redefines the nucleon convention at all points in space.

As before, it is possible to require that the theory be symmetric under the more demanding local gauge transformations and, as before, it is found necessary to introduce a massless gauge particle ρ to ensure the invariance of the Lagrangian:

$$\mathbf{G}^{SU(2)}(x)\mathcal{L}(N, \rho) \to \mathcal{L}(N^*, \rho^*).$$

The source of this new gauge particle is isospin, just as the source of the electromagnetic gauge field is electric charge. Because the nucleon may or may not change its electric charge in interaction, the new gauge particle must come in three charge states to do its job, (ρ^+, ρ^0, ρ^-). Furthermore, because electric charge is related to isospin by $Q = e[I_3 + (Y/2)]$, the ρ gauge particle carries its own isospin and so can act as its own source. This allows the charged gauge particle to interact with itself, in contrast to the neutral photons in QED (see Figure 21.1). The difference between the two is expressed by saying that QED is an Abelian field theory whilst the $SU(2)$ gauge theory is non-Abelian. The difference is a consequence of the mathematical structure of the gauge groups. The simple shift in phase in QED is said to be Abelian because a series of transformations can be performed in any order to produce the same effect as

one big transformation. The group of rotations in isospin space $SU(2)$ (which is also the group of ordinary spatial rotations in the 3-dimensional space of the everyday world) is non-Abelian because a series of transformations does depend on the order of operation.

Gauge invariance was first generalised to the isospin invariance of the Lagrangian by Yang and Mills in 1954 and Shaw in 1955, but further work was deemed nugatory as such invariance apparently required the existence of a charge triplet of massless spin-1 gauge bosons (the ρ particles) which do not exist. A few years later the ρ mesons were discovered and the possibility of a gauge theory for the strong interaction was briefly entertained. Unfortunately, the ρ mesons are massive bound states of two pions and cannot be considered as candidates for this fundamental rôle.

21.4 Gauge invariance and the weak interactions

The first step we must take in formulating a sensible theory of the weak interactions is to incorporate the basic laws of leptonic physics which we have already met. These are the separate laws of electron number and muon number conservation. It is natural, therefore, to group the leptons' wavefunctions into doublets of the same lepton-type:

$$l_e = \begin{pmatrix} \nu_e \\ e^- \end{pmatrix} \qquad l_\mu = \begin{pmatrix} \nu_\mu \\ \mu^- \end{pmatrix}.$$

If gauge theory is now to be used, it is necessary to identify some conservation law which will imply

Fig. 21.1. In QED, photons cannot interact directly, only with electrons (*a*). In $SU(2)$ gauge theory, the gauge particles can interact both directly and indirectly (*b*).

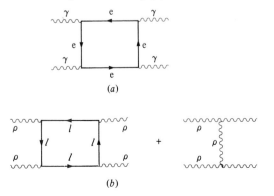

the existence of some symmetry of the Lagrangian. To discover this, we note the similarity of the above doublets to the isospin doublet of the nucleon containing the proton and the neutron. We can then carry this similarity further and propose that the weak interaction also is independent of the electrical charge of the particles on which it acts. The weak interaction 'sees' only a lepton and cannot distinguish between a neutrino and an electron.

This leads us to define 'weak isospin' in a fashion exactly analogous to the isospin of nucleons, and allows us to require that the weak interaction be invariant under rotations in this weak isospin space. So the Lagrangian must be invariant under the group of weak isospin rotations, denoted by $SU(2)^W$ to show it is acting on the leptons' wavefunctions,

$$\mathbf{G}^{SU(2)^W}\mathscr{L}(l_e, l_\mu) \to \mathscr{L}(l_e{}^*, l_\mu{}^*).$$

In exact parallel to the previous discussion, enforcing the more demanding local symmetry requires the introduction of massless gauge particles, W, to guarantee the invariance of the Lagrangian:

$$\mathbf{G}^{SU(2)^W}(x)\mathscr{L}(l_e, l_\mu, \mathbf{W}) \to \mathscr{L}(l_e{}^*, l_\mu{}^*, \mathbf{W}^*).$$

Here the physical reason for the existence of the W boson is to communicate between interacting leptons the locally defined convention governing the mixture of 'electron' and 'neutrino' constituting the lepton. Again, the gauge particle must be a charge triplet $(\mathbf{W}^+, \mathbf{W}^0, \mathbf{W}^-)$, and here we have something new for the weak interactions: the presence of the $\mathbf{W}^0$ particle allows *neutral-current* reactions in which a neutrino does not have to become an electron (Figure 21.2). However, this theory was proposed well before the discovery of neutral currents, and the prediction of a $\mathbf{W}^0$ seemed a

positive inconvenience. Also inconvenient was the fact that gauge invariance requires the gauge particles to be massless, yet the absence of experimental evidence for the W boson led inevitably to the conclusion that it must be heavy.

These two factors hampered the development of gauge theory for almost a decade. What was required was a mechanism by which the W bosons may be allowed to have mass whilst originating from a gauge-invariant (and thus possibly renormalisable) theory. Also, the observed absence of neutral currents withheld the experimental confirmation of the theory necessary for its credibility and its development.

Fig. 21.2. The neutral gauge particle allows weak neutral current reactions.

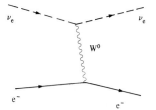

22

Spontaneous symmetry breaking

22.1 Introduction

The existence of asymmetric solutions to a symmetric theory is common to many branches of physics. Consider, for example, an ordinary magnet. Its magnetic field clearly defines a preferred direction in space (i.e. rotational symmetry is broken), but the equations governing the motions of the individual atoms in the magnet are entirely rotationally symmetric. How has this come about? The answer lies in the fact that the symmetric state is *not* the state of minimum energy (i.e. the ground state), and that in the process of evolving towards the ground state, the intrinsic symmetry of the system has been broken.

A simple mechanical example is the behaviour of a marble inside the bottom of a wine bottle (see Figure 22.1). The symmetric state obviously corresponds to the marble taking the central posi-

Fig. 22.1. The initial position of the marble is symmetric but not minimum energy. A small perturbation will cause the rotational symmetry to be broken and the system to assume the state of minimum energy.

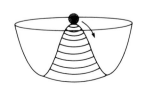 

tion on top of the hump: but this is not the state of least energy, as the marble possesses potential energy due to its elevation. A small perturbation will send the marble tumbling down into the trough, where the system will possess least energy, but will also be rotationally asymmetric. When the symmetry of a physical system is broken in this way by an asymmetric ground state, we say the system exhibits 'spontaneous symmetry breaking'.

22.2 Spontaneous breaking of global symmetry

We now want to apply these ideas to particle physics to see if a spontaneously broken gauge theory has anything to do with gauge boson mass. In this context, spontaneous symmetry breaking means that, although the Lagrangian is symmetric, the actual ground state is asymmetric. Recall from Section 4.4 that associated with every particle is an underlying quantum field; the particle is the quantum of the field. The ground state of the field – its vacuum state – is a state in which the field has its lowest possible energy and no particles are present (Section 4.9). For most fields, the energy is minimised when the average value of the field is zero; but for some, the energy is minimised only when it takes some uniform non-zero value.

Let us start by considering a hypothetical spinless particle consisting of two components:

$$\Phi = (\phi_1, \phi_2).$$

(This is analogous to considering the nucleon as a particle with two components: the proton and neutron, i.e. N = (p, n).) Let us now write down an hypothetical Lagrangian which specifies the interaction between ϕ_1 and ϕ_2. Suppose we choose the wine-bottle shape of the previous section to describe the interaction energy. Figure 22.2 shows such a choice for the interaction energy. The axes labelled ϕ_1 and ϕ_2 correspond to the average values of the associated quantum fields. For the interaction we have chosen, the energy is not a minimum at zero values of the fields, but around the circle defined by

$$\phi_1^2 + \phi_2^2 = R^2.$$

This equation defines the vacuum states of this theory, which are characterised by non-zero average values for ϕ_1 and ϕ_2. Despite this unusual property, the Lagrangian is still symmetric under transformations between ϕ_1 and ϕ_2, i.e. under rotations in the ϕ_1–ϕ_2 plane:

$$\mathbf{G}\mathscr{L}(\phi_1, \phi_2) \to \mathscr{L}(\phi_1{}^*, \phi_2{}^*)$$

where $\mathbf{G}$ is the rotation group in the plane. Note that for the moment we are dealing with global transformations: $\mathbf{G}$ acts in the same way at all points in space–time.

For definiteness, consider the particular vacuum state given by $\phi_1 = 0$ and $\phi_2 = R$. For reasons of convenience, we might want the average values of the fields to be zero in the vacuum state. In fact, this is necessary if we wish to use the mathematics of perturbation theory. Moreover, it can be arranged by a simple redefinition of the fields:

$$\phi_1' = \phi_1, \qquad \phi_2' = \phi_2 - R.$$

This simply corresponds to drawing new axes through the point R in Figure 22.2. If we write the Lagrangian in terms of the new fields, it should describe exactly the same physics (after all, all we have done is to make a redefinition):

$$\mathscr{L}(\phi_1, \phi_2) \equiv \mathscr{L}(\phi_1', \phi_2').$$

However, some interesting features arise in this redefined system. First, the Lagrangian is now *not* invariant under the original group $\mathbf{G}$ of transformations. This is not surprising because the interaction energy is not symmetric about the vacuum state. Secondly, the Lagrangian now describes ϕ_2' as a massive particle (mass proportional to R), and ϕ_1' as a massless particle. This is in contrast to the

original Lagrangian in which the concept of mass was rather ill-defined. The net result is that the global symmetry of the original Lagrangian has been broken and, as a consequence, one of the particles has been given a mass whilst the other remains massless. The interesting question now is whether or not particle masses in the real world originate in a similar fashion from some originally gauge-invariant (and hence possibly renormalisable) interaction.

Unfortunately, things are not of much use as they stand in this simple model. The presence of the massless spin-0 particle turns out to be a general consequence of this type of mechanism (a theorem proved by Cambridge physicist Jeffrey Goldstone in the early 1960s): whenever a *global* symmetry is spontaneously broken, a massless, spin-0 particle results. This particle is called a 'Goldstone boson'. This situation is unfortunate as no such massless, spinless particle appears to exist in the real world. Moreover, the particle which has developed a mass is nothing like a W boson (which has spin 1) and this remains massless. Our trouble seems to be increasing! Happily, however, all these difficulties can be resolved, as we shall now see.

22.3 Spontaneous breaking of local symmetry – the Higgs mechanism

Now take the original Lagrangian with the same wine-bottle-shaped interaction energy, and demand that it be invariant under *local* gauge transformations (i.e. under rotations in the $\phi_1\phi_2$ plane which vary from place to place). Then we know from Chapter 21 that we must introduce a gauge particle, which we shall denote by A, in order to maintain the invariance:

$$\mathbf{G}(x)\mathscr{L}(\phi_1, \phi_2, A) \to \mathscr{L}(\phi_1{}^*, \phi_2{}^*, A^*).$$

In this instance, the gauge particle is responsible for communicating the ϕ_1, ϕ_2 content of Φ from place to place. As before, we now redefine the fields so as to arrange that our axes pass through the point of minimum energy:

$$\phi_1' = \phi_1, \qquad \phi_2' = \phi_2 - R.$$

Rewriting the Lagrangian in terms of these redefined fields does not change the underlying physics, so we have

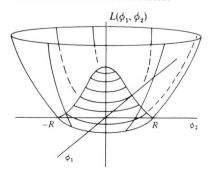

Fig. 22.2. The interaction energy chosen for the two components of a hypothetical quantum field. The state of minimum energy corresponds to a non-zero value for the field.

$$\mathcal{L}(\phi_1, \phi_2, A) \equiv \mathcal{L}(\phi_1', \phi_2', A).$$

It is in this last step that something remarkable occurs. The redefined ϕ_2' particle acquires, as before, a mass proportional to R, but astonishingly the massless Goldstone boson ϕ_1' disappears. Moreover, the formerly massless gauge particle now acquires a mass, again proportional to R.

What in fact happens is that the mathematical expressions describing the original massless gauge particle become mixed with R (the vacuum value of ϕ_2) in such a way as to create a mass term. At the same time, the Goldstone boson ϕ_1' becomes absorbed into the gauge particle in such a way as to lose its physical significance. Physicists say that the gauge particle 'eats' the Goldstone boson and thereby becomes massive.

The physical interpretation of all this is the following. The original Lagrangian describes a two-component particle, $\Phi = (\phi_1, \phi_2)$, and a massless vector gauge particle, A, consisting of two spin polarisation states. However, the redefined Lagrangian describes one massive spinless particle, ϕ_2', and one massive vector gauge particle, A', which, by virtue of its mass, now contains three polarisation states. The total number of physical degrees of freedom remains the same (i.e. four), but the hapless Goldstone boson has become the third polarisation state of the massive gauge boson. This looks more encouraging. The Goldstone boson has been avoided by using a *local* gauge symmetry, a step first taken by Peter Higgs of Edinburgh University and others in 1964. What is more, despite having started with a gauge-invariant theory, the gauge boson has acquired mass; this was the point of the entire exercise. The only price to be paid for this success is the presence of the massive spin-0 particle, ϕ_2' – the so-called 'Higgs boson' (see Table 22.1). Such a particle has not yet been detected, but its observation would lend great support to the idea of spontaneously broken gauge theories.

We have presented only a very simple example of how a local gauge symmetry may be spontaneously broken by the Higgs mechanism. This mechanism is quite general and can be applied straightforwardly to the gauge theory of the weak interactions, as we shall see in the next chapter.

Table 22.1. *The Higgs mechanism*

Spontaneous breaking of *local* symmetry avoids the unwanted Goldstone boson and generates a mass for the vector boson. Both before and after symmetry breaking, the total number of physical degrees of freedom is four.

Before	After
A	
ϕ_1 $\Big\}$ $\longrightarrow$	massive A'
ϕ_2 $\longrightarrow$	massive ϕ_2'

23

The Glashow–Weinberg–Salam model

23.1 Introduction

In 1967 and 1968 respectively, Steven Weinberg of Harvard and Abdus Salam of London independently formulated a unified theory for the weak and electromagnetic interactions, based in part on work developed previously by Sheldon Glashow, also of Harvard. The theory describes the interactions of leptons by the exchange of W bosons and photons, and incorporates the Higgs mechanism to generate the masses for the W bosons. Because the Lagrangian prior to spontaneous symmetry breaking (i.e. prior to the redefinition of the fields) is gauge-invariant, Weinberg and Salam conjectured, although were not able to prove, that the theory is renormalisable. The proof was demonstrated subsequently by Gerard 't Hooft of Utrecht, in 1971.

The idea of the model is to write down a locally gauge-invariant Lagrangian describing the interactions of leptons with massless W-gauge bosons, just as described in Chapter 21. Hypothetical Higgs fields are then introduced with a suitably chosen interaction Lagrangian which is added to that for the leptons. Following the redefinition of the Higgs fields, the Lagrangian now describes particles with mass but is no longer gauge-invariant under the same local transformations. Because the theory is renormalisable, the Feynman rules can be used to calculate finite answers for any physical quantities to any desired degree of accuracy.

We must take care during the spontaneous symmetry breaking to ensure that the photon remains massless whilst the W bosons are made massive. This is achieved by a sufficiently clever choice of the Higgs interactions such that, after redefinition of the fields, the Lagrangian is still invariant under some sub-group of the local gauge transformations. This subgroup is precisely the $U(1)$ gauge symmetry of QED, which describes the interactions of massless photons with charged particles.

23.2 Formulation

We have previously grouped the electron with the electron-neutrino as two different, weak isospin states of a single lepton wavefunction,

$$l_e = \begin{pmatrix} \nu_e \\ e^- \end{pmatrix}.$$

However, a straightforward grouping like this is unsatisfactory, for while the neutrino is massless and left-handed, the massive electron is both right- and left-handed. We have already mentioned that the weak interaction prefers its electrons to be left-handed, such that if the electron were actually massless (a state well-approximated by very relativistic electrons) it would act only on left-handed electrons. Therefore, let us split the electron wavefunction into separate right- and left-handed components and group them separately:

$$l_e = \begin{pmatrix} \nu_e \\ e^- \end{pmatrix}_L, \quad e^-_R.$$

In order to be able to do this consistently, the electron must be massless. So, we need to arrange for it to acquire a mass later. (This same separation is also made for the muon and all particles that 'feel' the weak force. They too must be massless at this stage.)

23.2.1 *Weak interaction charges and gauge symmetry*

We now wish to write down a Lagrangian describing the interaction of these leptons, and to introduce gauge bosons by requiring the Lagrangian to be invariant under certain gauge transformations. To do this, we need to know the generators of these transformations. Or, equivalently, we need to know the quantities (i.e. the charges) which are conserved by the interactions. These weak-interaction charges will differ before and after spontaneous symmetry breaking. Our hypothesis is that the present state of the world is the result of this symmetry breaking, and so we have relative freedom in choosing conservation laws prior to the breaking, provided that, after it, electric charge is conserved.

We have already identified weak isospin as a plausible candidate for a conserved charge for the weak interaction. The neutrino and the left-handed component of the electron form the weak isospin doublet, l_e, with $I_3^W = +\frac{1}{2}$ and $I_3^W = -\frac{1}{2}$ respectively. At this stage, prior to spontaneous symmetry breaking, the two components of the lepton doublet must be identical except for the value of the third component of their weak isospin, I_3^W. However, the electron has negative electric charge, whilst the neutrino is neutral. So, we must relate this electric charge difference to the difference in their I_3^W values. We can do this by introducing 'weak hypercharge' Y^W, which is defined by the equation:

$$Q = e(I_3^W + Y^W/2).$$

So, by awarding both ν_e and e_L^- a weak hypercharge of -1, the difference in their electric charges is given by the difference in their component I_3^W values. Furthermore, since I_3^W and Y^W are both conserved charges (before symmetry breaking), then electric charge will also be conserved.

The 'weak quantum numbers' of the leptons are summarised in Table 23.1. Note that because the right-handed component of the electron has no

Table 23.1. *The weak quantum numbers of the leptons*

Antileptons have opposite values of I_3^W, Y^W and Q.

	I^W	I_3^W	Y^W	Q
ν_e	$\frac{1}{2}$	$\frac{1}{2}$	-1	0
e_L	$\frac{1}{2}$	$-\frac{1}{2}$	-1	-1
e_R	0	0	-2	-1

weakly interacting partner, it must have weak isospin zero, $I^W = 0$ and $I_3^W = 0$ (since it must transform into itself under weak isospin rotations). Consequently, its weak hypercharge is directly related to its electric charge, and $Y^W = -2$.

We now demand that the interactions between leptons conserve weak isospin and weak hypercharge. We implement this by requiring the Lagrangian to be invariant under the $SU(2)_L^W$ group of weak-isospin transformations *and* under the $U(1)^W$ group of weak-hypercharge transformations (which correspond to simple shifts in the phase of the lepton wavefunction). So, the total symmetry group is $SU(2)_L^W \times U(1)^W$.

To realise these invariances under local (space–time dependent) transformations, we must introduce the appropriate gauge particles. Invariance under local rotations of weak isospin, requires the introduction of the gauge particle $W = (W^+, W^0, W^-)$. Then

$$\mathbf{G}^{[SU(2)_L^W]}(x)\mathscr{L}(l_L, W) \to \mathscr{L}(l_L{}^*, W^*).$$

Furthermore, to maintain invariance under shifts in the phase of the lepton wavefunction, we must introduce an additional gauge particle B, so that

$$\mathbf{G}^{[U(1)^W]}(x)\mathscr{L}(l_L, e_R, B) \to \mathscr{L}(l_L{}^*, e_R{}^*, B^*).$$

So, the total gauge invariance can be expressed as

$$\mathbf{G}^{[SU(2)_L^W \times U(1)^W]}(x)\mathscr{L}_1(l_L, e_R, W, B)$$
$$\to \mathscr{L}_1(l_L{}^*, e_R{}^*, W^*, B^*).$$

23.2.2 *Spontaneous symmetry breaking*

At this point, we add two further terms to the Lagrangian. Each involves the Higgs fields, which, like the left-handed leptons, take the form of a doublet:

$$\Phi = \begin{pmatrix} \phi^0 \\ \phi^- \end{pmatrix}$$

where the weak quantum numbers are the same as those of the left-handed lepton doublet. First, we add the term associated with the wine-bottle-shaped interaction energy. As this term must also be locally gauge invariant, it must also contain the gauge particles:

$$\mathscr{L}_2(\Phi, B, W).$$

Secondly, we may allow the Higgs fields to interact

with the leptons as we have yet to generate their masses:

$$\mathscr{L}_3(l_L, e_R, \Phi).$$

Then the local gauge invariance of the total Lagrangian $\mathscr{L}_1 + \mathscr{L}_2 + \mathscr{L}_3$ under the $SU(2) \times U(1)$ symmetry group is broken by the neutral Higgs component taking a non-zero vacuum value, $\phi^0 = R$, corresponding to the state of minimum energy. We must now redefine the Higgs field $\Phi \to \Phi'$ so that it is zero at the state of minimum energy:

$$\phi^{0\prime} = \phi^0 - R, \qquad \phi^{-\prime} = \phi^-.$$

After this redefinition, the Lagrangian must still describe the same physics, and so

$$\mathscr{L}_2(\Phi, B, W) \equiv \mathscr{L}_2(\Phi', B, W),$$
$$\mathscr{L}_3(l_L, e_R, \Phi) \equiv \mathscr{L}_3(l_L, e_R, \Phi').$$

Once the mathematical smoke has cleared following this redefinition, the following features emerge.

Weak isospin and weak hypercharge are no longer conserved charges since the $SU(2) \times U(1)$ gauge symmetry has been broken. However, it has been broken in such a way that the combination corresponding to electric charge (i.e. $Q = e(I_3^W + Y^W/2)$) is still conserved. This implies that the $U(1)$ gauge symmetry of QED remains unbroken, and the photon remains massless.

Gauge boson masses are generated by the mixing of R (the vacuum value of ϕ^0) with B and W in $\mathscr{L}_2$. However, as we have noted, the photon remains massless whilst the other gauge bosons become massive. We can see how this happens when we notice that neither the W-gauge particle of weak isospin, nor the B-gauge particle of weak hypercharge, can be identified with the electromagnetic gauge particle, the photon. Just as the electric charge is a mixture of weak isospin and weak hypercharge, the electromagnetic gauge particle is similarly a mixture of the neutral gauge particles of weak isospin W^0 and weak hypercharge B:

$$A = W^0 \sin \theta_W + B \cos \theta_W$$

where the weak angle, θ_W, is the parameter which adjusts the relative proportions of the two. The remaining portions of the W^0 and B wavefunctions

also mix together to produce another gauge particle:

$$Z^0 = W^0 \cos \theta_W - B \sin \theta_W.$$

This combination corresponds to the part of the weak interaction which has the same quantum numbers as the photon (i.e. zero electric charge). It is the neutral gauge boson corresponding to the weak neutral current mentioned previously, and is now denoted Z^0 to signify its origin as a combination of the two fundamental gauge fields.

By careful choice of the form of the interaction energy, the Z^0 can be given a mass, whilst the photon A remains massless. In obtaining a mass, the Z^0 absorbs a mixture of ϕ^0 and its antiparticle $\bar{\phi}^0$, which provides the third polarisation state of the massive particle. We are left with a real massive Higgs particle, ϕ', which comes from the remaining mixture of ϕ^0 and $\bar{\phi}^0$. The charged components of the W-gauge particles (i.e. W^- and W^+) obtain masses by absorbing ϕ^- and its antiparticle, ϕ^+. (See Table 23.2.) The values which emerge from the mathematics are:

$$M_{W^\pm} = \frac{38.5}{\sin \theta_W} \text{ GeV}, \qquad M_{Z^0} = \frac{M_{W^\pm}}{\cos \theta_W},$$

which clearly depend on the value of the weak angle, θ_W. This is a free parameter in this model and must be determined by experiment. The experimental value is given by $\sin^2 \theta_W \approx 0.23$, and so the model predicts masses of approximately 80 GeV and 90 GeV respectively. This explains why the W bosons were so difficult to detect. It was

Table 23.2. *The electroweak Higgs mechanism*

Spontaneous symmetry breaking in the electroweak model leads to three massive vector bosons and one massive Higgs boson. Both before and after symmetry breaking, the total number of physical degrees of freedom is 12.

Before	After
ϕ^-; W^- $\longrightarrow$	massive W^-
ϕ^+; W^+ $\longrightarrow$	massive W^+
$\phi^0, \bar{\phi}^0$; W^0, B $\longrightarrow$	$\begin{cases} \text{massive } \phi' \\ \text{massive } Z^0 \\ \text{massless } A \end{cases}$

not until 1983 that accelerators had enough energy to produce such massive particles.

Recall that in order to consistently split the electron wavefunction into left- and right-handed components, it needed to be massless (at least before symmetry breaking). After symmetry breaking, we find that l_e and e_R are mixed with R (the vacuum value of ϕ^0) in $\mathscr{L}_3$, and the correct mass is indeed generated for the electron. However, this part of the exercise does not have the same predictive power as the mass generation for the gauge bosons. Because we have incomplete knowledge of how the Higgs field interacts with leptons, we are free to choose appropriate coefficients in $\mathscr{L}_3$ so as to guarantee the correct electron mass.

23.3 Reprise

In presenting the basic structure of what is now the accepted theory of the 'electroweak' interactions, we have, for reasons of clarity, considered only the electron and its neutrino. However, *all* particles which 'feel' the weak force fit into the above scheme in a straightforward way. That is, all left-handed fermions form weak-isospin doublets with $I^W = \frac{1}{2}$ and $I_3^W = \pm\frac{1}{2}$, and all right-handed fermions form weak-isospin singlets with $I^W = 0$ and $I_3^W = 0$. So, for the muon we have

$$\begin{pmatrix} \nu_\mu \\ \mu^- \end{pmatrix}_L, \quad \mu_{\bar{R}}.$$

The same is true even for the quarks, but we postpone discussion of this until the next chapter.

This then is the Glashow–Weinberg–Salam model of the 'electroweak' interactions. It incorporates the successful theory of QED and provides a description of the weak force in terms of the exchange of massive vector bosons. It has, moreover, introduced a weak neutral current in a natural fashion, and the discovery of neutral-current reactions in 1973 was a great boost to the acceptability of the model. The model's cleverest feature, however, is the way it ensures the masslessness of the photon, whilst giving mass to the weak interaction gauge bosons $W^\pm$ and Z^0. This is achieved by the use of the Higgs mechanism and a suitable choice of Higgs fields. The gauge boson masses depend on the weak angle, θ_W, which is a parameter which must be measured from experiments

($\sin^2 \theta_W \approx 0.23$). In addition, the theory predicts the existence of a spinless Higgs particle ϕ', which has yet to be discovered. The production of $W^\pm$ and Z^0 bosons at CERN in 1983 with precisely the predicted masses was a great triumph for the electroweak model and indeed for the concept of non-Abelian gauge theory.

23.4 An academic postscript – renormalisability

Although the model was essentially formulated as above in 1967 and 1968, it was not enthusiastically received until its renormalisability had been demonstrated. As we have mentioned, the local gauge invariance of the Lagrangian prior to spontaneous symmetry breaking suggested that it would be, but it remained to be shown that the symmetry breaking itself did not spoil this property.

'T Hooft's proof of renormalisability essentially consisted of showing that the Feynman rules of the theory lead to mathematical expressions for the W-boson propagators, which avoid the problems associated with the use of massive spin-1 particles in perturbation theory (see Chapter 20). 'T Hooft showed that when the momentum flowing through a W-boson propagator is very large, then the mathematical expression of that propagator does not depend on the mass at all. As there is no mass dependence, there is no need for compensating momentum factors to ensure that the resulting probabilities are dimensionless numbers. It is these extra momentum factors which lead to the divergences (infinite values) when summing over all the internal momentum configurations of a complicated Feynman diagram; without them, the probabilities are finite and thus the theory is renormalisable.

24

Consequences of the model

24.1 Introduction

The Glashow–Weinberg–Salam model is now the undisputed theory of the weak (and electromagnetic) interactions. It was established over the decade 1973–83 by a series of experiments which have confirmed the model's predictions. First, in 1973, the discovery of neutral currents revealed a qualitatively new phenomenon, just as predicted by the model. Soon after, in the mid-1970s, new mesons were discovered which support the existence of a new quark type carrying the 'charm' quantum number – just as had been suggested by theorists attempting to describe the weak interactions of hadrons using the Glashow–Weinberg–Salam model. Next, in the late-1970s, various parity-violating effects were found to be in close agreement with the quantitative predictions of the model. Finally, and perhaps most impressively of all, the discovery at CERN of the $W^{\pm}$ and Z^0 bosons in 1983 with precisely the masses predicted by the theory established the gauge-theory structure of the weak interactions. In this chapter and the next we shall describe each of these in turn.

24.2 Neutral currents

In the discussion of neutrino–electron scattering in Chapter 17, we floated the possibility of the existence of neutral current processes by which, for example, an incoming neutrino may not have to turn into a charged lepton but could emerge with its identity unscathed. The emerg-ence of the neutral gauge boson Z^0 from the Glashow–Weinberg–Salam electroweak model provides a natural explanation for such neutral current phenomena, and their detection was seen in the early 1970s as being an important test for the validity of the model. Reactions which proceed by the neutral current are, for instance, elastic muon-neutrino–electron scattering or quasi-elastic muon-neutrino–proton scattering:

$$\nu_\mu + e^- \to \nu_\mu + e^-,$$
$$\nu_\mu + p \to \nu_\mu + p + \pi^0.$$

It was possible to check for the existence of these reactions only when neutrino beams were intense and energetic enough to permit detailed accelerator experiments to be performed. This became reality in 1973 when, much to everyone's surprise, neutral currents were found to be a significant effect, comparable in magnitude to the well-established charged currents. The process first seen was the second of the two mentioned above (see Figure 24.1), the magnitude of which, relative to its charged current version, was measured to be:

$$\frac{\sigma(\nu_\mu + p \to \nu_\mu + p + \pi^0)}{\sigma(\nu_\mu + p \to \mu^- + p + \pi^+)} = 0.51 \pm 0.25,$$

showing them to be effects of the same order. Since the first observation of this reaction, other neutral current processes have been observed as predicted by the electroweak model, including the elastic muon-neutrino–electron scattering mentioned above.

All these reactions allow us to determine the value of the weak angle θ_W, which is the parameter which essentially fixes the relative mixing of the weak and electromagnetic interactions. The value which represents the average of the different experiments conducted to date is about

$$\sin^2 \theta_W = 0.2324 \pm 0.0011.$$

As we saw in Chapter 23, the masses of the intermediate $W^{\pm}$ bosons depend directly on this quantity, which therefore predicts

$$M_{W^{\pm}} \approx 80 \, \text{GeV} \qquad M_{Z^0} \approx 90 \, \text{GeV}.$$

24.3 The incorporation of hadrons – charm

It is necessary also for the Glashow–Weinberg–Salem model to describe the weak

Fig. 24.1. (*a*) A neutral current reaction photographed in the Gargamelle bubble chamber at CERN, and its interpretation (*b*). (Photo courtesy CERN.)

(*a*)

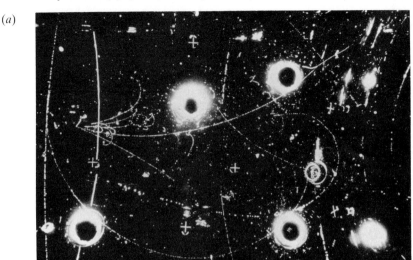

$$\nu_\mu + p \longrightarrow \nu_\mu + p + \pi^+ + \pi^- + \pi^0$$
$$\hookrightarrow \gamma$$

(*b*)

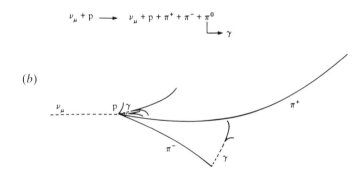

interactions of hadrons. In Chapter 18 we saw how these could be described using a weak interaction current written in terms of quarks. This form is the most convenient for investigating the consequences of the model. The charged current describing weak interactions where electric charge is changed is the same as before. It comprises a part which changes strangeness and a part which conserves strangeness, the relative importance of the two being regulated by the Cabbibo angle θ_C.

However, the model also requires an electrically neutral current, which, in the absence of further specifications, will similarly consist of a part which changes the strangeness of the particles

and a part which conserves it. An example of the strangeness-changing neutral current is given by the decay of the long-lived neutron kaon,

$$K_L^0 \rightarrow \mu^+ + \mu^-,$$

which is described in terms of quark interactions by the diagram Figure 24.2(*a*). Unfortunately, these processes, apparently allowed in the model, do not occur in the real world. Experimentally, the probability of this decay occurring is less than one part in one hundred million. So the model must be modified to ensure that these processes do not occur.

Several explanations were advanced in an

attempt to cure this problem. The most successful, later to be rewarded by striking experimental evidence, is the 'charm' scheme proposed in 1970 by the international triumvirate of Glashow, Iliopoulos and Maiani, commonly known as GIM. The key assumption in the scheme is the existence of a new 'charmed' quark, c. This quark is to carry a new quantum number, charm, just as the strange quarks carry strangeness. All the other quarks, u, d and s, are assigned zero charm. It is then possible to arrange that the unwanted strangeness-changing neutral current of Figure 24.2(*a*) is cancelled out by the corresponding diagram involving the charmed quark, shown in Figure 24.2(*b*).

In this way, it is possible to arrange for the disappearance of all the unwanted currents in reactions such as

$$K^+ \rightarrow \pi^+ + \mu^+ + \mu^-,$$
$$K^+ \rightarrow \pi^+ + e^+ + e^-.$$

The consequences of the GIM scheme are enormous. The existence of this fourth quark flavour implies the existence of whole new families of charmed mesons and charmed baryons, rather like

Fig. 24.2. Strangeness-changing neutral currents which can proceed by the intermediate quark process in (*a*) can be cancelled out of the model by additional quark processes involving the charmed quark in (*b*).

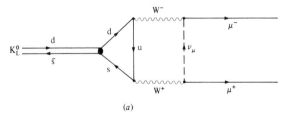

(*a*)

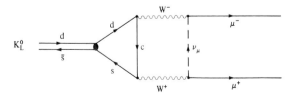

(*b*)

a repeat showing of all the strange particles. What is more, for the charmed quark diagram to cancel out the unwanted reactions effectively requires the mass of the charmed quark (and hence charmed particles) to be relatively small at about 1.5 GeV. So we should see charmed particles copiously produced in modern accelerators.

The charm scheme seemed to many physicists to be rather an 'expensive' way of solving the problem, i.e. having to introduce whole families of unknown particles just to alter the reaction rate of a small obscure class of reactions. However, this scepticism soon turned to amazement as the required particles tumbled out of the accelerators in the mid-1970s (see Chapter 36).

We mentioned in the previous chapter that the Glashow–Weinberg–Salam model treats all particles that 'feel' the weak force in precisely the same way. That is, all left-handed fermions form weak-isospin doublets with $I^W = \frac{1}{2}$ and $I_3^W = \pm\frac{1}{2}$, while all right-handed fermions form weak-isospin singlets with $I^W = 0$ and $I_3^W = 0$. This is true even for the quarks. For instance, for the up and down quarks we have

$$\begin{pmatrix} u \\ d' \end{pmatrix}_L, u_R, d_R.$$

Their weak quantum numbers are given in Table 24.1. Similarly, for the charmed and strange quarks we have

$$\begin{pmatrix} c \\ s' \end{pmatrix}_L, c_R, s_R.$$

Note that it is not d and s which appear in the above quark doublets, but d' and s'. This is because the

Table 24.1. *The weak quantum numbers of the quarks*

Antiquarks have opposite values of I_3^W, Y^W and Q.

	I^W	I_3^W	Y^W	Q
u_L	$\frac{1}{2}$	$\frac{1}{2}$	$\frac{1}{3}$	$\frac{2}{3}$
d_L	$\frac{1}{2}$	$-\frac{1}{2}$	$\frac{1}{3}$	$-\frac{1}{3}$
u_R	0	0	$\frac{4}{3}$	$\frac{2}{3}$
d_R	0	0	$-\frac{2}{3}$	$-\frac{1}{3}$

charged hadronic current has both strangeness-conserving and strangeness-changing components. Therefore, it is not d_L and s_L which couple to the charged $W^\pm$ bosons but the combinations

$$d_L' = s_L \sin \theta_C + d_L \cos \theta_C$$
$$s_L' = s_L \cos \theta_C - d_L \sin \theta_C$$

where θ_C is the Cabbibo angle.

24.4 Parity-violating tests of the Glashow–Weinberg–Salam model

In addition to the qualitatively new phenomena described in the previous two sections, the Glashow–Weinberg–Salam model also tells us the magnitude of the various parity-violating effects due to the weak force. The most convincing evidence in this area is provided in polarised electron–deuteron scattering in which all the electrons can be collided with their spins pointing in a specified direction. The experiment measures the difference in the scattering cross-sections between left- and right-handedly polarised electrons off polarised deuteron targets. Not only is the magnitude predicted correctly, giving a difference between cross-sections of about 0.01%, but the way this difference changes with the energy of the collision is also explained.

Another interesting parity-violating experiment is that concerning the interaction of light with matter at low energies. This provides a test of the model in a wholly new domain, and is thus that much stronger a test of its general worth. There are many variants of the experiment, but the basic idea is to shine a beam of polarised light (light whose electric field vector is aligned in a specific direction) through a vapour of metal atoms. The light is absorbed and re-emitted by the various transitions of the atomic electrons and, because of the effects of the weak interaction between the atomic nucleus and the electrons, this leads to a very slight, but well-defined, rotation of the plane of polarisation in a given direction. Such a given rotation, of course, implies a distinction between clockwise and anticlockwise and so is indicative of parity violation. Because the effect is slight and because there are uncertainties about using the Glashow–Weinberg–Salam model in the atomic environment, the experiments and their interpretation are extremely hard work. But after initial doubts and discrepancies, there is now a convincing consensus of experimental results supporting the model. The electroweak model also predicts novel effects in electron–positron annihilations (see Chapter 35) and these also provide support. The model is now our standard working understanding of the electromagnetic and weak interactions (sometimes now referred to as the unified electroweak force) and in recognition of this, Glashow, Weinberg and Salam were awarded the Nobel Prize in 1979.

25

The hunt for the W$^\pm$, Z^0 bosons

25.1 Introduction

The intermediate vector bosons of the weak force have been by far the most eagerly awaited particles of recent years. Their existence is crucial to the validity of the Glashow–Weinberg–Salam electroweak theory, and, by implication, to the acceptability of all the modern theories of the 'gauge' type. It was not surprising then that between 1976 and 1983 an enormous experimental effort was dedicated to their detection. But as we have seen, the masses of these particles are very large indeed. So just how were they detected?

The Z^0 boson decays into both quark–antiquark pairs and lepton–antilepton pairs. So, by the principle of microscopic reversibility of particle reactions, we may anticipate that sufficiently energetic collisions between quarks and antiquarks, or leptons and antileptons, will produce Z^0 bosons. The 'clearer' of the two possibilities is provided by lepton–antilepton annihilation, such as in the e$^+$e$^-$ experiments described in Part 8. The LEP collider at CERN, which was commissioned in 1989, was justified largely by this prospect, despite the inherent difficulties in the handling of very high energy e$^+$e$^-$ beams (see Section 35.2).

In the mid-1970s, physicists realised that an easier and less expensive means of searching for the W$^\pm$ and Z^0 bosons (although a less satisfactory environment in which to study them) was provided by quark–antiquark annihilation in collisions be-tween protons and antiprotons. This was despite the very messy final states resulting from the spectator quarks. In such reactions, one of the quarks inside the proton (uud) annihilates with an antiquark inside the antiproton ($\bar{u}\bar{u}\bar{d}$) to produce a W$^\pm$ boson if the pair is dissimilar (i.e. u$\bar{d}$ or $\bar{u}$d), or a Z^0 boson if the pair is similar (i.e. u$\bar{u}$ or d$\bar{d}$). This is illustrated in Figure 25.1.

In 1976, Carlo Rubbia, David Cline and their colleagues suggested converting a conventional 'fixed-target' proton accelerator into a proton–antiproton collider in order to provide the earliest possible opportunity for discovering these massive gauge bosons.

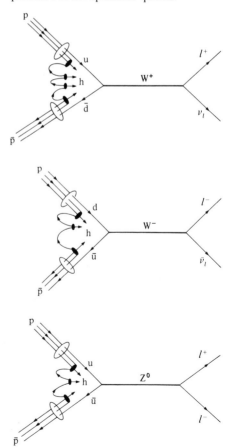

Fig. 25.1. The mechanisms for W$^\pm$, Z^0 boson production in p$\bar{p}$ collisions. In all cases h indicates the hadronic debris resulting from the presence of the spectator quarks.

25.2 The CERN pp̄ collider experiment

Following the general acceptance of the pp̄ idea, the accelerator chosen for the job was the super proton synchrotron accelerator (SPS) at CERN which, as of 1976, was one of the highest-energy machines in the world, able to accelerate protons to 400 GeV. The great beauty of the pp̄ idea is that, because the antiparticles (p̄) have the opposite charge to, but the same mass as, the particles (p), the accelerator configuration which accelerates protons in one direction will automatically accelerate antiprotons in the opposite direction. So the one-beam ring of the original SPS was made to accommodate the two counter-rotating beams of p and p̄. In the process of this conversion, each beam was designed for an energy of 270 GeV, giving a head-on collision energy of 540 GeV. This, of course, is far greater (over five times) than the thresholds for $W^{\pm}$, Z^0 production, but this is necessary as only a fraction of the energy will go into the quark–antiquark annihilations. The rest stays with the spectator quarks and gives rise to complicated, long-range hadron production.

Although the basic idea behind the experiment is simple, the reality is complicated by the absence of naturally occurring antiprotons. These must be painstakingly manufactured in preparatory particle collisions and stored until a sufficient number have been collected to form a beam of sufficient density (referred to as luminosity) for an observable rate of reactions to be possible. Achievement of this 'antimatter factory' is one of the wonders of modern physics and demonstrates convincingly the sophistication with which the experimentalists are now able to control elementary particle beams, through a veritable 'sphaghettiscape' of accelerators. See Figure 25.2.

Initially, protons are accelerated to 26 GeV in the 1959 proton synchrotron (PS) machine and are collided into a tungsten target. From the input of 10^{13} protons, approximately 20 million antiprotons of about 4 GeV energy emerge. These are then piped to the antiproton accumulator where an increasing number of such collision bunches are stored. When a bunch first enters the accumulator it is regulated or 'cooled' to remove all random components of the antiprotons' individual movements. This process is called 'stochastic cooling' and is achieved with a sophisticated control system

which detects any such deviation of the antiprotons' movements from the ideal orbit, flashes a signal across the diameter of the accumulator ring and 'kicks' the antiproton bunch back into shape by the application of a tailored magnetic pulse. All this in the time that it takes the antiproton bunch, travelling at practically the speed of light, to travel half-way around the ring! After about two seconds, the antiproton bunch is sufficiently cooled for it to be manoeuvered by magnetic fields into a 'holding' orbit in the accumulator whilst the next bunch is introduced and cooled, after which it too is manoeuvered to join the holding orbit. Over some two days about 60 000 injections are achieved to form a few orbiting bunches of about 10^{12} antiprotons in each. Eventually, the antiproton bunches are sent back to the PS to be accelerated up to 26 GeV, after which they are injected into the SPS. The PS also injects bunches of 26 GeV protons into the SPS in the opposite direction. The SPS can then accelerate the counter-rotating p and p̄ bunches each to 270 GeV when they can be collided through each other at specified interaction regions around the SPS, where various experiments observe the interactions. After this, the same pp̄ beam bunches can go on providing interactions over many hours.

Knowing the luminosities of the beam bunches (equivalent to approximately 10^{30} anti-

Fig. 25.2. High-energy plumbing. Both the proton synchrotron (PS) and the antiproton accumulator (AA) form an integral part of the CERN SPS pp̄ collider experiment. The intersecting storage rings (ISR) conduct separate experiments.

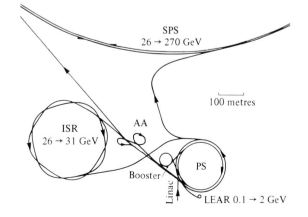

Fig. 25.3. The predicted production rate of $W^{\pm}$, Z^0 bosons as a function of collision energy, assuming the design luminosity of the CERN p$\bar{\text{p}}$ experiment.

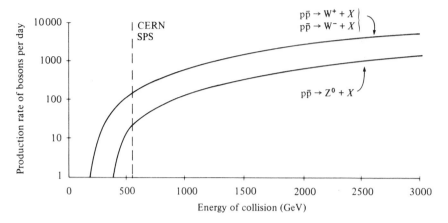

protons per square cm per second) and using the Glashow–Weinberg–Salam theory to calculate the probabilities of occurrence of the reactions producing the $W^{\pm}$, Z^0 boson allows us to estimate the rate of production of the bosons in the CERN p$\bar{\text{p}}$ collisions. As we can see in Figure 25.3, several hundred were expected each day. The problem then became one of finding the bosons amongst the debris of the collisions.

25.3 Detecting the bosons

Once produced, the $W^{\pm}$, Z^0 bosons are far too short lived to leave detectable tracks. As is often the case, their presence must be inferred from the behaviour of their decay products. The most significant of these for the $W^{\pm}$, Z^0 bosons are the charged leptons arising from the decays:

$$W^+ \to \mu^+ + \nu_{\mu},$$
$$W^- \to \mu^- + \bar{\nu}_{\mu}$$

and

$$Z^0 \to \mu^+ + \mu^-.$$

Several features of the charged lepton distributions emerging from p$\bar{\text{p}}$ collisions can provide the telltale signs of the $W^{\pm}$, Z^0 bosons.

Firstly, the presence of the parity-violating effect indicates the action of the weak force. The effect of angular momentum conservation and the unique-handedness of the neutrino and antineutrino requires an excess of positively charged lep-

tons emerging in the direction of the incoming antiproton beam (and, similarly, an excess of negatively charged leptons in the direction of the incoming proton beam). Although this is strong circumstantial evidence for the bosons, strictly speaking it indicates only the presence of the weak force and not specifically mediation via the $W^{\pm}$, Z^0 bosons.

This more demanding information is indicated by the momenta and energies of the emerging leptons. The production of the $W^{\pm}$, Z^0 bosons gives rise to a far higher proportion of leptons carrying a significant momentum perpendicular to the axis of the p$\bar{\text{p}}$ collision. By measuring the distribution of this transverse momentum of the emerging leptons, the presence of the bosons is manifestly obvious (see Figure 25.4(*a*)). Also, the decay of the Z^0 boson involves no 'invisible' neutrinos to carry off any of the energy and so gives rise to the additional distinguishing feature of a very sharp peak in the mass distribution of emerging lepton–antilepton pairs centred on the mass of the Z^0 (see Figure 25.4(*b*)).

During the latter part of 1982, the SPS p$\bar{\text{p}}$ collider achieved a sufficient luminosity to make feasible the discovery of the bosons at the rate of a few events per day. So two separate experiments using different detectors set out to find the bosons. The experiments were referred to as UA1 and UA2, respectively, denoting the different underground areas in which their detectors are located round the SPS ring. Each of the experiment's

Fig. 25.4. The anticipated effects of $W^{\pm}$, Z^0 bosons. In (a) we see the increased distribution of leptons emerging with high transverse momentum, p_T. In (b) the mass spectrum of charged lepton–antilepton pairs shows a peak at the mass of the Z^0.

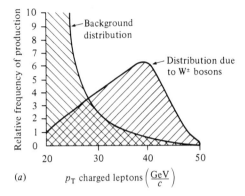

(a)

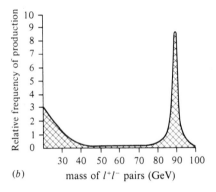

(b)

detectors were very massive assemblies of a variety of particle detection devices (2000 tonnes in the case of the UA1, 200 tonnes in UA2). The output of these detectors was fed directly into on-line computers, which allowed the subsequent reconstruction and analysis of the tracks of each event (see Figure 25.5(a) and (b)).

The experiments observed about 10^9 p̄p collisions of which 10^6 were recorded for subsequent analysis. The two experiments then applied various criteria to the observed collisions to find examples of the process,

$$p + \bar{p} \rightarrow W^{\pm} + X$$
$$\phantom{p + \bar{p} \rightarrow W^{\pm}} \rightarrow e^{\pm} + \nu.$$

The UA1 experiment sought just two separate classes of event: firstly, those with an isolated electron with large transverse momentum and, secondly, those events with a large fraction of transverse energy missing (carried off by the undetected neutrinos).

Starting from its initial sample of 140 000 events, the UA1 experimental group was able to identify those events containing the isolated electron with large transverse momentum by applying a series of exclusion conditions (such as demanding that the electron track have originated in the central detector, demanding that other tracks have low transverse momentum, and so on). Eventually, just five isolated electron events were identified.

Then, starting a new search on a sub-set of 2000 of the original events, the group searched for those events with missing energy (i.e. those events containing energetic neutrinos). This is done essentially by adding up the measured energies of all the observed tracks and checking the total against the known collision energy of 540 GeV. After another set of exclusion conditions just seven events were found, five of which were just those events with the isolated, high-energy electrons. So the UA1 experimental group were able to claim the discovery of five $W^{\pm}$ bosons. Furthermore, by adding up the measured energy of the electron track and the missing energy of the inferred neutrino track, the UA1 group was able to estimate the mass of the $W^{\pm}$ bosons from which these two particles had originated. The result was in excellent agreement with the prediction of the Glashow–Weinberg–Salam model.

The UA2 experimental group were able to perform a similar analysis of their sample of p̄p interactions, and were able to identify four candidate $W^{\pm}$ boson events and give an estimate of the mass which was also in good agreement with the prediction.

These discoveries of the $W^{\pm}$ bosons were announced in January 1983. On 1 June of the same year, the two experimental groups announced the discovery of the Z^0 boson, which was first identified in the process

$$p + \bar{p} \rightarrow Z^0 + X$$
$$\phantom{p + \bar{p} \rightarrow Z^0} \rightarrow e^+e^-.$$

This identification followed from another set of exclusion criteria in which the e^+e^- pair emerged back-to-back with equal and opposite high transverse momenta.

Carlo Rubbia, the driving force behind the experimental effort, and Simon van der Meer, the inventor of stochastic cooling, were jointly awarded the 1984 Nobel Prize in recognition of the major roles they played in these discoveries, which brought to a climax a decade of successful experiments verifying the gauge-theory framework of the Glashow–Weinberg–Salam electroweak model.

25.4 Epilogue

Since 1983, the massive electroweak gauge bosons have been produced in much greater numbers both at the CERN p$\bar{\text{p}}$ collider and also at the 'Tevatron' p$\bar{\text{p}}$ collider at Fermilab in Illinois. In 1989, two e^+e^- machines – LEP at CERN and SLC at Stanford – came into operation and began producing huge numbers of Z^0 bosons. This has permitted an accurate mass determination, and enabled physicists to refine their measurements of the width of the Z^0 peak (Figure 25.4(b)), hence establishing the particle's lifetime. These measurements had important implications for the total number of neutrino species, which in October 1989 was finally narrowed down to precisely three.

The final piece in the puzzle of the electroweak force remains the elusive Higgs boson. One process in which it might be discovered involves a virtual Z^0 boson radiating a Higgs particle before the Z^0 itself decays. Just how easy it will be to find the Higgs in such a process depends on how heavy it actually is. Its mass is not predicted by the electroweak theory, however we know from LEP that it must be heavier than about 45 GeV.

Fig. 25.5 (a) The 2000-tonne UA1 detector at the CERN proton–antiproton collider. (Photo courtesy CERN.) (b) Particle tracks heralding the discovery of the W boson in the UA1 detector. An electron with high transverse momentum (arrowed) emerges from the interaction point, and missing energy betrays the escape of an invisible neutrino. (Photo courtesy CERN.)

(a)

(b)

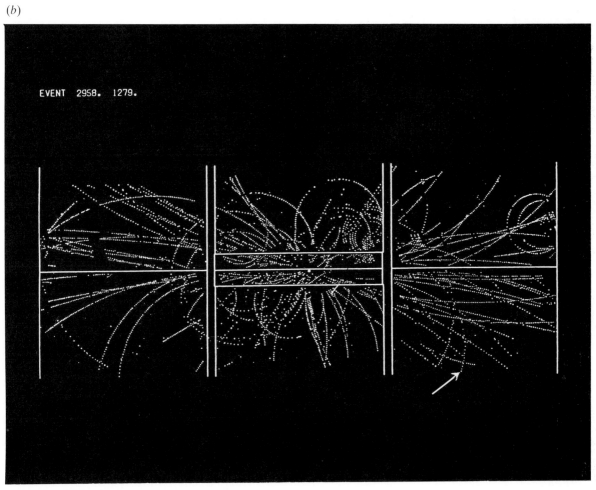

Part 6
Deep inelastic scattering

Deep inelastic processes

26.1 Introduction

Among the most important experiments of the last 25 years have been those which use the known interactions of the leptons to probe the structure of the nucleons. Their importance lies in the fact that they provided the first dynamical evidence for the existence of quarks, as opposed to the static evidence provided by the success of the internal symmetry scheme $SU(3)$.

The term *deep inelastic scattering* arises because the nucleon which is probed in the reaction nearly always disintegrates as a result. This is obvious from the momentum–wavelength relation for particle waves:

$$p\lambda = h. \tag{26.1}$$

The proton is approximately 10^{-15} m in diameter and so to resolve any structure within this requires that the probing particle wave has a smaller wavelength. The formula then gives the required momentum of the probe as being greater than 1 GeV/c, under the impact of which the target nucleon is likely to disintegrate.

Deep inelastic experiments divide into two classes, depending on the nature of the probe used, which in turn dictates the force involved. In electroproduction, electrons or muons are scattered off the target nucleon and the force involved is electromagnetic. The leading process of the scattering is that of single-photon exchange, which is assumed to be a sufficiently good description of the interaction (Figure 26.1(*a*)), although, in principle, more-complicated multi-photon processes may become significant as the energy of the collision becomes very large. The second class of experiment is called neutrinoproduction and, in this, neutrinos are scattered off the target nucleon by the weak nuclear force. The leading process is that of single-W-boson exchange, other more complicated processes being insignificant. Both charged and neutral currents may contribute (see Figure 26.1(*b*), (*c*)), but in practice it is the better-understood charged currents which are used in experiments. Indeed, this was necessarily the case in the early experiments, 1967–73, as the neutral currents had not then been discovered.

The main measurement of the experiments is the variation of the cross-section (the effective target area of the nucleon) with the energy lost by the lepton during the collision and with the angle through which the incident lepton is scattered. The energy lost by the lepton v is simply the difference between its incident and final energy:

$$v = E_i - E_f. \tag{26.2}$$

Fig. 26.1. (*a*) Electroproduction via photon exchange. (*b*) Charged current neutrinoproduction via $W^\pm$ exchange. (*c*) Neutral current neutrinoproduction via Z^0 exchange.

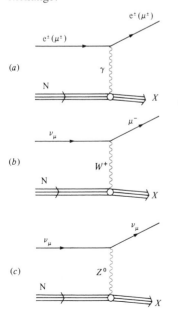

The angle through which the lepton is scattered is related to the square of the momentum transferred by the photon q^2 from the lepton to the nucleon by the formula

$$q^2 = 2E_i E_f (1 - \cos\theta). \tag{26.3}$$

These are the two main observables in deep inelastic scattering, which connect the data from experiments with our theoretical picture of the nucleon interior.

26.2 Two key ideas

Two ideas in particular played an important role in the development of the experiments and in our understanding of them. The two ideas, both put forward in 1969, are those of the parton model and of scaling.

The *parton model* was first put forward by Richard Feynman and is simply a formal statement of the notion that the nucleon is made up of smaller constituents: the partons. No initial assumptions about the partons are necessary, as it is the purpose of the experiments to determine their nature. But obviously we have at the back of our minds the identification of the partons with the quarks of $SU(3)$. However, we should not jump to the conclusion that *only* the familiar quarks will be sufficient to describe the composition of the nucleon. For instance, in addition to the proton's three quarks which are required by the internal symmetry scheme (the so-called *valence* quarks), it may be possible for virtual quark–antiquark pairs to emerge briefly from the vacuum by borrowing energy according to Heisenberg's uncertainty principle. The *sea* quarks may then form an additional material presence within the nucleon and provide a mechanism for the existence, albeit transient, of antimatter inside a 'matter' particle. Because they emerge in quark–antiquark pairs, the sea quarks will have no net effect on the quantum numbers of the nucleon, which are determined by the valence quarks. In addition to the quarks, we may be alert to the fact that quanta of the interquark force field may also be present inside the proton. Just as electrons interact by the exchange of photons, the quanta of the electromagnetic force, so quarks may interact by the exchange of quanta of their force field. These quanta have been called, rather simplistically, gluons, because they glue the quarks together.

Scaling is the name given to a phenomenon first predicted by the Stanford physicist James Bjorken. Stated simply, the prediction is that when the momentum carried by the probe becomes very large, then the dependence of the cross-section on parameters such as the energy ν and momentum-squared q^2, transferred by the photon, becomes very simple. In the parton model, the onset of this simple scattering behaviour has a straightforward interpretation. The complicated scattering of the probe off a nucleon of finite spatial extent is, at high momentum, replaced by the scattering of the probe off a point-like parton. The photon ceases to scatter off the nucleon *as a coherent object* and, instead, scatters off the individual point-like partons *incoherently*. We should expect this sort of behaviour to manifest itself when the wavelength of the probe is much less than the nucleon diameter, implying a probe momentum above about 1 GeV.

Observation of this scaling behaviour in 1969 immediately lent support to the parton model of the nucleon, although, as we shall see, the initial discovery was somewhat fortuitous. To understand the concepts of scaling and the parton model further, we must take a more detailed look at the processes involved. The importance of these ideas was recently acknowledged by the award of the 1990 Nobel Prize to the pioneers of these deep inelastic experiments, Jerome Friedman, Henry Kendall and Richard Taylor.

27

Electron–nucleon scattering

27.1 Introduction

Assuming that the electromagnetic interaction between the electron and the nucleon is dominated by the single-photon exchange mechanism, then the mathematics used to describe the reaction becomes relatively simple. To check the experimental observations, we want to derive a formula to explain how the cross-section varies with the energy transfer ν and momentum transfer squared q^2 of the intermediate photon. The formula is made up of factors associated with the different parts of the diagram in Figure 26.1(a). It consists of a factor describing the progress of the electron through the reaction (the lepton current), a factor describing the propagation of the virtual photon as a function of ν and q^2, and a factor describing the flow of the nucleon in the reaction including the complicated disintegration process (the hadron current). The factors describing the electron and the photon are well known from QED and present us with no problems. But the factor describing the hadron current is a complicated unknown, describing the evolution of nucleon structure during the reaction. This unknown can be characterised by a number of 'structure functions' of which we assume no prior knowledge and which are to be determined by the deep inelastic experiments (see Figure 27.1).

The form of the structure functions is discovered by writing down the most-general possible combinations of all the momenta appearing in the

Fig. 27.1. The formula describing the differential cross-section for electron–nucleon scattering with respect to the momentum transfer squared q^2 and the energy lost by the electron ν. The structure functions F_1 and F_2 essentially describe the shape of the nucleon target.

$$\frac{\mathrm{d}^2\sigma}{\mathrm{d}q^2\mathrm{d}\nu} = \frac{4\pi\alpha^2}{q^4}\frac{E_\mathrm{f}}{E_\mathrm{i}M_\mathrm{p}}\left[\frac{M_\mathrm{p}}{\nu}F_2(q^2,\nu)\cos^2\frac{\theta}{2} + 2F_1(q^2,\nu)\sin^2\frac{\theta}{2}\right]$$

reaction and then simplifying the result using general theoretical principles such as parity and time-reversal invariance.

The two separate functions of q^2 and ν that result – $F_1(q^2,\nu)$ and $F_2(q^2,\nu)$ – correspond to the scattering of the two possible polarisation states of the virtual photon exchanged: longitudinal and transverse. The longitudinal polarisation state exists only because of the virtual nature of the exchanged photon (because it temporarily has a mass). The virtual photon is 'off-mass-shell', meaning that $E = pc$ is violated, and implying that its mass is non-zero. On the mass shell when the virtual photon becomes real (massless), then the longitudinal polarisation state and its associated structure function disappear. The separate behaviour of the two structure functions can be determined from experiments because they are multiplied by coefficients involving different functions of the electron scattering angle. By observing the reaction at different values of this angle, the two behaviours can be separated out.

27.2 The scaling hypothesis

The scaling hypothesis mentioned previously is to do with these structure functions. It is important to realise that they are just numbers and have no physical dimension. The cross-section is usually given in units of area which are provided by the simple Rutherford scattering formula for elastic scattering. This has deep implications for the behaviour of the structure functions. If they are to have any dependence on physically dimensional quantities such as the energy ν or momentum-squared q^2 involved in the reaction, then these factors must have their physical dimensionality cancelled out to give structure functions in terms of pure numbers.

In low-energy elastic scattering (corresponding to $q^2 = 2\nu M_\mathrm{N}$), the photon effectively perceives the nucleon as a single extended object and the structure function essentially describes the spatial distribution of electrical charge on the nucleon. This leads to a dependence of the structure function on the momentum of the photon – but the dimensionality of the momentum in the structure function is cancelled out by factors of the nucleon mass:

$$\frac{\mathrm{d}\sigma}{\mathrm{d}q^2} = \frac{4\pi\alpha^2}{q^4}\cdot F\left(\frac{q^2}{M_\mathrm{N}^2}\right) \tag{27.1}$$

i.e.,

cross-section = unit of area × pure number

To signify this cancellation we say that the nucleon mass sets the 'scale' of reaction. It provides a scale against which the effect of the photon momentum can be measured.

By contrast, in very high energy, deep inelastic scattering (i.e. q^2, $\nu \to \infty$), the wavelength of the photon is so small that the existence of the complete nucleon is really irrelevant to the reaction: the photon interacts with only a small part of the nucleon and does so independently of the rest of it. This means that there is no justification for using the nucleon mass to determine the scale of the reaction. In fact, there is no justification for using any known mass or any other physically dimensional quantity to determine the scale of the deep inelastic regime. James Bjørken grasped the consequences of this abstract argument: if the structure functions are to reflect the dependence of the cross-section on the shape of the nucleon as seen by a photon of very high q^2 and ν, and if there exists no mass scale to cancel out the physical dimensions of these quantities, then the structure functions can only depend on some dimensionless ratio of the two. Choosing such a ratio as x,

$$x = \frac{q^2}{2M_\mathrm{N}\nu}, \tag{27.2}$$

then the scaling hypothesis is that the structure functions can depend only on it, and not on either or both of the quantities involved separately. So, as $q^2, \nu \to \infty$,

$$F_{1,2}^{\mathrm{eN}}(q^2, \nu) \xrightarrow[q^2, \nu \to \infty]{} F_{1,2}^{\mathrm{eN}}(x). \tag{27.3}$$

The scaling hypothesis becomes rather more accessible when it is combined with the parton model in which the nucleon is regarded as a simple collection of point-like constituents. A point has no dimension, and we are considering the scattering of a photon carrying infinitely high momentum (i.e. one which has a vanishingly small wavelength). In this situation, there are simply no physically significant quantities which are relevant to set the scale of the reaction. So quantities such as the energy and momentum transfer squared in the reaction can only enter into its description in

the form of pure numbers, which in turn implies a dimensionless ratio of the two.

Bjørken's choice of the 'scaling' variable x has a very significant interpretation. It turns out to be the fraction of the momentum of the nucleon carried by the parton which is struck by the photon. So the structure functions, which depend only on x, effectively measure the way in which the nucleon momentum is distributed amongst its constituent partons.

Figure 27.2(a) shows how the structure function $F_2^{\mathrm{ep}}(x)$ varies with x, as measured in early experiments at the Stanford Linear Accelerator Center. As can be seen, the shape implies that the majority of collisions occur with partons carrying a relatively small fraction of the nucleon momentum. Figure 27.2(b) shows a test of the scaling hypothesis that the structure function depends only on x, and not on q^2 (or ν) separately. At a given value of x, the structure function is found to be almost constant over a q^2 range from 1 to 8 GeV^2. Data such as this became available in the early 1970s. The apparent validity of the scaling hypothesis and the plausibility of its connection with the existence of point-like constituents within the nucleon led to a more detailed investigation of the structure functions to establish more information about the mysterious partons.

Fig. 27.2. Early data on scaling. In (a), $F_2(q^2, \nu)$ is a universal function of x for a range of values of q^2 (many experimental points are contained in the shaded area). In (b), $F_2(q^2, \nu)$ demonstrates approximate constancy over the range of q^2 measured in the early experiments.

27.3 Exploring the structure functions

One straightforward exercise is to compare the formula for electron–proton scattering with formulae derived from QED describing the electromagnetic interactions of electrons with other simple, electrically charged particles whose properties, such as spin, may be assumed. In this way it is possible to derive relationships between the structure functions depending on the similarities between the properties of the partons and those of the hypothetical particles assumed in the QED formulae. For instance, by comparing the formula of Figure 27.1 with the QED formula describing the scattering of an electron with an electrically charged particle of spin $\frac{1}{2}$, it is possible to derive the relationship between the structure functions:

$$2xF_1(x) = F_2(x). \tag{27.4}$$

Thus if, experimentally, the ratio $2xF_1/F_2$ is found to

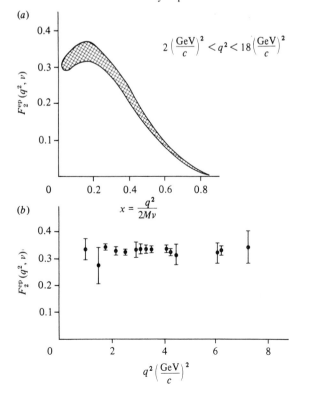

to be equal to one, this may be interpreted as evidence for the partons having spin $\frac{1}{2}$. Using similar formulae it is possible to show that if the ratio is observed to be zero, then this provides evidence for spin-0 partons. As can be seen from Figure 27.3, the evidence is firmly in favour of spin-$\frac{1}{2}$ partons.

Another lesson to be learned by comparing the deep inelastic formula with the simpler formula from QED is that the structure functions essentially measure the distribution of electric charge within the nucleon. In low-energy, non-relativistic physics, it is acceptable to speak of the distribution of charge over the spatial extent of the proton. But in high-energy physics this is better expressed in terms of the conjugate description of how the nucleon momentum is distributed amongst partons of various charges. If we say that the ith type of parton with charge Q_i has probability $f_i(x)$ of carrying a fraction x of the nucleon momentum, then it is possible to relate the overall structure functions of the nucleon to these individual parton momentum distributions,

$$\left.\begin{aligned} F_1^{eN}(x) &= \sum_{i\,\text{partons}} f_i(x)Q_i^2 \\ F_2^{eN}(x) &= x \sum_{i\,\text{partons}} f_i(x)Q_i^2 \end{aligned}\right\} \tag{27.5}$$

Obviously, if we integrate the structure functions over the fractional momentum carried by the partons, we should then expect to provide some measure of the total charge carried by the partons. In fact the most convenient relationship involves the sum of the squares of the parton charges:

$$\int_0^1 \frac{F_2(x)}{x}\,dx = \sum_i Q_i^2 \tag{27.6}$$

So, by investigation of relationships involving the structure functions, it is possible to find evidence for the spin of the partons and their charge assignments, which we will investigate further in Chapter 29.

Fig. 27.3. The ratio of the structure functions provides a test of the spin-$\frac{1}{2}$ assignment to the partons.

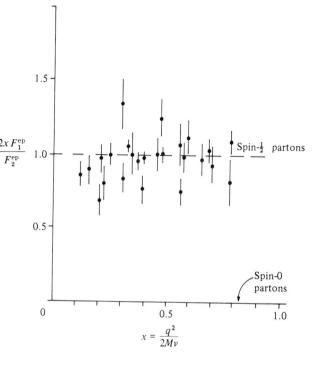

28

The deep inelastic microscope

28.1 Introduction

The physics behind the approach to scaling can be appreciated intuitively by regarding deep inelastic scattering as an extension of the ordinary microscope, whose successor the experiments quite literally are. The distance to which an ordinary microscope can resolve depends ultimately on the wavelength of the light scattered from the object under view. The shorter the wavelength, the smaller the distances which can be resolved. The high-energy photon exchanged between the electron and the nucleon in deep inelastic experiments is simply the logical development of the microscope technique. The origin of the scaling phenomena becomes clear by considering a succession of snapshots of a virtual photon–nucleon collision.

When the momentum carried by the photon is low, its wavelength is relatively long compared with the dimensions of the nucleon. It will not be able to resolve any structure and will effectively see the nucleon as a point. In this case, structure functions are irrelevant, being represented simply by the nucleon charge in the Rutherford formula for the scattering of two point charges (see Figure 28.1(*a*)).

With higher momentum, the photon will have a wavelength comparable to that of the nucleon. The photon will begin to resolve the finite spatial extent of the nucleon and the structure functions will depend in a non-trivial fashion on

the momentum carried by the photon, modifying the Rutherford scattering cross-section (Figure 28.1(*b*)).

Eventually, with high-momentum transfer the photon will have a very short wavelength and may resolve the internal structure of the nucleon. Scattering off point-like constituents within the nucleon is indicated when the structure function assumes a simple dependence on some dimensionless 'scaling variable'. In fact, under certain circumstances the structure functions are simply replaced by a constant equal to the sum of the squares of the charges of the constituents, which multiplies the simple Rutherford cross-section (Figure 28.1(*c*)). This reversion to the simplicity of point-like scattering after a relatively more complicated transitional phase is taken to indicate that we have broken through to a new, more basic level of matter within the nucleon.

28.2 Free quarks and strong forces

An essential consequence of the validity of the scaling hypothesis, in which the deep inelastic probe scatters *incoherently* off the individual partons, is that the partons are essentially free from

Fig. 28.1. An illustration of the approach to scaling as the virtual photon wavelength becomes much smaller than the nucleon diameter. The wavelengths become progressively smaller in (*a*), (*b*) and (*c*).

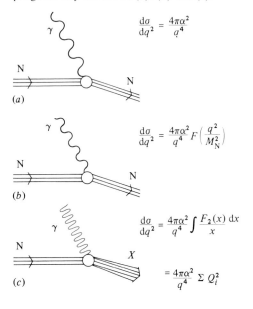

mutual interactions over the space–time distances of the probe–parton interaction. This has important consequences for the nature of interparton forces.

To understand this more fully, we can associate one interaction time with the duration of the probe–parton interaction τ_1 and another time to characterise the duration of interquark forces τ_2. Obviously, for the nucleon to have any sort of collective identity, the nucleon must exist for $\tau_{\text{life}} > \tau_2$, so that the partons can become aware of each others' presence. As a rough guide we may estimate the interaction times by the following prescriptions:

$$\tau_1 = \frac{\text{wavelength of probe}}{\text{speed of light}} \approx \frac{\hbar}{v},$$

$$\tau_2 = \frac{\text{interquark distance}}{\text{speed of light}} < \tau_{\text{life}}.$$

If $\tau_2 \lesssim \tau_1$, then the interparton forces will have transmitted the effects of the probe collision to all the partons inside the nucleon within the probe's interaction time. In this case, the probe will not be scattering off the individual partons but off the entire nucleon instead. However, in deep inelastic scattering in which the probe wavelength is very small, $\tau_1 \ll \tau_2$, and therefore the probe interaction is completed well before the interparton forces have had time to relay the event to the rest of the nucleon. So the probe–parton interaction occurs well within the lifetime of the nucleon.

For a short time, subsequently, the nucleon exists in an uncomfortable state: one of its partons has been struck hard and flies off with the high momentum imparted by the probe, but the other partons know nothing of this and continue to exist in a quiescent state (see Figure 28.2). This situation cannot last long, otherwise the struck parton would eventually appear as an isolated particle separated from the rest. This has never been observed. Instead, it is thought that final-state interactions come into effect between the struck parton and the others, turning the energy of the collision into the observed hadrons (Figure 28.2(c), (d)).

An important feature of the parton model is the assumption that the cross-section for deep inelastic scattering can be calculated simply by summing over the individual probe–parton interactions and that the complicated final-state interactions become important only over longer space–time distances (i.e. greater spatial separations and longer interaction times).

If we think particularly in terms of the strong interactions between quarks, the overall picture emerging from the success of the parton model is

Fig. 28.2. When a deep inelastic probe strikes a parton (*a*) and (*b*), it flies off with large momentum (*c*) until some confining mechanism pulls extra parton–antiparton pairs from the vacuum, creating new particles (*d*).

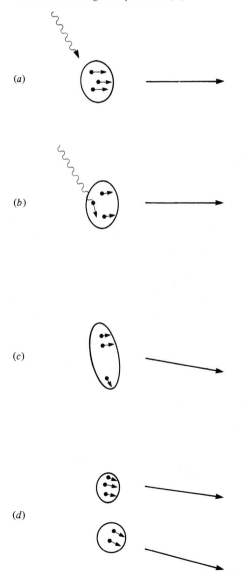

that of a force whose strength varies with distance. At the short distances probed by the deep inelastic reactions (say 10^{-17} m) the interquark force is weak and the quarks are essentially free. But as the distance between two quarks grows to the nucleon diameter (10^{-15} m), the force also grows, confining the quarks, perhaps permanently, within the observed hadrons. Eventually, any energy expended in trying to separate the quarks will be sufficient to pull a new quark–antiquark pair from the vacuum, so allowing the production of a new hadron, but preventing the emergence of the individual quarks.

The tendency of the interquark forces to become weaker at small distances is known as *asymptotic freedom*. As we shall see in Part 7, the discovery that this property can be explained naturally in non-Abelian gauge theory was a significant breakthrough in our attempt to understand quark dynamics. The other tendency of the interquark forces, to become increasingly strong as the quarks are separated, is known as confinement. As discussed also in Part 7, a satisfactory description of this phenomenon remains an outstanding problem.

29

Neutrino–nucleon scattering

29.1 Introduction

Just as in electron– or muon–nucleon scattering the exchanged photon acts as a probe of the electromagnetic structure of the nucleon, so in neutrino (or antineutrino)–nucleon scattering, the exchanged W boson probes the distribution of 'weak charges' within the nucleon. For this purpose, the two most important processes are the charged current inclusive reactions (see Figure 26.1). An important feature of these reactions is that they are able to distinguish between the partons and the antipartons of the target nucleon. This is because the space–time structure of the weak interaction ensures that target partons of differing helicities are affected differently. In the relativistic limit, in which the rest mass of a particle is regarded as being negligible, the parton and antiparton helicities are opposite, so they will interact with the W-boson probe differently. Also, because the W-boson probe is electrically charged, the target parton must be able to absorb the charge. As we shall see, this rules out the participation of some types of parton, making the weak interaction a more selective probe of the nucleon's interior than the indiscriminate photon.

29.2 Neutrino experiments

Although these weak interaction experiments are theoretically more illuminating than their corresponding electromagnetic counterparts, the practical difficulties of dealing with neutrinos tend to spoil their potential.

The electrically neutral, weakly interacting neutrinos cannot be directed by electric and magnetic fields, as can electrons; and building a usable neutrino beam is a complicated process. Firstly, a primary beam of protons is accelerated to a high energy and is made to collide with a stationary target such as a piece of iron. From these collisions, a host of secondary particles, mainly mesons, will emerge in the general direction of the incident proton beam, although with somewhat less energies. These secondary mesons can then decay into neutrinos or antineutrinos and various other particles by decays such as

$$\pi^{\pm} \rightarrow \mu^{\pm} + \nu_{\mu} \text{ (or } \bar{\nu}_{\mu}\text{)}.$$

Because the muon-decay mode of the mesons is generally the most common, it is mainly muon-type neutrinos which make up the beam. Finally, the neutrinos are isolated by guiding the secondary beam through a barrier of, perhaps, 0.5 km of earth. Only the weakly interacting neutrinos can pass through this amount of matter and so the beam emerging from the far side of the barrier is a pure neutrino beam with a typical intensity of about 10^9 particles per cm^2 per second. Unfortunately, the initial proton–target collisions and the subsequent decays of the secondary mesons mean that the resulting energy of the neutrino beam is rather uncertain and often must be inferred by adding up the energies of the products of the neutrino–nucleon interactions under study.

To obtain a reasonable rate of interactions, the neutrino beam must be passed through a very massive target. For instance, the Gargamelle bubble chamber at CERN contains about ten tonnes of some heavy liquid such as freon to ensure a satisfactory rate of reactions. Because the beam consists of both neutrinos and antineutrinos, both sorts of reaction will occur during the same experiment. The two can be distinguished by observing the charge of the outgoing muon in any particular reaction.

29.3 The cross-section

For $\nu_{\mu}N$ scattering, the cross-section may be written down in a fashion similar to that used for $e^{\pm}N$ scattering, by combining various factors describing the different sub-processes which go to make up the collision. Referring back to Figure 26.1, we can see that these factors must include one to describe the transformation of the incoming neutrino into a muon by emitting the W boson (the lepton current); one describing the propagation of the W boson; and one describing the disintegration of the target nucleon under the impact of the W boson (the hadron current). Also analogous to the case of $e^{\pm}N$ scattering is that the behaviour of the factors is well known apart from that describing the hadron current. The lepton current and the propagation of the W boson are well known from the gauge theory of the weak interactions – which is well approximated by the simpler Fermi theory at these energies. But, as before, the unknown form of the hadron current must be characterised by a number of structure functions whose nature is the job of the experiments to discover (see Figure 29.1).

The format of the structure functions is found as before by writing down all the possible combinations of momenta involved in the reaction and then appealing to general principles to simplify the result. In contrast to the electromagnetic force, the weak force does not respect parity invariance and so this simplifying influence cannot be

Fig. 29.1. The formula describing the differential cross-section for (anti)neutrino–nucleon scattering with respect to momentum transfer squared q^2 and energy lost by the neutrino ν. Three structure functions F_1^W, F_2^W and F_3^W (functions of q^2 and ν in general) are needed to describe the shape of the nucleon target (i.e. the way in which weak charge is distributed over the nucleon momentum).

$$\frac{d^2\sigma}{dq^2 d\nu} = \frac{G_F^2}{2\pi} \frac{E_{\mu}}{E_{\nu(\bar{\nu})}} \left[2F_1^W(q^2, \nu) \sin^2 \frac{\theta_{\mu}}{2} + F_2^W(q^2, \nu) \cos^2 \frac{\theta_{\mu}}{2} \pm F_3^W(q^2, \nu) \frac{(E_{\mu} + E_{\nu(\bar{\nu})})}{M_N} \sin^2 \frac{\theta_{\mu}}{2} \right]$$

used in neutrino–nucleon scattering. Because of this, a third weak structure function is introduced (F_3^W) which enters with a different sign in the formula, depending on whether neutrinos or antineutrinos are being scattered. This is the manifestation of the effect mentioned earlier by which the parity-violating weak interaction will distinguish between matter and antimatter involved in the reactions as a result of helicity effects. In general, the structure functions all depend on q^2 and v separately. It is interesting to note that it is *only* through the structure functions that these quantities enter the description of the reactions at all.

29.4 The scaling hypothesis

Although first described in connection with the electromagnetic interactions, the scaling hypothesis is equally valid for the weak force. In this case, all the dimensionality of the cross-section is contained within the Fermi coupling constant G_F (remember that this was the trouble which provided one of the major motivations for the development of weak interaction field theory). As a result, the structure functions must be pure numbers. In the absence of any 'scaling factor' to cancel out the dimensionality of q^2 and v, the structure functions $F_{1,2,3}^W$ cannot depend on them as individual quantities, but only on some dimensionless ratio of the two:

$$F_{1,2,3}^W(q^2, v) \xrightarrow[q^2, v \to \infty]{} F_{1,2,3}^W(x), \qquad (29.1)$$

where the ratio x is the same as before. The structure functions can be measured directly as in $e^\pm N$ scattering and the scaling hypothesis tested. The general shape of the weak structure functions is much the same as that of the electromagnetic example illustrated in Figure 27.2, but because the parameters of the neutrino beam are that much more uncertain than of the electron or muon beam, the experimental errors are much larger, thereby providing a weaker test of the scaling hypothesis.

However, the scaling hypothesis does predict that a far more obvious characteristic will hold in neutrino–nucleon scattering. As mentioned earlier, the cross-section does not depend on q^2 or v apart from through the structure functions. If, then, this dependence is removed by the scaling hypothesis, it means that the cross-section will not

depend on these quantities at all. In this case, it is possible to integrate the formula for the cross-section over all possible values of q^2 and v in a very simple fashion to obtain the total cross-section for neutrino– or antineutrino–nucleon scattering.

$$\sigma^{v(\bar{v})N} = \int \frac{d^2\sigma}{dq^2\,dv}\,dq^2\,dv \propto \frac{G_F^2 M E_{v(\bar{v})}}{\pi}. \quad (29.2)$$

The resulting scaling prediction for neutrino–nucleon scattering is that the total cross-section should rise linearly with the energy of the incident neutrino. The slope of the rise is given by constants which will be different for neutrino or antineutrino reactions, because of the different signs in front of F_3 in the formula of Figure 29.1. Experimental measurements of the total cross-sections are consistent with the linear rise with energy predicted by the scaling hypothesis and its interpretation in terms of point-like partons carrying both electric and weak charges (see Figure 29.2).

The difference in the slopes of the energy dependencies of v and $\bar{v}$ scattering is measured to be about a factor of 3. This factor can be easily

Fig. 29.2. Total neutrino–nucleon and antineutrino–nucleon cross-sections plotted against energy support the scaling hypothesis by exhibiting a linear increase.

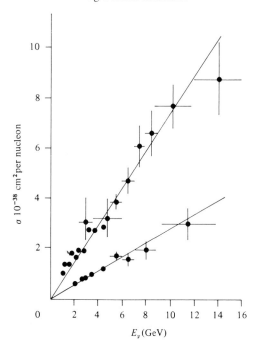

Fig. 29.3. (*a*) Neutrino–parton scattering: spins cancel. There are no restrictions on the directions of the emerging particles. (*b*) Antineutrino–parton scattering: spins add. Restrictions limit the angles of emergence of the outgoing particles.

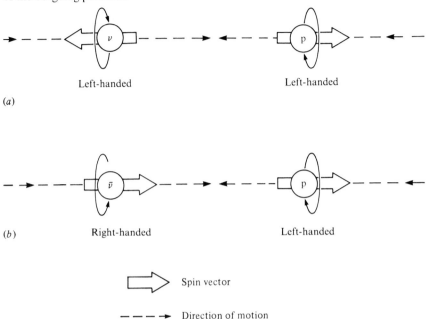

understood in terms of the underlying neutrino–parton interactions. Recall that neutrinos exist exclusively with left-handed spin, and antineutrinos exclusively with right-handed spin (a situation which is possible only because they are massless). Moreover, inasmuch as the mass of the partons may be neglected (i.e. in the relativistic limit), they too are solely left-handed (assuming spin-$\frac{1}{2}$ partons). This follows from the fact that in the Glashow–Weinberg–Salam theory, only left-handed fermions take part in charged current weak interactions (see Part 5). Because partons predominate over antipartons in the nucleon, the neutrino– and antineutrino–parton collisions can be distinguished by the way in which these spins add up. Consider Figure 29.3. If the two colliding spins cancel each other out, as in the case of neutrino–parton scattering, then no restrictions are placed on the angles of emergence of the outgoing particles. If, on the other hand, the two colliding spins add up, as in the case of antineutrino–parton scattering, then the existence of non-zero angular momentum in the system restricts the permitted angles of emergence of the

outgoing particles. This means that the cross-section of antineutrino–parton scattering is reduced relative to that of neutrino–parton scattering. This is because the integration over q^2 to obtain the total cross-section is equivalent to summing over all possible angles of emergence (bearing in mind the definition (26.3)) which are more restricted when the angular momentum is non-zero. The mathematics predicts a factor of three between νN and $\bar{\nu}$N scattering just as observed.

Neutrino–nucleon scattering provides an independent test of the scaling hypothesis and of the parton model. What is now possible is the comparison of muon–nucleon and neutrino–nucleon scattering to establish that the electromagnetic and weak interactions 'see' the same partons.

Bearing in mind that we believe the electromagnetic and weak interactions to be just the different manifestations of the same 'electroweak' force, we certainly expect that this should be the case. Also, we should like to compare the properties of the partons with those of the quarks of $SU(3)$.

30

The quark model of the structure functions

30.1 Introduction

To be more specific about the content of the nucleon (i.e. about what makes up the structure functions), the approach adopted is to assume that the partons have the same properties as the quarks and then to work out the consequences for electron–nucleon and neutrino–nucleon scattering independently. Following this, it is possible to compare the results obtained to see if they are compatible. So we examine, in turn, the electron's eye view and the neutrino's eye view of the structure functions.

30.2 Electromagnetic structure functions

We have mentioned earlier the interpretation of the structure functions as the distribution of the squares of the parton charges within the nucleon, according to the fraction of momentum x carried by the parton, given by formula (27.5):

$$
\left.
\begin{aligned}
F_1^{eN}(x) &= \sum_i f_i(x) Q_i^2, \\
F_2^{eN}(x) &= x \sum_i f_i(x) Q_i^2.
\end{aligned}
\right\} \tag{27.5}
$$

In the simple four-flavour quark model, the quarks and their charges are:

$$u(\tfrac{2}{3}e),\ d(-\tfrac{1}{3}e),\ s(-\tfrac{1}{3}e),\ c(\tfrac{2}{3}e)$$

where c denotes the fourth charmed quark of the

GIM scheme with two-thirds of the charge on the electron. Assuming these values, we can write out the explicit quark content of the structure functions for both the proton and the neutron. In doing so, we must allow for the presence of both quarks and antiquarks from the vacuum sea. This gives, for the proton:

$$
\begin{aligned}
F_1^{ep}(x) = [&(\tfrac{2}{3})^2(f_u(x) + f_{\bar{u}}(x)) \\
+ &(\tfrac{1}{3})^2(f_d(x) + f_{\bar{d}}(x)) \\
+ &(\tfrac{1}{3})^2(f_s(x) + f_{\bar{s}}(x)) \\
+ &(\tfrac{2}{3})^2(f_c(x) + f_{\bar{c}}(x))].
\end{aligned} \tag{30.1}
$$

The expression for the second structure function $F_2^{ep}(x)$ is just the same as above, only multiplied by x, and the corresponding expressions for the neutron structure functions, $F_1^{en}(x)$ and $F_2^{en}(x)$, can be obtained by interchanging $f_u(x) \leftrightarrow f_d(x)$ in the above expressions, as the distribution of up quarks inside the proton is equivalent to the distribution of down quarks inside the neutron.

The total fractional momentum carried by any particular sort of quark can be obtained simply by integrating over its momentum distribution. For instance, the total share of the momentum carried by the up quarks and antiquarks in the proton is given by:

$$
P_u = \int_0^1 x(f_u(x) + f_{\bar{u}}(x))\, dx. \tag{30.2}
$$

Similar expressions will obtain for other varieties of quark. The integrals involved are all contained within the integrals over the total structure functions which are measured in the experiments as the area under the distribution of Figure 27.2(*a*). This quantity gives a linear combination of the momentum shares of all the possible constituent quarks:

$$
\left.
\begin{aligned}
&\int dx\, F_2^{ep}(x) \\
&\quad = (\tfrac{4}{9}P_u + \tfrac{1}{9}P_d + \tfrac{1}{9}P_s + \tfrac{4}{9}P_c) \\
&\quad = 0.18 \text{ (experiment)}, \\
&\int dx\, F_2^{en}(x) \\
&\quad = (\tfrac{1}{9}P_u + \tfrac{4}{9}P_d + \tfrac{1}{9}P_s + \tfrac{4}{9}P_c) \\
&\quad = 0.12 \text{ (experiment)}.
\end{aligned}
\right\} \tag{30.3}
$$

In the above formula, it is fair to assume that the total fraction of the proton's momentum carried by the strange and charmed quarks is negligible. This assumption leaves us with two equations and two unknowns which may be solved to give:

$$P_u = 0.36, \qquad P_d = 0.18. \tag{30.4}$$

Thus the total fractional momentum carried by the up quarks is measured to be twice that carried by the down quark, which supports the quark model's picture of the proton as (uud).

However, the measurements also indicate that the total fractional momentum carried by the quarks is only one-half of the total proton momentum. The interpretation of this is that the other half is carried by neutral gluons which are the quanta of the strong nuclear force between the quarks. Because these gluons are electrically neutral, they do not experience the electromagnetic force and so show up only as missing momentum in the overall accounting for the proton.

30.3 Weak interaction structure functions

The last section showed that the quark model can lend a great deal more detail to our picture of the structure functions. But the picture cannot be filled in completely because the photon transferred in electron–nucleon scattering differentiates between the quarks only by virtue of their electrical charge. To make further distinctions it is necessary to use the W-boson probe of the weak interaction. It is straightforward to establish which of the W-boson–quark interactions are possible, and which of those possible are dominant.

Because of lepton-number conservation in charged current reactions, the neutrino must turn into a negatively charged muon and emit a positively charged W boson, and the antineutrino must turn into a positively charged muon emitting a negatively charged W boson. In principle, the W^+ boson can collide with a down quark, thereby changing it to an up, or it may collide with a strange quark, also turning it into an up. But the W^+ boson cannot be absorbed by an up quark, as this would result in a quark of charge $\frac{5}{3}e$, which does not exist. The possible reactions can be simplified further because the weak interactions which change strangeness, such as

$$W^+ + s(-\tfrac{1}{3}e) \to u(\tfrac{2}{3}e),$$

are much smaller than the strangeness-conserving

weak interactions. This is just the Cabbibo hypothesis mentioned in Chapter 18. Because of it, we may ignore all the strangeness-changing reactions which may occur in principle. In addition to the interactions with quarks mentioned so far, the W bosons can also interact with the antiquarks from the vacuum sea. Combining all these considerations, we can summarise all the significant neutrino–quark interactions which are possible in neutrino–nucleon scattering:

$$\left.\begin{array}{ll}
(1) & \nu_\mu + d(-\tfrac{1}{3}e) \to \mu^- + u(\tfrac{2}{3}e); \\[4pt]
(2) & \bar\nu_\mu + u(\tfrac{2}{3}e) \to \mu^+ + d(-\tfrac{1}{3}e); \\[4pt]
(3) & \bar\nu_\mu + \bar d(\tfrac{1}{3}e) \to \mu^+ + \bar u(-\tfrac{2}{3}e); \\[4pt]
(4) & \nu_\mu + \bar u(-\tfrac{2}{3}e) \to \mu^- + \bar d(\tfrac{1}{3}e).
\end{array}\right\} \tag{30.5}$$

We may now proceed to write the cross-section for, say, neutrino–proton scattering as a sum of the neutrino–quark cross-sections involved.

In doing this, it is usual to express the differential cross-section, not as before expressed as varying with the momentum squared q^2 and energy ν carried by the intermediate W boson, but instead expressed as varying with the momentum fraction of the target carried by struck quark x and the fraction of the incident neutrino energy carried across by the W boson y. Mathematically, this requires us to make the following transformation:

$$\left.\begin{array}{c}
\dfrac{d^2\sigma}{dq^2\,d\nu} \to \dfrac{d^2\sigma}{dx\,dy}, \\[10pt]
\text{with} \\[10pt]
x = \dfrac{q^2}{2M_N\nu} \quad \text{and} \quad y = \dfrac{\nu}{E_i}.
\end{array}\right\} \tag{30.6}$$

This is simple to do, after which the neutrino–proton cross-section can be written in terms of the neutrino–quark cross-sections (1) and (4) of (30.5), the proportions of the two being determined by the distribution of down quarks and antiup quarks inside the proton. This gives:

$$\frac{d^2\sigma}{dx\,dy} = f_d(x)\frac{d^2\sigma}{dx\,dy}(\nu_\mu + d \to \mu^- + u)$$

$$+ f_{\bar u}(x)\frac{d^2\sigma}{dx\,dy}(\nu_\mu + \bar u \to \mu^- + \bar d).$$

$$\tag{30.7}$$

The neutrino–quark scattering is of a very simple point-like kind, and so the cross-sections just provide the usual factors for point-like scattering. Writing the above expression in terms of these factors leaves us with the differential cross-section for deep inelastic scattering, but with the structure functions expressed directly in terms of the distributions of quarks within the target nucleon:

$$\frac{d^2\sigma}{dx\,dy} = \frac{2MEG_F^2}{\pi}[xf_d(x) + x(1 - y^2)f_u(x)].$$

$\uparrow$ neutrino–quark point-like scattering $\uparrow$ structure functions expressed as quark distributions   (30.8)

These expressions may be derived for ν and $\bar{\nu}$ scattering off both protons and neutrons. The average of the two gives the scattering of ν and $\bar{\nu}$ off a general nucleon target N consisting of a mixture of protons and nucleons. The ratio of $\bar{\nu}$N scattering to νN scattering allows the cancellation of point-like scattering factors leaving only a ratio of the quark distributions. By integrating over the variables x and y, the ratio of the total cross-sections is expressed in terms of the total fractional momentum carried by each species of quark:

$$\frac{\sigma^{\bar{\nu}N}}{\sigma^{\nu N}} = \frac{(P_u + P_d) + 3(P_{\bar{u}} + P_{\bar{d}})}{3(P_u + P_d) + (P_{\bar{u}} + P_{\bar{d}})}. \quad (30.9)$$

If quarks only are present inside the target nucleon, then this ratio will be $\frac{1}{3}$, and if antiquarks only are present, then the ratio will be 3. The measured value of 0.37 ± 0.02 suggests a very slight presence of antiquarks from the sea.

The presence of the factors of 3 in the above ratio arises from the allowed helicity states of the νN collisions as described earlier, the integration over the y variable effectively being the integration over the allowed angles of νq scattering, thus favouring neutrino–quark scattering over neutrino–antiquark scattering (and vice versa for antineutrino scattering).

30.4 Electron and neutrino structure functions compared

The quark content of the neutrino scattering structure functions may be obtained simply by comparing the differential cross-section expressed in terms of quark distribution functions (30.7), with the same quantity expressed in terms of the

structure functions. This comparison provides equalities such as:

$$\left.\begin{array}{l} F_2^{\nu p} = 2x(f_d(x) + f_{\bar{u}}(x)), \\ F_3^{\nu p} = -2(f_d(x) - f_{\bar{u}}(x)), \\ F_2^{\nu n} = 2x(f_u(x) + f_{\bar{d}}(x)), \\ F_3^{\nu n} = -2(f_u(x) - f_{\bar{d}}(x)). \end{array}\right\} \quad (30.10)$$

By comparing the quark content of the neutrino structure functions above with the quark content of the electron structure functions in (30.1), it is possible to predict a numerical relation between the neutrino and electron–nucleon scattering structure functions. The resulting relationship is:

$$F_2^{\nu N}(x) = \tfrac{18}{5}F_2^{eN}(x). \quad (30.11)$$

Experimentally, the relationship is found to hold very well, as illustrated in Figure 30.1. The significance of this is that the quark content of the target nucleon has been verified independently by two separate interactions and is thus that much more credible.

Fig. 30.1. The constant relationship between electron and neutrino scattering structure functions is verified by superimposing the two.

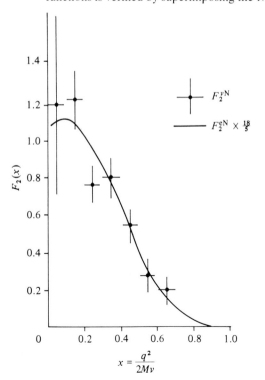

$F_2(x)$ versus $x = \dfrac{q^2}{2M\nu}$

$F_2^{\nu N}$

$F_2^{eN} \times \frac{18}{5}$

30.5 Sum rules

We can learn more about the roles played by the various quarks inside the proton by using expressions (30.1) and (30.10) to relate particular quark distribution functions to combinations of the observed structure functions. When integrated over the fractional momentum variable of the quark distributions, these relationships can reveal the total fractional momentum carried by each species of quark:

$$
\left.
\begin{aligned}
&\int x(f_s(x) + f_{\bar{s}}(x))\,dx \\
&= \int (9F_2^{eN}(x) - \tfrac{5}{2}F_2^{\nu N}(x))\,dx \\
&= 0.05 \pm 0.18, \\
&\int x(f_u(x) + f_d(x))\,dx \\
&= \tfrac{1}{2}\int (F_2^{\nu N}(x) - xF_3^{\nu N}(x))\,dx \\
&= 0.49 \pm 0.06, \\
&\int x(f_{\bar{u}}(x) + f_{\bar{d}}(x))\,dx \\
&= \tfrac{1}{2}\int (F_2^{\nu N}(x) + xF_3^{\nu N}(x))\,dx \\
&= 0.02 \pm 0.03.
\end{aligned}
\right\} \quad (30.12)
$$

The numbers show that, as expected, the strange quarks and the various sorts of antiquarks carry only a few per cent of the proton's total momentum. So the total contribution of 'sea' quarks is small. As discovered in electron–nucleon scattering, the up and down quarks together carry about half of the proton's total momentum. Again the missing half of the proton's momentum is ascribed to the neutral gluons which are not affected by the weak interactions.

When we integrate the quark distribution functions (or, correspondingly, the structure functions) over the fractional momentum variable x, it is possible to derive 'sum rules' relating these quantities to physically significant numbers. For instance the Gross–Llewellyn Smith sum rule measures the difference between the numbers of quarks and antiquarks in the target. As expected from the quark model, this number is measured to be approximately 3, as illustrated in Figure 30.2.

$$
\tfrac{1}{2}\int_0^1 (F_3^{\bar{\nu}}(x) + F_3^{\nu}(x))\,dx = N_q - N_{\bar{q}} = 3.
$$

We have spent some time examining the structure functions of deep inelastic scattering, as it is these which have provided us with a direct look inside the proton at its constituent quarks. All the observations are compatible with the standard model of spin-$\tfrac{1}{2}$ quarks with fractional charges. The big surprise is that these quarks carry only half of the proton's total momentum – the remainder presumably being carried by the gluons. The next step is to proceed to see if we can learn something of the dynamics of the interactions between quarks and gluons.

30.6 Summary

Deep inelastic lepton–nucleon scattering has been an extremely useful tool for probing the structure of hadrons. What we have learnt from these experiments may be stated succinctly as follows. We now know that the nucleon contains point-like constituents (partons), as evidenced by the approximate scale invariance of structure functions: $F(\nu, q^2) = F(x)$. That these partons have spin $\tfrac{1}{2}$ is clear from the observed relation between the electromagnetic structure functions: $2xF_1(x) = F_2(x)$. Furthermore, the behaviour of electroweak cross-sections strongly suggests the identification of these spin-$\tfrac{1}{2}$ partons with fractionally charged quarks, which account for just one half of the nucleon momentum (the remainder being due to the gluon constituents).

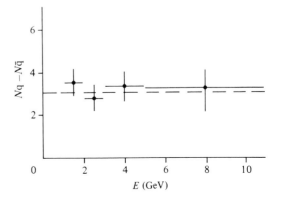

Fig. 30.2. The Gross–Llewellyn Smith sum rule measures the difference between the numbers of quarks and antiquarks in the nucleon.

Part 7
Quantum chromo-dynamics – the theory of quarks

31

Coloured quarks

31.1 Introduction

The results of deep inelastic scattering experiments are able to tell us a lot about the nature of the quarks.

- The scaling behaviour of the cross-sections indicates scattering off point-like quarks with relatively weak interactions between them at short distances.
- The ratio of structure functions F_1^{eN}/F_2^{eN} supports the assignment of half-integer spin for the quarks.
- The comparison of structure functions in electron and neutrino scattering reactions supports the assignment of fractional charges to the quarks.
- The momentum sum rules in both electron– and neutrino–proton scattering suggest that quarks carry only about half of the total proton momentum. The other half is thought to be carried by neutral gluons, the quanta of the interquark force field.

This wealth of information on the structure of the proton was discovered between 1968 and the mid-1970s and represents an experimental triumph similar to the 1911 scattering experiments of Geiger, Marsden and Rutherford, which established the nuclear picture of the atom. In both cases, experimental observation led the way to the development of theories describing the phenomena.

Just as Bohr's early quantum theory of the atom had been advanced to describe Rutherford's discoveries, so quantum chromodynamics (QCD) was put forward as a description of the behaviour of the quarks inside the proton. Pressing the analogy further, just as Bohr's description of the atom was an extension of the quantum theory propounded earlier by Planck, so QCD is an application of the ideas of gauge field theory developed in the 1960s.

QCD was proposed in 1973 by Fritzsch, Leutwyler and Gell-Mann (the last of whom, appropriately enough, was one of the original proponents of quarks in 1963), although a similar idea had been put forward in 1966 by Nambu. The basic idea is to use a new charge called *colour* as the source of the interquark forces, just as electric charge is the source of electromagnetic forces between charged particles.

31.2 Colour

Soon after the proposal of the quark model, it was realised that the suggested quark content of some particles clashed with one of the most fundamental principles of quantum mechanics. The Pauli exclusion principle states that no two fermions (particles with half-integer spin) within a particular quantum system can have exactly the same quantum numbers. However, the proposed contents of some particles consist of no less than three identical quarks. For instance, the doubly charged, spin-$\frac{3}{2}$ resonance Δ^{++} must consist of three 'up' quarks, all with their spins pointing in the same direction (Figure 31.1(a)). Similarly, the famous Ω^- particle, the discovery of which first confirmed the validity of the $SU(3)$ flavour scheme, must consist of three strange quarks (Figure 31.1(b)). These two examples seem to contradict the rules of quantum mechanics.

One immediate way out of this problem is to suggest that the quarks are not fermions at all but, instead, are spinless or integer-spin bosons. However, it was realised early on that only fermionic quarks can account for the spins of the observed hadrons and subsequent observations, say, of structure functions in deep inelastic scattering, have always supported this conjecture. In fact, the mathematical statement of the Pauli exclusion principle deals in terms of the symmetry of the

wavefunction which is the total description of the quantum-mechanical system. The statement that no two fermions can have exactly the same quantum numbers in a particular system is equivalent to the statement that the wavefunction describing a system of fermions must be antisymmetric (i.e. it must change sign) on the interchange of any two of the constituent fermions. The wavefunction which describes a hadron made up of three quarks consists of at least three factors: one describing the positions of the quarks; one describing the spins of the quarks; and one describing the flavours of the quarks. The product of these three factors gives the overall wavefunction:

$$\psi_{\text{TOTAL}} = \psi_{\text{SPACE}} \times \psi_{\text{SPIN}} \times \psi_{\text{FLAVOUR}}.$$

For particles such as the Δ^{++}, all quarks have the

Fig. 31.1. The quark content of the Δ^{++} in (a) and the Ω^- in (b) consists of three identical quarks, which would seem to contradict Pauli's exclusion principle. The introduction of colour distinguishes the quarks and preserves the principle.

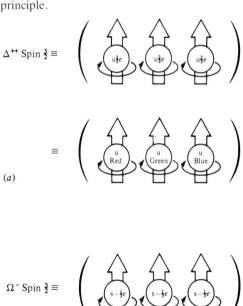

same flavour, and so the flavour factor of the wavefunction is obviously symmetric under the intercharge of any two quarks. The same is true of the spin factor because all quark spins are the same. Because the spins of the quarks add up to give the total overall spin of the particle, it means that there is no orbital angular momentum belonging to the three quarks. This implies that the quarks are positioned symmetrically so that the space factor is symmetric under the interchange of any two quarks. As all the individual factors are symmetric, the total wavefunction must be symmetric and the combination of the three quarks seems to violate the Pauli exclusion principle.

In 1964, Greenberg and, later, Han and Nambu, suggested that the quarks would have to carry another quantum number which would distinguish otherwise identical quarks and so satisfy the demands of the Pauli exclusion principle. This new quantum number they called colour, although it should be stressed that this new property has nothing to do with the normal meaning of the word colour; it is just a label. The total wavefunction will now be multiplied by a new 'colour factor':

$$\psi_{\text{TOTAL}} = \psi_{\text{SPACE}} \times \psi_{\text{SPIN}} \times \psi_{\text{FLAVOUR}} \times \psi_{\text{COLOUR}}$$

The colour hypothesis is that each of the otherwise identical quarks has a different colour assigned to it and this makes the colour factor, and so the total wavefunction, antisymmetric under interchange of two quarks. The quark model is thus reconciled with the Pauli exclusion principle at the expense of introducing a new quantum number to differentiate between the quarks. Because there are three quarks inside the proton, three quark 'colours' are needed to distinguish them uniquely, say red, green and blue. Each of these labels the three quarks inside the Δ^{++} and the Ω^- as shown in Figure 31.1(*a*) and (*b*). So the net effect of the introduction of colour is to triple the number of quarks; each of the flavour must come in three colours. This is illustrated in the quark table (Table 31.1).

Obviously, tripling the number of the supposedly fundamental quarks is rather against the spirit of the model, which attempts to make do with as few basic components as possible. So the above arguments for colour, although theoreti-

Table 31.1. *A table of the flavours and colours of quarks*

Flavour	Colour		
	Red	Green	Blue
$u(\frac{2}{3}e)$	u_r	u_g	u_b
$d(-\frac{1}{3}e)$	d_r	d_g	d_b
$s(-\frac{1}{3}e)$	s_r	s_g	s_b

cally compelling, had to be demonstrated by more-direct means before the colour quantum number became established as the physical reality on which a theory of quarks could be based. Happily, these more-direct means are readily evident.

One piece of evidence supporting the hypothesis that quarks come in three colours is provided by the decay of the neutral π^0 meson into two photons (see Figure 7.2). In the quark model, this decay rate is calculated by adding up all the possible varieties of quarks which can act as intermediate states in the decay. The experimental decay rate can be matched by the theory to within a few per cent if the quarks are assumed to come in three different colours. If only one colour of quark is allowed then the answer turns out to be a factor of 9, too small (the number of quarks entering the decay rate formula as a square).

A second piece of evidence for the existence of three different colours of quark is provided by electron–positron annihilations. In these events, the electron and positron annihilate each other to produce a virtual photon loaded with energy. This virtual photon may then decay into either a muon and an antimuon or into a shower of hadrons. The shower of hadrons is the end result of the initial production of a quark–antiquark pair which subsequently transforms into the conventional hadrons (see Figure 31.2).

The ratio of the cross-sections for these events is a very significant quantity which will be discussed further in Part 8. Suffice it to say at this stage that the ratio is proportional to the sum of the squares of the quark charges, and that only if each flavour of quark comes in three different colours does the number predicted by the quark model agree with the number observed experimentally.

So three separate pieces of evidence attest to

the physical reality of the colour quantum number: quark 'spectroscopy' (i.e. how quarks build up the known hadrons); the π^0 decay; and e^+e^- annihilations. Two immediate questions arise from this reality. Firstly, if quarks carry these colour charges, is it possible to discover observable hadrons which also carry them? Secondly, given the existence of colour, what is its purpose? Can it form the basis of a theory of quark interactions?

31.3 Invisible colour

The observability of colour and the quark structure of matter are intimately linked. In fact, we will see that the introduction of colour corresponds to a formal method of categorising allowed quark structures. We saw in Part 2 how the observed hadrons fit into the multiplets generated by treating the up, down and strange quarks as the elements of the fundamental representation of the symmetry group $SU(3)$. However, we saw also that only very specific combinations of the fundamental representation (the quark *flavour* triplet) could generate the correct multiplets for the observed hadrons (see Table 31.2).

The introduction of colour provides a way of categorising which combinations of quarks and

Table 31.2. *Certain combinations of the fundamental quark flavour triplets generate the observed multiplets*

Fundamental representation	Possible combinations	Multiplets generated	Seen
$q = \begin{pmatrix} u \\ d \\ s \end{pmatrix}$ 'flavour'	$q \times \bar{q}$	**1 + 8**	✓
	$q \times q \times q$	**1 + 8 + 8 + 10**	✓
	$q \times q$	**3* × 6**	✗

antiquarks are allowed to exist. The first step to realise is that the three colours of any one flavour of quark can be taken as the elements of the fundamental representation of a new symmetry group $SU(3)_C$. The methods of group theory then allow us to combine these fundamental representations (the quark colour triplet) into colour multiplets representing the different ways of combining all the colours of the quarks (see Table 31.3). The mathematics of $SU(3)_C$ is exactly the same as that for $SU(3)_F$, although it must be appreciated that these $SU(3)_C$ colour multiplets represent the various combinations of colour for a given flavour of quark and are completely distinct from the $SU(3)_F$ flavour multiplets. Just as it was necessary earlier with flavour multiplets to establish which of them could be taken to represent the observed hadrons, it is now necessary to repeat the exercise for the colour multiplets to see which of these are permitted.

The original purpose of the introduction of colour was to ensure that the combinations of quarks representing hadrons are multiplied by factors which are antisymmetric under the inter-

Fig. 31.2. An electron–positron pair annihilates into a virtual photon which may then disintegrate into a muon–antimuon pair, as in (*a*), or into a quark–antiquark pair (which then transforms into hadrons), as in (*b*).

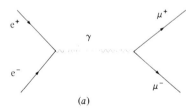

(*a*)

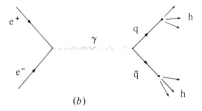

(*b*)

Table 31.3. *Combinations of quark colour triplets generate multiplets*

Fundamental representation	Possible combinations	Multiplets generated	Seen
$q = \begin{pmatrix} r \\ g \\ b \end{pmatrix}$ 'colour'	$q \times \bar{q}$	**1 + 8**	?
	$q \times q \times q$	**1 + 8 + 8 + 10**	?
	$q \times q$	**3* × 6**	?

change of two of the quarks (colours). By using mathematical analysis it is possible to show that some of the multiplets are antisymmetric, just as required, whilst others are symmetric, and so do not help us. The most obvious and simplest anti-symmetric multiplet is the singlet. This observation is then elevated to a hypothesis for explaining the observed quark structure:

All observed hadrons are colour singlets **1**.

Under this hypothesis, qq̄ and qqq combinations are allowed because the series of colour multiplets generated includes a singlet. Also allowed are qqq̄q̄ and qqqqq̄. These theoretically allowed combinations are referred to as exotic hadrons or baryonium states and they have long been the subject of experimental searches. Following several false alarms and tentative detections, their existence awaits experimental verification. Also under this hypothesis, combinations such as qq or qqqq are not allowed because those combinations of the fundamental representation of the colour group (quark *colour* triplet) do not generate a singlet combination. Experimentally, there has been no suggestion of their existence.

To summarise, let us recap the two parallel descriptions of quark structure, those of flavour and colour. Each combination of quarks generates a set of flavour multiplets and a set of colour multiplets. For certain of these combinations of quarks, the flavour multiplets will correspond to the observed hadrons and these combinations are identified as those which generate a colour singlet. All the observed hadrons are thought to be in colour singlet states. This scheme is summarised in Table 31.4.

The statement that all the observed hadrons are colour singlets is equivalent to saying that they are colourless. Just as flavour singlet states can carry no net electric charge or strangeness, the colour singlet states can carry no net colour. This can be understood simply by examining the colour combinations of the allowed quark structures:

$$q\bar{q} = \sqrt{\tfrac{1}{3}}(r\bar{r} + g\bar{g} + b\bar{b}),$$
$$qqq = \sqrt{\tfrac{1}{6}}(rgb - grb - rbg + gbr + brg - bgr).$$

The laws of quantum mechanics forbid us to say exactly what colour any one quark is at any one

time; all we can say is that there is a certain probability of it being red or green or blue. However, what we can say is that in the singlet state of a qq̄ combination the colour of the quark is exactly cancelled by the anticolour of the antiquark, and that in the singlet state of a qqq combination all the

Table 31.4. *The colour and flavour multiplet structure of various quark–antiquark combinations and their postulated observability*

(a) q × q̄ — quark antiquark forms mesons like π⁰, ρ, K, ψ, ...

COLOUR

$\begin{pmatrix}r\\g\\b\end{pmatrix} \times \begin{pmatrix}\bar{r}\\\bar{g}\\\bar{b}\end{pmatrix}$

FLAVOUR: $\begin{pmatrix}u\\d\\s\end{pmatrix} \times \begin{pmatrix}\bar{u}\\\bar{d}\\\bar{s}\end{pmatrix} \Rightarrow \begin{matrix}1\\+\\8\end{matrix}$

$\Downarrow$ **1 + 8**

$\begin{pmatrix}\text{Yes} & \text{No}\\ \text{Yes} & \text{No}\end{pmatrix}$ Colour states

(b) q × q — diquark states are not observed

COLOUR

$\begin{pmatrix}r\\g\\b\end{pmatrix} \times \begin{pmatrix}r\\g\\b\end{pmatrix}$

FLAVOUR: $\begin{pmatrix}u\\d\\s\end{pmatrix} \times \begin{pmatrix}u\\d\\s\end{pmatrix} \Rightarrow \begin{matrix}3^*\\+\\6\end{matrix}$

$\Downarrow$ **3* + 6**

$\begin{pmatrix}\text{No} & \text{No}\\ \text{No} & \text{No}\end{pmatrix}$ Colour states

(c) q × q × q — three quarks form a baryon like the proton, neutron, Ω, Δ, ...

COLOUR

$\begin{pmatrix}r\\g\\b\end{pmatrix} \times \begin{pmatrix}r\\g\\b\end{pmatrix} \times \begin{pmatrix}r\\g\\b\end{pmatrix}$

FLAVOUR: $\begin{pmatrix}u\\d\\s\end{pmatrix} \times \begin{pmatrix}u\\d\\s\end{pmatrix} \times \begin{pmatrix}u\\d\\s\end{pmatrix} \Rightarrow \begin{matrix}1\\+\\8\\+\\8\\+\\10\end{matrix}$

$\Downarrow$ **1 + 8 + 8 + 10**

$\begin{pmatrix}\text{Yes} & \text{No} & \text{No} & \text{No}\\ \text{Yes} & \text{No} & \text{No} & \text{No}\\ \text{Yes} & \text{No} & \text{No} & \text{No}\\ \text{Yes} & \text{No} & \text{No} & \text{No}\end{pmatrix}$ ← Colour → states colours mix in equally to provide a 'white' baryon, i.e. one with no net colour quantum number. Under this scheme, the colour of the quarks is permanently hidden from us because all the allowed quark structures are colourless, colour singlets. The confinement of the quarks within the hadrons can accordingly be restated as the confinement of the colour quantum number.

At this point it is perhaps worth making clear that the confinement of quarks and of colour is really only a hypothesis which is occasionally re-examined, and quite properly so. In 1976, when some anomalous events were detected in electron–positron annihilations, Pati and Salam suggested that they may be the signals of unconfined quarks emerging freely and decaying into leptons rather than undergoing their forced transformations into hadrons. In fact, it was later realised that these anomalous events signalled the production of a new heavier brother of the muon. But, at least for a time, the free quark hypothesis was tenable.

In experiments carried out in the late 1970s at Stanford University, William Fairbank and his collaborators claimed to have detected fractional electric charges on supercooled niobium balls. The experiments were modern versions of Milikan's oil-drop experiment (which first established the value of the electric charge, e). However, these results have never been confirmed elsewhere. In fact, all the evidence from accelerator experiments and searches for pre-existing quarks suggests that free quarks do not exist. For example, analysis of sea water indicates that if free quarks do exist, they are so rare that there is less than one for every 10^{24} nucleons. Despite this, it is by no means certain that free quarks and/or coloured mesons and baryons will never be seen.

Worthy of mention in passing is that the effects of new flavours are easily incorporated into the above picture and do not affect the colour of the quarks at all. The only effect of a new flavour is to generate bigger flavour multiplets to accommodate the greater number of quark combinations possible when an extra degree of freedom is present. The number of quark colours is always three, because that is the number required to distinguish the three valence quarks in each baryon.

32

Colour gauge theory

32.1 Introduction

The fundamental idea of QCD is that the colour 'charges' of the quarks act as the sources of the strong, so-called 'chromodynamic' force between quarks, just as electric charge acts as the source of the electromagnetic force between electrically charged particles. As the quarks carry both colour and electric charge, they experience both the strong and electromagnetic forces, as well as the more feeble, weak and gravitational interactions. However, the chromodynamic force is by far the strongest in most regions of interest and so we are justified in examining it in isolation from the others.

In the terms of classical physics, the colour charges may be thought of as giving rise to a chromostatic force, just as electrical charges give rise to the electrostatic force given by Coulomb's inverse square law. Although there are great similarities between the two cases, the colour force will be a good deal more complicated. In chromodynamics there are three different colour charges between which the force must be attractive to bind together the three different colours inside each baryon, and the force between colour and anti-colour must also be attractive to bind together the quark and antiquark in each meson. However, despite these complications the analogy with the theory of electrodynamics is worth pursuing as far as possible.

Any theory of quarks, like any other fundamental theory, must be compatible with the laws of quantum mechanics and relativity and the most common approach to achieve this is relativistic quantum field theory. Both QED and the Glashow–Weinberg–Salam theory of the electroweak interactions are examples of a particular class of quantum field theory, namely a gauge theory. As this class has enjoyed such striking success in these other two areas, it must have seemed a reasonable candidate for describing the new chromodynamic forces as well.

    The essence of gauge theory is to explain the origin of the fundamental forces in terms of a symmetry. As we saw in Chapter 21, in QED the form of the interaction between electrons is dictated by the demand that the description of electron behaviour be invariant under arbitrary local redefinitions of the phase of the electron wavefunction. This invariance is a reformulation of the law of conservation of electric charge. The redefinition of the phase can be expressed mathematically by the action of certain symmetry transformations associated with the group $U(1)$. The resulting formulation of QED is summarised in Figure 32.1.

32.2 The formulation of QCD

    It will be useful to bear in mind the step-by-step summary of the formulation of QED because the formulation of QCD is remarkably similar. The first step is to identify the symmetry thought to be the fundamental origin of the colour forces. This comes readily to mind.

    We have already seen that the quark colour triplet may be used as the fundamental representation of the symmetry group $SU(3)_C$ and that the colour multiplet structure generated by this group gives an acceptable way of categorising the known hadrons (i.e. all are in colour singlet states). The fundamental symmetry of the colour force may then be taken as invariance under the redefinition of quark colours. This is eminently consistent with the philosophy of quantum theory because, just like the phase of the electron wavefunction, colour is believed to be unobservable and so no real phenomena should depend on the convention which defines it.

    The redefinition of quark colours is achieved by applying the $SU(3)_C$ group of transformations

Fig. 32.1. A summary of the formulation of QED.

(1) A wavefunction describes the propagation of an electron.

(2) The Lagrangian describes the wavefunctions of two electrons in interaction.

$$\mathscr{L}(\psi_1, \psi_2)$$

(3) Gauge invariance demands that in any theory of electrons the Lagrangian must be invariant under the redefinition of the phase of the electron wave. This is represented by the action of some group of transformations on the Lagrangian.

$$\mathbf{G}\psi \to \psi^*, \quad \mathbf{G}\mathscr{L}(\psi_1, \psi_2) \to \mathscr{L}(\psi_1^*, \psi_2^*)$$

(4) Furthermore, it is possible to require that this be true independently at each point in space. This is called local gauge invariance.

$$\mathbf{G}(x)\psi \to \psi^*, \quad \mathbf{G}(x)\mathscr{L}(\psi_1, \psi_2) \nrightarrow \mathscr{L}(\psi_1^*, \psi_2^*)$$

(5) For the Lagrangian to remain invariant under this last operation, a new gauge field must be introduced.

$$\mathbf{G}(x)\,\mathscr{L}(\psi_1, \psi_2, A) \to \mathscr{L}(\psi_1^*, \psi_2^*, A^*)$$

(6) This gauge field communicates between the two electron fields their locally defined phase conventions.

(7) Stated in more familiar language, the electromagnetic field mediates the force between two electrons. In quantum theory, this is described as the exchange of photons, the quanta of the electromagnetic field.

to the quark colour triplet. Suppose we define the quark colours initially by the multiplet

$$q \equiv \begin{pmatrix} r \\ b \\ g \end{pmatrix}.$$

We may then choose to change our colour scheme by applying an $SU(3)_C$ group transformation to this triplet. This will have the effect of mixing up the colours to provide three different combinations each with different proportions of r, g and b:

$$\mathbf{G}^{SU(3)}q \rightarrow \begin{pmatrix} c_1(r\,b\,g) \\ c_2(r\,b\,g) \\ c_3(r\,b\,g) \end{pmatrix} \equiv \begin{pmatrix} v \\ y \\ o \end{pmatrix}.$$

However, it is perfectly possible to label these new combinations as new colours, say violet, yellow and orange. The underlying physical requirement is that the theory describing the quark interactions does not depend on whatever 'colour coding' is chosen. In fact, the colour coding of the quarks may be different at each point in space and this will require the Lagrangian to be *locally* gauge invariant under the application of the $SU(3)_C$ group of transformations.

Just as in the other gauge theories, this requires the introduction of a new gauge field to communicate the local colour conventions from place to place. The quanta of this new colour gauge field are massless spin-1 gauge particles. These are the 'gluons' which mediate the chromodynamic forces between quarks. Because the interaction between two quarks corresponds to an interaction between two colour states, gluons must come in colour multiplets which correspond to the colour combinations allowed by group theory. More precisely, since a quark and an antiquark can annihilate into a gluon, the colour quantum numbers of gluons should correspond to a combination of those of quarks and antiquarks. That is, the colour on a gluon must come from the combination of a quark colour triplet $\mathbf{3} = (r, g, b)$ with an antiquark anticolour triplet $\mathbf{3}^* = (\bar{r}, \bar{g}, \bar{b})$:

$$\mathbf{3} \times \mathbf{3}^* = \mathbf{1} + \mathbf{8}$$

and, indeed, gluons form a colour octet $\mathbf{8}$ with quantum numbers such as red–antigreen ($r\bar{g}$) and blue–antired ($b\bar{r}$). It is now clear how, in any QCD reaction, redness, greenness and blueness are conserved. For instance, in Figure 32.2, we show how a red quark can emit a red–antigreen gluon thereby transforming itself into a green quark.

In QED the photon is electrically neutral and so does not act as a source of electromagnetic

Fig. 32.3. (a) In QED, photons cannot interact directly, they must dissociate into an e^+e^- pair to do so at all. (b) In QCD, the coloured gluons can interact with each other directly.

QED

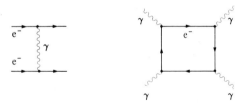

QCD

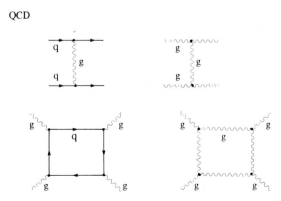

Fig. 32.2. Quarks interact by gluon exchange. The colour quantum numbers flowing into the gluon are equivalent to those of a q$\bar{q}$ pair.

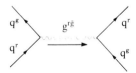

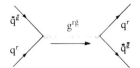

fields. This means that photons cannot interact amongst themselves directly. The only way they can interact is for each to dissociate into a virtual electron–positron pair or other charged particle pair which may then do the interaction for them. However, this has a much lower probability of occurrence than the direct interaction between electrons. In QCD, the gluons carry colour and so give rise to their own colour fields. This means that they can interact amongst themselves directly. The two cases are illustrated graphically in Figure 32.3.

Fig. 32.4. A summary of the formulation of QCD.

---

(1) A wavefunction describes the propagation of a quark.

$$\psi \qquad q \longrightarrow$$

(2) The Lagrangian describes the wavefunctions of two quarks in interaction.

$$\mathscr{L}(\psi_1, \psi_2) \equiv$$

$$q_1 \qquad q_2$$

(3) Gauge invariance demands that the Lagrangian must be invariant under the redefinition of the quarks' colour code.

$$\mathbf{G}^{SU(3)_c}\,\psi \to \psi^*, \quad \mathbf{G}^{SU(3)_c}\,\mathscr{L}(\psi_1, \psi_2) \to \mathscr{L}(\psi_1^*, \psi_2^*)$$

(4) This gauge invariance may be required to hold locally

$$\mathbf{G}^{SU(3)_c}(x)\psi \to \psi^*, \quad \mathbf{G}^{SU(3)_c}(x)\,\mathscr{L}(\psi_1\,\psi_2) \nrightarrow \mathscr{L}(\psi_1^*\,\psi_2^*)$$

(5) The Lagrangian can remain invariant under this local group only if a new, self-interacting gauge field is introduced

$$\mathbf{G}^{SU(3)_c}(x)\mathscr{L}(\psi_1, \psi_2, \tilde{A}) \to \mathscr{L}(\psi_1^*, \psi_2^*, \tilde{A}^*)$$

(6) This gauge field communicates between the quarks their locally defined colour coding. In more-familiar terms, the quanta of the colour gauge field, the gluons, mediate the strong force between the quarks and also between themselves.

This difference between the two theories is a very fundamental one and has far-reaching consequences of great importance. The reason for the difference boils down to the number of charges in the theory. As we saw in Part 5, QED is an Abelian gauge theory with a single charge Q, whereas the more complicated electroweak charges mean that the Glashow–Weinberg–Salam model is a non-Abelian gauge theory, so that the net result of two gauge transformations depends on the order in which they are performed. Similarly, in QCD the colour charges give rise to a non-Abelian gauge theory with the symmetry group $SU(3)_C$. The formulation of QCD is summarised in Figure 32.4.

The picture of the hadrons painted by this theory is one of continual interchange of gluons between the constituent quarks which, as a result, continually keep changing colour but in such a fashion as to always maintain the hadron in its colour singlet state (see Figure 32.5).

One interesting consequence of the presence of gluon self-interaction in QCD is the possibility of the existence of particles made only of glue with no quarks. These are referred to as glueballs or gluonium states and are possible because when two colour octets are combined, a colour singlet state always results in addition to non-allowed

Fig. 32.5. The quarks inside the hadrons are bound together by continual gluon exchange.

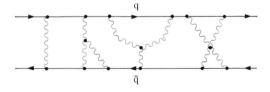

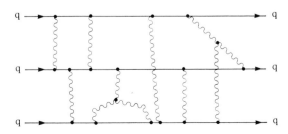

colour multiplets (see Figure 32.6). Other combinations of more than two gluons are also possible, giving rise to the possibility of a whole spectrum of gluonia. Their existence is currently the subject of lively debate, with some authors claiming as evidence of their presence certain irregularities observed in the quark model of the hadron spectrum.

Fig. 32.6. The non-Abelian nature of QCD allows the formation of particles made from gluons only.

$$8 \otimes 8 = 1 + 8 + 8 + 10 + 10 + 27$$

33

Asymptotic freedom

33.1 Introduction

What we have done so far, formulating a gauge theory of the colour force in analogy to the theory of the electromagnetic interaction, is all very well – but is it correct? Does it explain any of the features of the strong interaction which are observed in the real world? Only by this test, and not by its theoretical elegance or any other criterion, may it be accepted as correct. In particular, we are interested to see if the forces resulting from gluon exchange correspond to the behaviour of the strong interaction as observed in deep inelastic scattering. In these experiments, we saw that when the distances probed were very small (i.e. when the momentum transferred by the probe was very high) then the force between quarks is surprisingly weak and they behave rather like free particles. On the other hand, no free quark has ever been observed, so we may be sure that, over long distances, the force between quarks becomes increasingly strong.

Another, and as it turns out, related theoretical question is whether or not we can actually perform any meaningful calculations in QCD. In QED, calculations of quantities of physical interest are possible because the increasingly complicated higher-order processes become decreasingly important. This is due to the smallness of the electron–photon coupling constant $(\alpha = \frac{1}{137})$. However, in QCD, the strength of the chromodynamic forces may require the quark–gluon and gluon–gluon couplings to be large (greater than one) and this would mean that the increasingly complicated higher-order processes become increasingly important. In this case, it is impossible to use the same mathematical techniques of perturbation theory to calculate quantities of physical interest.

Resolution of both the experimental and theoretical points mentioned above hinges on the far-from-obvious physical reality that the intrinsic strength of a force (the size of the coupling constant concerned) depends on the distance from which it is viewed. This dependence is in addition to the conventional spatial variation in the strength of forces as given, say, by the inverse square laws of classical physics.

A good example of this phenomenon is provided by the electromagnetic force. In classical physics the electrostatic force between two charged particles is given by Coulomb's inverse square law:

$$F = K \frac{N_1 e \cdot N_2 e}{r^2}.$$

In this formula, the intrinsic 'strength' of the force is fixed by the numerical value of the constant electric charge e. But when the distance separating the two particles becomes very small, then classical physics is no longer adequate and quantum-mechanical effects must be taken into account.

These quantum-mechanical effects can be described as the polarisation of the vacuum by a sea of virtual electron–positron pairs in the environment of an electric charge. In the region of an electron, say, the virtual positron is attracted towards it and the virtual electron is repelled away from it. This leads to a cloud of virtual positive charge shielding the 'bare' negative charge of the real electron (see Figure 33.1(a)). The effect of this is that from a distance the effective negative charge is much reduced compared to its 'bare' value. The electric charge appearing in Coulomb's law is this shielded, effective charge.

The quantum-mechanical shielding effect is known as the 'renormalisation' of the bare electri-

Fig. 33.1. (a) Virtual electron–positron pairs shield the 'bare' electric charge at very short distances. This effect can be calculated by evaluating Feynman diagrams such as that shown in (b).

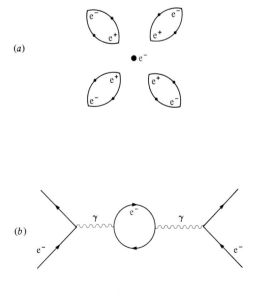

Fig. 33.2. (a) The apparent strength of an electrical charge as a function of the distance from which it is viewed. (b) The apparent strength of the colour charge on a quark.

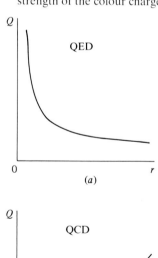

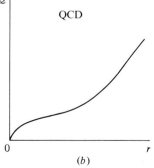

cal charge and it can be calculated by evaluating the probabilities of occurrence of the various quantum-mechanical processes such as the one illustrated in Figure 33.1(*b*). In fact, the probabilities associated with these processes are infinite! This suggests that the bare charge is also infinite, but negative, so that the infinities cancel, leaving a finite value *e* for the classical electric charge.

We can carry out a simple thought-experiment to investigate the quantum-mechanical behaviour of electric charge by considering the scattering of two electrons off each other at increasingly high energies. As the distance of approach decreases, the electrons begin to penetrate each other's virtual charge cloud and to experience each other's more negative bare charge. The change in the effective value of the electric charge is shown against distance of approach in Figure 33.2.

A parallel phenomenon exists in QCD. Just as the vacuum can be considered to be a sea of virtual electron–positron pairs, so too can it be considered as a sea of virtual quark–antiquark pairs and gluons (see Figure 33.3(*a*)). The 'bare' colour charge on a single quark may then be shielded by the polarisation of this vacuum sea of virtual quarks, antiquarks and gluons. The resulting renormalisation of the bare colour charge may be calculated by evaluating the corresponding probabilities of occurrence of the various quantum-mechanical processes, such as those shown in Figure 33.3(*b*). The essential new feature in the QCD case is the presence of gluon shielding, possible because of gluon self-interactions. Whereas in QED, the single variety of electron–positron shielding leads to a decrease in the effective electric charge compared with its bare charge, the presence of the gluon shielding effect in QCD provides a greater, opposite effect and leads to an increase in the effective colour charge relative to the original bare charge. Conversely stated, the effective strength of the colour charge on a quark appears to decrease as the distance from which it is viewed decreases (see Figure 33.2(*b*)).

This phenomenon is qualitatively similar to the effect noticed in deep inelastic scattering: when the quarks are close together, the chromodynamic forces between them are weak; as the distance between them increases, so too do the forces. The term coined to denote this behaviour is 'asymptotic freedom', to denote the fact that when the interquark distances probed become asymptotically small (or, as in the original formulation, when the momentum of the deep inelastic probe becomes asymptotically high), then the chromodynamic forces disappear and the quarks become, effectively, free particles.

This remarkable property of QCD was discovered in 1973 by H. David Politzer of Harvard and independently by David Gross and Frank Wilczek of Princeton. Immediately, the feasibility of developing a field theory for the 'strong' interaction received a tremendous boost. For not only does the picture of the forces presented resemble their behaviour as observed in experiment, but also the demonstration that the 'strong' interaction coupling constant can, under certain circumstances, be small allows for the application of the traditional methods of perturbation theory to calculate quantities of physical interest.

Fig. 33.3. (*a*) Virtual quark–antiquark pairs and gluons combine to 'antishield' the colour charges on the quarks. This effect is described by the Feynman diagrams shown in (*b*).

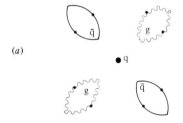

(*a*)

(*b*)

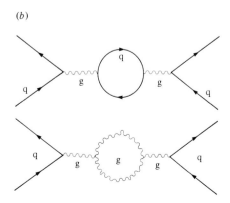

The position of QCD as the candidate theory of strong interactions was strengthened still further in 1974 when Gross and Wilczek went on to prove mathematically that only non-Abelian gauge field theories (of which QCD is an example) can give rise to asymptotically free behaviour. Also, they showed that this behaviour is possible only if there are a limited number of fermions in the theory (no more than 16 quark flavours in QCD) and if there are no Higgs particles involved in any form of spontaneous breaking of the $SU(3)_C$ symmetry. Thus if we decide that asymptotic freedom is a desirable feature of the theory of chromodynamic forces, certain other possibilities are decided for us automatically.

At this stage, it is desirable that we strengthen the credibility of QCD still further by the demonstration of the validity of the use of perturbation theory in calculating the effects of processes involving quarks and gluons. Fortunately, an example is close at hand.

33.2 Violations of scaling

The description of deep inelastic scattering in Part 6 represents a first attempt to gain some understanding of the interior of the proton. In this capacity it served us very well. For not only were we able to interpret the deep inelastic experiments as the first dynamical evidence for the existence of the point-like quarks within the proton, but the

implications of the experiments for the interquark forces provided us with the basic material for the formulation of QCD. Armed with this new theory we may now re-examine deep inelastic scattering and provide a more detailed explanation of the structure functions describing the constituents of the nucleon.

If the quarks were truly free particles then each would carry a third of the proton momentum (if we assume there are three valence quarks inside the proton). This would give the very simple form of proton structure function shown in Figure 33.4(a). However, we know that this is not the whole truth as the quarks are confined to within the dimensions of the proton; thus the uncertainty in their position can be no greater than $2r_p$. By Heisenberg's uncertainty principle this means that the uncertainty in their momentum must be at least

$$\Delta p \gtrsim \frac{\hbar}{2r_p} \approx 200 \text{ MeV}/c$$

and so the structure function tends to be smeared out as observed in Figure 33.4(b).

A better understanding of deep inelastic scattering is possible when fuller data on the behaviour of the deep inelastic structure functions are examined. These data reveal that, far from being a constant shape for all values of momentum transferred (the original scaling hypothesis), the structure functions vary with it in a very well-

Fig. 33.4. The expected behaviour of a nucleon structure function, for (a) free quarks, and for (b) confined quarks.

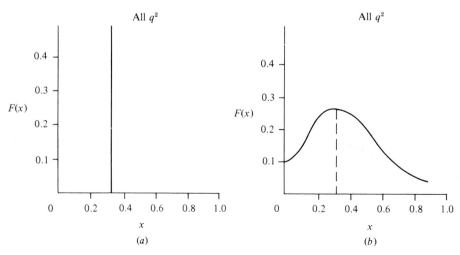

(a)

(b)

Fig. 33.5. (*a*) The violation of scaling behaviour; the nucleon structure functions vary systematically with momentum transfer squared. Values of *x* quoted are in fact the mid-points of ranges centred on those values. (*b*) The violation of scaling behaviour; the pattern of the variation of the nucleon structure functions.

(*a*)

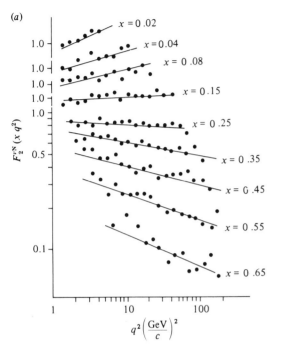

(*b*)

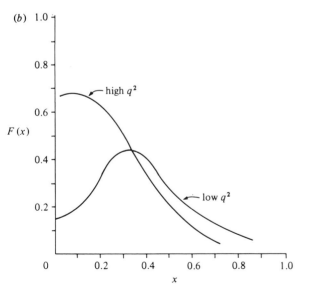

defined fashion. This is shown in Figure 33.5(*a*) and (*b*). Interestingly, this diagram illustrates why scaling was at first believed to be more exact than it really is. In the early experiments, the structure functions were examined for variations over only a limited range of q^2, predominantly in the mid-*x* regions – where there genuinely is no variation. The important variations in q^2 occur at low and high *x* values. Because of this, scaling was credited with more importance than its due. The fuller data show that it is not constancy of the structure functions which is important, but their variation.

The variation of the structure functions is such that at low values of *x* they increase with increasing momentum transfer, and that with high values of *x* there is a compensating decrease. This means that, as the momentum of the probe increases, it becomes more likely to hit a quark carrying a small fraction of the total proton momentum and less likely to hit a quark carrying a large fraction. This rather complicated behaviour can be understood by the application of the 'deep inelastic microscope technique' to the QCD picture of the proton.

As we have said, if there are no interquark forces then each valence quark will carry a third of the momentum of the proton. The corresponding structure function is shown in Figure 33.6(*a*). However, to confine the quarks inside the proton, we know that there must be some interquark forces – even if they do weaken in effect as the distance resolved by the probe decreases to less than the proton diameter. In QCD, chromodynamic forces are mediated by the exchange of gluons between the quarks. This continual exchange of gluons transfers momentum between the quarks, so smearing out the deep inelastic structure function (Figure 33.6(*b*)). As the momentum of the probe increases and the distance it resolves decreases, it begins to see the detailed quantum-mechanical sub-processes of QCD in the environment of the struck quark. For instance, what to a longer-wavelength probe may have appeared to be a quark may be revealed to a shorter-wavelength probe as a quark accompanied by a gluon (Figure 33.6(*c*)). What is more, the total momentum of the quark as measured by the long-wavelength probe must now be divided between the quark and the gluon, leaving the quark with a lower fraction of

Fig. 33.6. The shorter the wavelength of the probe used, the more constituents are seen, each with a smaller fraction of the nucleons' total momentum.

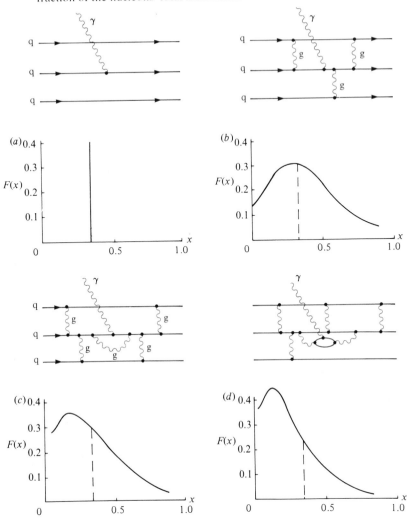

the total proton momentum. So, as the momentum of the probe increases, the average fraction of the total proton momentum carried by the quarks appears to decrease – just as observed in Figure 33.5. As the momentum of the probe increases still further and its resolving distance becomes more minute, it may see the gluon radiated by the valence quark dissociating into a quark–antiquark pair from the vacuum sea. So there will appear to be even more quarks carrying very low fractions of the total proton momentum (Figure 33.6(*d*)).

Using QCD, it is possible to calculate the probabilities of occurrence of these various quantum-mechanical sub-processes and to derive the way the structure functions vary with the momentum of the probe. Unfortunately, this is a fairly complicated business and there is no direct way of comparing the predictions of QCD with the behaviour of the structure functions as described. Comparison is of course possible using rather more sophisticated descriptions of the structure functions and the observed behaviour is completely consistent with the predictions of QCD.

34

Quark confinement

34.1 Introduction

The fact that a single quark has never been observed has for years been the single greatest puzzle of elementary particle physics. No matter how energetically protons are collided together in the enormous accelerators at CERN and elsewhere, no quarks are seen to emerge in the debris. Many other varieties of particles are produced, but never any fractionally charged particles which may be identified with the quarks. This means that the forces which bind the quarks together are much stronger than the forces of the collision – which means that they are enormously strong. As an indication, we may note that the energies which bind the electrons into their atomic orbits are of the order of a few electronvolts. The energies binding the protons and neutrons in the nucleus are of the order of a few million electronvolts. Pairs of protons have been collided at energies of hundreds of thousands of millions electronvolts and still no quarks are observed, which means the chromodynamic force between them must be at least that strong.

Not surprisingly, other more bizarre quark hunts have met with no success. Attempts have been made to detect the existence of fractional electric charges in all manner of materials from oysters (because they filter a large amount of sea water) to moon dust, with no convincing record of success. Because of the very delicate nature of the experiments, which are basically modern variants of Milikan's oil-drop experiment, fractional charges are sometimes reported. But none of these have yet gained general acceptance.

These basic experimental facts have led theorists to conjecture that quarks may be permanently confined within hadrons as a result of the fundamental nature of the chromodynamic force. In contrast to Abelian QED which gives rise to Coulomb's inverse square law of electrostatic attraction, it may well be that the non-Abelian nature of QCD gives rise to a confining force which does not decrease with increasing distance. In fact, the corollary of asymptotic freedom is that the effective strength of the chromodynamic force increases as the quarks are drawn apart, a phenomenon known as 'infra-red slavery'. It is not yet known whether QCD gives rise to infra-red slavery or whether, after a period of rising, the chromodynamic force tends to a constant strength or even decreases as the quarks are separated. If the force does eventually begin to drop off, then the quarks will eventually be separable and confinement only a temporary phenomenon, apparent because accelerator energies are not yet high enough. The various possibilities are shown in Figure 34.1.

The major hindrance to a straightforward examination of the confinement problem is the difficulty in developing a mathematical description of strong forces. The method of perturbation theory used in QED and in the asymptotically free regime of QCD is valid only because the forces are weak. Attempts have been made to develop other

Fig. 34.1. Possible behaviours of the chromodynamic force at large distances.

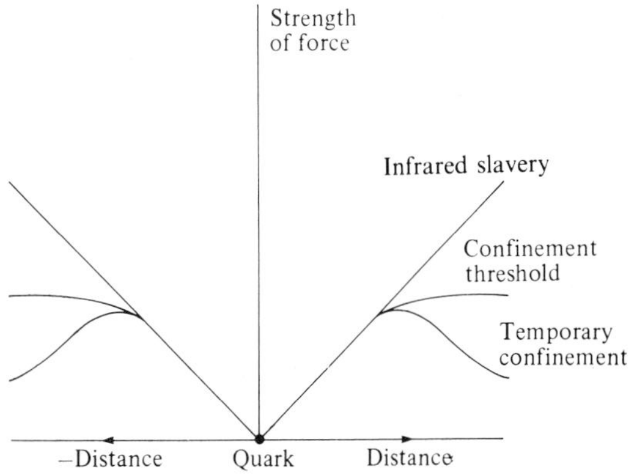

methods such as one which divides the space–time continuum into a lattice of discrete points (so-called lattice gauge theory). But no totally satisfactory description has been achieved.

Instead, we will have to content ourselves with an intuitive picture of how the non-Abelian nature of QCD may give rise to the confinement mechanism. As usual, we start off with the familiar case of electrodynamics. The field lines joining two charges spread out to infinity in a spherical fashion. As they are drawn apart the field lines become more spread out. Because the density of field lines at any point is related to the strength of the electrostatic force at that point, this means that the force decreases as the separation increases (see Figure 34.2).

Consider now what may happen in QCD to the chromodynamic force between the quark and antiquark in a meson. The chromodynamic field lines would like to spread out like the electrodynamic ones, but because the non-Abelian nature of QCD gives rise to self-interactions of the gauge field, the field lines are drawn together instead. This is illustrated in Figure 34.3 by field lines forming a 'flux-tube' between the quarks. As the

quarks are separated, the field lines do not spread out but, instead, are drawn out into a tube in which the density of chromodynamic force lines may be constant. This would lead to a constant force existing between the quarks like a perfectly elastic band. Eventually, as we put more and more work into increasing the separation of the quarks, the system will gain enough energy to promote a virtual quark–antiquark pair from the vacuum sea into physical reality. This will give rise to the creation of a new meson. So the energy we expend in attempting to separate the $q\bar{q}$ pair has, in fact, resulted in the production of another meson, just as occurs in the high-energy collisions!

34.2 Quark forces – hadron forces

Having seen how QCD may provide an acceptable picture of the interquark forces, it is worth pausing to relate this picture to that of the forces between the observable hadrons, mentioned briefly in Part 2. These are the forces which bind the protons and neutrons together in the nuclei and, when the hadrons are in collision at

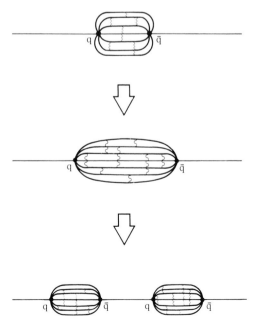

Fig. 34.3. Colour force lines between quarks are collimated into a tube-like shape and do not spread out as the quarks are separated. Eventually a single tube will split into two when the force applied has completed enough work.

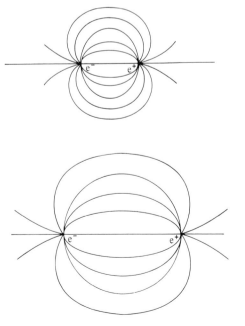

Fig. 34.2. Electric field lines spread out as the electric charges are separated.

high energies, produce the numerous secondary particles.

These forces are now seen as the 'van der Waals' forces between hadrons. The van der Waals forces between atoms are the very feeble residual electrodynamic effects remaining after the electrons and nucleus have formed a net electrically neutral atom (Figure 34.4). Analogously, the van der Waals forces between hadrons are the chromodynamic effects remaining between colour singlet states once their colour constituents are bound together. Unlike the electrodynamic case, however, there is no guarantee that these 'secondary' forces will be weaker than the 'primary' inter-quark chromodynamic forces. This is because they are predominantly long-range phenomena, which is in the strong-coupling regime of QCD.

Hadron collisions can be divided into either of two main classes. In the first are the diffractive collisions which are, in effect, glancing blows between the colliding particles. In Part 2 we saw that these form the vast majority of hadron collisions and that they can be described, albeit in a less than fundamental manner, by the Regge theory of collisions. In the QCD picture, these long-range collisions are complicated affairs involving multiple-gluon exchange with many sub-processes occurring (Figure 34.5). Because the forces are strong, there is no well-established method of describing the quark and gluon behaviour in these collisions.

Indeed, there is no great motivation for examining these collisions at the level of the details of quarks and gluons, as we are unlikely to be able to deduce much about the fundamental nature of the forces from such a complicated event. It is as if we were to attempt to study the electromagnetic force by observing collisions between complex atoms!

In the second main class of hadron collisions are the non-diffractive or 'head-on' events. Because they are much rarer than the diffractive events, they tended to be rather ignored prior to the development of QCD. Another reason for this perhaps was that they could not be described properly by Regge theory! However, at the levels of quarks and gluons, non-diffractive collisions are rather simple and so have become one of the centres of attention in the quest to understand

Fig. 34.5. Diffractive (glancing) hadron–hadron collisions are the result of complicated multi-quark and multi-gluon processes. In (*a*) the reggeon (R) exchange picture is shown. In (*b*) the QCD picture of the quark and gluon sub-processes is shown.

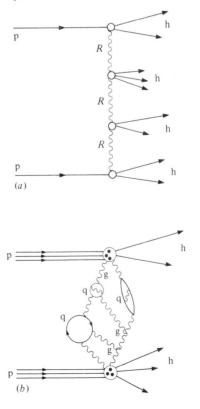

Fig. 34.4. The analogy between the van der Waals force between atoms and the long-range colour force between the observed hadrons.

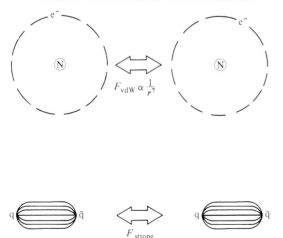

Fig. 34.6. Head-on hadron–hadron collisions are described by simple quark and gluon processes, such as one-gluon exchange, which give rise to jets of hadrons emerging from the collisions.

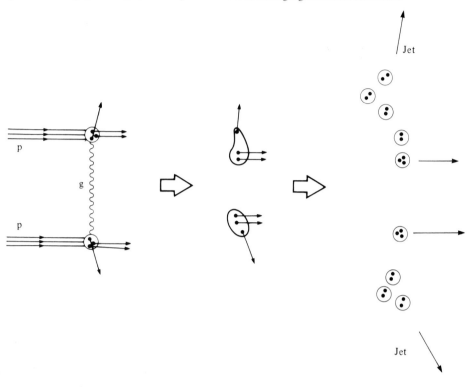

more of the interquark forces. Because the hadrons collide head-on, it means that the quarks in each collision will approach each other very closely. The collisions are then thought to proceed predominantly by the exchange of a single gluon between two passing quarks, all the others acting as passive spectators (Figure 34.6). The result is that the interacting quarks are knocked violently sideways out of their parent hadrons. Of course, they do not emerge as free particles (the confinement mechanism dresses them up as hadrons), but the result is a jet of hadrons emerging along the directions of motion of the original quarks. These 'high transverse momentum' jets were observed in the early 1980s at the CERN p$\bar{\text{p}}$ collider.

Jets occur also in other classes of high-energy collisions, such as electron–positron annihilations, where there are no complications due to spectator quarks. These events are altogether cleaner, as we will see in Part 8. But it is encouraging to note common phenomena in two very different circumstances, as this suggests a common, fundamental origin which we take to be the underlying dynamics of quarks and gluons.

Part 8
Electron–positron collisions

| **35** |

| ***Probing the vacuum*** |

35.1 Introduction

The primary means of studying the fundamental constituents of matter and their interactions is through performing scattering experiments. This has been the case since the very beginning of particle physics. One class of scattering experiments which has, over the past two decades, been extremely fruitful involves collisions between electrons and positrons. These e^+e^- experiments have yielded a great deal of information about the nature of the strong, weak and electromagnetic forces, and have played a major role in establishing the 'standard model' of particle physics: QCD and the Glashow–Weinberg–Salam electroweak theory.

An attractive feature of e^+e^- experiments is that because the electron and the positron are antiparticles, they often annihilate into a 'vacuum' state of pure energy. All the quantum numbers of the initial particles cancel and so we avoid the inhibiting effects of some conservation laws. The energy resulting from the annihilation (in the form of a virtual photon or Z^0) is then free to sample the hidden, negative-energy content of the vacuum. It can be regarded quite literally as the photon of Figure 4.2 boosting a particle from its negative-energy sea to create, for instance, another e^+e^- pair or a $q\bar{q}$ pair. In this way e^+e^- experiments are ideal reactions in which to look for new particles. By bringing beams of electrons and positrons into collision, physicists have also been able to study the couplings of leptons and quarks to the photon and Z^0, and hence the electroweak properties of these fermions, as well as the nature of the strong interaction between hadrons.

There is a further convenient feature of these reactions. If an electron and positron are collided head-on, with equal and opposite momentum, then the centre of mass of the reaction is absolutely stationary, and the total centre-of-mass collision energy is just the sum of the energy of each particle. (This is in contrast to accelerating electrons into stationary targets where the centre of mass is moving, and the total centre-of-mass collision energy is much less than the energy of the electron.) Furthermore, because the centre of mass is stationary in head-on collisions, the angular distribution of created particles can be measured directly and significant asymmetries detected that much more easily.

Because the e^+e^- pair can annihilate to a vacuum state, it means also that the reactions are very clean. There is no debris surviving from the initial state to mask interesting new effects or to confuse the experiments' detectors. This is in contrast to the deep inelastic experiments in which the photon–quark interaction has to take place in the presence of spectator quarks and leptons.

35.2 The experiments

Naturally, the many benefits of e^+e^- collisions are balanced by some penalties. The basic requirement is to collide bunches of electrons with bunches of positrons head-on and this requires comparatively elaborate accelerator technology. The most serious drawback is the numbers of e^+ and e^- available in the bunches. The available flux limits the rate at which reactions can be observed, and this in turn limits the accuracy of the measurements obtained.

Another limitation is that the range of phenomena open to study depends upon the total energy of the collision. The total energy is just the sum of the e^+ energy and the e^- energy and, as the collision is head-on, all of it is available to create the mass of new particles. But whilst the e^+e^- bunches are kept in orbit they emit their energy in the form of synchrotron radiation at a rate which rises rapidly with their energy (as the fourth power) and with the tightness of the bends (as the

Table 35.1. *Important e^+e^- storage rings*

Accelerator	Location	Start	Maximum energy (centre of mass)
SPEAR	Stanford, USA	1972	8 GeV
DORIS	DESY, Hamburg, Germany	1973	12 GeV
PETRA	DESY, Hamburg, Germany	1978	45 GeV
CESR	Cornell, USA	1979	12 GeV
PEP	Stanford, USA	1980	30 GeV
TRISTAN	Tsukuba, Japan	1987	64 GeV
SLC	Stanford, USA	1989	100 GeV
LEP	CERN, Geneva, Switzerland	1989	200 GeV

inverse of the radius of curvature). So, to achieve very high energies, it is necessary to build very large rings and to input a lot of radio frequency (r.f.) power to replenish the lost energy. So the search for new particles has led to the construction of increasingly more powerful accelerators. The progress of this construction in the past two decades is summarised in Table 35.1.

At the current frontier of e^+e^- experiments are the Large Electron–Positron (LEP) collider at CERN and the Stanford Linear Collider (SLC) at SLAC in California. But historically, it was the SPEAR ring at SLAC which provided such important advances during the mid-1970s and which may thus serve as the best illustration of e^+e^- experiments (see Figure 35.1).

Electrons are created at the end of the two-mile-long linear accelerator tube. After some initial acceleration by electric fields, some of the electrons can be collided with a target to produce general particle debris from which the positrons can be filtered. Bunches of electrons and positrons are then accelerated down the tube by alternating electric fields and are then injected in opposite directions into the storage ring SPEAR. The bunches can be stored and accelerated in orbit in this ring for several hours by magnetic fields and r.f. power cavities, during which time e^+ and e^- bunches can be made to pass through each other at specified interaction regions. The individual e^+e^- interactions are studied by many different types of

Fig. 35.1. An aerial view of the Stanford Linear Accelerator Center. This is the home of the historic SPEAR electron–positron ring and the much larger PEP collider. The 3-kilometre linear accelerator now forms part of the novel Stanford Linear Collider (SLC). In the SLC, bunches of electrons and positrons are boosted down the linear accelerator and guided around separate arcs, before being brought into collision. (Photo courtesy SLAC.)

detectors, which are usually cylindrical or spherical distributions of sensors wrapped around the interaction region.

35.3 The basic reactions

It is useful to classify the various possibilities that can occur during an e^+e^- collision. Firstly, there are the purely electromagnetic processes. 'Bhabba scattering' is the name given to elastic $e^+e^- \to e^+e^-$ scattering. This can occur by either of the two Feynman diagrams of Figure 35.2(a) corresponding to the possibilities of photon exchange, and of annihilation into a virtual photon with subsequent reproduction of an e^+e^- pair.

However, the simplest electromagnetic process is muon-pair production $e^+e^- \to \mu^+\mu^-$ (Figure 35.2(b)), as this can occur only through the single-photon annihilation mechanism. The single

photon must be virtual, as we remarked in Chapter 4, as it is impossible to conserve both energy and momentum. The energy of the initial two-electron state is always greater than $2m_e$ but its momentum is zero; whereas the energy momentum relationship for real photons is $E = pc$.

If the energy of the e^+e^- pair is greater than $2m_\mu$ then the virtual photon can promote a muon pair from the negative-energy sea in the same way as it can reproduce an e^+e^- pair. Another possibility is the production of two real photons (Figure 35.2(c)). This is allowed as the photons can emerge with equal and opposite momentum, so both energy and momentum can be conserved simultaneously.

Accurate measurement of these electromagnetic effects allows us to test the validity of QED at very high energies. This is done usually by measuring the angular distribution of particles emerging from the collision and comparing results with predictions. The correctness of QED has been verified up to the highest accelerator energies. This implies that the leptons are indeed fundamental point-like particles (any possible substructure must be smaller than 10^{-18} m in size).

The second class of e^+e^- collisions comprises those in which hadrons emerge in the final state and which indicate that the strong interaction is involved somewhere (see Figure 35.3(a)). One of the most significant quantities in particle physics is the ratio R of the cross-section for $e^+e^- \rightarrow$ hadrons to that for $e^+e^- \rightarrow \mu^+\mu^-$, measured as the energy of the collision varies:

$$R = \frac{\sigma(e^+e^- \rightarrow \text{hadrons})}{\sigma(e^+e^- \rightarrow \mu^+\mu^-)}.$$

The significance of the ratio R is that it compares a reaction we understand very well (muon-pair production) with the class of reactions we wish to understand (hadron production), thus providing a very useful guide to our thinking about the unknown. Also, the ratio R is relatively straightforward to observe experimentally. Only two

charged 'prongs' are seen to emerge in muon-pair production whereas, almost invariably, more emerge from a hadronic final state. So the ratio can be obtained by dividing the number of events detected with more than two prongs by the number with only two prongs, as measured during a given experiment.

Surprisingly, the ratio R is constant over large energy ranges, indicating that the complicated hadronic state is produced in much the same way as the simple muon pair. The virtual photon is probing the negative-energy sea of hadrons contained in the vacuum instead of electrons or muons. We will see how this can be given a clear interpretation in terms of quarks in the next chapter. Suffice it at this stage to note that the hadrons cannot have been produced by the e^+e^- pair annihilating into a virtual gluon, as the leptons have no colour and so have no connection with gluons whatsoever.

Before passing on, we must finally identify a third class of e^+e^- reaction involving the weak force. This results because the e^+e^- *do* carry weak isospin, which allows them to annihilate into a virtual Z^0 boson. In fact, as we discussed during our look at the Glashow–Weinberg–Salam model, the photon γ and the Z^0 boson are simply the rather dissimilar quanta of the unified electroweak force. Thus we might expect weak interaction effects to come into the picture somewhere.

Fig. 35.2. Possible electromagnetic effects following an e^+e^- collision. (*a*) Bhabba scattering. (*b*) Muon-pair production. (*c*) Two-photon production.

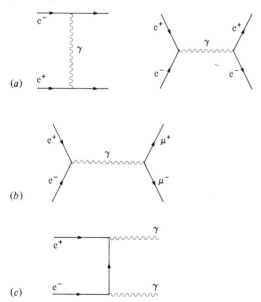

Fig. 35.3. Non-electromagnetic effects: (*a*) hadron production in the final state (the blob); (*b*) annihilation into a Z^0 boson.

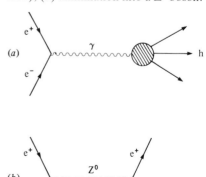

The virtual Z^0 boson is free to explore the negative-energy content of the vacuum just like the photon (see Figure 35.3(b)). As a result we should expect to see some uniquely weak interaction effects (such as parity violation) creeping in at higher energies. This we shall discuss further in Chapter 38. Until then, however, we shall ignore these very slight effects. Most of our attention will focus on hadron production and the ratio R.

36

Quarks and charm

36.1 Introduction

The observation of scaling in deep inelastic scattering provides firm evidence for the interaction of the photon with point-like quarks inside the observed hadrons. So when we come to explain the process $e^+e^- \to$ hadrons, the most likely picture is that of the virtual photon interacting with quarks rather than directly with complete hadrons (Figure 36.1(a)). The photon promotes a quark–antiquark ($q\bar{q}$) pair from the vacuum, giving the quark and antiquark a kinetic energy depending on the initial collision energy. The q and $\bar{q}$ must separate with equal and opposite momentum to maintain the net momentum of zero and in so doing are 'dressed up' into hadrons by the, as yet unknown, quark-confinement mechanism (Figure 36.1(b)). This may be viewed as the potential energy of the long-range attractive force between q and $\bar{q}$ being used to promote extra $q\bar{q}$ pairs from the vacuum.

36.2 The quark picture

Because the confinement stage of the process *always* occurs (at least assuming the permanently confined quark hypothesis), it enters the calculation of the process $e^+e^- \to$ hadrons only as a final probability of one multiplying the underlying process $e^+e^- \to q\bar{q}$. As the quarks are observed to

Fig. 36.1. $e^+e^- \rightarrow$ hadrons proceeds by a $q\bar{q}$ intermediate state, shown in (*a*). The transformation of this state into the observed hadrons involves the creation of more $q\bar{q}$ pairs (*b*).

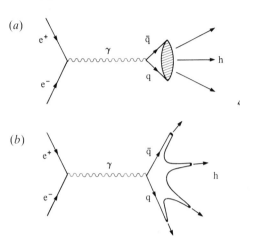

(*a*)

(*b*)

be point-like and spin $\frac{1}{2}$, the process $e^+e^- \rightarrow q\bar{q}$ is very similar to the process $e^+e^- \rightarrow \mu^+\mu^-$, the only difference being that the charges on the quarks are only some fraction of that on the muons. This explains the constancy of the ratio R mentioned earlier and displayed in Figure 36.2. The fundamental dynamics of the two processes are the same, so giving an R constant with energy, but their magnitudes differ by an amount equal to the ratio of the squares of the charges involved (see Figure 36.3). As several species of quark will be able to act as intermediaries to the creation of hadrons, and as the charge on the muon is 1, then R is equal to the sum of the squares of the quark charges.

Now the significance of R is gloriously apparent. It is a directly observable quark-counting opportunity which provides a measure of the number of quarks and their properties. For instance, in the simplest quark scheme with just three quark flavours, namely up ($\frac{2}{3}e$), down ($-\frac{1}{3}e$) and strange ($-\frac{1}{3}e$), the value of R is predicted to be

$$R_{uds} = (\tfrac{2}{3})^2 + (-\tfrac{1}{3})^2 + (-\tfrac{1}{3})^2 = \tfrac{2}{3}.$$

As we mentioned in Chapter 31 on QCD, the advent of the colour degree of freedom triples the

Fig. 36.2. The ratio R of the total hadronic cross-section to $\sigma(e^+e^- \rightarrow \mu^+\mu^-)$ as a function of the cm energy, E.

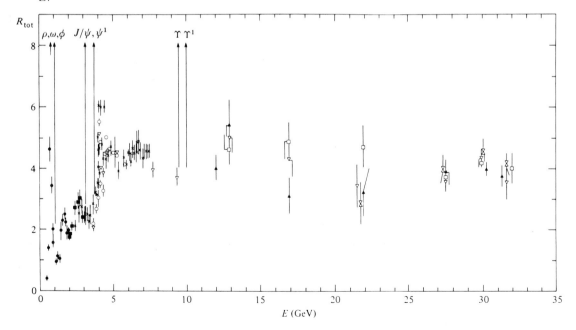

Fig. 36.3. The value of the ratio R is equal to the sum of the squares of the quark charges.

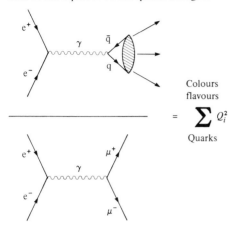

number of quarks and so predicts the correct value at low energy, $R = 2$.

We have now explained the constancy of the ratio R but have not so far mentioned the very pronounced spikes which punctuate the picture. These shapes are highly reminiscent of the phenomena of resonance particles in Chapter 9 and this in fact is just what they are. At certain energies of the e^+e^- collision, the $q\bar{q}$ pair into which the photon transforms will have just the correct mass to stay intact as a single-meson resonance. This is signalled by a large increase in the probability of the event occurring compared with non-resonant 'background' $q\bar{q}$ production at neighbouring energies and this leads to the observed spikes in the cross-section. After its brief existence, the resonance particle will then decay by its usual mechanisms into the final-state hadrons observed.

The resonance particles produced are a select sub-set of the hundreds which have been observed in hadron–hadron reactions. The sub-set is defined by the quantum numbers of the virtual photon from which the resonances transform: spin 1, zero charge and strangeness. This defines the allowed quark content of the meson as being that of the quark–antiquark combinations in the vector nonet ($= \mathbf{8} + \mathbf{1}$) of $SU(3)$ flavour symmetry.

36.3 The advent of charm

In what was undoubtedly the most sensational experimental surprise of the 1970s, an extra-ordinary new resonance spike was discovered in $e^+e^- \rightarrow$ hadrons at a collision energy of 3.096 GeV, followed quickly by the discovery of a similar spike at 3.687 GeV and a subsequently turbulent rise in the value of the ratio R to a new plateau (see Figure 36.2). The new resonance was denoted the ψ (psi) by Burton Richter and his colleagues at SLAC, who observed the particle in e^+e^- annihilations, but it was also seen simultaneously as a resonance production phenomenon in the reaction $p + p \rightarrow e^+e^- + X$ at Fermilab by Samuel Ting and his team, who denoted it J. For their discovery both Richter and Ting were awarded the Nobel Prize in 1976. Subsequently, J/ψ has become the accepted symbol for the particle at 3.097 GeV and ψ' for that at 3.687 GeV.

After a brief period of speculation, the correct interpretation of the J/ψ emerged. What had happened was that the increasing energy of the e^+e^- collision had become sufficiently large to create a new flavour $q\bar{q}$ pair. It had boosted a new heavier type of quark from its negative-energy sea in the vacuum. The J/ψ and ψ' were bound-state mesons consisting of the $q\bar{q}$ pair and, at energies above the threshold of its production, the pair could contribute to the ratio R, thus accounting for its observed rise.

This new flavour, called charm, had been anticipated in advance by the GIM theorists attempting to explain the behaviour of hadrons in the Glashow–Weinberg–Salam theory of the weak force. As we saw in Section 24.2, it was put forward as an explanation for the absence of strangeness-changing neutral currents. Also, it was able to complete an aesthetically pleasing matching between the numbers of fundamental leptons and fundamental hadrons. With the advent of charm, it became possible to group the leptons and the quarks into two generations, the second being simply a massive repetition of the quantum numbers of the first (see Table 36.1).

But the discovery of the charmed quark automatically implied the existence of a horde of new particles corresponding not only to the excited states formed from the $c\bar{c}$ pair, but also to all possible combinations of the charmed quark with the up, down and strange quarks in both mesonic and baryonic configurations and their excited states. In short, the $SU(3)$ flavour symmetry of the

Table 36.1. *The charmed quark completes two generations of the fundamental-quarks and leptons*

Generation	First	Second	Charge
Quarks	u	c	$\frac{2}{3}$
	d	s	$-\frac{1}{3}$
Leptons	e^-	μ^-	-1
	ν_e	ν_μ	0

hadrons is enlarged to $SU(4)$. Mesons which combine the charmed quark with the up or down antiquarks are denoted the D mesons. These mesons carry explicit charm (i.e. have a non-zero charm quantum number), just as the K mesons carry strangeness. This is in contrast to the J/ψ itself which, being a $c\bar{c}$ combination, has the charm of its quark cancelled by the anticharm of its antiquark. There are also mesons consisting of both charmed and strange quarks and antiquarks,

Fig. 36.4. The hexadecimet (16-plet) of spin-0 mesons generated by $SU(4)$ flavour symmetry. The familiar nonet of $SU(3)$ flavour is the middle $C = 0$ plane.

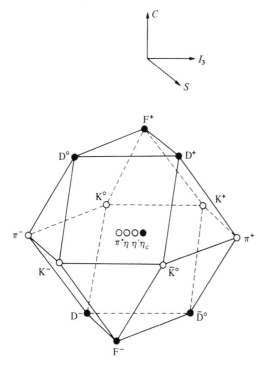

which thus possess both charm and strangeness. These mesons were initially designated $F^{\pm}$, but have been recently renamed $D_s^{\pm}$. All possible spin-0 mesons are contained in the hexadecimet of $SU(4)$ flavour symmetry, which is illustrated in Figure 36.4. Similarly, there are also new baryons containing both charmed and strange quarks.

With the prospect of such a feast of new particles, both experimenters and theorists busied themselves in the 1970s in confirming the anticipated picture. It often became a race between experimental teams to be the first to detect a particularly tricky candidate, whilst theorists vied with each other to predict the masses and properties of the particles as accurately as possible. This led to a rapid advance in our understanding of quark behaviour and in the formulation of QCD.

36.4 Psichology

What remained a puzzle for some time was the extraordinary size of the resonance (its formation being some 3000 times more probable than the production of a non-resonant $q\bar{q}$ pair at neighbouring energies), and its extreme narrowness. The ψ has a width of only 0.002% of its mass compared with the ρ width of 20% of its mass. By Heisenberg's uncertainty principle this means that the ψ has a lifetime much longer than that generally associated with hadrons.

This unusual narrowness can be explained in terms of the inhibition of its preferred decay modes because of the masses of the charmed mesons. Its preferred decay mode would normally be expected to be into the charmed mesons D^+D^- or $D^0\bar{D}^0$, proceeding by a quark line diagram rather like that for the decay of the ρ^0 into $\pi^+\pi^-$ or $2\pi^0$ (see Figure 36.5(a)). However, the ρ^0 decay can proceed only because the ρ^0 mass of 0.77 GeV is larger than the mass of the two-pion state, 2×0.135 GeV. The J/ψ is so narrow because its mass is *less* than that of two charmed mesons. These mesons were detected well after the discovery of the J/ψ with a mass of 1.86 GeV, thereby requiring a particle with a mass of at least 3.72 GeV to produce them.

The J/ψ can decay only by rather sophisticated means. The $c\bar{c}$ pair has to annihilate itself into a state of three gluons which must then transform themselves into the observed hadrons by the

Fig. 36.5. The obvious decay mode of the mesons (*a*) is not possible because the charmed mesons are too massive. The more complicated decay into non-charmed hadrons (*b*) takes longer.

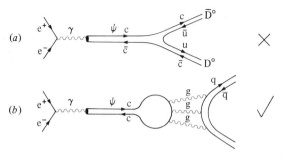

mechanism of colour confinement similar to that practised by the quarks mentioned earlier (see Figure 36.5(*b*)). Intermediate states of one or two gluons are prohibited by the conservation laws of colour and C parity respectively.

The *J*/*ψ* is but the most obvious of a whole family of mesons consisting of a c̄c pair. Some differ only in mass, the differences being due to the increased radial excitation energy of the c̄c pair. The *ψ′* at 3.687 GeV is the lightest of these and, like the *J*/*ψ*, is very narrow because it too lies below the DD̄ threshold. The *ψ″* is next at 3.77 GeV but this is a hadron of normal width (at 0.7% of mass) as it lies above the DD̄ threshold (but only just!). Above this there are a number of other states.

However, some of the heavier c̄c mesons have different spin, parity (**P**) and charge conjugation (or C-parity, **C**) assignments from those of the *J*/*ψ*. These are denoted *χ* and are produced when one of the heavier members of the *ψ* family emits a photon, thereby allowing the c̄c pair to change its quantum numbers (Figure 36.6).

Fig. 36.6. If a heavy *ψ*-like state emits a photon, c̄c mesons with new quantum numbers are created.

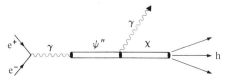

A simplified version of the entire c̄c family (often called 'charmonium') is shown in Figure 36.7. Note that η_c with spin 0 is the lightest c̄c meson. To discover all these 'secondary' c̄c states, it is necessary to observe the energies of photons emerging from a process such as that in Figure 36.6. If one energy is preferred above all those possible, this is taken to indicate the mass difference between the heavy *ψ*-like particle (the energy of the e^+e^- collision) and the secondary c̄c state with different spin or parity assignments.

To achieve this prodigiously detailed particle-hunting task, experimenters at SLAC built a novel photon detector nicknamed the crystal ball. This consists of a spherical array of sodium iodide crystals pointing towards its centre which is colocated with the interaction region. The sodium iodide crystals are monitored by photomultipliers which can measure the energy deposited in the crystal by an incident photon (see Figure 36.8).

Readers familiar with atomic physics will recognise the pattern of Figure 36.7 as being very similar to the energy level structure of the hydro-

Fig. 36.7. The experimentally observed spectrum of c̄c mesons resulting from the different values possible for the spin and the orbital angular momentum of the constituent quarks. In this notation, S, P and D refer respectively to measured orbital angular momentum equal to 0, ℏ and 2ℏ.

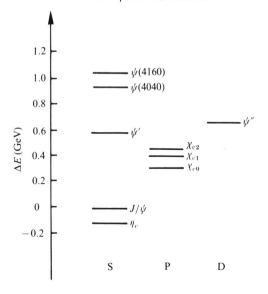

gen atom. This similarity is understandable because the c and c̄ have bound themselves together into an exotic sort of elementary particle atom. Recognition of this phenomenon provided an enormous opportunity for particle physicists be-

cause such an atomic arrangement of the relatively heavy charmed quarks can be described by well-understood non-relativistic quantum mechanics. The force between the quarks can be formulated as a potential acting in the vicinity of a colour charge,

Fig. 36.8. The crystal ball detector at SLAC. Numerous photomultiplier tubes bristle from the surface of the spherical container. They are monitoring the sodium iodide crystals mounted in the interior which detect the photons originating from the interaction point. (Photo courtesy SLAC.)

Fig. 36.9. The spectrum of energy levels expected from the familiar electric potential is shown in (*a*). In (*b*), is shown the spectrum generated by the proposed form of the interquark potential. It is a much closer match to the observed spectrum.

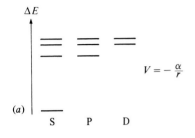

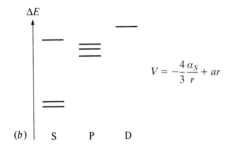

just as in classical electrodynamics an electric potential surrounding an electric charge gives rise to Coulomb's force law between charges.

The particular form of the potential will determine the splitting of the energy levels or, in the $c\bar{c}$ case, the mass differences between mesons. As these can be measured experimentally with great accuracy, this can be used to provide a detailed picture of the force between the quarks.

The form of the potential arising from a colour charge which is found to give the most satisfactory match to the spectrum of mass levels is one which combines a simple Coulomb law at short ranges (one corresponding to single gluon exchange in the asymptotically free regime) with an attractive potential rising linearly with range at longer ranges, giving rise to the ever-increasing forces of quark confinement. The theoretical pattern of $c\bar{c}$ mass states generated by this potential is

Fig. 36.10. A magnified bubble chamber photograph of charmed particles decaying. In the top half, a positively charged charmed meson (track entering from left) decays into three other charged particles. In the lower half, an invisible, neutral charmed meson decays into a pair of charged particles.

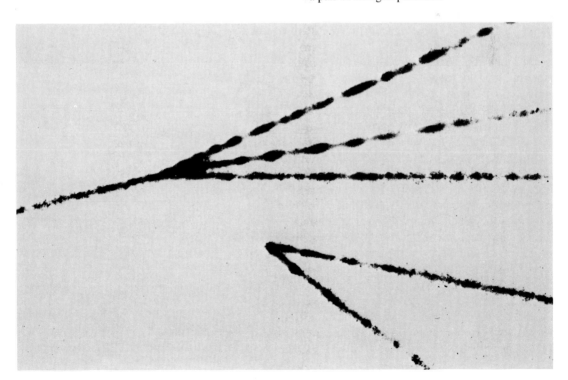

shown in Figure 36.9(*b*) and, in comparison, the energy levels of positronium, the bound states of e^+e^- arising from the Coulomb potential between the two electric charges, is shown in Figure 36.9(*a*).

So the masses of the c c̄ mesons (sometimes referred to as the spectrum of charmonium) provide direct support for the QCD picture of interquark forces containing both asymptotic freedom at short ranges and confining forces at longer ranges.

36.5 Charmed particles

For several years after the discovery of the J/ψ, experimenters sought the scores of particles which should be expected to carry explicit charm and their various excited states with everincreasing spins. These were a lot harder to dig out of the experiments, as they could be found only by searching amongst the final-state hadrons for particular combinations at given masses. When a lot of debris is present in the final state and when the decays of the sought-for particle are uncertain, this is a tricky business. Eventually, a respectable roll call of the particles was built up which supports their categorisation by $SU(4)$ flavour symmetry.

Like the strange particles, charmed particles decay by the strong force, emitting pions until they arrive at the lowest-mass charmed state. Charm is conserved by the strong force and so this state, for example a D meson, is obliged to decay by the weak force. This it does by emitting a virtual W boson which changes the flavour of the emitting quark. This is an extension of the Cabbibo hypothesis of Chapter 18. The charmed quark will prefer to turn into a strange quark rather than an up or down, and this is signalled by the presence of a high proportion of strange particles amongst the decay products of the Ds. These strange particles must then decay either to non-strange mesons or directly into leptons by another weak interaction process. Thus the decay of the charmed particle is a complex laboratory of weak decays involving as many as three in succession (see Figure 36.10).

In summary, the discovery of charm has enabled us to find out a great deal about the strong force between quarks, as carried by the gluons of QCD, by studying the spectrum of c c̄ mesons. The decay of the charmed D and F (i.e. D_s) mesons has confirmed our understanding of the weak decays of hadrons as contained in the Glashow–Weinberg–Salam theory.

Another generation

37.1 Introduction

Soon after physicists had digested the consequences of the ψ mesons and the charm scheme, the discovery of yet another particle threatened them with elementary particle indigestion. In an experiment similar to Ting's discovery of the J/ψ, Leon Lederman and his team at Fermilab discovered a new particle in the reaction,

$$p + N \rightarrow \mu^+ \mu^- + X.$$

Lederman and his colleagues observed that this reaction was enhanced slightly for a $\mu^+ \mu^-$ pair mass of 9.46 GeV compared to its generally declining probability over the neighbouring range, see Figure 37.1. This was taken as the signal of a new, very massive meson resonance consisting of yet another flavour of quark bound to its antiquark. The new meson is denoted by upsilon, Υ, and its new constituent, the bottom quark, b (after a spirited but doomed effort on the part of a romantic school to call it beauty).

37.2 The upsilon

This interpretation was by no means certain at the beginning and the pN experiment is by no means an ideal reaction in which to study the particle. This is because the hadronic debris X confuses the final state, and the fact that the very massive $\mu^+ \mu^-$ pairs are relatively rare makes it difficult to obtain accurate statistics. If its interpretation were correct, then it should be produced

also in $e^+ e^-$ annihilations exactly like the J/ψ and so this was the obvious way to examine it in more detail. The trouble was that with its mass at 9.46 GeV, the Υ lay above the energy range of the SPEAR ring at SLAC and *below* the range provided by the new PETRA ring opened at DESY in 1978. Doubtless, the high-energy planners thought that no divine guiding hand would deal such a low card as to stick a particle between 8.4 and 10 GeV. However, it was vital that the Υ be investigated in the uncluttered environment of $e^+ e^-$ annihilations and so the energy range of the DORIS ring (PETRA's predecessor at DESY) was tweaked to give just enough energy to reach the Υ.

The $e^+ e^-$ experiments confirmed that the Υ was indeed a $(b\bar{b})$ bound state and confirmed also the existence of its radially excited relative, Υ' at 10 GeV and Υ'' at 10.40 GeV (Figure 37.2). The width of the states was much harder to establish than that of the J/ψ as the energy resolution of the storage ring is not as accurate at the very end of its energy range as in the middle. The best value for the Υ width is about 0.005% of its mass, which indicates that it too, like the J/ψ, has its preferred decay mode (into explicit bottom mesons) suppressed. It too must annihilate the bottom of its quark with the anti-bottom of its antiquark into a

Fig. 37.1. The $\mu^+ \mu^-$ mass spectrum in pN collisions, containing the telltale bump of the upsilon.

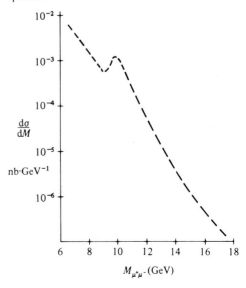

Fig. 37.2. Evidence for the Υ and Υ′ from the total cross-section for $e^+e^- \rightarrow$ hadrons.

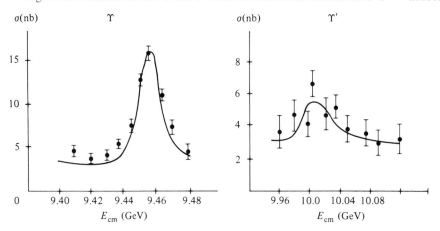

state of three gluons which will then transform into non-bottom hadrons. The full significance of this will be seen later in Chapter 38. From measurement of the Υ width, it is possible to deduce that the most likely charge of the bottom meson is $-\frac{1}{3}$, which establishes it as a more massive successor to the down and strange quarks. The spacing of the masses of the Υ and Υ′ can be calculated in the same way as the spectrum of ψ states. The experimental value observed supports the form of the interquark force as described by the potential of Figure 36.9(b).

The existence of yet another flavour of quark of course means that there must exist an entire new family of mesons with explicit bottom for all the various values of isospin, strangeness and charm discussed previously. The $SU(4)$ flavour symmetry is enlarged to $SU(5)$ so that the basic multiplet of spin-0 mesons is now expanded from the hexadecimet of Figure 36.4 to a 25-plet. Similarly, baryons with non-zero bottom will augment all the baryonic multiplets. Detection of explicit bottom particles is even harder than that of naked charm as they are much more massive and thus require high-energy collisions. These will contain more debris in the final state from which the suspected decay products of the bottom mesons must be sorted. Despite these difficulties experiments have detected explicit bottom particles, as shown in Figure 37.3. Bottom particle spectroscopy has provided confirmation of the quark dynamics formulated in the context of the charm spectrum.

In some ways, just the existence of the Υ meson and bottom quark is of more significance than the details of its properties. For there is no place for the bottom quark in the first two generations. This suggests that it is the herald of a third generation containing yet another quark (denoted, naturally, the top quark) and a new lepton and its neutrino. Indeed, simultaneous with the discovery of the Υ, evidence for a new lepton was already mounting.

Fig. 37.3. Experimental evidence for the production of explicit bottom hadrons. An excess of electron production at the beam energy of an Υ state suggests that it is decaying into bottom mesons which then produce the electrons in their own weak decays.

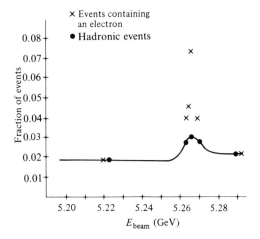

37.3 The tau heavy lepton

In 1975, at the time of the e^+e^- charm experiments, a team of physicists led by Martin Perl, also working on the SPEAR ring at SLAC, reported the existence of 'anomalous μe' events occurring in e^+e^- reactions. They suggested that these might signal the existence of a new heavy lepton, denoted τ. The 'anomalous μe events' are reactions of the form

$$e^+e^- \rightarrow e^{\pm}\mu^{\mp} + \text{missing energy}$$

and the suggested origin of the final state is that of the separate electronic and muonic decays of the new intermediate pair of heavy leptons (Figure 37.4).

It took some time to establish the truth of Perl's suggestion, due to several complicating factors. The most serious of these was that the energy threshold for the production of the $\tau^+\tau^-$ pair is approximately 3.6 GeV (implying a mass for the τ of about 1.8 GeV). This of course, is very close to the threshold of 3.72 GeV required for the production of a charmed meson pair $D^0\bar{D}^0$. As we know, these must decay by the weak interaction and so can quite easily be confused with tau heavy-lepton production and decay. However, in the case of charmed mesons, one would generally expect other hadronic tracks to be present. It is extremely unlikely that charmed-particle decays will give rise to the final state detected by Perl.

The problems were in ensuring that absolutely no other charged particles had been produced and had slipped past the detectors, or that the electrons and muons detected were in fact hadrons confusing the detectors (misidentification is always possible).

Eventually, Perl was able to place his identification beyond doubt, the final evidence for this being the production of the μe states *below* the threshold for charmed meson-pair production (possible because of the slightly lesser mass of the τ). The evidence for this production is shown in Figure 37.5, which shows the growth of the process away from its theoretical threshold. The shape of the energy dependence also establishes the spin of the τ to be $\frac{1}{2}$, like that of the electron and the muon, and in contrast to the spin-0 D mesons. Since its confirmation, all the evidence has supported the identification of the τ as being a very massive copy of the muon (which is itself simply a massive copy of the electron). Like all leptons, the τ does not experience the strong force as do the quarks. However, the τ does have one new feature compared with the muon and the electron. Because of its large mass, it can decay into hadrons and this means that it can add up to one unit to the ratio R, as defined previously, above its production threshold, over and above the value predicted by the charges of the quarks.

Fig. 37.4. Production of a $\tau^+\tau^-$ heavy-lepton pair in e^+e^- annihilation gives rise to an 'anomalous' μe final state.

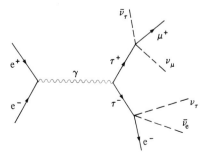

Fig. 37.5. The growth of $\tau^+\tau^-$ production from threshold as signalled by the ratio of candidate events (those containing an electron and some other charged particle only, eX) to known $\mu^+\mu^-$ production. Note that J denotes the spin of the τ.

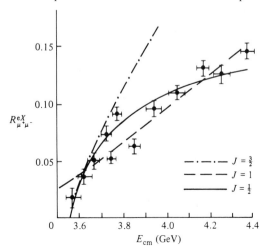

Table 37.1. *The generations of fundamental particles are almost complete. Only the top quark is missing*

Generation	First	Second	Third	Charge
Quarks	u	c	$\widehat{t}$	$\frac{2}{3}$
	d	s	b	$-\frac{1}{3}$
Leptons	e^-	μ^-	τ	-1
	ν_e	ν_μ	ν_τ	0

Apart from this new feature, the τ behaves exactly like its less-massive relatives during interactions. Despite its mass, it shows no deviation from point-like behaviour down to the current experimental limit of 10^{-18} m, and provides us with no hints that the leptons themselves may be composites of even smaller particles.

Further experiments have determined that, just like the electron and the muon, the tau has its own tau-type neutrino, and that tau number is conserved during weak interactions. The advent of the tau-neutrino completes the third generation of fundamental leptons, as shown in Table 37.1. However, as far as the quarks go, there is still one undiscovered member of the third generation, viz. the top quark. Few physicists doubt its existence, but it has so far proven very elusive. It is generally expected to be heavier rather than lighter, and most estimates put its mass in the region of about 100 to 250 GeV.

It is an amusing historical diversion to note that the first generation was begun in 1897 by J. J. Thomson with his discovery of the electron and concluded between 1963 (with Gell-Mann's introduction of the u and d quarks) and, let us say, 1968 (with their shadows being detected in deep inelastic experiments). The second generation was begun in 1937 with Anderson's (mis)identification of the muon and completed in 1974 with the evidence for the charmed quark. The third generation began in 1975 with Perl's first evidence for the tau and is still awaiting completion in the discovery of the sixth, top quark. It will be surprising if this generation takes anything like the length of time needed to fill in the other two!

38

e^+e^- today and tomorrow

38.1 Introduction

The Stanford Linear Collider (SLC) in California and the Large Electron–Positron (LEP) storage ring at CERN in Switzerland are the newest and most powerful of the world's e^+e^- colliders. In 1989, both machines began producing vast numbers of Z^0 gauge bosons with the aim of making precision measurements of the electroweak theory. Both groups were quickly rewarded with the almost simultaneous announcement, in October of the same year, that measurements of the Z^0 decay width were precise enough to establish that there are only *three* generations of light fermions. But despite the glamour attached to these huge accelerators, some of the credit for this important result belongs to their predecessors.

In the 1980s in particular, three e^+e^- colliders contributed a great deal to our understanding of the standard model: PETRA at DESY, near Hamburg in Germany; PEP at SLAC in California; and TRISTAN, located near Tsukuba in Japan. PETRA was commissioned in 1978 and finally shut down in 1986. Figure 38.1 shows its basic layout and how previous generations of storage rings were used as pre-accelerators. Each of the five experiments around the ring consisted of a very large array of detectors of various types wrapped around the interaction regions, with each configuration tailored for the particular experiment. As many as nine separate institutions were involved in the experimental collaborations. PEP

Fig. 38.1. The layout of the PETRA ring at DESY.

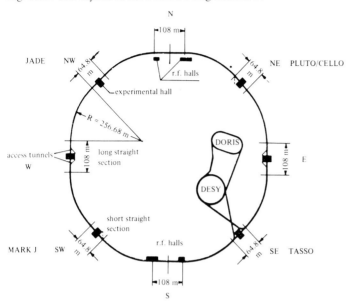

was completed in 1980 and is a rather similar machine, with a circumference of 2.2 km, just 0.1 km smaller than PETRA, and a maximum collision energy of 30 GeV. TRISTAN, on the other hand, has a circumference of just over 3 km, and a maximum collision energy of 75 GeV. Commissioned in 1986, it has extended many of the PEP and PETRA studies to higher energies.

38.2 Hadronic processes

Electron–positron collisions in which hadrons emerge in the final state are an important class of the reactions studied at PETRA, PEP and TRISTAN. Measurements of R (i.e. the ratio of $e^+e^- \rightarrow$ hadrons to $e^+e^- \rightarrow \mu^+\mu^-$) have now been extended up to almost 60 GeV. We noted earlier that the ratio R is a constant given by the sum of the squares of the quark charges. However, this is only the simplest approximation to its value. According to QCD, it is possible that, at high energies, one member of the outgoing quark–antiquark pair will radiate a gluon. As the energy is increased, this process becomes increasingly more probable and spoils the constancy of R above about 40 GeV. Figure 38.2 shows the process concerned and the effect it has on the value of R. At slightly higher

energies, there are also modifications to R coming from the electroweak theory. In particular, processes in which the incoming e^+e^- pair annihilate into a virtual Z^0 (instead of a photon) become important. Nevertheless, the experimental results are perfectly compatible with both of these modifications, and further strengthen our faith in QCD and the electroweak theory.

38.2.1 *Jets*

With PETRA and PEP, new evidence became available to support the reality of the quark and gluon sub-processes which were believed to mediate the process $e^+e^- \rightarrow$ hadrons. This evidence was in the form of jets of hadrons, which are the footprints of emerging quarks and gluons. Near the threshold energy for the production of a heavy quark–antiquark pair (e.g. $b\bar{b}$), the quark and antiquark are produced almost at rest. So, when they fragment into hadrons, they do so isotropically, i.e. the hadrons emerge in all directions. However, as the beam energy is increased, the quark and antiquark are produced with very large momenta, moving in opposite directions. The fragmentation into hadrons then takes place, preferentially along the direction of motion of the quark and antiquark, resulting in jets

Fig. 38.2. Gluon radiative corrections to q̄q production, and the variation in R predicted as a result by QCD.

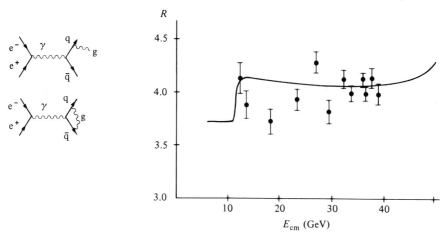

of hadrons which become more and more collimated as energy is increased (see Figure 38.3(*a*)).

An important property of these hadronic jets is that the jet axes correspond to the directions of the outgoing quarks. This is significant because the directions in which the quarks emerge depend upon their spin. Thus the measurement of the angular distribution of jet axes provides a means of determining quark spin. For any particular event, the jet axes can be precisely determined from the observed hadron tracks with the aid of a computer. Figure 38.3(*b*) shows the inferred angular distribution of jets from a large number of events. The result is a curve proportional to $(1 + \cos^2 \theta)$, where θ is the angle between the incoming beam and a particular jet. This is just what is expected for spin-$\frac{1}{2}$ quarks, thus confirming our earlier spin assignment.

As we have mentioned, QCD predicts a host of more complex processes which become increasingly significant at higher energies. One such process involves the outgoing quark or antiquark radiating a gluon, which forms a separate jet of its own. The result is a three-jet event like that shown

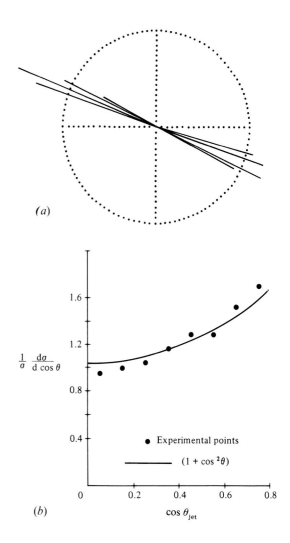

Fig. 38.3. (*a*) An obvious two-jet event in high-energy e⁺e⁻ annihilation, resulting from the emergence of a quark–antiquark pair. (*b*) The angular distribution of jet axes emerging from e⁺e⁻ collisions follows the $(1 + \cos^2 \theta)$ law expected from the production of intermediate spin-$\frac{1}{2}$ quarks.

Fig. 38.4. A three-jet event in e^+e^- annihilation resulting from a quark–antiquark pair plus a radiated gluon.

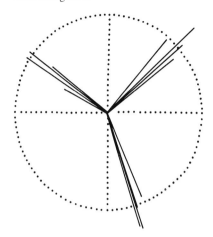

in Figure 38.4, which provides firm dynamical evidence for the reality of gluons. Furthermore, measurements of the angular distribution of the jet axes confirm that gluons are indeed spin-1 particles.

Another sort of three-jet event also supports the existence of gluons. We have seen previously that the only way in which the psi meson (J/ψ) can decay is via three gluons. But, because its mass is comparatively small, there is little hope of discerning any jet structure. However, the heavier upsilon meson (Υ) also decays into three gluons and detection of the resultant jets is a far more feasible proposition. These jets are not as obvious as the true high-energy jets discussed above, but a clear three-jet structure can indeed be discerned.

38.3 The weak force in e^+e^-

In all the processes described above, it was assumed that the electron–positron pair annihil-ates into a virtual photon. In fact, another possibility is that the pair may annihilate into a virtual Z^0 boson, which is then able to probe the 'weak' (as opposed to electromagnetic) content of the vacuum. But because the real Z^0 is very massive (91 GeV), the Z^0s produced at PETRA, PEP and TRISTAN are very virtual (or, 'far off mass-shell', to use the technical phrase). This means that their effects are diluted by the dominant photon-annihilation channel. However, as the collision energy is increased, the virtual photon is moved further off its mass-shell (i.e. $E = pc$ is more ser-iously violated), whilst the virtual Z^0 is moved closer to its mass-shell (i.e. $E^2 = p^2c^2 + M_{Z^0}^2c^4$ is less seriously violated). So, the weak interaction effects of the Z^0 become increasingly more import-ant compared with the electromagnetic effects of the photon. In larger accelerators, such as LEP or the SLC, the collision energy can actually be set equal to M_{Z^0}, and real Z^0 bosons are produced at rest. At this energy, weak interaction effects are dominant in shaping the gross features of the final state.

However, even at PETRA, weak interaction effects could be discerned. As we have seen, the primary distinguishing feature of the weak force is parity violation. For the process $e^+e^- \rightarrow \mu^+\mu^-$, this means that the outgoing muons are not distrib-uted symmetrically about the interaction regime, as would be the case if only the electromagnetic force were present. In fact, the electroweak theory predicts that slightly more μ^+s should emerge in one hemisphere than in the other (with the reverse asymmetry for the μ^-s, as in Figure 38.5). This prediction was confirmed at PETRA in 1981. Other weak interaction effects have since been observed and found to be in complete agreement with the Glashow–Weinberg–Salam model.

Fig. 38.5. The annihilation of the e^+e^- pair into a Z^0 boson gives rise to an observed asymmetry in the distribution of similarly charged muons.

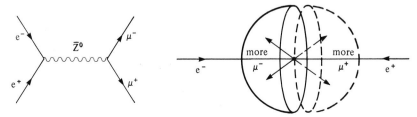

38.4 The SLC

In the early morning of 11 April 1989, a novel particle accelerator at the Stanford Linear Accelerator Center in California recorded the first ever Z^0 produced in an e^+e^- collider (see Figure 38.6). However, more remarkable than the event itself is the machine that produced it: the Stanford Linear Collider, or SLC.

From its conception some ten years earlier, the SLC was always considered a daring project. Far from relying on proven accelerator technology, it necessitated the development of ambitious new technologies, which caused more than just a few headaches for Burton Richter, the direc-

tor of SLAC. Unlike LEP and other conventional storage-ring colliders, the SLC is a single-pass collider: the two beams are lost after one collision. The machine employs the 2-mile-long Stanford linear accelerator to accelerate bunches of electrons and positrons to 50 GeV, before they are steered around separate arcs and brought into a head-on collision at a combined energy of 100 GeV (Figure 38.7). Surrounding the collision point is a state-of-the-art particle detector known as the SLAC Large Detector, a view of which is shown in Figure 38.8.

Despite the ambitious scope of the project, there were very good reasons for choosing this

Fig. 38.6. A computer reconstruction of the first Z^0 boson detected at the SLC. (Photo courtesy SLAC.)

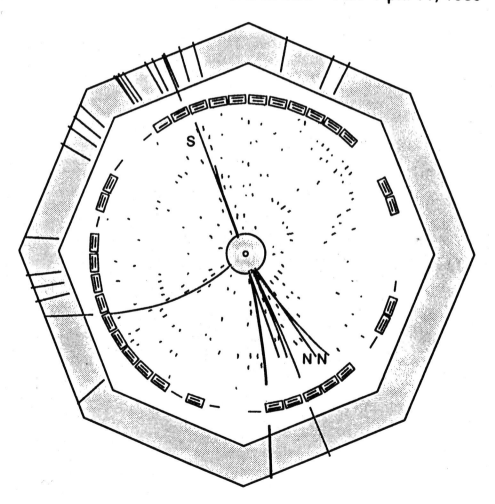

Run 17723 Event 1493 First Z at SLC 7:37 April 11, 1989

novel design. Linear accelerators are on the whole free from the problem of synchrotron radiation, which saps energy from charged particles moving along curved paths. With storage rings, this energy loss accumulates with every revolution, and very high energy machines, such as LEP, must have extremely large rings to minimise the loss. However, with the SLC, there is only a small, one-off energy loss as the beams are steered into collision. Hence, a linear accelerator offered a smaller and cheaper alternative. Secondly, the fact that a linear accelerator was already available, provided SLAC with the opportunity to complete the SLC in time to compete with LEP. In the event, they got there with four months to spare.

Fig. 38.7. The layout of the SLC at Stanford.

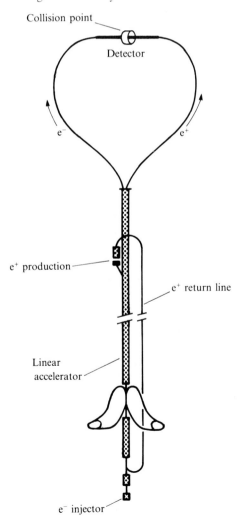

38.5 LEP

Since August 1989, the attention of the particle physics community has been focussed on the largest scientific instrument ever built. With a circumference of 27 kilometres, the Large Electron–Positron (LEP) collider, straddling the Franco–Swiss border at CERN on the outskirts of Geneva, took six years and 1.2 billion Swiss francs to complete.

The 3.8-metre-diameter LEP tunnel, which houses the particle beams, lies some 100 to 150 metres below ground level, and is inclined at an angle of 0.8° to the horizontal, in order to minimise the depth it penetrates under the Jura mountains (Figure 38.9). The enormous scale of the project can be best appreciated from Figure 38.10, in which the LEP and SPS rings are superimposed on an aerial view of the surrounding countryside. A ring 27 km in size is necessary in order to keep the energy loss through synchrotron radiation to a minimum. The rate at which a charged particle loses energy by synchrotron radiation is proportional to $(E/m)^4/R^2$, where E and m are the particle's energy and mass, and R is the radius of the ring. With beams of 50 GeV (half the design energy), LEP loses some 400 MeV per revolution; but when the energy is doubled to 100 GeV per beam the energy loss will amount to over 3 GeV per revolution.

Electrons and positrons are injected into the LEP ring after having been pre-accelerated to 20 GeV by the SPS. Then, circulating in opposite directions, the e⁺s and e⁻s are guided around the ring by 3328 bending magnets, and kept in tight, pencil-like beams by a further 1272 focusing magnets. At four equally-spaced points, the counter-rotating beams pass through each other, initiating a series of e⁺e⁻ collisions which are monitored by four huge detectors each weighing in excess of 3000 tonnes. The four detectors, OPAL, ALEPH, DELPHI and L3, built and operated by international collaborations which include non-member states such as the USA, incorporate many different features designed to complement each other.

The construction of LEP was an organisational, technological and engineering miracle. That the project was completed on time and on budget is truly remarkable. A fitting testimony to this miracle was the 'pilot run' – the first operation with counter-rotating e^+e^- beams – which took place on 14 August 1989. With the individual beam energies set at 45.5 GeV, it took only 16 minutes before the first Z^0 boson was observed! By contrast, the SLC took almost two years to produce its first Z^0, owing to unforeseen difficulties with unproven accelerator technology.

Currently LEP is in its first stage, running at a maximum beam energy of 55 GeV, leading to maximum centre-of-mass collision energies of 110 GeV. However, over the next few years, the

Fig. 38.8. A beam's-eye view of the SLAC Large Detector which surrounds the collision point of the SLC. (Photo courtesy SLAC.)

Fig. 38.9. A schematic view of the LEP collider at CERN.

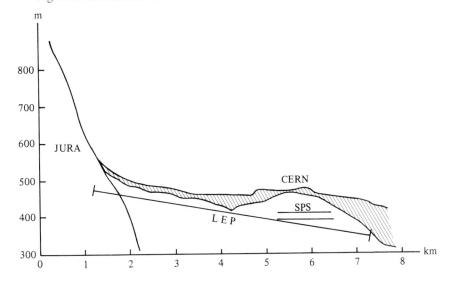

Fig. 38.10. Aerial view of CERN with the Jura mountains in the background. The two large circles, which are respectively 27 km and 7 km in circumference, indicate the line of the LEP and SPS accelerators. (Photo courtesy CERN.)

128 copper r.f. accelerating cavities will gradually be replaced by superconducting niobium r.f. cavities, enabling the beam energy to be doubled. In the meantime, LEP is being used essentially as a Z^0 factory, facilitating very accurate determinations of its mass and decay width.

38.6 Outlook

Let us conclude this discussion with a summary of what we have learnt from e^+e^- collisions over the past two decades. Most spectacular perhaps was the discovery of charm and of the whole host of associated new particles which were identified as quark–antiquark bound states. This was followed by the confirmation of the existence of the bottom quark and its associated $b\bar{b}$ bound states, which were the first manifestations of the third generation of fermions (quarks and leptons). Shortly afterwards, the τ lepton was identified as the leptonic member of the third generation.

In addition to telling us much about the electroweak properties of quarks and leptons, e^+e^- collisions have also provided a great deal of information about hadronic structure and the strong interactions. First, the constancy of the ratio R provided further evidence for partons as the point-like constituents of hadrons. Secondly, the angular distribution of events with two hadronic jets offered proof that the charged partons have spin $\frac{1}{2}\hbar$. Thirdly, the value of R can only be understood if the charged partons are quarks with fractional charges. And finally, the angular distribution of three-jet events provided additional evidence for the existence of vector (i.e. spin-1) gluons.

While the world's e^+e^- colliders continue to make ever more precise measurements of the detailed predictions of the standard model, particle physicists await a more exciting and important confirmation of the theory: the discovery of the elusive Higgs boson and top quark. Yet even if both of these particles lie beyond the energies accessible to current accelerators, we should not despair. Many important discoveries in physics (e.g. of charm) were at the time great surprises, so the probing of ever-increasing energies always offers the possibility of truly significant, yet wholly unexpected discoveries.

Part 9
Research in progress

| | **39** |
| | *Grand unified theories* |

39.1 Introduction

The success of the Glashow–Weinberg–Salam model of the electroweak force, and the evidence supporting QCD, have reduced the scope of particle physics to two sets of particles (the leptons and the quarks) and two forces (the electroweak and the strong), ignoring gravity for the time being. Together, the electroweak theory and QCD constitute what has become known as 'the standard model' of particle physics. The standard model has been remarkably successful in accounting for all experimental phenomena. There is no indication whatsoever that the theory is wrong, or breaks down at any level. However, it is not expected to be the ultimate theory of the microworld because of its great complexity and because of the many questions it leaves unanswered. These objections seem to suggest that there may be deeper symmetries underlying the standard model, leading perhaps to the unification of the strong and electroweak interactions into a single 'grand unified theory', or GUT. The distinct forces we observe would then be merely different low-energy manifestations of a single grand unified force.

Even at low energies, there is a strong resemblance between quarks and leptons. They are point-like objects, are arranged in three generations, and are treated in exactly the same way by the electroweak interactions. In addition, the fundamental forces between them share a common description in terms of gauge theories, which, moreover, seem to predict that the strengths of the interactions converge to the same value at very short distances (or very high energies). These observations hint at both a unified description of quarks and leptons, and a unified formulation of the strong and electroweak interactions – i.e. a grand unified theory.

As well as the obvious aesthetic appeal, there are also practical reasons for wishing to formulate a unified theory. Foremost is the desire to explain the quantisation of electric charge and why the charge on the electron should be *exactly* opposite to that on the proton (and so an exact multiple of the $\frac{1}{3}$-charges on the quarks). Another reason is to reduce the number of 'free' parameters which must be introduced into the theoretical picture. In the separate theories, 19 parameters (charges, masses, mixing angles, etc.) must be introduced and their values determined experimentally. In a unified theory many of these will be determined by self-consistency within the theory.

The general approach to the unified theory is suggested by the success of the principle of gauge invariance applied to both weak and strong forces. Also, the unification of the very dissimilar electromagnetic and weak forces by the use of spontaneous symmetry breaking suggests that this too will play a role in our grand unified theory. Let us briefly recap this mechanism.

The introduction of Higgs bosons allows the electroweak gauge bosons to be given different masses ($M_{W^\pm} = 80$ GeV, $M_{Z^0} = 91$ GeV, $M_\gamma = 0$). At interaction energies much higher than the gauge boson masses (say 10^4 GeV), the masses are unimportant and all the gauge bosons are easily produced. This means that the forces they mediate all appear equally important and the $SU(2) \times U(1)$ electroweak symmetry is manifest. At energies comparable to the gauge boson masses (about 10^2 GeV), the masses can no longer be neglected and it becomes difficult to produce the $W^\pm$ and Z^0.

Finally, at energies less than the gauge boson masses, the $W^\pm$ and Z^0 cannot be produced (because of energy conservation) except as virtual particles. Here, the short-range nature of the weak force is apparent: the uncertainty principle means that energy and momentum conservation can only be violated over a short distance. So, at distances greater than 10^{-16} cm (corresponding to energies below 10^2 GeV), the weak force is much weaker than the electromagnetic force, and the electroweak symmetry is broken. Only the Abelian electromagnetic symmetry remains.

39.2 The structure of a GUT

We have seen that the correct gauge bosons for the electroweak force are generated by requiring that the Lagrangian describing the interactions of leptons be invariant under the $SU(2) \times U(1)$ group of transformations. Similarly, the gluon structure of QCD is generated by requiring the quark Lagrangian to be invariant under the $SU(3)_C$ group of transformations. So, the gauge-boson structure of a GUT will be generated by requiring the total Lagrangian describing the interaction of quarks *and* leptons to be invariant under some grand symmetry group which contains $SU(2) \times U(1)$ and $SU(3)_C$ as sub-groups.

There are several possible choices for the grand symmetry group and each gives rise to a distinct particle structure of its own. The simplest is the theory based on $SU(5)$ symmetry. In this theory the basic entity is a 5-component vector (denoted **5**) containing the right-handed components of each of 3 colours of the down quark, the positron and the electron antineutrino (which exists only in a right-handed state). The right-handed components of the remaining particles of the first generation are contained in a 10-component multiplet (which is denoted **10***), with the left-handed states contained in conjugate multiplets **5*** and **10** (see Figure 39.1). The second and third generations reside in multiplets of the same

Fig. 39.1. The five- and ten-component multiplets of the simplest $SU(5)$ grand unified theory. These are repeated for the second and third generations.

$$\mathbf{5} = [(d^r, d^g, d^b), e^+, \bar{\nu}_e]_R \; ; \mathbf{10^*} = [(u^r, u^g, u^b), (\bar{u}^r, \bar{u}^g, \bar{u}^b), (d^r, d^g, d^b), e^-]_R$$

$$\mathbf{5^*} = [(\bar{d}^r, \bar{d}^g, \bar{d}^b), e^-, \nu_e]_L \; ; \mathbf{10} = [(\bar{u}^r, \bar{u}^g, \bar{u}^b), (u^r, u^g, u^b), (d^r, d^g, d^b), e^+]_L$$

$SU(5)$ structure: $(\mathbf{5} + \mathbf{10}^*) + (\mathbf{5}^* + \mathbf{10})$. An $SU(5)$ transformation on the multiplets corresponds to the redefinition of what is a quark and what is a lepton. Requiring the Lagrangian to be invariant under the local group of $SU(5)$ transformations, then, leads to the introduction of a 24-plet of gauge bosons. Of these, 12 are familiar (the photon, $W^\pm$, Z^0 and 8 gluons). The remaining 12 are new bosons with charges $\pm\frac{4}{3}$, $\pm\frac{1}{3}$ and denoted X; these carry new forces which can transform quarks into leptons and vice versa (see Figure 39.2).

Even at this stage, charge quantisation can be seen as a natural consequence of this $SU(5)$ symmetry. Firstly, the symmetry demands that the average charge of the basic entity be zero and, secondly, that the charge can change only by amounts carried by the gauge bosons (i.e. by strict multiples of $\frac{1}{3}e$). The theory thus accommodates the relationship between quark and lepton charges.

To the present day, no force transforming quarks into leptons or leptons into quarks has ever been observed. This is explained by the spon-

Fig. 39.2. The $SU(5)$ symmetry of the GUT causes transformations between the members of the basic **5** multiplet (the right-handed components of the three colours of the down quark, d, the positron and the antineutrino). The gauge bosons include the eight gluons (g) of the QCD $SU(3)_C$ theory and the γ, $W^\pm$, Z^0 bosons of the $SU(2)_L \times U(1)$ electroweak theory.

	d_R^{red}	d_R^{green}	d_R^{blue}	e_R^+	$\bar{\nu}_e$
d_R^{red}	g^0, γ, Z^0	g^{r+g}	g^{r+b}	$x_{-\frac{4}{3}}^{red}$	$x_{-\frac{1}{3}}^{red}$
d_R^{green}	g^{g+r}	g^0, γ, Z^0	g^{g+b}	$x_{-\frac{4}{3}}^{green}$	$x_{-\frac{1}{3}}^{green}$
d_R^{blue}	g^{b+r}	g^{b+g}	g, γ, Z^0	$x_{-\frac{4}{3}}^{blue}$	$x_{-\frac{1}{3}}^{blue}$
e_R^+	$x_{\frac{4}{3}}^{red}$	$x_{\frac{4}{3}}^{green}$	$x_{\frac{4}{3}}^{blue}$	γ, Z^0	W^+
$\bar{\nu}_e$	$x_{\frac{1}{3}}^{red}$	$x_{\frac{1}{3}}^{green}$	$x_{\frac{1}{3}}^{blue}$	W^-	Z^0

taneous breaking of the $SU(5)$ symmetry by a suitable multiplet of Higgs fields, denoted H. By this means the correct masses for the photon and the $W^\pm$, Z^0 bosons are generated, whilst the masses of the X bosons turn out to be $M_X \approx 10^{15}$ GeV. These super-heavy particles lie many orders of magnitude beyond the energy ranges of any conceivable accelerator and even beyond the high energy-densities in any known region of the universe. (However, they were present in great abundance in the first 10^{-35} seconds after the Big Bang when the temperature was higher than 10^{28} Kelvin, which is equivalent to 10^{15} GeV). So, they are unlikely ever to be observed directly. Also, their enormous mass means that the forces they mediate will be very unimportant indeed compared with those forces whose quanta are exchanged freely at the energies of interaction concerned.

The grand unified theory therefore incorporates a hierarchy of spontaneous symmetry breaking which reflects the transition from a very high energy simplicity to the low-energy complexity we observe. At energies well above 10^{15} GeV, all gauge bosons (including the Xs) can be produced freely and all forces are apparent; quarks transform into leptons as easily as they change colours. The grand $SU(5)$ symmetry is manifest. At an energy of about 10^{15} GeV, the $SU(5)$ symmetry breaks down to separate $SU(3)$ and $SU(2) \times U(1)$ symmetries and the grand unified force separates into the strong colour force and the electroweak force, whilst the 'new' quark–lepton transforming force becomes unimportant. For energies between 10^4 and 10^{15} GeV, the strong and electroweak forces exist with little interaction between quark and lepton sectors. At about 10^2 GeV, the $SU(2) \times U(1)$ symmetry becomes broken, reflecting the separation of the electroweak force into the distinct weak and electromagnetic forces.

This picture of the unification of forces also incorporates the variation in the strengths of charges, depending on the distance from which they are viewed. Recall that the discovery of asymptotic freedom in QCD was the first indication of the importance of this effect. The colour charge of a quark is spread out on the sea of virtual gluons and quark–antiquark pairs surrounding the quark. So the closer the quark is approached (the higher energy with which it is probed), the less the colour charge seems to be. A similar, but less pronounced, weakening affects the strength of the weak charge on a lepton, whilst, as we mentioned in Chapter 32, the Abelian nature of electromagnetism gives rise to an increase in the effective strength of the electric charge as it is approached.

The relative strengths of the forces, as measured at current experimental distances (shown in Table 5.1), and the forms of variations, as expressed as above, lead to the proposition that there is a particular distance at which all three forces have the same strength. In fact, it is possible to calculate this distance, and in $SU(5)$ theory the answer is remarkably small: 10^{-29} cm. Consequently, the energy required to probe this distance is extremely large, about 10^{15} GeV. This is just the energy at which the grand symmetry group $SU(5)$ becomes relevant and X bosons freely mediate quark–lepton interconversions. The variation in the strengths of the forces with distance is shown in Figure 39.3: from the unique strength of the single unified force experienced at energies above 10^{15} GeV, to the three different strengths associated with the three separate forces observed below 10^2 GeV.

Fig. 39.3. Charge shielding and antishielding effects in quantum theory predict the equality of coupling constants (force strengths) at very high energies, about 10^{15} GeV (or very small distances, about 10^{-29} cm). The discontinuity at 10^{-16} cm is due to the dissociation of the unified electroweak force into electromagnetism and the weak nuclear force (which does not act over distances greater than 10^{-16} cm).

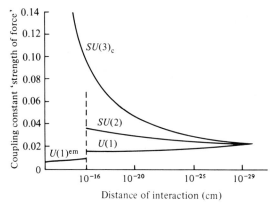

Grand unified theories also permit the calculation of other quantities, which provide an indication of the success of the unification concept. Firstly, the unification of coupling strengths allows the ratio of the $U(1)$ coupling constant to the $SU(2)$ coupling constant in the electroweak theory to be calculated, thus determining the weak mixing angle. In the minimal $SU(5)$ theory this gives $\sin^2 \theta_W = 0.214$, which should be compared with the current experimental value of 0.2324 ± 0.0011. Clearly, these values do not quite agree; but that they are so close is a major triumph for grand unification. In fact, one can construct slightly more complicated GUTs with more fields and/or different grand symmetry groups (also incorporating $SU(3)_C$ and $SU(2) \times U(1)$) which fit the experimental value extremely well. Secondly, the ratio of the mass of the bottom quark to that of the tau lepton can be calculated to be about 3, compared with the experimental value of about 2.5.

These calculations form the basis for our confidence in the grand unified approach. But the validity of the approach hinges on experimental detection of one or more of the unique consequences of the theory, yet to be observed.

39.3 Consequences of grand unification

The most dramatic consequence of grand unification is that baryon number is no longer necessarily conserved. A quark (baryon number $\frac{1}{3}$), for example, can be transformed into a lepton (baryon number 0) or antiquark (baryon number $-\frac{1}{3}$). This raises the possibility of the decay of the proton – a particle which had previously been regarded as absolutely stable, being the least massive of the baryons. Proton decay can occur when a u quark emits a virtual X boson to transform into a ū antiquark, the X boson being absorbed by the d quark which transforms into a positron (see Figure 39.4). In this way, both the proton and the neutron can decay to a pion and a positron:

$$p \rightarrow \pi^0 + e^+,$$
$$n \rightarrow \pi^- + e^+.$$

If we then go on to note that the neutral pion decays into two photons, and that the positron may annihilate with any nearby electron again into photons, it leads to the apocalyptic conclusion that the eventual fate of matter is to transform into radiation, and so act like some cosmic torchlight battery.

However, the fact that none of us burns up like a comet on our daily round gives some comfort. It is a very slow process. It is possible to calculate the decay rate of the proton in any grand unified theory. The minimal $SU(5)$ GUT we have been discussing predicts a value for the mean lifetime of the proton of about 10^{31} years. Bearing in mind the age of the universe (i.e. since the Big Bang) is a mere 10^{10} years, the proton is comfortably long-lived. But, despite this fantastically long life, if enough protons are available then we may be fortunate enough to see the odd one decay – perhaps one or two a year in a typical detector. For example, there are about 5×10^{31} protons and neutrons in 100 tonnes of matter, and so according to $SU(5)$ we might expect to see about five instances of proton decay per year for every 100 tonnes of matter.

In recent years, several experiments have been mounted in attempts to detect such decays. Generally, the experiments are set up deep underground in disused mines in an attempt to filter out all spurious events resulting from cosmic rays. The

Fig. 39.4. The decay of a proton into a positron and a pion proceeds by the exchange of a superheavy X boson between u and d quarks.

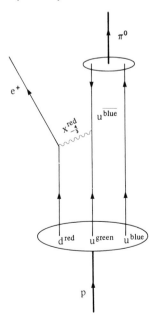

experiments consist of monitoring a large volume of matter, looking for spontaneous reactions which might signal proton decay. However, to date the various groups operating detectors in places as far flung as the USA, Japan, India and Europe have failed to come up with a single unambiguous event. This has led to an experimental lower limit on the proton lifetime of $\tau(p \rightarrow \pi^0 e^+) > 4 \times 10^{32}$ years. This limit, together with the measured value of $\sin^2 \theta_W$, seems to rule out the simplest $SU(5)$ GUT. However, there are many other GUTs which predict correctly $\sin^2 \theta_W$ and a value of the proton lifetime consistent with experiment. So, although the minimal GUT may be dead, the concept of grand unification is still very much alive.

Another consequence of grand unified theories is that neutrinos may, after all, possess a very small mass. In the Glashow–Weinberg–Salam theory of the electroweak force, the unique-handedness of the neutrinos and the particular form of Higgs fields introduced to accomplish the spontaneous symmetry breaking ensure that the neutrinos have no mass. This is true also in the simplest $SU(5)$ GUT. However, it is possible to formulate acceptable GUTs, with a more elaborate configuration of Higgs fields and a larger set of basic particles (such as a right-handed neutrino) which result in a very small mass for the neutrino of between 10^{-3} and 10 eV.

If such a mass is present, then it is possible that neutrinos may oscillate between electron-, muon- and tau-types, thus violating the previously sacrosanct law of lepton-type number conservation. This effect is similar to the K^0–$\bar{K}^0$ mixing we discussed in Chapter 15. That is, the particular mass states of neutrinos may not be identical to the eigenstates of the weak force. This would lead to an oscillation between different neutrino types as a beam of neutrinos propagates through space, just as the K^0 beam in Chapter 15 was seen to transform spontaneously into $\bar{K}^0$.

Because of this effect, one would expect a beam of electron-type antineutrinos (such as emerge from a nuclear reactor) to oscillate into muon-type antineutrinos and thereby give rise to a decrease in the frequency with which electron-type reactions $\bar{\nu}_e + p \rightarrow e^+ + n$ can be observed in the vicinity of the reactor (i.e. relative to rates expected in the non-oscillating neutrino theory). Several experiments are currently looking for neutrino oscillations, and report somewhat contradictory results. The general feeling of the experimental community is, however, that as yet there is no convincing evidence for such oscillations, but neither can they be entirely ruled out.

Although neutrino oscillations imply the existence of a finite mass for the neutrino, the converse is not necessarily true. The neutrino may have a finite mass without giving rise to oscillations. Other types of experiment can attempt to observe the neutrino mass more directly by very accurate measurements of the masses of all the other products of nuclear β-decay. A Russian experiment, running since 1980, still reports a finite mass for the neutrino, but this has not yet been corroborated elsewhere. In fact, this experiment is in conflict with many others which put an upper limit on the mass of the electron neutrino of $m_{\nu_e} < 10\,\text{eV}$.

The possible existence of a very small mass of the neutrino may sound at first to be of rather marginal interest but, in fact, the cosmological consequences of such a discovery are extremely profound. The present number density of neutrinos throughout the universe is about 110 of each type per cubic centimetre. Even a small (but non-zero) mass would mean that neutrinos contribute significantly to the mass density of the universe and therefore play a role in determining the ultimate fate of the cosmos (see Section 42.3). This is only the beginning of a series of connections between particle physics and cosmology which we shall meet in more detail in Chapter 42.

Grand unified theories also predict the existence of super-heavy magnetic monopoles, slightly heavier than X bosons with a mass of $10^{16}\,\text{GeV}$. (Magnetic monopoles will be described in detail in Section 44.4.) In addition to their electromagnetic interactions, GUT monopoles may possibly act as a catalyst in baryon-violating processes and so, for example, speed up proton decay.

39.4 Baryogenesis

In this section, we shall foreshadow a little of the close interrelationship between particle physics and cosmology that will be described in Chapter 42. We shall examine a possible explanation of

why the universe is made of matter and not anti-matter.

It now seems likely that there are no large concentrations of antimatter in the universe, and almost certainly none within our local cluster of galaxies. The question arises naturally of how this has come about, assuming (the only philosophically attractive option) that in the instant of the Big Bang no divine guiding hand arbitrarily decided on one or another (i.e. assuming the initial state to be matter–antimatter symmetric). By matter we mean, of course, ordinary baryonic matter; so, we must explain how a net excess of baryons over antibaryons was generated. In 1967, the Russian physicist (and political dissident) Andrei Sakharov showed that the generation of a net baryon number requires: (1) baryon number violation; (2) **CP** and **C** violation (otherwise, the rates of reactions producing quarks and antiquarks will be equal); and (3) non-equilibrium (otherwise in equilibrium, **CPT** conservation requires the number of baryons and antibaryons to be equal). Grand unified theories certainly have the requisite baryon number violation. If **CP** and **C** violation are also present and if the universe goes through a stage in which reactions occur out of equilibrium, then a non-zero baryon number can be generated. This process is called 'baryogenesis'. Although still very speculative, a convincing chain of reasoning can be advanced.

At a time less than 10^{-35} s after the Big Bang, the temperature of the fledgling universe would have been higher than 10^{28} K, corresponding to an average energy of the material particles more than $10^{15}\,\text{GeV}$ ($\approx M_X$). In this regime the super-heavy gauge bosons, X, and their antiparticles, $\bar{X}$, would have been produced with ease in particle collisions and the equal populations of X and $\bar{X}$ would have remained in thermal equilibrium (i.e. as many would be produced in collisions as would be annihilated):

$$X + \bar{X} \leftrightarrow \text{matter and radiation.}$$

But, as the universe expanded it also cooled. Very soon the temperature would have fallen below M_X and the X and $\bar{X}$ bosons could no longer be produced, as the average collision energy would have been too low. By the same token, they would have been unable to annihilate, as the expansion of the universe destroyed equilibrium. That is, the universe expanded faster than the bosons could interact (see Figure 39.5). Then, the large numbers of X and X̄ bosons would have begun to decay. Because of the presence of the **CP**-violating effect described in Chapter 15, there is no guarantee that the average value of the baryon number of the states into which the Xs decayed would be exactly opposite to that of the states into which the X̄s decayed.

So the **CP**-violating decays of the X and X̄ bosons could generate a net baryon number for the universe from an initial state consisting of equal numbers of Xs and X̄s. As the average energy of the particles in the universe would then have continued to fall, baryon-number violating processes would have become increasingly insignificant and the net baryon number would thus have become 'frozen'. This is generally stated as a ratio of the net number density of baryons n_B to the number density of cosmological Big-Bang photons n_γ (≈ 400 per cubic centimetre). The observed value of this ratio, namely

$$\eta = \frac{n_B}{n_\gamma} = (4 \pm 1) \times 10^{-10},$$

can be reproduced in many GUTs.

Although we have examined an appealingly simple picture of baryogenesis, it should be realised that the application of GUTs to the early universe is highly speculative. We have ignored many possible unknowns. Amongst these, for instance, are the roles which may have been played by the super-heavy grand Higgs bosons (which are a necessary part of all GUTs), primordial black holes (see Section 40.4) and magnetic monopoles. In fact, in many grand unified theories, it is the super-heavy Higgs bosons, H, which play the most important part in baryogenesis; however, the above discussion still applies if we replace X and X̄ by H and H̄.

39.5 Summary

The concept of grand unification is an important one. It offers the attractive prospect of a unified description of the strong, weak and electromagnetic interactions in terms of a single universal force obeying a single grand unified symmetry group. The three separate forces that we observe are merely different low-energy (or long-distance) manifestations of the grand unified force. Clearly, GUTs cannot be the final answer as they do not incorporate gravity. However, their many appealing features – charge quantisation, baryogenesis, prediction of $\sin^2 \theta_W$, etc. – suggest that they may well be signposts pointing to a more complete unified theory.

Fig. 39.5. The number density of X, X̄ bosons relative to the number density of photons. When T falls below M_X, the universe's expansion causes a departure from equilibrium and the X and X̄ cannot annihilate. As the universe cools further, they begin to decay in order to restore equilibrium. (Note: when we write $T = M$, we mean that the temperature is such that the average energy of relativistic particles is Mc^2.)

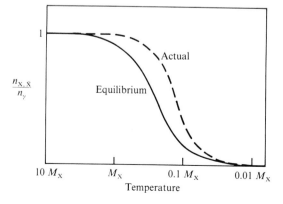

40

Quantum gravity

40.1 Introduction

In Chapter 5, we noted that gravity is a universal force: it is felt by every particle by virtue of its mass and energy. Gravity is also the feeblest of the four known forces by very many orders of magnitude. Nevertheless, it possesses two special properties which give it a very important and conspicuous role. First, it is long range; and secondly, it is always attractive. Together, these properties make the force of gravity cumulative, so that the very weak gravitational forces between the individual particles of, for example, the sun and the earth add up to a very considerable attraction. By contrast, the other three forces are either (1) short range, or (2) sometimes attractive and sometimes repulsive so that the effects tend to cancel out. It is for this reason that gravity, essentially on its own, determines the large-scale evolution of the universe.

General relativity and quantum mechanics are the two great physical theories of this century, and any attempt to formulate a theory of quantum gravity must incorporate them in a fundamental way. However, as they stand, these theories describe two extreme and very different physical regimes. Whereas general relativity describes physics on the largest possible scale – the scale of the universe – quantum theory describes physics on the smallest possible scale – the scale of elementary particles. So, what is the motivation for wishing to unite them in a common framework? The

answer is simple: there are many instances in which both general relativity and quantum theory are equally important and a common framework is essential. One obvious such instance is in the very early universe, immediately after the Big Bang (see Chapter 42). The observed expansion of the universe suggests that in the remote past the universe was very much smaller than it is today (i.e. the typical distance between galaxies was very much smaller). Indeed, naive extrapolation backwards in time shows that, some 10 or 20 billion years ago, the distance between galaxies was zero. At that time, the density of the universe and the curvature of space–time were infinite, implying a total loss of predictability. So, general relativity predicts that there is a point in space–time where the theory itself breaks down. This point is called the Big-Bang singularity. However, before we get back as far as the singularity, there was a time when all the matter in the universe was so close together that the typical cosmological distance was the same as the distances characteristic of quantum processes. At this time, the scales of macroscopic and microscopic physics coalesce, and a unified formulation of quantum theory and gravity becomes essential.

In attempting to quantise gravity along the lines of the other forces, Newton's gravitational constant, G, plays the role of a coupling constant analogous to the fine-structure constant ($\alpha = e^2/\hbar c$) in QED. However, unlike α, which is dimensionless, G has dimensions. In 1899 Max Planck pointed out that a new fundamental length scale, l_{Pl}, and a new fundamental time scale, t_{Pl}, can be formed from the constants G, $\hbar$ and c (see Table 40.1). These fundamental units, called the Planck length and Planck time respectively, define the scales at which both quantum and gravitational effects are important. So, when the length or time scales of interest fall below the Planck values, then a full quantum theory of gravity must be invoked. Note how extremely small these scales are, and how extremely large the associated Planck energy, Planck mass and Planck temperature are.

The early universe is not the only situation in which quantum gravitational effects play a role. According to general relativity, gravity couples to all forms of matter and energy – including gravitational energy. So, gravity couples to gravity itself

Table 40.1. *The Planck units*

Note that k_B is Boltzmann's constant, which relates energy to
absolute temperature on the Kelvin scale.

Planck length	$l_{Pl} = (G\hbar/c^3)^{1/2}$	$= 1.62 \times 10^{-33}$ cm
Planck time	$t_{Pl} = (G\hbar/c^5)^{1/2}$	$= 5.39 \times 10^{-44}$ s
Planck mass	$M_{Pl} = (\hbar c/G)^{1/2}$	$= 2.17 \times 10^{-5}$ g
Planck energy	$E_{Pl} = (\hbar c^5/G)^{1/2}$	$= 1.22 \times 10^{19}$ GeV
Planck temperature	$T_{Pl} = (\hbar c^5/G k_B^2)^{1/2}$	$= 1.42 \times 10^{32}$ K

(cf. the self-couplings of the $W^{\pm}, Z^0$ and gluons which are described by non-Abelian gauge theories). This non-linear nature means that some quantum-gravitational effects are possible at all length and time scales. A great deal of effort has been devoted to investigating such effects far away from the Planck scales, where a semi-classical approach to quantum gravity can be employed.

40.2 Towards a quantum theory of gravity

As mentioned in Section 5.2, no one has succeeded in formulating a consistent quantum theory of gravity. There are subtle and profound problems in reconciling general relativity with quantum mechanics. At the heart of these difficulties are the different roles ascribed by these theories to the concepts of space and time. The conceptual problems are moreover compounded by many technical complexities to do with the symmetries and non-linearity of general relativity.

40.2.1 *General relativity as a gauge theory*

There have been many (unsuccessful) attempts to quantise general relativity along the lines of non-Abelian gauge theories. The essence of these attempts can be summarised as follows.

In the absence of gravity, the laws of physics are invariant under both the Lorentz transformations of special relativity (Section 2.5) and space–time displacements (the laws of physics are the same at different points in space and do not change with time). These transformations are collectively called global Poincaré transformations and together constitute the global Poincaré group (Section 6.2). However, we can go one step further and demand, analogously to the procedure in gauge theories, that the laws of physics not change

under *local* Poincaré transformations which are different at each space–time point. For example, one such transformation consists in shifting each space–time point by an arbitrary amount, so that a regular grid drawn in space–time would become horribly distorted. Then, in order to straighten out the distortions resulting from the local Poincaré transformations, we are led to introduce the concept of the gravitational field. This is analogous to the way in which we were led to introduce the electromagnetic field in order to compensate for the change in the electron's wavefunction under a local gauge transformation in QED and thus restore the gauge symmetry. So, the gravitational field communicates between the different grids (i.e. coordinate systems) which may be chosen at different space–time points.

In this way, general relativity can be regarded as a gauge theory corresponding to the local Poincaré symmetry group. However, it is a gauge theory of a very different kind as it involves a space–time (as opposed to an internal) symmetry group and the gauge quanta are spin-2 bosons (gravitons).

40.2.2 *Semi-classical gravity*

In spite of the difficulties mentioned above, it is possible to develop a semi-classical description of some quantum gravitational effects. We begin by considering classical general relativity. Space–time can be decomposed into a fixed background space–time plus small gravitational disturbances (gravitational waves):

$$(\text{space–time}) = \begin{pmatrix} \text{background} \\ \text{space–time} \end{pmatrix}$$
$$+ \begin{pmatrix} \text{small gravitational} \\ \text{disturbances} \end{pmatrix}.$$

As an example, let us return to the analogy of space–time as a stretched rubber sheet. The shape of the sheet when still is analogous to a background space–time. Tapping the sheet lightly causes vibrations, small ripples, which move over the rubber. These ripples are like gravitational waves travelling in a background space–time. As long as the ripples are small, the rubber sheet keeps roughly the same overall shape.

Recall that Einstein's equations (Section 5.2.1) relate the geometry of space–time (which is actually what we call gravity) to mass and energy (which are sources of gravity). It seems natural, therefore, to reinterpret space–time in such a way that the gravitational waves (which are a form of energy and hence a source of gravity) are included with the sources on the right-hand side of Einstein's equations. The geometry of space–time is therefore specified by the background space–time which remains on the left-hand side. Hence we may rewrite Einstein's equations schematically as

$$\begin{pmatrix} \text{geometry of} \\ \text{background space–time} \end{pmatrix}$$
$$= 8\pi G \times \begin{pmatrix} \text{mass and energy,} \\ \text{including gravity waves} \end{pmatrix}.$$

So far, this is all classical. We shall now take account of the quantum nature of matter (including gravitational waves). That is, we shall consider the sources of gravity – matter and gravitational waves – as quantum fields. The background space–time will continue to be treated as classical. Then we have a semi-classical (and semi-quantum) theory, which might be regarded as a starting point for a full quantum theory of gravity. However, this theory leads to Feynman diagrams which are infinite and, in contrast to QED, cannot be renormalised (see Section 4.8). That the theory is not renormalisable can be traced to the fact that the gravitational coupling, G, is not dimensionless. Nevertheless, if quantum gravitational effects are limited to no more than one loop involving virtual gravitons (Figure 40.1), then this truncated theory provides a useful framework for many calculations. This is analogous to Fermi's current–current theory, which, despite being unrenormalisable, gives an adequate description of the weak interactions at low energies. Of course, it ignores

higher-order quantum gravity effects, but this should be a valid approximation in most cases provided the relevant length and time scales are much greater than the Planck values.

Let us summarise the picture we have built up. The graviton is a massless spin-2 particle which is the quantum of the gravitational field, and hence mediator of the force of gravity. It propagates, along with other quantum fields, in a classical background space–time which satisfies Einstein's equations. Furthermore, as general relativity interprets gravity in terms of space–time geometry, gravitons carry with them small disturbances to the classical background space–time. Despite being an incomplete formulation of quantum gravity, this approach has provided important insights and yielded many interesting results which are expected to form part of the full theory. Examples include the creation of particles in curved space–time, and Stephen Hawking's celebrated discovery that black holes radiate particles.

40.3 Particle creation in curved space–time

Particle creation in curved space–time is an interesting illustration of the effects that gravity can have on quantum fields. In particular, particles can be 'created' by a gravitational field that is changing with time, such as the gravitational field of an expanding universe. Recall our discussion in Section 4.4, in which we considered quantum fields in terms of an infinite number of harmonic oscillators (masses attached to springs). We

Fig. 40.1. Some quantum gravity contributions to e^+e^- scattering involving virtual gravitons propagating in a fixed background space–time.

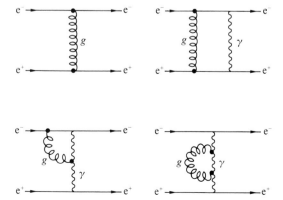

imagined harmonic oscillators at every point in space–time. Now, when the space–time curvature changes as it does in an expanding universe, this is equivalent to effecting a change in the physical properties of the oscillators. Suppose an oscillator is initially in its ground state (i.e. state of minimum energy), undergoing zero-point oscillations (see Section 4.9). Then if one of its physical properties (e.g. the mass, or the stiffness of the spring) is changed, the zero-point oscillations must readjust themselves. After this readjustment is complete, there is a non-zero probability that the oscillator is no longer in its ground state, but in a state of excited vibration corresponding to a particle.

As one might expect, particle production is greatest where the curvature is greatest and changing most rapidly. Moreover, the lightest particles are produced most easily, as they require less energy. But the energy of the created particles does not appear out of nothing: it comes from the gravitational field, that is, from the curvature of space–time itself.

Studies of particle production in various space–times have revealed that the concept of a 'particle' is not universal. It is observer-dependent. Particles may be registered by certain particle detectors and not by others. In particular, the state of motion of a particle detector can have a bearing on whether or not particles are observed. For example, even in flat space–time, a detector which is accelerating detects particles, even when an identical unaccelerating detector does not. In fact, a uniformly accelerated observer detects a thermal distribution of particles at a temperature of

$$T = \frac{\hbar}{2\pi k_B c} a,$$

where a is the acceleration and k_B is Boltzmann's constant. This should not be too surprising. By the principle of equivalence, the effects of acceleration and gravity are identical. So, it is as though the accelerating observer is sitting in a gravitational field, the curvature of which modifies the oscillators associated with any quantum field. Hence, what this observer calls a zero-point oscillation is completely different from what an unaccelerated observer calls a zero-point oscillation. In particular, zero-point oscillations for the latter

correspond to excited states of vibration (i.e. particles) for the former.

40.4 Hawking radiation from black holes

Black holes are massive classical objects predicted by the theory of general relativity. A black hole can be defined as a region of space–time from which nothing – not even light – can escape. Consequently, they look completely black and can only be observed by virtue of the effects of their strong gravitational fields on neighbouring matter. The boundary of a (non-rotating) black hole of mass M is a spherical surface called the event horizon, situated at a radius of

$$r_{Schw} = \frac{2GM}{c^2}. \tag{40.1}$$

This is called the Schwartzchild radius.

Black holes are believed to be the final evolutionary stage in the life of very massive stars. As an example, imagine a very massive star, many times the mass of the sun. For the first billion or so years of its life, this star shines brightly, with heat and light generated by nuclear fusion reactions. The energy released in the fusion processes creates sufficient pressure to stop the star from collapsing under its own gravity. This is a situation of hydro-static equilibrium in which the inward gravitational force is exactly balanced by the outward pressure. However, once the star exhausts all its nuclear fuel, the pressure cannot be maintained and surrenders to gravity: the star collapses. During collapse, the gravitational field at the star's surface increases enormously. Once the radius has fallen to the Schwartzchild value, it is so strong that even light cannot escape: a black hole has been born.

For stars 10 times more massive than the sun, the Schwartzchild radius is approximately 30 km. We now have much observational evidence to support the belief that black holes of about this size exist in binary star systems in which a visible star orbits an unseen companion (a black hole?). Much more massive black holes – 10^8 times the mass of the sun and with a Schwartzchild radius of 300 million km – may act as the power source at the centre of quasars. There is also the possibility that black holes very much lighter than the sun may also exist. Such black holes could not have formed by gravitational collapse – they are much too light. However, they might have formed under the very high temperatures and pressures prevailing in the early universe, shortly after the Big Bang. Such a 'primordial' black hole with the mass of a billion tonnes would have a Schwartzchild radius of about 10^{-13} cm, the size of an atomic nucleus! This is small enough to suggest that quantum-gravitational effects associated with primordial black holes may be significant – as indeed they are.

40.4.1 *Entropy and thermodynamics*

When matter falls into a black hole, its mass increases and, as a consequence of equation (40.1), so does the Schwartzchild radius. Since nothing can escape from beyond the event horizon, its surface area cannot decrease. This is reminiscent of a thermodynamic quantity called *entropy* (see box, p. 186), which is a measure of the degree of disorder in a system, or equivalently, of the lack of knowledge of the state of the system. One of the most fundamental laws in physics, the second law of thermodynamics, states that entropy cannot decrease. This similarity was the first hint of a connection between black holes and thermo-dynamics. It led to the suggestion, in 1972, that the area of the event horizon is a measure of a black hole's entropy. As matter (which carries entropy) falls into a black hole, the horizon's area increases by an amount which is, in fact, greater than the entropy of the in-falling matter. Not only does the black hole swallow up the entropy of the matter itself, but it also swallows up all information about the particular state it was in. For example, whether the matter was in the form of a star, or a cloud of gas, or lumps of rock makes no difference to the black hole. This information is lost and so the total entropy increases.

In fact, there is a mathematical theorem (called the 'no-hair' theorem) which states that, when a star collapses into a black hole, it quickly approaches a state specified by only three quantities: its mass, its angular momentum and its charge. No other details of the star survive. So, in the collapse we lose virtually all knowledge of the state of the star, and entropy increases by an enormous amount. For example, the entropy of a black hole of solar mass is about 10^{54} joules per degree, whereas the entropy of the sun is only about 10^{35} joules per degree.

However, there seemed to be a fatal flaw in the argument. If a black hole has entropy, then it ought to have a temperature, and hence it should emit radiation. But, according to classical physics, *nothing* at all can escape from a black hole. So, how can it possibly have a temperature?

40.4.2 *Hawking radiation*

Then, in 1974, came one of the most surprising and potentially one of the most fundamental results in modern physics. Stephen Hawking was investigating the behaviour of quantum fields in the vicinity of a black hole and found, to his surprise, that indeed black holes seem to emit particles at a steady rate, until they eventually evaporate together. Furthermore, the spectrum of the outgoing particles is precisely thermal: particles and radiation are emitted just as if the black hole were a hot body at a temperature of

$$T_H = \frac{\hbar c^3}{8\pi k_B G M}. \tag{40.2}$$

In this expression k_B is Boltzmann's constant, M is the mass and T_H is called the Hawking temperature. Hence, the identification of the surface area of a black hole's event horizon with its entropy was complete. Table 40.2 gives the Hawking temperatures and approximate lifetimes of black holes of various sizes.

To appreciate how it is possible for black holes to radiate particles, recall the discussion of quantum vacuum in Section 4.9. According to the uncertainty principle, there are, even in empty

Table 40.2. *Properties of black holes of various sizes*

Note that the mass of the sun is $M_\odot = 2 \times 10^{27}$ tonnes

Mass	Schwartzchild radius	Hawking temperature	Lifetime
10^9 tonnes	10^{-13} cm	10^{11} K	10^{10} years
$1 \times M_\odot$	3 km	10^{-7} K	10^{66} years
$10 \times M_\odot$	30 km	10^{-8} K	10^{69} years

space, zero-point quantum fluctuations associated with all quantum fields. These fluctuations may be thought of as pairs of virtual particles and antiparticles which constantly materialise out of the vacuum, briefly separate, and then annihilate. (See Section 33.1.)

In the vicinity of a black hole, the strong tidal forces may lead to one member of a virtual particle–antiparticle pair falling into the hole, leaving the other without a partner with which to annihilate. If the latter does not suffer the same fate as its partner, it becomes a real particle and appears to be emitted by the black hole (Figure 40.2). However, as this is a real process, energy and momentum must be strictly conserved. So, one can view the process in the following way. One member of the particle–antiparticle pair has positive energy and escapes to infinity, while the other enters the black hole with negative energy relative to infinity. The black hole absorbs the negative energy, thus reducing its mass by precisely the amount that the positive-energy particle carries away. The entropy of the black hole is decreased by the reduction in its mass, but this is more than compensated for by the entropy carried away by the emitted particle. Hence, overall, entropy is increased. Furthermore, we would expect the smaller the black hole, the stronger the tidal

forces, the more likely it is for a virtual particle to fall into the hole, and therefore the greater the rate of emission. This is borne out in Table 40.2, where we see that smaller black holes have higher temperatures.

As a black hole radiates, it loses mass and, by equation (40.2), gets hotter. As it becomes hotter, it radiates more particles, loses more mass, etc. There comes a time in this runaway particle emission when the rate of mass loss is so great that the notions of thermal equilibrium and of a fixed background space–time are no longer valid. What happens after this time is not entirely certain. However, it is likely that in the final tenth of a second the black hole should release some 10^{23} joules ($\approx 10^{33}$ GeV) of energy – mostly in γ-ray photons. This is equivalent to a 10^6-megatonne thermonuclear explosion!

One final consequence of black hole radiation is the violation of baryon number and other global quantum numbers. A black hole formed from collapse of a star forgets its baryon number and radiates a thermal spectrum of equal numbers of baryons and antibaryons.

40.5 Quantum cosmology

In the previous sections, we have seen how a semi-classical formulation of quantum gravity leads to interesting quantum effects when applied to black holes and expanding universes. It is apparent that quantum mechanics can modify the predictions of general relativity in very profound and surprising ways. Physicists have been encouraged by these results to conjecture that quantum mechanics might offer a way around the apparently unavoidable conclusion that the universe began from a singularity. Perhaps the true theory of quantum gravity will reveal how the Big-Bang singularity is 'smoothed out' by some quantum-gravitational uncertainty principle. If this is true, then space–time may have been perfectly smooth at the Big Bang and the laws of physics may have been valid even at the beginning of time (see Figure 40.3). In this section, we consider an approach which attempts to transcend the limitations of the semi-classical formulation and describe the large-scale quantum properties of the universe as a whole. This approach is based on

Fig. 40.2. Hawking radiation can be explained in terms of virtual particle–antiparticle pairs in the vicinity of a black hole.

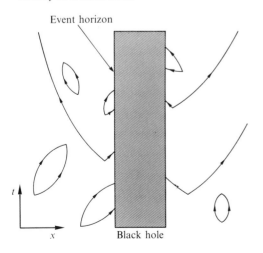

Event horizon

t

x

Black hole

Richard Feynman's formulation of quantum mechanics in terms of a 'sum over histories'.

40.5.1 *The sum over histories*

Let us put gravity aside for a moment to examine Feynman's formulation. Consider a particle which moves from an initial point **y** at an initial time t_i to a final point **x** at a final time t_f. Classically, it follows a well-defined path, or *history*, determined by Newtonian physics. Quantum mechanics, however, tells us that this cannot be the case, since a precisely determined path is inconsistent with the uncertainty principle. We must, therefore, consider paths other than the classical one (Figure 40.4). According to Feynman, a particle does not necessarily move along the classical path, but may take *any* arbitrary path in space–time. So, a particle has an infinite number of possible histories. Corresponding to every path, Feynman assigned a quantum-mechanical amplitude which is a measure of how probable it is for that path to be taken. Then the probability that the particle passes through a particular point at a particular time is found by adding up the amplitudes associated with all the paths passing through that point. This is Feynman's 'sum over histories'.

The reader might be wondering what the connection is with the wavefunction of the particle. Well, we can find the particle's wavefunction at a given time t_0 by simply performing the sum over histories:

$$\psi(\mathbf{x}, t_0) = \sum_{\text{histories, } H} m[H] \tag{40.3}$$

where $m[H]$ is the quantum-mechanical amplitude associated with a history H from an initial point **y** in the infinite past ($t_i = -\infty$) to the point **x** at a time $t_f = t_0$. This is precisely equivalent to solving Schrödinger's wave equation, and, hence, the two formulations of quantum mechanics are equivalent.

Because of technical problems in evaluating the sum over histories, quantum cosmologists usually work with 'imaginary' time. Imaginary time, τ, is related to 'real' time, t, by the formula $\tau = it$, where i is the square root of -1. That is, $i = \sqrt{-1}$, or $i^2 = -1$. In imaginary time, the invariant space–time interval we met in Section 2.8.3, is given by

$$s^2 = x^2 + (c\tau)^2 \tag{40.4}$$

since $\tau^2 = -t^2$. Note that there is now no minus sign distinguishing time from space – they both appear on an equal footing. So, time has become space-like. A space–time with this property is called Euclidean after Euclid, the father of geometry. (The above equation may remind the reader of the relationship between the squares of the sides of a right-angle triangle – a theorem accredited to Pythagoras, but based on Euclidean geometry.)

Fig. 40.3. Space–time can be schematically represented by the surface of a cone, the apex of which corresponds to the Big-Bang singularity, (*a*). At any given time, the universe corresponds to the circle formed by a horizontal slice through the cone. However, quantum mechanics may imply that there is no Big-Bang singularity, (*b*).

Fig. 40.4. All paths contribute to the probability amplitude for a particle to go from **y** (at time t_i) to **x** (at time t_f).

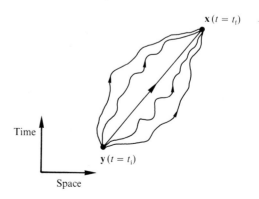

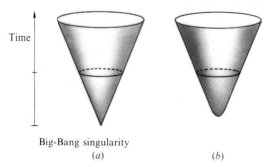

Hence, in imaginary time, four-dimensional space–time becomes four-dimensional Euclidean space.

Physicists hold greatly diverging views on the significance of imaginary time. Some argue that it is merely a mathematical trick, or a convenient tool devoid of any physical significance. Others suggest that imaginary time is the true physical quantity. In fact, the whole question of the meaning of time in quantum mechanics in general is very controversial. Many physicists hold that time is only a semi-classical concept and cannot – even in principle – be extended to the domain of quantum gravity. In this domain, imaginary time may perhaps be the appropriate concept.

40.5.2 *The wavefunction of the universe*

In quantum cosmology, the above ideas are applied not to a particle, but to the universe as a whole. Working in curved Euclidean space (i.e. with imaginary time), the probability that the universe has a particular space–time geometry at a particular time is given by the sum over all possible histories of the universe which lead to that geometry. The classical path, or classical history, is the one specified by Einstein's equations, in the same way that the classical path of a particle is that given by Newton's equations.

Just as one can consider the wavefunction of a particle, so can one also consider the wavefunction of the entire universe, an object which completely specifies its quantum state. (There are a host of interpretational and philosophical difficulties associated with this concept, which take us well into the realm of metaphysics and beyond the scope of this discussion.) The wavefunction of the universe is also given by a sum over histories similar to equation (40.3), and satisfies a Schrödinger-like wave equation which is called the Wheeler–De Witt equation. This equation is, like the sum over cosmic histories, extremely complicated. Every solution to this equation describes a universe. However, we are interested in one particular solution – the one corresponding to our universe – and we must impose suitable boundary conditions on the wavefunction to obtain this solution. That is, to obtain the wavefunction for *our* universe, we must impose boundary conditions

describing its initial state. Unfortunately however, the appropriate boundary conditions are not known.

One proposal, put forward by Stephen Hawking and James Hartle, is that the boundary condition appropriate to our universe is that it should have no boundary. In other words, Euclidean space–time should be like the surface of a sphere (but with two extra dimensions to make it four-dimensional), that is, finite in extent, but without a boundary or an edge. This is possible because in Euclidean space–time, space and time are on equal footing. The no-boundary proposal removes the need for choosing (and justifying) particular boundary conditions. It also suggests that, at least in imaginary time, the Big-Bang singularity is not a singularity at all. This is because a sphere – even a four-dimensional one – is a smooth object with no beginning; it has no sharp points or holes at which the laws of physics might break down. On the other hand, a singularity on the sphere would constitute a boundary to space–time.

However, even if the no-boundary proposal is correct, to find the wavefunction of the universe we are still confronted with the problem of solving the Wheeler–De Witt equation, or equivalently, performing the sum over cosmic histories. Both involve considering an infinite number of degrees of freedom and are presently far beyond our capabilities. Instead, physicists consider only a finite subset of all these degrees of freedom, which, albeit unjustified, at least enables the equations to be solved. It is still too early to tell if this approach will be ultimately successful. The simple solutions which have been found are still too crude and unrealistic. They, nevertheless, provide much-needed insight into this fascinating field, whose frontiers lie well within the realm of yesteryear's metaphysics.

40.6 Summary and outlook

The marriage of gravity and quantum theory in a consistent framework has yet to be achieved, yet we are confident of certain ingredients which should go into such a theory. Firstly, the marriage of the two theories leads to an entirely new set of physical units – the Planck units – based on combinations of the gravitational constant, G, and Planck's quantum constant, $\hbar$. These fundamental units define the scales of length, time and energy at which quantum-gravitational effects become strong and a consistent theory of quantum gravity must be invoked. However, far away from these scales, a semi-classical approach suffices, in which gravity can be interpreted as specifying a classical background space–time together with gravitons which, like other quantum fields, propagate in that background. This has led to the discovery of particle production in curved space–time and Hawking radiation from black holes.

However, the semi-classical picture is not ultimately credible. Physicists feel that the properties of space–time should emerge from the theory – not be imposed as a background structure. Space–time should be an active ingredient and not play the passive role of a background. It follows that the true quantum theory of gravity should predict that the structure of space–time itself be radically altered on the scale of the Planck length. Roger Penrose of Oxford University goes even further and argues that such a theory must also be intrinsically time asymmetric, so that **T** invariance is broken at a fundamental level. Regardless of whether or not this is so, it seems clear that this field, more than any other, requires the input of radically new concepts and ideas.

Entropy

Entropy is essentially a measure of disorder. Consider two containers filled with gas molecules of two types, say A and B. If the containers are connected, the molecules mix. The total system evolves from an ordered state (A and B separated) into a more disordered state (A and B mixed). That is, the system evolves from a state with less entropy into a state of greater entropy. Note that the disordered state can be realised by many more arrangements of the individual molecules than the ordered state. The former, therefore, has a far greater probability of occurring.

Equivalently, entropy is a measure of our lack of knowledge about a system. If a system is in a disordered state, we clearly know less about it than if it is in an ordered state. In the disordered state, the molecules may be in any one of a large number of possible arrangements; in the ordered state, the molecules may be in any one of a much smaller number of possible arrangements. The units of entropy are those of (energy)/(temperature).

41.1 Introduction

In Chapter 39, we saw how grand unified theories can provide a credible framework for the unification of the electroweak and strong forces. This is one possible way in which the theory of the standard model may be modified at high energies. Another possible way is suggested by supersymmetry – a new, as-yet unproven symmetry which, unlike any other, relates particles of different spin.

In attempting to divine the course that physics might take beyond the realm of the standard model, physicists have constructed supersymmetric versions of the Glashow–Weinberg–Salam model, QCD and even GUTs. The advantages of these formulations are twofold. Firstly, supersymmetry solves a fundamental problem: why the electroweak mass scale ($M_{W^\pm}$, $M_{Z^0} \approx 10^2$ GeV) is so very different to the grand unification mass ($M_X \approx 10^{15}$ GeV), or the Planck mass ($M_{Pl} \approx 10^{19}$ GeV). This is often referred to as the 'hierarchy problem'. If the standard model is a part of some larger, more fundamental theory – such as a grand unified theory, or quantum gravity – then the masses of the $W^\pm$ and Z^0 bosons should be predicted by the theory. Moreover, for lack of any reason forbidding it, we would expect these masses to turn out to be very large as well, i.e. $M_{W^\pm}$, $M_{Z^0} \approx M_X$ or M_{Pl}.

Of course, it is conceivable that the numbers may just work out to generate values as small as 10^2 GeV – which is 10^{17} times smaller than M_{Pl}.

But, without a specific reason for it, this would be an incredible accident, requiring coupling constants to be 'fine-tuned' to 17 decimal places!

Supersymmetry, however, leads to delicate cancellations in the computation of these masses in an entirely natural way. Hence, the enormous hierarchy between the electroweak scale and the GUT or Planck scale is an uncontrived feature of supersymmetric models.

The second motivation for supersymmetry is its intimate connection with gravity. If supersymmetry is promoted to a *local* gauge symmetry, then the theory automatically incorporates Einstein's theory of general relativity! Theories with local supersymmetry are, for this reason, called supergravity theories.

41.2 Supersymmetry

Supersymmetry differs from all other symmetries in that it relates two classes of elementary particles which are so fundamentally different: fermions (i.e. particles with spin $\frac{1}{2}$, $\frac{3}{2}$, $\frac{5}{2}$, etc.) and bosons (i.e. particles with spin 0, 1, 2, etc.). These had previously been regarded as two wholly different entities, and so the symmetry linking the two, which emerged in the early 1970s, was entirely novel and unexpected.

According to supersymmetry, every 'ordinary' particle has a companion particle – its superpartner – differing in spin by half a unit, but with otherwise identical properties. Furthermore, the strengths of the interactions of the superpartners are identical to those of the corresponding ordinary particle. The superpartners are given names according to the following rules, examples of which appear in Table 41.1. For the bosonic superpartners of fermions, the prefix 's' is added to the fermion name. So, for example, the spin-$\frac{1}{2}$ electron, e, has a spin-0 superpartner called the *s*electron, ẽ. For the fermionic superpartners of bosons, the suffix 'ino' is added to the boson name. Hence, the spin-1 photon γ, for example, has a spin-$\frac{1}{2}$ superpartner called the photino, $\tilde{\gamma}$. Supersymmetric partners are usually denoted by a tilde (˜) above the appropriate symbol.

In Figure 41.1, we show how the electromagnetic coupling of a photon to two electrons (Figure 41.1(*a*)) is accompanied by similar couplings of the photon to two *s*electrons (Figure 41.1(*b*)) and of a

Table 41.1. *Superpartner nomenclature*

If supersymmetry is exact, a superpartner has identical properties to its corresponding ordinary particle, except for spin.

Particle	Spin	Superpartner	Spin
fermions:		sfermions:	
quark	$\frac{1}{2}$	squark	0
lepton	$\frac{1}{2}$	slepton	0
bosons:		bosinos:	
Higgs	0	Higgsino	$\frac{1}{2}$
gauge boson	1	gaugino	$\frac{1}{2}$
graviton	2	gravitino	$\frac{3}{2}$

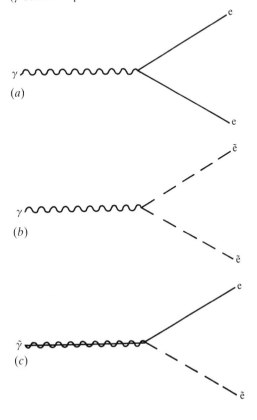

Fig. 41.1. In supersymmetric QED, these three electromagnetic couplings are of equal strength. ($\tilde{\gamma}$ denotes a photino, $\tilde{e}$ denotes a selectron.)

photino to an electron and selectron (Figure 41.1(c)). More generally, the interactions in a supersymmetric version of the standard model can be found from the standard couplings by replacing any two particles by their superpartners. So for example, in a similar way, the coupling of a W boson to an electron and neutrino is accompanied by couplings of (1) the W boson to a selectron and sneutrino, (2) a Wino to a selectron and neutrino, and (3) a Wino to an electron and sneutrino. Other combinations are not possible because an even number of spin-$\frac{1}{2}$ particles must be involved to conserve angular momentum.

41.2.1 *Implications for experiments*

If supersymmetry were an exact, unbroken symmetry, then we would see sleptons with the same masses as the leptons, squarks with the same masses as the quarks, and gauginos with the same masses as the gauge bosons. However, no such particles have ever been observed, and supersymmetry, therefore, if it is a true symmetry of particle physics, must be broken. The breaking of supersymmetry allows superpartners to be heavier than the corresponding particle, which explains why they have not yet been seen. But it also jeopardises the elegant solution to the hierarchy problem. If supersymmetry were exact, then the very large contributions to the masses of the $W^{\pm}$ and Z^0 coming from very large mass scales (M_X or M_{Pl}) naturally cancel. However, once supersymmetry is broken, this cancellation is spoiled – but not totally. The contribution is now proportional to the difference in mass between particles and their superpartners. So in order for the electroweak scale to be naturally (i.e. without fine tuning) at about 10^2 GeV, this mass difference should be less than about 10^3 GeV:

$$|m^2_{\text{particle}} - m^2_{\text{superpartner}}| < (10^3 \, \text{GeV})^2.$$

Hence, supersymmetry predicts a host of new particles with masses of at most 10^3 GeV – just the mass range accessible to the new generation of accelerators.

In many supersymmetric models there is a new conservation law. This law is associated with a quantum number called **R**-parity, which distinguishes ordinary particles from superpartners. The **R**-parity of a particle is determined by its

Table 41.2. **R**-*parity assignments*

Ordinary particles have even **R**-parity, while superpartners have odd **R**-parity. (B, L and s refer respectively to baryon number, lepton number and spin.)

Particle	$3B + L + 2s$	**R**	Superpartner	$3B + L + 2s$	**R**
quark	2	$+1$	squark	1	-1
lepton	2	$+1$	slepton	1	-1
photon	2	$+1$	photino	1	-1
$W^{\pm}, Z^0$	2	$+1$	Wino, Zino	1	-1
gluon	2	$+1$	gluino	1	-1
Higgs	0	$+1$	Higgsino	1	-1
graviton	4	$+1$	gravitino	3	-1

baryon number, lepton number and spin: **R** is equal to $+1$ if $3B + L + 2s$ is an even number, and -1 if it is odd. Hence, **R**-parity is a quantum number similar to parity (space inversion) which we met in Chapter 6. Furthermore, it is even (**R** $= +1$) for all ordinary particles and odd (**R** $= -1$) for all superpartners (see Table 41.2). Conservation of **R**-parity has two immediate consequences. Firstly, superpartners cannot be produced on their own; they must be produced in pairs (the initial **R**-parity is even, so the final **R**-parity must also be even). Secondly, a superpartner must decay into states with an odd number of superpartners (the initial **R**-parity is odd, so the final must also be odd). Both aspects of this conservation law are illustrated in Figure 41.2.

A direct result of **R**-parity conservation is that the lightest superpartner must be a stable particle, since there are no lighter superpartners into which it can decay. The identity of the lightest superpartner depends on the specific supersymmetric model one is using, but it is generally expected to be a combination of the neutral Higgsino, photino and Zino. Of these, the most favoured candidate in many models is the photino. The lightest superpartner may – like the massive neutrino – have very important implications for cosmology. With a mass of 10–10^3 GeV, it may constitute a significant proportion of the total mass density of the universe (see Chapter 42).

41.2.2 *Supersymmetric GUTs*

Supersymmetric grand unified theories are similar in structure to ordinary GUTs. However,

they involve new light particles (the superpartners plus some extra Higgs) which modify the prediction of both the X-boson mass and the weak mixing angle. In the minimal supersymmetric $SU(5)$ GUT, we find a slightly higher value for the mass of the X bosons of $M_X \approx 10^{16}$ GeV. The value of $\sin^2 \theta_W$ is marginally consistent with the experimental value – a slight improvement on conventional GUTs.

The inclusion of supersymmetry also leads to some new mechanisms for proton decay. In contrast to non-supersymmetric models in which the most important decay is into a pion and positron, the dominant decay in supersymmetric models is

Fig. 41.2. **R**-parity conservation means that superpartners are produced in pairs, and all except the lightest must decay into an odd number of superpartners. This diagram is an example of how a $p\bar{p}$ collision produces a squark and gluino which subsequently decay. We assume that the photino is the lightest superpartner.

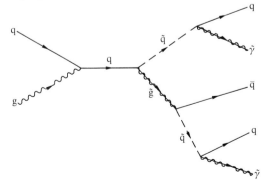

into a kaon and muon-type antineutrino. This decay takes place through a box diagram involving the exchange of a Wino and Higgsino (Figure 41.3). In the minimal $SU(5)$ GUT, this leads to a prediction of the proton lifetime of 10^{26}–10^{31} years, depending on the unknown masses of the superpartners. Unfortunately, this is short compared with the current experimental limit of $\tau(p \to \bar{\nu}_\mu K^+) > 7 \times 10^{31}$ years, which therefore rules out the minimal supersymmetric $SU(5)$ model. However, there are still a whole host of supersymmetric GUTs (with different matter multiplets and/or different GUT symmetries, e.g. E_6, or $SO(10)$) which are consistent with this value, and which, moreover, predict acceptable values for $\sin^2 \theta_W$.

Fig. 41.3. The decay of the proton into a kaon and muon antineutrino proceeds via Wino and coloured Higgsino exchange with intermediate squarks.

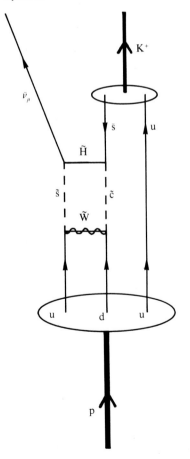

41.3 Supergravity

In the last chapter, we stressed how little we know of the quantum nature of gravity. A consistent theory of quantum gravity has yet to be found. However, for the practical purposes of calculation, we require a field theory describing both gravity and the other interactions, which, although not fully consistent, still has some predictive power. Supergravity is such a theory.

One of the most remarkable features of supersymmetry is that the transformations which change the spin of a particle also shift it in space and time. In particular, two (global) supersymmetry transformations lead to a global space–time displacement, i.e. a global Poincaré transformation. Now, following the gauge-theory approach, let us demand that supersymmetry be a *local* symmetry and that physics be unchanged under *local* supersymmetry transformations at each space–time point. Then local supersymmetry must include the local Poincaré symmetry group, and therefore, by the argument in Section 40.2.1, the theory must incorporate gravity. For this reason, local supersymmetry is called supergravity.

In order to make supergravity invariant under the local 'super-Poincaré' symmetry group (Poincaré group + supersymmetry), we must now introduce a super-gravitational field. This field includes, in addition to the graviton, a further gauge particle: the *gravitino*. The gravitino is the superpartner of the graviton and, unlike any other gauge particle, is a fermion with spin equal to $\frac{3}{2}$. At first sight, this is worrying! We certainly do not want to introduce any new forces which modify the successful predictions of general relativity. Luckily, however, forces corresponding to fermionic quanta have no classical counterpart: they are a purely quantum effect. This is because they obey the Pauli exclusion principle, which stops large numbers of fermions doing the same thing, and so prevents any long-range macroscopic force from building up.

So, what does this new field do? We have already seen how the quantum-mechanical probability that a scattering will occur is found by adding up all the possible exchanges of gauge fields. Hence, in supergravity, we have to add all the possible gravitino exchanges to all the graviton exchanges which we already know about. The result is remarkable. In many of the calculations, the infinite answers which result from graviton exchange are exactly cancelled by new infinities resulting from gravitino exchange! For the first time, we have removed some of the infinities which have plagued quantum gravity for so long. Unfortunately, it is not the case that all infinities are banished by supersymmetry. Despite the great promise it showed initially, supergravity has provided only a first step (albeit one of great importance) towards a renormalisable quantum theory of gravity.

41.3.1 *Extra space–time dimensions*

The everyday world we perceive is manifestly four-dimensional: three spatial dimensions plus time. However, one particular supergravity theory, which at one stage showed great promise as a unified theory of the fundamental interactions, required that space–time have not four but eleven dimensions. That this was considered a serious physical theory raises an obvious question: how can extra space–time dimensions be reconciled with our perception of only four? The short answer is that these extra dimensions are hidden. More specifically, they are very tightly curled up like an extremely small ball – so small that they cannot be seen.

The notion of unobservably small dimensions can be understood in terms of a simple two-dimensional example. Consider a garden hosepipe. Its surface is essentially two-dimensional, like the surface of a cylinder. However, when viewed from a great enough distance, it appears to be only a one-dimensional line, because its diameter is too small to be resolved. But moving closer we see that each point along this 'line' is really a tiny (one-dimensional) circle. Similarly, according to eleven-dimensional supergravity theory, each point in the four-dimensional space–time which we observe is really a tiny seven-dimensional ball, with a diameter of about the Planck length. We say that these extra dimensions are *compact* (see Figure 41.4).

The idea of extra compact dimensions is not a new one. In 1919, Theodor Kaluza suggested that space–time was five-dimensional with the fifth dimension curled up into a tiny circle. By considering general relativity in five dimensions, he attempted to formulate a unified theory of the two fundamental interactions known at that time: gravity and electromagnetism. The electromagnetic force emerged from the theory by virtue of the fifth compact dimension. We can appreciate how this comes about in the following way.

To specify a point (or event) in Kaluza's five-dimensional space–time requires five coordinates: the usual four, x, y, z, t, plus an additional fifth coordinate, θ. It is straightforward to see that this extra coordinate is actually an *angle*: θ locates the position of the event on the circle which corresponds to the extra dimension. Because the radius of the circle is so small, the value of θ is completely unobservable. Consequently, the laws of physics should be invariant under shifts in θ – even under local shifts which are different at different space–time points. This behaviour is reminiscent of that of another angular quantity, namely, the phase of the electron wavefunction. And the invariance under local shifts in θ is just the same sort of invariance we met in the gauge theory of QED (Section 21.2). Hence, a local coordinate transformation applied to the compact dimension corresponds precisely to a local gauge transformation, and implies the existence of a local $U(1)$ gauge symmetry.

This still leaves us with one question: where does the gauge boson come from? To answer this, we should recall that gravitons can be regarded as

Fig. 41.4. If a circle is associated with every point along a line, the result is a two-dimensional surface – a cylinder. In a similar way, an extra, hidden dimension may take the form of a tiny circle associated with every point in four-dimensional space–time.

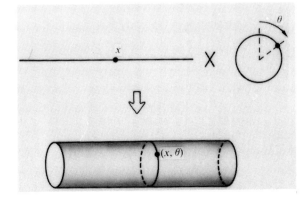

gravitational waves which take the form of small distortions to space–time. This is true even in five dimensions and therefore applies to Kaluza's theory. Moreover, the distortions associated with the four ordinary space–time dimensions, x, y, z, t, correspond as before to spin-2 gravitons. However, the distortions associated with the compact dimension, θ, are new and can be shown to correspond to the presence of spin-1 gauge bosons.

Hence, in Kaluza's theory we see a close relationship between space–time symmetries and internal symmetries. Another way of viewing this relationship is to think of the five-dimensional Poincaré group (the gauge symmetry of gravity in five dimensions) as being decomposed into the four-dimensional Poincaré group plus a local $U(1)$ gauge symmetry. In this way, the theory describes both gravity and electromagnetism.

Kaluza's theory illustrates a very general phenomenon: extra compact dimensions lead to local gauge symmetries. Moreover, several such dimensions can give rise to non-Abelian gauge symmetries. In the eleven-dimensional supergravity theory, for example, seven dimensions are compact and various non-Abelian gauge symmetries are possible, depending on the particular seven-dimensional shape these dimensions assume. It was hoped that this would prove to be a fully unified theory of the four interactions, with the known particles and gauge symmetries emerging from the seven compact dimensions. Unfortunately, however, this hope was dashed by a number of problems, including non-renormalisability. But perhaps the most devastating was the prediction of the same $W^{\pm}$ boson couplings for both left- and right-handed particles, in clear conflict with experiment.

41.3.2 *Supergravity and particle physics*

Despite the failure of eleven-dimensional supergravity, simple supergravity in four dimensions offers an attractive (albeit imperfect) framework in which to consider gravity and the other forces. However, on its own, it does not provide a realistic description of particle physics. To do so, it must be combined with a suitable gauge theory. Combining four-dimensional supergravity with $SU(3)_C \times SU(2) \times U(1)$, for example, yields a supersymmetric version of the standard model

coupled to gravity, or with $SU(5)$ yields a supersymmetric GUT coupled to gravity. These are theories with a total gauge symmetry group given by $\mathbf{G} \times$ super-Poincaré group, in which we can calculate some of the effects of quantum gravity on particle physics. But they remain only approximate theories. Nevertheless, this should not be too much of a worry, as long as the effects in which we are interested are at energies much smaller than the Planck energy and at distances much bigger than the Planck length (Table 40.1).

As in theories with global supersymmetry, local supersymmetry must be broken in order to agree with experiment. The spontaneous breaking of this local symmetry is very similar to the spontaneous breaking of a local gauge symmetry (see Chapter 22). However, in this case, the breaking of supersymmetry does not give rise to a massless Goldstone boson, but a massless Goldstone *fermion*! Furthermore, because the supersymmetry is local, the Goldstone fermion is absorbed by the gravitino, and the gravitino becomes massive. This is similar to what happens in a spontaneously broken gauge theory: the Goldstone boson is absorbed by the massless gauge boson to form a massive gauge boson. This is called the 'super-Higgs' mechanism.

Once supergravity has been spontaneously broken, the theory then looks exactly like a theory with approximate global supersymmetry. Hence, the properties of spontaneously broken supergravity models are virtually identical to those of global supersymmetric models; but the former have the added advantage of providing a framework in which to treat gravity.

Grand unified theories with local supersymmetry (i.e. supergravity GUTs) offer a tantalising first step towards a completely unified theory of all four of the fundamental interactions. Such a theory, if it exists, is often referred to as a 'theory of everything', or TOE. At present, the best candidate for a theory of everything is the superstring theory (see Chapter 43). One feature of the superstring theory is that, at relatively low energies (less than E_{Pl}) and large distances (greater than l_{Pl}), it becomes equivalent to a supergravity GUT. The precise character of this grand unified theory is not known at present; however, there are several promising possibilities.

42

Particle physics and cosmology

42.1 Introduction

Recently, physics has witnessed the convergence of two of its most fascinating and most fundamental branches: elementary particle physics and cosmology. These two subjects, dealing with the universe on the smallest and largest possible scales, are now thought to be inextricably intertwined within the framework of the Big-Bang theory of the origin of the universe. This intimate interrelationship between particle physics and cosmology is revealed in the profound implications each discipline holds for the other.

According to the Big-Bang theory, the universe began some 10^{10} years ago from a space–time singularity, a single point of infinite energy-density and infinite space–time curvature. The act of creation – the Big Bang – was an enormous explosion from which an extremely hot and dense, rapidly expanding universe came into being. The early universe was a thick, hot primordial 'soup', filled with a great abundance of elementary particles of every kind, its evolution governed by the fundamental forces between them.

Consequently the early universe was also the ultimate particle accelerator. Its extremely high temperature and high density offer an unrivalled opportunity to probe physics beyond the reach of terrestrial accelerators and test ideas such as grand unified theories, supersymmetry and superstrings.

42.2 Big-Bang cosmology

There are just three observations that form the basis of Big-Bang cosmology. The first of these is that of the expansion of the universe, which was first discovered in 1929 by Edwin Hubble. He observed that distant galaxies are moving away from us, and moreover, the farther away a galaxy is, the faster it is receding. This discovery is embodied in the equation known as Hubble's law:

$$v = Hr$$

where v is the galaxy's recessional velocity, r is its distance from us, and H is a constant of proportionality called Hubble's constant. We now know that H is not strictly constant but decreasing very slowly with time. Its present value is a little uncertain, owing to the difficulties in measuring galactic distances. The accepted range of values is

$$H = 100h \text{ km s}^{-1}/\text{Mpc} \quad (0.4 \leqslant h \leqslant 1),$$

that is, between 40 and 100 kilometres per second for every megaparsec in distance. (A megaparsec is given by $1 \text{ Mpc} = 3 \times 10^6$ light-years $= 3 \times 10^{24}$ cm). So, a typical galaxy 1 megaparsec away will be moving away from us at a speed of between 40 and 100 km s^{-1}. A galaxy 10 megaparsecs away will be receding at 10 times this speed.

The second principal observation is that of the cosmological abundance of the light elements. In the late 1940s, George Gamow and his collaborators explained these observed abundances in terms of an early universe which was very hot and dense. The light elements, they proposed, were synthesised when the universe was at an absolute temperature of 10^9 K (on the Kelvin scale 0 K $= -273°$C). This temperature is equivalent to a thermal energy per particle of about 0.1 MeV. (It is often convenient to express temperatures in electronvolts. Note that $1 \text{ eV} = 1.2 \times 10^4$ K). This process is called 'nucleosynthesis' and accounts only for the light elements – heavier elements were formed much later inside stars and distributed throughout the cosmos by supernova explosions.

The third principal observation is that of the cosmic microwave background radiation, discovered by chance in 1965 by Arno Penzias and Robert Wilson. This radiation, in which we are constantly bathed from all directions, is a 'residue' of the hot universe. However, it now has a temperature of only 2.7 K, owing to the cooling effect of the universe's expansion.

42.2.1 *Friedmann models*

It was Einstein who should have predicted an expanding universe. However, he ignored the fact that he was unable to find a static cosmological solution to general relativity, and so modified the theory by introducing a new term, a 'cosmological constant', into his equations. A cosmological constant acts as a repulsive antigravity force which is not connected with the presence of matter: it corresponds to an energy in empty space. It is a property of space–time itself, and, Einstein argued, exactly balances the gravitational attraction of all the matter in the universe. The net result is a static cosmological model.

In 1922, working with Einstein's unmodified equations, the Russian Alexandre Friedmann (and much later Howard Robertson and Arthur Walker independently) considered expanding cosmologies based on two assumptions: the universe is (1) isotropic (i.e. looks the same in all directions); and (2) homogeneous (i.e. looks the same from every point in the cosmos). These assumptions give rise to the so-called Friedmann models which seem to describe our universe. Although on smallish scales the universe appears very different in different directions, on very large length scales (much greater than the distances between galaxies) it is indeed remarkably uniform and isotropic. Distant galaxies are distributed more or less uniformly. Moreover, the cosmic microwave background radiation is uniform to within a few parts in 10^5, indicating that the universe was even more isotropic in the past.

An expanding universe raises the question of whether the expansion will continue for ever or eventually end. In the framework of Friedmann models, the answer depends on (1) how fast the universe is expanding, and (2) how much matter there is. If the mass/energy density of the universe is greater than a certain critical value, then gravitational attraction will eventually overcome the expansion and the universe will collapse. If on the other hand, the density is less than this critical value, expansion will continue *ad infinitum*. The critical density is

$$\rho_{\text{crit}} = \frac{3H^2}{8\pi G} = 2 \times 10^{-29} h^2 \text{ g/cm}^3$$
$$= 1 \times 10^4 h^2 \text{ eV/cm}^3$$

and is equivalent to ten hydrogen atoms per cubic metre throughout the universe. Current observations suggest that the density of the universe is between about $0.1\rho_{\text{crit}}$ and $2\rho_{\text{crit}}$ (see Section 42.3.2). So its ultimate fate hangs in the balance, awaiting more precise measurements.

General relativity is above all about geometry (see Figure 42.1). If the density is greater than the critical density, then space (not space–time) is positively curved like the surface of a sphere, and the universe is said to be 'closed', expanding for a certain time before contracting again. But if the density is less than critical, then space is negatively curved like a saddle, and the universe is said to be 'open', expanding forever. Finally, if the density just so happens to be exactly equal to the critical

Fig. 42.1. Open (*a*), flat (*b*) and closed (*c*) Friedmann cosmologies, showing how the size of the universe changes with time. The spatial curvature is such that the three angles of a triangle add up to less than 180° in an open universe and to greater than 180° in a closed universe.

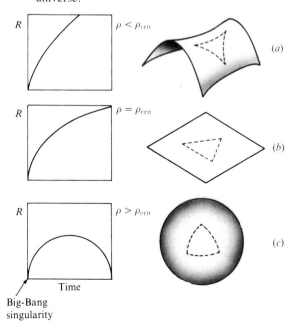

Big-Bang singularity

density, then space is not curved at all but 'flat' (but space–time is still curved).

These Friedmann models form the basis of standard Big-Bang cosmology. Which one describes our universe depends on the actual rate of expansion (i.e. H) and the density (i.e. ρ), neither of which is known with accuracy. However, despite their vastly different predictions for the eventual fate of the cosmos, these models nevertheless paint very similar pictures of the universe at early times.

42.2.2 *Chronology of Big-Bang cosmology*

According to Hubble's law, the universe is expanding in such a way that any two points are separating at a velocity proportional to the distance between them. This expansion is summarised in the behaviour of the cosmological scale factor, R: all cosmological distances increase with R. Furthermore, the temperature decreases in inverse proportion:

$$T \propto \frac{1}{R}.$$

(Note that this means that there are points so far apart that they are separating faster than the speed of light! There is no contradiction here: this is perfectly consistent with general relativity. Locally, special relativity remains valid and c remains the limiting velocity.) Now, let us chart the universe's evolution from a time shortly after the Big Bang.

- $t \approx 10^{-43}$ s, $T \approx 10^{32}$ K $\approx 10^{19}$ GeV: The universe emerges from the Planck era in which quantum gravity was dominant; it is (?) described by a grand unified theory. The energy density is dominated by very relativistic particles and is falling as $\rho \propto 1/R^4 \propto T^4$.

- $t \approx 10^{-35}$ s, $T \approx 10^{28}$ K $\approx 10^{15}$ GeV: The grand unified symmetry is broken. What we know as our observable universe – a region which is today some 10^{10} light years (or 10^{28} cm) in size – is, at this time, contained in a region of space only a millimetre across. A little later, baryogenesis leads to the cosmological excess of matter over antimatter (Section 39.4).

- $t \approx 10^{-10}$ s, $T \approx 10^{15}$ K $\approx 10^2$ GeV: Electroweak symmetry breaking takes place. The presently observable universe is contained in a region 10^{14} cm in size.

- $t \approx 10^{-5}$ s, $T \approx 3 \times 10^{12}$ K ≈ 300 MeV: QCD becomes confining: free quarks combine to form hadrons.

- $t \approx 10^{-2}$ s, $T \approx 10^{11}$ K ≈ 10 MeV: The universe consists mainly of photons, electrons, positrons, neutrinos and antineutrinos. There are small numbers of protons and neutrons which undergo rapid interconversions at this high temperature:

$$\nu_e + p \leftrightarrow e^+ + n$$
$$\bar{\nu}_e + n \leftrightarrow e^- + p$$

Our observable universe is one light year (or 10^{18} cm) in size. The density is over a billion times that of water.

- $t \approx 0.1$ s, $T \approx 3 \times 10^{10}$ K ≈ 3 MeV: At this temperature it is much easier for the heavier neutrons to turn into lighter protons than vice versa. There are 1.5 times more protons than neutrons.

- $t \approx 1$ s, $T \approx 10^{10}$ K ≈ 1 MeV: Neutrinos and antineutrinos begin to behave as free particles. They 'decouple' from the rest of matter and evolve independently. Electrons and positrons begin to annihilate into photons, increasing the temperature of photons relative to neutrinos: $T_\gamma = 1.4 T_\nu$.

- $t \approx 10^2$ s, $T \approx 10^9$ K ≈ 0.1 MeV: The universe is almost entirely made up of photons, neutrinos and antineutrinos, with a small number of electrons and nucleons. There are now six times as many protons as neutrons. Our observable universe is about 100 light years (or 10^{20} cm) in size. The density of the universe is 40 times that of water.

- $t \approx 3$ or 4 minutes, $T \approx 8 \times 10^8$ K: Nucleosynthesis begins. During nucleosynthesis, all free neutrons and some free protons are synthesised into the nuclei of light elements: chiefly deuterium (^{2}D); helium

(^{3}He and ^{4}He); and lithium (^{7}Li). Within a few hours the synthesis is completed, leaving 24% helium by weight and 76% hydrogen (i.e. unused protons), plus smaller amounts of other light elements. However, the universe is still mostly photons and neutrinos.

- $t \approx 10^4$ years, $T \approx 10^5$ K: The energy density becomes dominated by non-relativistic matter and now only falls as $\rho \propto 1/R^3 \propto T^3$.

- $t \approx 10^5$ years, $T \approx 4000$ K: Electrons combine with nuclei to form electrically neutral atoms. With the disappearance of charged particles the universe becomes transparent (there are now no charged particles to scatter photons). In particular, the cosmic microwave radiation was last scattered at this time. Optical and radio astronomy cannot see back beyond this time.

- $t \approx 10^9$–10^{10} years, $T \approx 10$ K: Galaxies form.

- $t \approx 10^{10}$ years, $T \approx 2.7$ K: Today. Size of observable universe is 10^{10} light-years (or 10^{28} cm).

Such is the career of our universe to date. Fortunately for us, the next ten billion or so years look much more relaxing!

42.3 The cosmic connection

The description of 'standard' Big-Bang cosmology in the previous section highlights the important role played by particle physics in the evolution of the early universe. The properties of elementary particles – their masses and interactions – determine the overall rate of cosmological expansion, the temperature as a function of time, and the major chain of events in the universe's history. Thus the implications of particle physics for cosmology are extremely profound. However, as we have already emphasised, cosmology has equally profound implications for our current understanding of particle physics. An idea of the depth of this cosmic connection can be gained by a closer look at the physics of nucleosynthesis.

The light elements – ^{2}D, ^{3}He, ^{4}He, ^{7}Li – are synthesised via nuclear reactions in the first 10^2–10^4 seconds. Within the context of the Big-Bang model, the primordial abundances depend on only one free parameter: the nucleon energy density, or equivalently, the ratio of the number of baryons to photons, $\eta = n_B/n_\gamma$. By comparing the primordial abundances predicted from the Big Bang with those inferred from observation, η can be constrained:

$$\eta = (4 \pm 1) \times 10^{-10}.$$

The number density of photons in the universe is known from the temperature of the cosmic microwave background radiation to be about 400 photons/cm^3. Hence, the above value for η translates into a constraint on the present energy density in baryons:

$$0.10\rho_{\text{crit}} \lesssim \rho_B \lesssim 0.12\rho_{\text{crit}}.$$

Even if η were known exactly, there is a large uncertainty in ρ_{crit} owing to our ignorance of Hubble's constant. So nucleosynthesis tells us that nucleons account for merely 1–10% of the density necessary to close the universe. Therefore if the universe is closed ($\rho > \rho_{\text{crit}}$), then something other than nucleons must completely dominate it.

Nucleosynthesis also places strong constraints on the total number of neutrino types through the observed helium abundance. The amount of primordial helium synthesised depends on the ratio of protons to neutrons at the beginning of nucleosynthesis ($T \approx 10^9$ K). This is because virtually all the neutrons which are present (but only some of the protons) end up as helium. If additional light neutrinos exist, then their presence would increase the universe's expansion rate (relative to its value for three neutrinos). So, the universe would more rapidly cool to the temperature at which nucleosynthesis begins. This, in turn, means that there would be less time in which neutrons could be converted into protons before nucleosynthesis begins; so more neutrons would be present and, therefore, more helium produced. Since the helium produced in stars since nucleosynthesis is small, the observed abundance leads to

$$N_\nu = 3 \pm 1,$$

indicating there is *at most* one further neutrino

awaiting discovery, and hence at most one further generation of elementary particles. For many years, this was the best constraint available. It was only in October 1989 that accelerator experiments at CERN and Stanford were able to improve on this result and establish that there are precisely three light neutrinos.

By a different argument, cosmological bounds can also be found for possible non-zero neutrino masses. At high temperatures, neutrinos (and antineutrinos) are in thermal equilibrium with the radiation and matter in the primordial soup. However, as the temperature falls, the interaction rate between neutrinos and the soup decreases. At some temperature, T_{dec}, the interaction rate falls below the expansion rate, and the universe expands faster than the neutrinos can interact. Thereafter they are essentially free particles, decoupled from the rest of matter and radiation. By demanding that the total energy in neutrinos today be less than the critical density, i.e. $\rho_\nu \lesssim \rho_{crit}$, we find the following constraints on their mass. For each species of heavy neutrinos:

$$m_\nu \gtrsim 2 \text{ GeV}.$$

For the total mass of all light neutrinos:

$$\sum_i m_{\nu_i} \lesssim 92 h^2 \text{ eV},$$

where h accounts for the uncertainty in Hubble's constant $(0.4 \leqslant h \leqslant 1)$, and which implies $\Sigma m_{\nu_i} \lesssim 15$–$92$ eV. This is to be compared with the laboratory bounds on ν_e, ν_μ and ν_τ of 10 eV, 250 keV and 35 MeV. Clearly the cosmological bounds are very restrictive indeed compared with current laboratory bounds.

42.3.1 *Exotic relics from the Big Bang*

These sorts of considerations can be extended to all stable, weakly interacting, massive particles (WIMPs), which may have survived as relics from the Big Bang. Most of these bounds depend on assumptions concerning the details of specific grand unified theories and the masses of other undiscovered particles. The bound on the photino, $\bar{\gamma}$ (the superpartner of the photon), for example, assumes that (1) it is the lightest superpartner, (2) it is stable, and (3) the sfermions (squarks and sleptons) all have equal mass. Then if the sfermion mass is 100 GeV, we have $m_{\bar{\gamma}} > 5$ GeV. On the other hand, the axion (see Section 44.2) is an example of a particle which is very tightly constrained relatively independently of assumptions. Assuming it exists, then the requirement that its density be less than critical imposes the bound $m_a \gtrsim 10^{-5}$ eV. Moreover, in order that it not upset the evolution of stars and supernovae (see Section 17.2), it must also satisfy $m_a \lesssim 10^{-3}$ eV, leaving a remarkably small window of allowed axion masses.

42.3.2 *Dark matter*

We have seen that cosmology has many implications for the properties of elementary particles. The question raises itself: what evidence is there that these (mostly hypothetical) particles play an important role in cosmology? The answer may lie in what astrophysicists call 'dark matter'.

There is now overwhelming evidence that luminous matter – matter that can be seen by astronomers in stars and galaxies – may account for less than 10% of the total mass in the universe (see Table 42.1). So 90% of the matter in the universe may be dark! The best documented evidence for dark matter comes from spiral galaxies. The pull of gravity, measured in terms of the speed at which stars rotate within a spiral galaxy, does not decrease with distance as it should (Figure

Table 42.1. *Cosmological density estimates from various sources*

The characteristic scale on which each estimate is made is given in megaparsecs (1 parsec = 3.26 light-years). Note that h is the uncertainty in Hubble's constant, and is a number between 0.4 and 1.

Source	Scale (Mpc)	ρ/ρ_{crit}
Luminous parts of galaxies	$0.02h^{-1}$	0.01
Galactic haloes + groups of galaxies	0.1–$1h^{-1}$	0.02–0.2
Clusters and superclusters	3–$30h^{-1}$	0.2
Cosmological tests	$3000h^{-1}$	0.1–2
Cosmological inflation	$>3000h^{-1}$ (?)	1

42.2). This suggests that five to ten times more mass is present than is observed. This non-luminous matter would appear to take the form of a roughly spherical halo surrounding the galaxy (Figure 42.3). On much larger scales, it has been known since as early as 1933 that the total gravitational field necessary to hold clusters of galaxies together is typically ten times that provided by visible component galaxies.

But this may not be the full story. There may, in fact, be dark matter which does not clump around galaxies and galactic clusters, but spreads itself thinly throughout the universe. This may provide the necessary mass for the universe to be critical, $\rho = \rho_{crit}$. One compelling argument that $\rho = \rho_{crit}$ is provided by cosmological inflation (see Section 42.5). However, because of the nucleosynthesis constraint, $\rho_B \lesssim 0.12\rho_{crit}$, this dark matter cannot be baryonic.

Certainly the dark matter associated with galaxies and clusters may be baryonic. However, there are problems with what form it takes, and with reconciling galaxy formation with the isotropy of the cosmic microwave radiation. So, we are left with the possibility that some of the WIMPs mentioned above may fit the bill. Clearly, depending on the mass of each of the candidates, the universe may contain varying amounts of non-baryonic dark matter. Such exotic dark matter is generally divided into two general classes: hot and cold, depending on the temperature at which it decouples from the rest of the matter and thereafter evolves freely. Hot dark matter, such as light neutrinos ($m_\nu \approx 30\,\text{eV}$), decouples whilst relativistic (i.e. $T_{dec} \gg m$). Cold dark matter, such as heavy neutrinos ($m_\nu \approx 1\,\text{GeV}$), photinos or axions, decouples when non-relativistic (i.e. $T_{dec} < m$).

Several experiments are currently underway worldwide in the hope of detecting cold dark matter in the halo of our galaxy. Experiments can be broadly split into two groups: those looking for WIMPs such as the photino or heavy neutrino, and those looking for the axion. Such experiments pose enormous technical challenges. However, a positive result would have important and far-reaching consequences. It would provide insights not only into the nature of particle physics beyond the standard model, but also into the cosmology of the Big Bang, the formation of galaxies and the ultimate fate of the universe.

42.4 Supernova 1987A

We have traced the profound relationship between particle physics and cosmology back some 10^{10} years to the very early universe. However, even today the connection between particle physics and events taking place on astronomical scales is apparent. On 23 February 1987 the bright-

Fig. 42.3. Astronomers can explain the motion of stars in spiral galaxies only if there is a spherical halo of dark matter.

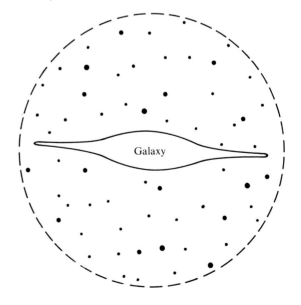

Fig. 42.2. Typical rotation curve for stars in a spiral galaxy suggesting the presence of dark matter.

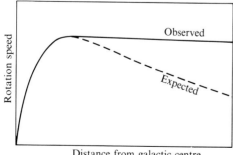

est supernova in nearly four-hundred years appeared in a nearby galaxy called the Large Magellanic Cloud. The Large Magellanic Cloud is a satellite of our own Milky Way galaxy, situated only some 50 kiloparsecs (about 160 000 light-years) away. The progenitor (pre-supernova star) was a massive blue supergiant of some 15 to 20 solar masses known as Sanduleak $-69°$ 202. This supernova, designated SN1987A, was the closest since 1604, and has provided many new results and a host of unusual features. The most dramatic was the first detection ever of the neutrino burst which occurs when the core of a massive star collapses to a neutron star.

SN1987A was a type II supernova, the result of the evolution of stars of about eight solar masses or more. The very large mass implies a very high temperature, so that once the hydrogen in the core of the star has been used up, other fusion reactions begin there. First the helium produced by hydrogen fusion is fused into carbon, then carbon into oxygen, and progressively neon, silicon and finally iron. Fusion stops with iron, ^{56}Fe, because this element has the greatest (i.e. most negative) binding energy per nucleon (see Figure 5.6 in Section 5.4). Beyond ^{56}Fe, fusion does not actually release any energy: in fact, energy must be supplied. Towards the end of its life, therefore, the star is

characterised by an onion-like structure, with an iron core surrounded by layers of progressively lighter elements. Iron accumulates in the core until its mass exceeds 1.4 solar masses. At this point, the weight of the star becomes too much for the core to support, and the latter collapses into a neutron star as the atomic electrons in the iron are 'squeezed' into the atomic nucleus, converting protons into neutrons and releasing a huge burst of neutrinos:

$$e^- + p \rightarrow n + \nu_e.$$

When the core density reaches nuclear density (10^{14} g/cm^3), the nuclear forces become strong enough to stop collapse and the core 'bounces', producing an enormous shockwave which propagates outward and blows a huge shell of matter off into space.

Calculations show that regardless of the specific details of the star, the total energy released is enormous:

$$E_{\text{tot}} \approx 2 \times 10^{46} \text{ joules}.$$

Of this, the total light output and kinetic energy of ejected matter amounts to only 10^{44} joules. So, more than 99% of the total energy comes out in invisible form – as neutrinos. Some 10% of this 99% is carried away by electron neutrinos from the above 'neutronisation' of the core, and the remainder by neutrino–antineutrino pairs of all types which are produced thermally (Figure 42.4).

Two experimental groups, IMB (a US collaboration from Irvine, Michigan and Brookhaven) in the USA and Kamioka (a Japanese–US collaboration) in Japan reported detecting neutrinos at the time of the supernova. Figure 42.5 shows these events in a plot of energy versus time, and

Fig. 42.4. Neutrino emission from a supernova occurs in two stages: (1) neutronisation, in which 10^{57} protons convert into neutrons; (2) thermal $\nu\bar{\nu}$ emission, in which approximate thermal equilibrium is established.

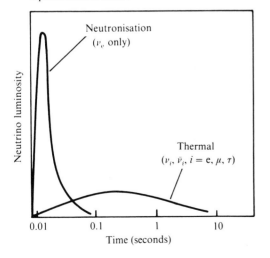

Fig. 42.5. Energies of the IMB and Kamioka neutrino events plotted against arrival time.

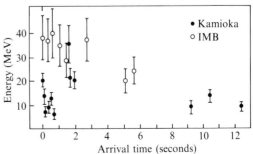

agrees well with theory. (Actually, the detectors could only detect electron antineutrinos via the reaction $\bar{\nu}_e + p \rightarrow n + e^+$.) These results also imply constraints on neutrino properties independently of those given above. These constraints, $N_\nu \lesssim 7$ and $m_{\nu_e} < 20\text{--}25$ eV, are consistent with other bounds, but offer no improvement.

42.5 Inflation

The Big-Bang theory provides a highly successful account of the evolution of the universe from early times. However, it fails to address a number of important questions, which were in the past dismissed as being mere initial conditions:

(1) Why, after 10^{10} years, is the density so close to the critical density? The fact that the present density is between $0.1\rho_{crit}$ and $2\rho_{crit}$ implies that its value 10^{-10} seconds after the Big Bang was equal to the critical density to within one part in 10^{25} (i.e. to 25 decimal places)! Why this was so very close to the critical value, or in other words, why space was so flat, is difficult to fathom.

(2) Why is the universe so uniform on large scales? The cosmic microwave background radiation is uniform in temperature to within a few parts in 10^5, even over regions separated by vast distances. However, such regions are so far apart that there has not been enough time since the Big Bang for even light to have travelled between them. It is therefore difficult to see why they should happen to have the same temperature.

(3) What is the ultimate origin of the relatively small-scale irregularities we see as galaxies?

Cosmologists believe that the answer to these questions may lie in the notion of cosmological 'inflation', which was proposed by Alan Guth in 1981. This involves a period in the history of the universe during which it underwent extremely rapid expansion, i.e. inflation. According to this hypothesis, the cosmological scale factor, R, grew exponentially and, in a minute fraction of a second, the size of the universe increased by 10^{30} times!

According to Guth, this enormous expansion is the result of a (temporary) cosmological constant. Recall that Einstein, in constructing his static model of the universe, introduced a cosmological constant to provide a repulsive force to counteract the gravitational attraction of matter. In his model the two forces were delicately balanced. However, if the cosmological constant is large enough, then the repulsion overwhelms the attraction and the universe expands rapidly: it inflates.

Scalar fields, such as Higgs fields, can provide a natural mechanism for inflation. Suppose such a field has the same wine-bottle-shaped interaction energy as in Chapter 22. Then the state of minimum energy (i.e. the vacuum state) corresponds to a non-zero average value for the field. However, at the very high temperatures prevailing in the early universe, this interaction energy is modified (Figure 42.6(a)) such that the average value of the field is actually zero: this is the state of minimum energy at high temperatures. Then as the universe cools, the preferred configuration becomes the low-temperature vacuum state in which the symmetry is broken. So, the field finds itself in an unstable situation at zero with a large potential energy (Figure 42.6(b)). This potential energy is not associated with the presence of actual particles, but corresponds to a constant energy locked in the field and evenly distributed throughout space. It therefore corresponds to a cosmological constant, and a period of inflation begins. As inflation proceeds, the value of the field moves slowly away from zero towards the low-temperature vacuum state, like a ball rolling down a hill. Because the potential energy curve is almost flat near zero, this motion is initially very slow, allowing the universe to inflate a considerable amount. Inflation finally stops as the field rolls quickly down the rest of the hill, converting potential energy into particles in the process.

A variation on this mechanism is called chaotic inflation. In this picture, large quantum fluctuations cause the scalar field to take random (chaotic) values in different regions of the universe. In some regions, these values correspond to a large potential energy and inflation ensues; in others there is little or no potential energy and hence no inflation. According to chaotic inflation, what we now see as the observable universe was contained entirely within one of the inflating regions.

Inflation provides very appealing answers to

Fig. 42.6. (*a*) The interaction energy at high temperatures. Note that the vacuum value of the scalar field is zero. (*b*) The interaction energy at low temperatures leading to inflation. The scalar field begins in the high-temperature vacuum state with a large potential energy, and slowly evolves toward the low-temperature vacuum state. During this evolution, the non-zero potential energy acts as a cosmological constant and inflation ensues.

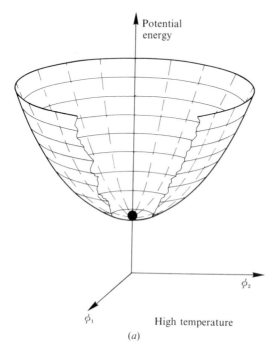

(*a*)

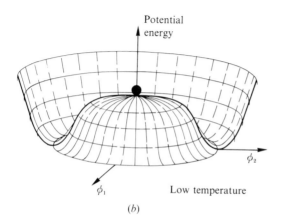

(*b*)

the three questions posed above, by demonstrating how a very large number of very different possible initial conditions could have evolved into the present universe. For example, whatever the value of the density of the universe before inflation, afterwards it is extremely close to critical. The rapid expansion causes space to become flatter in the same way the surface of an inflating balloon becomes flatter. Certainly, if the balloon is as large as the earth, it is flatter than when it was the size of a cricket ball. A flat universe has precisely critical density. Secondly, the uniform temperature of the cosmic microwave radiation is also explained by inflation. Regions which were thought to have always been far apart were actually very close just before inflation. So, it is quite reasonable that they have the same temperature after a period of inflation. Thirdly, quantum fluctuations in the field responsible for inflation are a natural source of fluctuations in density which ultimately lead to the formation of galaxies. Inflation also explains why we do not see exotic objects like magnetic monopoles – the expansion dilutes them to unobservable levels.

42.6 Summary

The intimate connection between particle physics and cosmology allows us to (1) describe the early history of the universe in terms of the known or suspected behaviour of the elementary particles; and (2) constrain the parameters of both observed and hypothetical particles from cosmological observations. Furthermore, this connection has established how the processes of the quantum field theory (e.g. spontaneous symmetry breaking) may have been responsible for determining the currently observed large-scale structure of the universe.

43

Superstrings

43.1 Introduction

Since 1984, much of the physics community has been brimming with excitement over a new fundamental physical theory – the theory of superstrings – which offers the promise of a totally unified formulation of the four fundamental interactions, a 'theory of everything'. This theory seems to be completely free of the two crippling diseases that inflict so many quantum field theories: divergences (infinite quantities), and anomalies (quantum-mechanical inconsistencies). So compelling is the theoretical framework of superstrings that particle physicists and even mathematicians speak of its great elegance and beauty, and its vast richness, which, they argue, are by themselves sufficient to make it worthy of extensive study. However, not all physicists share this view. Some argue that the whole concept of strings is a misleading one, both physically and philosophically. Nevertheless, superstrings have certainly captured the imagination of vast numbers of people and generated an enormous number of papers in fields as diverse as particle physics, statistical physics and pure mathematics.

According to the theory of superstrings, the fundamental constituents of the material world are not point-like elementary particles, but tiny one-dimensional objects, which we can think of as strings. These strings are truly one-dimensional: they have a length, which is approximately equal to the Planck length ($l_{Pl} = 1.62 \times 10^{-33}$ cm), but

no thickness. Like the string of a violin, they can vibrate in many different ways. The different modes of vibration (at frequencies determined by the tension of the string) correspond to the different elementary particles observed in nature. From our discussion of quantum mechanics in Chapter 3, we saw that particles and waves are dual aspects of the same phenomenon: sometimes the concept of a particle is appropriate, and sometimes it is the notion of a wave which is more apt. Here, we have the connection between superstrings and the so-called elementary particles. What we call elementary particles can be understood as different modes of vibration (or different types of waves) of a single string. In other words, the difference between an electron and a muon, for example, is more or less due to different patterns of vibration. The frequency of a particular mode of vibration determines the energy of the particle and hence its mass.

If superstring theory is correct, then the standard laws of physics are only approximations – albeit very good ones – to a more complete and richer set of laws which are valid at all scales of length and energy. In particular, superstring theory incorporates a consistent quantum theory of gravity, valid even for distances smaller than the Planck length. It is in the extended, one-dimensional nature of the strings that the essential difference between this and conventional theories lies. This is crucial to the consistency of the entire approach.

43.2 General properties of strings

We shall now discuss some of the properties of strings, as distinct from those of point particles. We begin by noting that we can envisage two broad classes of strings: those with ends (open strings), and those without consisting of tiny loops (closed strings). Internal gauge symmetries can be associated with open strings through conserved charges which sit at the endpoints. So among the 'particles' associated with the vibrational states of open strings are the spin-1 gauge bosons. By contrast, the states of closed strings include the spin-2 graviton.

Particle states of both open and closed strings correspond to particular patterns of vibration called 'standing waves'. Standing waves are formed when two travelling waves of the same frequency and wavelength, but moving in opposite directions, are combined (Figure 43.1). They are characterised by the presence of 'nodes', i.e. points of the string which do not move as it vibrates. Moreover, the wavelengths and frequencies are quantised: the only possible standing waves are those which exactly fit the length, l, of the string. Therefore, for an open string, e.g. a violin string, standing waves can only have wavelengths equal to $2l/N$ ($N = 1, 2, 3, \ldots$). Similarly for a closed string: the wavelength must fit around the string an integer number of times and the allowed wavelengths are l/N.

When a point particle moves through space–time, it follows a geodesic (a path of minimum length) and sweeps out a one-dimensional curve which is referred to as its *world-line*. The world-line represents the particle's history in space–time.

Fig. 43.1. Standing waves on strings of length l. Standing waves result from two travelling waves of the same frequency and wavelength which are moving in opposite directions.

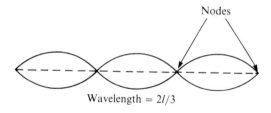

Wavelength = $2l/3$

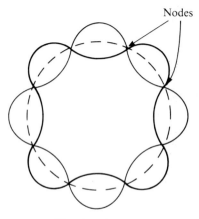

Wavelength = $l/4$

Fig. 43.2. World-sheets for (*a*) open and (*b*) closed strings.

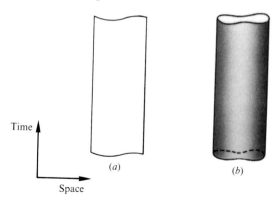

However, when a string propagates through space–time, it sweeps out a two-dimensional surface which, by analogy, is called its *world-sheet*. The world-sheet represents the string's space–time history, and just as a particle moves along a path of minimum length, a string moves along a surface of minimum area. The world-sheet of an open string is a long strip, while that of a closed string is a cylinder (Figure 43.2). If we cut a slice through either of these world-sheets, we get a line or a loop, representing the position of the string at one particular time.

Strings can also undergo interactions with other strings. For example, two strings can join together to form a single string: for open strings the world-sheet looks like an inverted Y; for closed strings the world-sheet looks like a pair of trousers (Figure 43.3). Similarly, the ends of a single open string may join to form a closed string. Hence, in any theory of open strings, there are also closed strings; however the converse is not true.

Fig. 43.3. Two strings join to produce a single string.

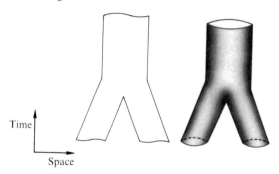

43.3 Brief history

The history of superstrings dates back to 1968 when Gabriele Veneziano came across a formula – unrelated to quantum field theory – which described several aspects of hadronic physics. If was subsequently revealed that this formula corresponds to a description of hadrons in terms of vibrating strings. Loosely speaking, the strings were what bound the quarks together inside a hadron. This theory became known as the bosonic string theory, because the modes of vibration correspond to particles with integer spin. Then in 1971, Pierre Ramond, André Neveu and John Schwarz formulated a closely related theory, which included fermions and was the precursor of the modern superstring theory. However, both of these early string theories had severe problems as theories of hadrons. Perhaps the most striking of these was the fact that they could be rendered consistent with quantum mechanics only if space–time has many more dimensions than the four we see. For the bosonic string theory, the dimensions of space–time must be 26; for the fermionic version (and also for modern superstrings), space–time must have 10 dimensions. Other problems included the prediction of tachyons – particles that travel faster than light and make the theory inconsistent – and of massless spin-1 and spin-2 hadrons which are not observed.

This latter problem seemed to suggest that string theory should be interpreted not as a theory of hadrons, but as a theory of gravity and the other fundamental interactions. Although this was an attractive idea, the difficulties associated with tachyons and extra space–time dimensions remained major obstacles. Furthermore – and much more devastating – such a theory seemed to be plagued by anomalies (quantum-mechanical inconsistencies which destroy gauge invariance).

Superstrings – offspring of the 10-dimensional fermionic string theory – were born in 1976. Incorporating supersymmetry, they immediately resolved the problem of tachyons: since (as it turns out) tachyons do not have superpartners, they cannot be present in the theory. Then in August 1984, Michael Green, of Queen Mary College in London, and John Schwarz, of the California Institute of Technology, showed that the 10-dimensional superstring theory is completely free of quantum-mechanical anomalies, but *only* if the associated gauge group is either $SO(32)$ or $E_8 \times E_8$. This very important result quickly revived the interest in superstrings. Quite suddenly a large number of physicists were working on the subject, and it was not long before a promising new version of superstring theory emerged.

The 'heterotic' superstring theory is a theory of closed strings, which is nevertheless associated with a gauge symmetry. In sharp contrast to open strings with gauge charges at the endpoints, here the gauge charges are 'smeared' over the entire heterotic string. Its construction is unusual, to say the least. It combines the desirable elements of the superstring (fermions and supersymmetry) with the simplicity of the bosonic string. Vibrations (waves) can travel around any closed string in two directions, but the unusual feature of the heterotic string is that the waves moving in each direction are completely different. In fact, the clockwise-moving waves are the waves of the 10-dimensional superstring, whereas the waves moving anticlockwise are those of the original 26-dimensional bosonic string. Particles, therefore, correspond to stationary-wave patterns (Figure 43.1) formed by the superposition of these very different travelling waves. So, the heterotic string is a hybrid of two quite different theories. But how can this make sense? We cannot have a different number of space–time dimensions for the same string!

Combining the two theories in this way entails making a straightforward mathematical connection. However, it is the physical interpretation of this connection which is crucial. We noted above that the bosonic string is consistent only in 26 dimensions. More correctly, we should have said that the theory is consistent only if there are 26 degrees of freedom, each of which behaves in the same way as a space–time dimension. Saying it this way suggests that we may interpret some of these degrees of freedom not as space–time dimensions, but as *internal* degrees of freedom. So, for the part of the heterotic superstring theory which is associated with the bosonic string, only 10 of the original 26 dimensions are really space–time dimensions; the other 16 correspond to internal degrees of freedom. Furthermore, the latter 16 are unobservable and the theory, therefore, should be invariant under local transformations of these internal degrees of freedom. This implies the existence of a local gauge symmetry, and is analogous to the way in which a gauge symmetry emerges from the compact (space–time) dimension in Kaluza's theory in Section 41.3.1. Although the principle is the same, the difference is that Kaluza's theory involves a real space–time dimension which is compact, whereas here we are dealing with internal degrees of freedom. So, the appearance of a gauge symmetry in the heterotic string is owed to the different space–time dimensions associated with the two string theories from which it is constructed.

Furthermore in this theory, particles correspond to stationary waves formed when travelling waves of the bosonic string are superimposed on those of the superstring. They therefore derive certain properties from each of the separate string theories. For example, a particle acquires its gauge properties from the anticlockwise-moving waves of the bosonic string, and its supersymmetry properties from the clockwise-moving waves of the superstring.

43.4 Quantum properties

So far, we have only considered the string as a classical (non-quantum) object. The quantum-mechanical properties of strings are very different to those of point particles, and it is worth while looking at them in a little detail. Before doing so, however, we need to make a small digression.

43.4.1 *Topology*

String theory, and indeed many other areas of physics, makes much use of a branch of mathematics called 'topology'. Topology has to do with smooth, gradual, continuous change. It is a kind of geometry in which lengths, angles and shapes can be continuously deformed. For example, a square can be continuously deformed into a circle by pushing in the corners and rounding the sides. So, a square and a circle are topologically equivalent. Similarly, a sphere of soft clay can be smoothly moulded into a bowl or vase – these are all topologically equivalent. But a sphere of soft clay cannot be smoothly moulded into a cup without tearing a hole for the handle. Thus a sphere and a cup are topologically different. However, a cup is topolo-

gically equivalent to a torus (i.e. a ring), since both have precisely one hole.

43.4.2 *The sum over histories*

In Section 40.5, we mentioned Richard Feynman's formulation of quantum mechanics in terms of a sum over histories. According to Feynman, when a particle moves through space, it does not necessarily move along the classical path (geodesic) between its initial and final states. Instead, there is an infinite number of possible paths (or histories) it may follow, and associated with each is a measure of the probability that that particular history is chosen. The very same formulation can be applied to superstrings.

In the case of strings, the classical path between initial and final string states is simply the surface (world-sheet) of smallest area. The sum over histories is a sum over all the possible surfaces which connect the initial string state to the final string state. The different surfaces can be thought of as quantum fluctuations associated with the world-sheet. But there is something surprising in the application of this approach to strings. The sum is over *all* possible connected surfaces. In particular, it includes all the surfaces formed by stretching, pulling, twisting and otherwise deforming (without tearing) the classical world-sheet. That is, we sum over all surfaces 'topologically' equivalent to the classical world-sheet. So, included in the sum are surfaces with very long, thin tentacles. As shown in Figure 43.4, these tentacles can be interpreted as very small closed strings that appear from out of the vacuum and join on to the original string, or as closed strings which break off from the original and then disappear into the vacuum. Hence the sum over histories automatically includes the interactions of the string with the background space–time in which the string moves. This is true even for free (non-interacting) strings (but certainly not for point particles). It is because of these interactions that the space–time in which the string moves is so tightly constrained in string theories.

If the strings can interact with each other, then we must consider quantum–string effects as well. Analogously to virtual sub-processes in quantum field theory, such as a propagating photon being converted into a virtual electron–

Fig. 43.4. Quantum mechanics implies that even a free string interacts with space–time. The sum over histories includes world-sheets corresponding to strings materialising from the vacuum and joining up with the string, or breaking off and disappearing into the vacuum.

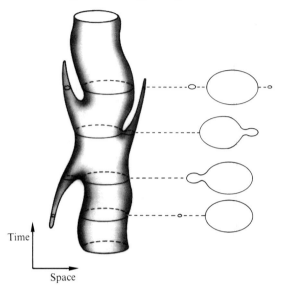

Fig. 43.5. A 'one-loop' string diagram.

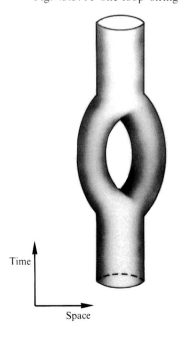

positron pair which then recombine into a photon (Chapter 4), a propagating string may split into two virtual strings which then recombine into a single string (Figure 43.5). Quantum processes of this kind must also be taken into account. The sum over histories must therefore include all surfaces which are topologically equivalent to the world-sheet in Figure 43.5, which has a hole through it. Note that this world-sheet is *not* topologically equivalent to the classical world-sheet. In fact, any number of such sub-processes can occur as the string propagates, and we must also sum over all surfaces with two, three and more holes.

More generally, if a given set of initial strings interact with each other, we can ask what is the probability that a particular final state will be produced. This probability can be evaluated in terms of world-sheet diagrams, which are the Feynman diagrams for strings. These show all the possible joinings and splittings of the world-sheets (i.e. all possible topologies), summed over histories for each case (i.e. summed over all smooth deformations). The sequence of all possible string Feynman diagrams is remarkably simple compared with that for point particles. By way of illustration, consider the elastic scattering of two strings shown in Figure 43.6. The initial strings can simply join and then separate. The Feynman diagram for this – the first in the sequence – is topologically equivalent to the surface of a sphere. Another possibility is for the single intermediate string to split into two strings which then rejoin into one. The Feynman diagram for this is topologically equivalent to the surface of a torus (i.e. the surface of a ring). Alternatively, the intermediate string may split and rejoin, and then split and rejoin again, corresponding to a diagram with two holes. The sequence is continued by simply adding more and more holes.

43.5 Superstrings and particle physics

Elegant and fascinating as the properties of superstrings are, they are worthless if the theory cannot describe the world we observe. At present, the most promising theory from the point of view of particle physics is based on the heterotic superstring with gauge symmetry group $E_8 \times E_8$. In this section, we explain how the standard model might be obtained from this theory. Although there is a great deal of uncertainty connected with many aspects of how this may be achieved, we shall concentrate on one particular approach, that pioneered by Philip Candelas, Gary Horowitz, Andrew Strominger and Edward Witten.

If superstring theory is a serious physical theory then it is necessary to explain why we only see four space–time dimensions and not the ten that superstrings would have us believe exist. From our discussion of extra dimensions in Section 41.3.1, the answer is obvious: six of these dimensions are hidden. In particular, they are very tightly curled up like an extremely small six-dimensional ball which is so minute that it cannot

be seen. So, according to superstring theory, at each point in the four-dimensional space–time which we observe, there is a tiny ball of compact dimensions, with a diameter of about the Planck length.

We saw in the previous section that even free strings interact with the vacuum. These interactions give rise to certain constraints on the properties of space–time, which must be satisfied if the theory is to be consistent. For example, the dimension of space–time must be equal to ten. Furthermore, there are exceptionally strong constraints on the way in which the extra six dimensions can be curled up. They can only be curled up

Fig. 43.6. String–string scattering. Interactions are represented by world-sheets with 0, 1, 2, etc. holes. The surfaces are continuous except at A, B, C and D which correspond to the initial and final strings.

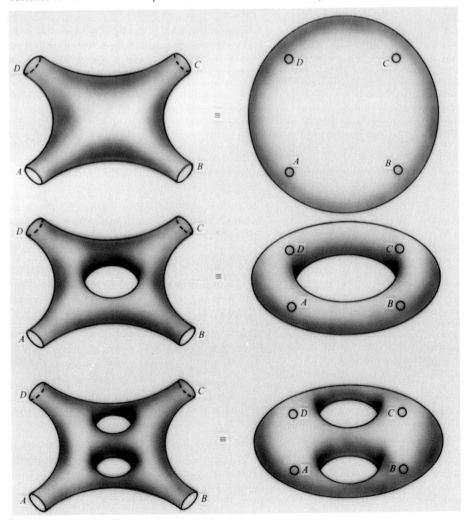

in the form of either a special six-dimensional 'space' called a Calabi–Yau space (named in honour of two mathematicians), or a generalisation of this called an orbifold. A Calabi–Yau space is a very complicated object indeed; an orbifold is a six-dimensional space with conical spikes on its surface.

At distance scales much greater than the Planck length, strings will effectively behave as point particles (since their stringy structure cannot be resolved) and, furthermore, the extra curled-up dimensions can be neglected (since they too cannot be resolved on these scales). So, we can describe particle physics in terms of an 'effective' quantum field theory of particles in 4 dimensions. Remarkably, the topological properties of the compact dimensions determine many of the details of low-energy ($<E_{Pl}$) or long-distance ($>l_{Pl}$) particle physics. For instance, it turns out that the number of generations of light fermions can be predicted from topology. So far, only one Calabi–Yau space has been found which predicts three generations, and this has become the object of considerable study.

In addition, the curvature of a Calabi–Yau space leads to a non-zero average value of some of the gauge fields. Furthermore, other gauge fields may get trapped around holes in this compact space. (Recall that these gauge fields arise from the 16 internal degrees of freedom coming from the bosonic string.) Both of these effects imply that the compact dimensions instead of enhancing the gauge symmetry as we might expect, actually break $E_8 \times E_8$ down to the smaller group $\mathbf{G} \times E_8$. The E_8 part of this group corresponds to gauge interactions which are quite separate from those we observe; none of the known particles is influenced by the E_8 gauge bosons except through gravitational interactions, which we know to be universal. As it constitutes a part of particle physics which is completely independent of the part we know, matter formed from particles associated with this E_8 has been dubbed shadow matter. Shadow matter is real matter; it just happens to have purely exotic gauge interactions. Hence, the symmetry group $\mathbf{G}$, the identity of which depends on the precise details of the compact space, describes all the known particles and forces, plus some unknown ones too.

So, at energies below E_{Pl} or distances greater than l_{Pl}, the heterotic string theory effectively leads to an ordinary grand unified theory with symmetry group $\mathbf{G}$. One promising example is $SU(3)_C \times SU(3) \times SU(3)$. Local supersymmetry is still present, and so the effective theory is really a supergravity GUT like those we mentioned in Section 41.3. Hence, the connection between superstrings and ordinary ('low-energy') particle physics is through the framework of supergravity GUTs.

43.6 Summary and outlook

The theory of superstrings – in particular, heterotic strings – is a fundamentally new physical theory, which offers the promise of a totally unified and consistent formulation of the four fundamental interactions. According to the theory, the basic building blocks of the world are tiny one-dimensional strings, whose different patterns of vibration correspond to the different elementary particles. Superstrings exist only in ten dimensions, six of which must be curled up to an unobservably small size. At distance scales much greater than the Planck length, or energies much lower than the Planck energy, the tiny strings effectively behave as point particles and the extra curled-up dimensions can be neglected. In this situation, particle physics can be described in terms of an 'effective' theory which is just an ordinary grand unified theory coupled to supergravity.

There are several outstanding problems in this programme. Firstly, physicists crave a better understanding of the process of compactification of extra dimensions. Why are six dimensions compact? Why not five, or seven? Why should a particular compact space be chosen over others? Recently, versions of the theory have been devised in which these extra six dimensions are not space–time dimensions at all, but internal degrees of freedom. The notion of compact dimensions is then not needed. Secondly, it is extremely difficult to extract concrete, testable predictions from superstrings. If a theory cannot be tested experimentally, what use is it? It is in search of the answers to these and other questions that particle physicists worldwide are currently devoting a great deal of effort.

44

The latest ideas

44.1 Introduction

So far in Part 9 we have looked at several promising theories which may describe some of the new physics which lies beyond the current experimental frontiers of high-energy physics. These theories, which developed from plausible (but untested) principles of symmetry, economy and elegance, incorporate, as they must, the highly successful standard model. However, we need not look quite so far for new and interesting physics. There are many aspects of the standard model which are poorly understood and may hold some surprises in the years ahead.

In this chapter, we shall briefly review some of the modern (and not-so-modern) ideas connected with some of this poorly understood physics. Many of these ideas involve modifications and/or minor extensions to the standard model. All are speculative to some degree; some much more so than others. Our survey is by no means exhaustive; rather our aim is to give a flavour for some of the current issues which have yet to be resolved.

44.2 Axions and the strong CP problem

One of the outstanding problems associated with the standard model is understanding why the strong interactions are invariant under **CP**. This is because, associated with the strong interactions, there are potentially large **CP**-violating effects coming from two sources: QCD and the Glashow–Weinberg–Salam electroweak model. The former

refers to the **CP** violation present in the QCD Lagrangian, and the latter to that associated with quark masses. The net **CP** violation is measured in terms of a parameter, usually denoted θ, which satisfies

$$|\theta| = |\theta_{\mathrm{QCD}} + \theta_{\mathrm{GWS}}| < 10^{-9}.$$

So, the two contributions appear to cancel and the strong interactions are observed to be **CP**-invariant. This raises certain questions which cannot be answered within the framework of the standard model and which constitute the 'strong **CP** problem': What determines the value of θ? Why is it so small?

Axions provide a theoretically attractive solution to this problem. In this solution, θ is interpreted not as a parameter, but as the average value of a spinless quantum field – the *axion*. The value of θ, therefore, is determined dynamically as the state of minimum energy of the axion field. In fact, $\theta = 0$ corresponds to the state of minimum energy, and thus provides an elegant solution to the strong **CP** problem.

The axion solution requires the existence of a *global* symmetry, which is called the Peccei–Quinn symmetry. By a mechanism similar to that discussed in Section 22.2, this Peccei–Quinn symmetry is actually spontaneously broken. We learnt in Section 22.2 that associated with a spontaneously broken global symmetry there is always a massless Goldstone boson. In fact, the Goldstone boson associated with the broken Peccei–Quinn symmetry is none other than the axion itself.

However, this picture is slightly complicated by the fact that the Peccei–Quinn symmetry is not quite an exact global symmetry; but it is a very good approximation. We can think of the wine-bottle-shaped potential energy of Section 22.2 as being very slightly tilted. Consequently, the axion is almost – but not quite – massless. Hence, it is actually a 'pseudo-Goldstone boson'. In fact, we saw in Section 42.3.1 that cosmological and astrophysical considerations constrain the axion's mass to be $m_{\mathrm{a}} = 10^{-5}\text{–}10^{-3}\,\mathrm{eV}$. Despite this exceptionally small value, the axion interacts only weakly with normal matter, which may explain why it has not yet been observed. In addition to direct searches, several experiments designed to detect cosmic axions which might be trapped in our galactic halo are currently in progress.

44.3 Technicolour theory

The theory of technicolour attempts to replace the as yet unproven Higgs mechanism of spontaneous symmetry breaking by a method of mass generation resulting from the dynamics of new 'technicolour' forces, introduced especially for the purpose.

We have seen how in the electroweak theory, the weak nuclear force is mediated by massive $W^\pm$ and Z^0 gauge bosons. These particles acquire mass by absorbing some components of a (so far hypothetical) Higgs field introduced for just this purpose. The remaining component of the Higgs field corresponds to a real elementary particle – the Higgs boson – which should be observable in experiments, but has yet to be discovered. Unfortunately, if the Higgs boson is a truly elementary particle, then there appears to be some difficulty in understanding the Higgs sector of the theory. One way to avoid this is to drop the idea that Higgs particles are elementary, and consider the possibility that they are composites of still more-elementary particles. Technicolour theory is an attempt to describe this possibility, and, as its name suggests, borrows its approach from QCD (colour theory).

Technicolour theory proposes the existence of a family of new elementary particles called 'techni-fermions'. These are spin-$\frac{1}{2}$ particles which carry a new technicolour charge and so act as the sources of new technicolour forces. By dynamics presumably similar to the (poorly understood) dynamics of QCD, the techni-forces bind the techni-fermions into techni-mesons. Some of these techni-mesons would be absorbed by the $W^\pm$ and Z^0 bosons, giving them a mass, whilst others would remain as observable particles. However, in the electroweak theory, the Higgs field also generates the masses of the quarks and leptons, and it is in attempting to provide this service that the technicolour theory runs into difficulties. It would seem to require the existence of a direct four-fermion interaction (two fermions interacting with two techni-fermions), which is known to be inconsistent. (This was the original reason for the introduction of the W gauge bosons.) To avoid this prob-

lem, it is necessary to introduce a family of massive techni-bosons to mediate the techni-forces between fermions. But these techni-bosons must themselves have their masses generated in a fashion similar to that used for the W bosons. So we are led to the so-called 'extended technicolour theory'. Although this theory may provide an acceptable explanation of mass generation, it also predicts a plethora of techni-mesons which have not been seen in accelerator experiments.

Despite the fact that extended technicolour is fraught with many problems to do with unobserved techni-mesons and flavour-changing neutral currents, the basic concept remains a plausible alternative to the Higgs mechanism. The fundamental idea is a very general one, but the actual realisation of a hypothetical composite Higgs boson may be quite different to the technicolour picture. Nevertheless, the new generation of accelerators should help to clarify the issue as they begin to probe energies in the TeV region.

44.4 Magnetic monopoles and solitons

In one of the classic papers of theoretical physics, published in 1931, Dirac predicted the existence of magnetic monopoles. (In the preface to this paper, and almost in passing, he finally identified the holes in the sea of negative-energy electrons as antielectrons rather than protons, as had been his original conjecture.) His motivation for writing the paper was the attempt to explain the origin of the quantisation of electric charge. This he could do only by postulating the existence of a magnetic charge g which is related to the familiar electronic charge e by the relation,

$$2eg = \hbar cn$$

where n is an integer.

Although startlingly strange, the idea of a magnetic monopole is theoretically attractive. Immediately, the existence of a source of the magnetic field allows Maxwell's equations describing electromagnetic fields to be written in a symmetric form with both electric and magnetic fields arising from both electric and magnetic charges. The experimental absence of magnetic monopoles was assumed to be due to their mass and to the strength of the force between monopoles preventing their individual observation. The subject continued to attract theoretical attention (albeit at a rather meagre level) and the magnetic monopole gradually became known as 'a well-known, undiscovered object' which was subject to routine searches in the succeeding generations of higher-energy particle accelerators, in cosmic ray experiments and in bulk matter – all to no avail.

Modern interest in the possibility of magnetic monopoles revived suddenly in 1974 with the prediction of such objects on the basis of modern gauge theory. This was made independently by Gerard 't Hooft and by Alexander Polyakov of the Landau Institute in Moscow. They discovered that magnetic monopoles should exist as the so-called 'soliton solutions' in certain gauge theories which undergo spontaneous symmetry breaking.

44.4.1 *Solitons*

Soliton solutions are configurations of fields which have finite energy, and which are both *localised* and *stable*. They are an entirely general consequence of classical physics and are well known in subjects such as hydrodynamics. For instance, a normal water wave is initially localised and has finite energy. But it is not stable. As it propagates across the surface of the water it spreads out and dies away. This is because of the dispersion of the wavelengths (i.e. different wavelengths will travel through the medium at different velocities). However, in very particular circumstances, the effects of dispersion can be cancelled out by some non-linear effects between the wave and the medium, giving rise to a non-dissipative, 'soliton' wave. In contrast to the normal dissipative water waves, a soliton wave will not disperse but will propagate as a stable disturbance. This is the origin of tidal bores of which that on the River Severn is perhaps the most famous.

The condition for the existence of soliton solutions in field theory is related to the vacuum structure of the theory and is well illustrated by the example of a two-dimensional rubber sheet which we may think of as our field. The field has two possible vacuum states (states of minimum energy) corresponding to either of the two sides being face up. When only one vacuum is present, the sheet is flat. Any finite energy disturbance will propagate through the sheet as a wave and will eventually disperse (see Figure 44.1). But when two vacua are

present, both sides are face up and a twist must join together the different vacuum regions. Although this twist may move about, it can never disperse while the two vacua remain. This then is simple soliton: a localised concentration of a finite amount of energy joining two distinct, but equivalent, vacua. This simple example illustrates that the existence of solitons of this type requires (1) more than one vacuum state and (2) boundary conditions which involve combinations of the different vacua (e.g. the twist has different boundary conditions at each end of the sheet). Such solitons are called *topological*.

As is possible to visualise, solitons (twists in the sheets) and antisolitons (twists in the other direction) can collide and annihilate each other, giving rise to a normal, dissipative-type wave. It is also possible for solitons to collide and pass through each other, maintaining their shape and energy through the collisions. Thus soliton solutions to classical field theory can exhibit very particle-like behaviour, which we would normally expect to describe using quantum field theory. The added intriguing fact is that solitons are extended objects and so may help us describe, in a natural

Fig. 44.1. A rubber sheet lying flat can have two vacua when either side is uppermost (*a*). When only one vacuum is present (*b*), any disturbance will dissipate. When both vacua are present (*c*), a twist must interpolate between them. The twist cannot disperse for topological reasons.

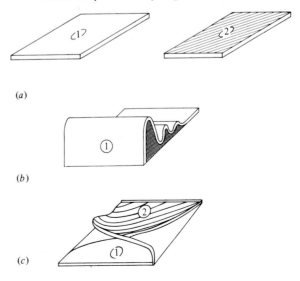

(*a*)

(*b*)

(*c*)

way, particles of finite size, in contrast to the invariably point-like particle behaviour modelled in quantum field theory.

The modern excitement began with the discovery of soliton solutions in $SU(2)$ gauge theory. In this case, multiple vacua occur because of the Higgs field necessary for spontaneous symmetry breaking. As we saw in Chapter 22, the form of the potential energy of the Higgs field is such that a state of minimum energy exists for any values of the two-component field $\Phi = (\phi_1, \phi_2)$ which satisfy the equation

$$\phi_1^2 + \phi_2^2 = R^2$$

(Actually, the 't Hooft–Polyakov monopole solution requires a triplet of Higgs fields, $\Phi = (\phi_1, \phi_2, \phi_3)$, so that the vacuum states of the theory form a sphere rather than a circle. However, the soliton is easier to visualise if we suppress the third component and consider only ϕ_1 and ϕ_2.)

During spontaneous symmetry breaking, the system of fields evolves towards the state of minimum energy. This corresponds to the selection of one particular vector orientation in the space defined by ϕ_1 and ϕ_2. Thus there is a potentially infinite number of the orientations of the vector Φ in the space defined by ϕ_1 and ϕ_2, which give rise to the state of minimum energy. The solitons in this theory are the kinks in the gauge field which connect the regions of different orientation of the Higgs field. Further work reveals that these particular gauge field kinks contain an isolated unit of magnetic charge and may be extremely massive, say about 10^3 GeV.

Solitons in this theory can also give rise to elaborate effects involving angular momentum. It is possible to show that, in the presence of a soliton, bosons may combine to form a fermion – a previously impossible occurrence. Additionally, it is possible for a soliton to divide a fermion into two half-fermions, also previously impossible.

44.4.2 *Monopoles*

Magnetic monopoles are also predicted by GUTs. In these theories, the monopoles are formed on the spontaneous breaking of the grand unified gauge group which, as we have seen, occurs typically at energies around 10^{15} GeV. From this it is possible to show that GUT monopoles must have enormous masses in the region of 10^{16} GeV,

or approximately 10^{-8} g. Just as in the case of the grand unified X-gauge particles, there is no possibility of producing such particles in any conceivable accelerator. But also as with the X-particles, it transpires that GUT monopoles should have been produced copiously in the early universe some 10^{-35} seconds after the Big Bang. It is, to say the least, an extremely non-trivial problem for the theories to explain the apparent absence of relic GUT monopoles and their effects. For instance, one might plausibly expect them to gather at the earth's core where they would most certainly give rise to various geomagnetic effects which have not been observed. Despite an enormous research effort worldwide using a variety of different detection techniques, magnetic monopoles remain elusive. There was a reported sighting of a single monopole in February 1982 by Blas Cabrera of Stanford University. But the lack of plausible candidates since then has cast doubt on this event. Furthermore, the existence of the galactic magnetic field implies that there cannot be very many monopoles in our galaxy. However, cosmological inflation (Section 42.5) offers a plausible explanation for why there may be fewer monopoles than GUTs predict: any monopoles would have been diluted to unobservably low densities as the universe inflated.

Moreover, as we have indicated, there are many possible GUTs and super-gravity theories which differ in the symmetry groups governing their Lagrangians and in their particle contents. It is possible that many soliton solutions exist for each of these theories and so give rise to a wide range of entities with a diverse range of properties. This so-called 'non-perturbative structure' of the modern gauge theories presents a whole new area of theory, undeveloped until recently, and holds out some intriguing possibilities. For instance, in some theories objects called dyons are discovered, which carry both electric and magnetic charge. It is conjectured that these dyons could provide some sort of model for the quarks, which would then be bound together by magnetic strings connecting their magnetic charges.

44.4.3 *Sphalerons and instantons*

Although there are no solitons in the electroweak theory, there are related objects called *sphalerons*. Sphalerons also correspond to localised field configurations of finite energy, but, unlike solitons, they are unstable. In fact, their name derives from the Greek word *sphaleros*, meaning unstable. The electroweak sphaleron is expected to have a mass of about 10 TeV ($= 10^4$ GeV).

There is a simple mechanical analogy which illustrates the nature of a sphaleron. Consider a bead threaded on a closed loop of wire. In general, gravity pulls the bead down to the bottom of the loop: this is the position of minimum energy analogous to the vacuum state in a field theory. But there is one point, viz. the top of the loop, from which the bead will not slide down to the bottom. At this point, the wire is horizontal and gravity has no effect. Although classically the bead can remain in this position forever, any small perturbation will send it sliding down to the state of lowest energy. The position of the bead at the top of the loop is analogous to a sphaleron field configuration. Any small perturbation (e.g. from quantum fluctuations) will cause the sphaleron to decay as the field theory moves towards the configuration of lowest energy, i.e. the vacuum state. So a sphaleron is essentially an unstable classical particle.

Another leading conjecture is that, in the situation of multiple vacua, a system of fields can effectively undergo quantum-mechanical tunnelling from one vacuum to another (by a mechanism similar to the tunnelling of electrons through potential barriers at the junctions of semiconductors). We can think of this as the vacuum configuration of the gauge fields flexing into an equivalent, but distinct, alternative vacuum state. This tunnelling is described by a family of soliton-like solutions that are transient in time. For this reason they are called *instantons*. Instantons are not particles and have no direct physical interpretation. Rather, they represent vacuum fluctuations of gauge fields which may lead to some observable forces on nearby particles such as quarks. On the basis of this, some theorists have attempted to explain various meson-mass anomalies as resulting from the effects of instantons on the interquark forces. Others have put forward instantons as yet another mechanism which may provide the understanding of quark confinement. But despite such suggestions, the physical importance of instantons is still far from obvious. Such ideas are still very speculative. However, even now it is certain that

the modern gauge theories have a rich structure (and, one day possibly, consequences?) far beyond the original reasons for their introduction.

44.5 Strange matter

One example of the surprises that even well-known and long-established physics can hold is provided by the possible existence of strange matter. Strange matter is a hypothetical form of ultra-dense nuclear matter containing roughly equal numbers of up, down and strange quarks. In 1984, Edward Witten of Princeton University conjectured that quark matter of this kind may have been produced in the hot and dense conditions prevailing in the early universe when the temperature was 100–200 MeV, or about 10^{12} K. He furthermore suggested that strange matter may well be stable enough to have survived until the present day in the form of 'nuggets' of 0.01–10 cm in size with a density of 10^{15} g/cm^3 and a mass of 10^9–10^{18} g.

Ordinary nuclear matter consist of only two quark flavours – u and d – which are confined within nucleons in groups of three. However, in quark matter there are no nucleons (or baryons) in which the individual quarks are confined. The quark wavefunctions are completely delocalised and the quarks move freely throughout the volume of a quark nugget.

At first sight, it may appear that quark matter cannot be stable in view of the fact that ordinary nuclei do not spontaneously transform into this hypothetical matter. However, this does not quite follow. Although it is true that ordinary nuclear matter is more stable than the corresponding quark matter, the addition of strange quarks may tip the balance the other way. Certainly, the addition of strangeness does not help to stabilise nuclear matter, since strange baryons are heavier than their non-strange counterparts. But adding strange quarks to (exclusively u and d) quark matter should lower the energy due to the Pauli exclusion principle (Section 3.10). This is because, for a given number of quarks, there will be fewer identical fermions if strange quarks are present. If this decrease in energy is greater than the increase in energy due to strange quarks being heavier than either u or d, then strange quark matter may well be more stable than nuclear matter.

Nuggets of strange matter might have survived as relics from the Big Bang and may account for much of the dark matter in the universe. Being so tightly bound in this very dense state, the quarks in such nuggets would not have participated in nucleosynthesis. So, the constraint on the total energy density of baryons in Section 42.3 does not apply.

Strange matter may also be produced in the conditions of extreme temperature and pressure created by a supernova explosion. Weak interaction processes, such as $u + d \rightarrow u + s$, may be responsible for the formation of 'strange stars', instead of neutron stars, after such an event.

45

The beginning of the end?

45.1 Introduction

History has an uncanny knack of repeating itself. The last 20 years have echoed the progress of the 1920s in the way that both theory and experiment have advanced in a complementary fashion to provide a convincing understanding of the microworld at a level several orders of magnitude smaller than could be understood 50 years earlier. Whereas in the 1920s, energies of a few eV probed atomic distances of 10^{-8} cm, we now use energies of hundreds of GeV to probe distances as small as 10^{-16} cm.

As a result of our recent successes, we now believe we have a reasonable understanding of the world on this scale. It has been the path towards this understanding which we have attempted to describe in this book. Having trod the path, our understanding of the microworld can be summarised meaningfully in just a few short sentences (which is always a good sign!).

We believe that the fundamental constituents of the material world are 24 elementary spin-$\frac{1}{2}$ particles (fermions). Of these, there are: six leptons,

$$\begin{pmatrix} e \\ \nu_e \end{pmatrix}, \quad \begin{pmatrix} \mu \\ \nu_\mu \end{pmatrix}, \quad \begin{pmatrix} \tau \\ \nu_\tau \end{pmatrix}$$

which exist as free, point-like particles; and 18 quarks (6 flavours times 3 colours),

$$\begin{pmatrix} u \\ d \end{pmatrix}_{r,g,b}, \quad \begin{pmatrix} c \\ s \end{pmatrix}_{r,g,b}, \quad \begin{pmatrix} t \\ b \end{pmatrix}_{r,g,b}$$

which seem to exist only in certain combinations, specifically, as baryons (qqq) and mesons (q$\bar{q}$). Furthermore, the six leptons and six flavours of quarks together seem to fall into 3 generations in the pattern indicated above by the parentheses. Each successive generation is simply a heavier version of particles with the same basic quantum numbers.

We believe that the forces between these fundamental particles are mediated by the exchange of spin-1 gauge bosons, the properties and interactions of which are determined by local gauge symmetries. In particular, the electromagnetic force is mediated by the massless photon, the weak nuclear force is mediated by the massive $W^{\pm}$ and Z^0 bosons, and the strong nuclear force is mediated by eight massless coloured gluons. Moreover, these forces are well described by the theories of quantum chromodynamics (QCD) and the Glashow–Weinberg–Salam electroweak model, which together constitute what is called the standard model of particle physics.

45.2 Where to next?

The standard model is a highly successful theory, which accounts for all current experimental data. But are there any surprises in store for us? Where will particle physics take us in the 1990s? Grand unification is an appealing theoretical possibility, but there is, as yet, no concrete evidence in its favour. Moreover, the simplest GUTs, such as $SU(5)$, predict no essentially new phenomena between the energy scales of $100\,\text{GeV}$ (where electroweak unification takes place) and 10^{15}–$10^{16}\,\text{GeV}$ (where grand unification occurs). Although we are now probing the former scale with enormously complicated and expensive accelerators, the latter scale appears to be beyond our direct reach. Cosmology can be of some help by providing us with a picture of the universe at very early times, when the temperature was so high that the grand unified symmetry was exact. However, there are many uncertainties associated with this approach and its usefulness is, unfortunately, ultimately limited.

On the other hand, the baroque school of physics can point to a wide range of alternatives to the simplest GUTs, including technicolour, supersymmetry, supergravity, and superstring theories. If any of these possibilities proves to be pointing in the right direction, then we should, in the years to come, see the discovery of new phenomena, new particles and new forces in the would-be desert above $100\,\text{GeV}$.

But, this is a rather complacent mood in which to leave this subject. As we have noted, history has the knack of repeating itself. In the closing years of the nineteenth century, no one could have thought that the slight irregularities in the theory of black-body radiation would have led to an entirely new conception of the world in terms of quantum theory. Likewise, few physicists pondering the apparent constancy of the speed of light would have made the connection with $E = mc^2$. Even when quantum theory and special relativity were well established, the prediction of antimatter resulting from the synthesis of the two was wholly novel. Moreover, in the 1930s, few can have realised that the discoveries of the positron and the muon, and the predictions of the pion and the neutrino, were the first heralds of a new order of matter. So, on the basis of history, the only surprise will be if there are no surprises.

45.3 A Theory of Everything?

Physics, like all science, is concerned with the explanation and prediction of physical phenomena. It has always progressed by drawing together seemingly disparate phenomena into one framework. The physicist begins with a selection of empirical facts out of which he constructs a model. But the process does not stop there. Other classes of facts lead to other models, which are eventually unified and subsumed into a single overarching theory. The theory goes much further than the initial models, furnishing a deeper understanding of nature, and possessing, by this same fact, far greater predictive power. At a later stage, this theory, along with others like it, is itself subsumed into a larger, more unified theory, and the process continues. Figure 45.1 illustrates this process of unification, taking as an example the development of theories of the fundamental forces. (Note that we have emphasised logical, rather than historical, connections.)

The question naturally arises: will this process ever end? Is there an ultimate theory towards which we are inevitably progressing? Is there a Theory of Everything?

The Theory of Everything – if it exists – would be much more than just a catalogue of physical laws. It would constitute a truly unified description of the material universe, weaving an intricate web of interconnections between its component parts, each one essential to the overall consistency of the whole. Based on just a few plain assumptions, the Theory of Everything would be utterly compelling in structure, symmetry and elegance. And, in addition to elucidating the properties and interactions of the elementary particles, it would account for such things as the observed geometry of space–time, and why there are precisely four space–time dimensions.

Fig. 45.1. The road to unification of the fundamental forces.

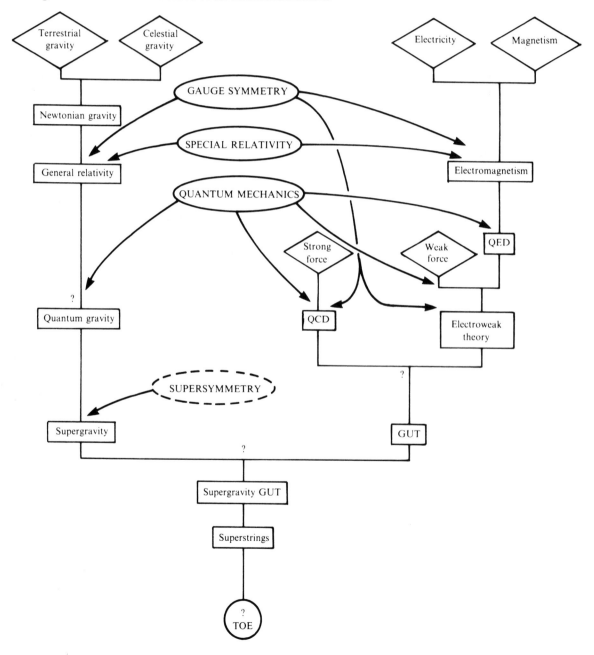

Unfortunately, history is littered with examples of reputable physicists who rashly announced that the ultimate Theory of Everything was at hand, and that the end of physics was imminent. Such claims should be greeted with a good measure of scepticism. There is much we have not even begun to understand about the universe, and it is likely that scientists in centuries to come will look back and view the quest for a Theory of Everything as we today view certain past attempts at unification, as having been both simplistic and premature.

45.4 Write-it-yourself

Historical trends may be the only guide to the long-term future, but the next few years promise to be exciting ones. With particle accelerators like LEP and the SLC up and running, and others like the US 'Superconducting Super Collider' (SSC) now under construction, the short-term outlook for particle physics is intriguing. We can look forward to the discoveries of the Higgs boson and the top quark, not to mention the possible detection of a whole host of hypothetical exotic particles. So, perhaps an appropriate way to complete this account of particle physics would be for you, the reader, to keep an up-to-date record of future discoveries for yourself. Table 45.1 gives you a pro-forma guide to anticipated events which should certainly be announced in the daily press and scientific magazines as they come to light.

It is an interesting exercise to review the progress of the subject since the completion of the first edition in early 1983. At that stage, the first few candidate $W^{\pm}$s and Z^0s had just been detected. Since then, the existence of these massive gauge bosons has been established beyond doubt. But apart from that major discovery, little in the table has changed over the past seven years. None of the other potential discoveries has become any firmer, and no new possibilities have been added. This is in sharp contrast to the theoretical side of the subject, in which we have seen tremendous advances, particularly in the field of superstring theory. It will be fascinating to see how particle physics progresses in the next seven years.

Table 45.1. *A guide to forthcoming discoveries in particle physics*

Topic	Status: (a) Discovered by (b) Date (c) Current status	Crucial measurement	Significance	Reference
$W^{\pm}$, Z^0 bosons	(a) CERN teams $p\bar{p}$ collider (b) January/May 1983 (c) Confirmed at Tevatron, SLC and LEP.	$M_{W^{\pm}} = 80$ GeV $M_{Z^0} = 91$ GeV	Confirms electroweak gauge theory.	Ch. 25
Glueball	(a) (b) (c) Candidate events; require confirmation. Good odds.	$M_g \approx 1$–2 GeV, spin 0, 2	Supports QCD.	Ch. 32
Higgs bosons	(a) (b) (c) Possible at LEP or SSC.	$M_H \gtrsim 45$ GeV	Confirms spontaneous symmetry breaking in gauge theories.	Ch. 23
Top quark	(a) (b) (c) Possible at LEP or SSC.	$M_t \gtrsim 90$ GeV	Completes third generation of elementary fermions.	Ch. 37

Continued

Table 45.1 (*cont.*)

Topic	Status: (a) Discovered by (b) Date (c) Current status	Crucial measurement	Significance	Reference
Proton decay	(a) (b) (c) No unambiguous event. Many experiments running. Long odds.	$\tau_{\mathrm{p}} > 4 \times 10^{32}$ years	Dramatic evidence for validity of GUTs.	Ch. 39
Magnetic monopole	(a) (b) (c) Candidate event claimed. More needed. Then independent confirmation. Long odds.	Existence Mass? Electric charge?	Electric and magnetic charge symmetry predicted by Dirac. Structure of gauge field configurations.	Ch. 44
Neutrino oscillations and mass	(a) (b) (c) Oscillations perhaps seen in reactor experiment. Mass perhaps detected in nuclear β decay. Long odds.	Presence. $M_{\nu_{\mathrm{e}}} < 10\,\mathrm{eV}$	Structure of GUTs. Eventual fate of the universe.	Ch. 39
Free quarks	(a) (b) (c) Fractional charges claimed. Must be confirmed elsewhere. Quarks or not? Long odds.	Existence	Would confuse all current prejudices.	Chs 31, 34
Super-symmetric particles	(a) (b) (c) Possible any time. Long odds.	Existence $M > 50\text{–}100\,\mathrm{GeV}$	Supersymmetric particles indicate some hope of understanding gravity.	Ch. 41
Technicolour particles	(a) (b) (c) Possible any time. Long odds.	Existence	Dynamical symmetry breaking. Composite Higgs.	Ch. 44
Gravitational waves (gravitons)	(a) (b) (c) Possible any time. Detectors most probably not yet sensitive enough. Very long odds.	Existence	Supports general relativity.	Chs 5, 42

Appendices

Units and constants

Energy: The most common unit of energy in the microworld is the electronvolt, eV. This is defined as the energy possessed by an electron after it has been accelerated through a potential difference of one volt.

$$1 \, \text{eV} = 1.602 \times 10^{-19} \, \text{J}$$

$$1 \, \text{keV} = 10^3 \, \text{eV}; \qquad 1 \, \text{MeV} = 10^6 \, \text{eV}$$

$$1 \, \text{GeV} = 10^9 \, \text{eV}; \qquad 1 \, \text{TeV} = 10^{12} \, \text{eV}$$

Mass:

$$\text{electron} \quad m_e = 9.109 \times 10^{-31} \, \text{kg}$$
$$= 0.511 \, \text{MeV}/c^2$$

$$\text{proton} \quad M_p = 1.673 \times 10^{-27} \, \text{kg}$$
$$= 938.27 \, \text{MeV}/c^2$$

Charge:

$$\text{electron} \quad e = 1.602 \times 10^{-19} \, \text{C}$$

Speed of light:

$$c = 2.998 \times 10^8 \, \text{m/s}$$

Planck's constant $\left(\hbar = \dfrac{h}{2\pi} \right)$:

$$\hbar = 1.055 \times 10^{-34} \, \text{J s}$$
$$= 6.582 \times 10^{-22} \, \text{MeV s}$$

Fine structure constant $\left(\alpha = \dfrac{e^2}{\hbar c} \right)$:

$$\alpha = \frac{1}{137.036}$$

Glossary

Abelian group A mathematical group of transformations with the property that the end result of a series of transformations does not depend on the order in which they are performed.

absolute temperature Temperature measured on the Kelvin scale: 0 Kelvin = −273.15° Celsius. Absolute temperature is directly related to (kinetic) energy via the equation $E = k_B T$, where k_B is Boltzmann's constant. So, a temperature of 0 K corresponds to zero energy, and room temperature, 300 K = 27°C, corresponds to an energy of 0.025 eV.

alpha (α) particles Particles first discovered in radioactive α decay, and later identified as helium nuclei (two protons and two neutrons bound together).

amplitude *See* quantum-mechanical amplitude.

angular momentum The rotational equivalent of ordinary momentum, being mass × angular velocity. It is a vector quantity directed along the axis of rotation. In quantum mechanics, (orbital) angular momentum is quantised in integer multiples of $\hbar$. This corresponds classically to only certain frequencies of rotation being allowed.

antiparticles Particles predicted by combining the theories of special relativity and quantum mechanics. For each particle, there must exist an antiparticle with the opposite charge, magnetic moment and other internal quantum numbers (e.g. lepton number, baryon number, strangeness, charm, etc.), but with the same mass, spin and lifetime. Note that certain neutral particles (such as the photon and π^0) are their own antiparticles.

asymptotic freedom A term used to describe the observed *decrease* in the intrinsic strength of the colour force between quarks as they are brought closer together. At asymptotically small separations, the quarks are virtually free. This is in contrast to the electromagnetic force whose intrinsic strength increases as two charged particles approach each other.

axion A hypothetical spin-0 particle with a very small mass of 10^{-5}–10^{-3} eV. It was postulated in order to provide a natural solution to the 'strong **CP** problem'. See Section 44.2.

baryogenesis The process by which the universe's net baryon number was generated. This explains why the universe is made predominantly of baryons and not antibaryons. See Section 39.4.

baryon The generic term for any strongly-interacting particle with half-integer spin in units of $\hbar$ (e.g. the proton, neutron and all their more massive excited resonance states).

beta (β) particles Particles first discovered in radioactive β decay – later identified as electrons.

Big-Bang theory The most widely accepted theory of the origin of the universe. It asserts that the universe began some 10^{10} years ago from a space–time point of infinite energy density (a singularity). The expansion of the universe since that time is akin to the expansion of the *surface* of an inflating balloon: every point on the balloon's surface is moving away from every other point. So, microbes living on the surface see their two-dimensional world expanding, yet there is no centre to the expansion which is everywhere uniform.

boson Any particle with integer spin: 0, $\hbar$, 2$\hbar$, etc.

Cabbibo angle (θ_C) The measure of the probability that one flavour of quark (u) will change into other flavours (d or s) under the action of the weak force.

CERN The European Laboratory for Particle Physics (formerly the Conseil Européen pour la Recherche Nucléaire), located near Geneva in Switzerland. Here, the resources of the European member nations are pooled to construct the large particle accelerators needed for high-energy experiments. The major facilities at CERN include the Super Proton Synchrotron (SPS) and the Large Electron–Positron (LEP) collider.

charm The fourth flavour (i.e. type) of quark, the discovery of which in 1974 contributed both to the acceptance of the reality of quarks and to our understanding of their dynamics. The charmed quark exhibits a property called 'charm' which is conserved in strong interactions.

colour An attribute which distinguishes otherwise identical quarks of the same flavour. Three colours – red, green and blue – are required to distinguish the three valence quarks of which baryons are composed. It must be stressed that these colours are just labels and have nothing to do with ordinary colour. Colour is the source of the strong force which binds quarks together inside baryons and mesons, and so the three colours (r, g, b) can be thought of as three different colour charges analogous to electric charge.

cosmological constant A term added by Einstein to the gravitational field equations of his theory of general relativity. Such a term would produce a repulsive antigravity force at very large distances and would correspond to energy locked up in the curvature of space–time itself. There is, at present, no evidence for the existence of a cosmological constant (although one may have existed in the past).

cosmological principle The hypothesis that the universe is isotropic and homogeneous on very large distance scales.

coupling constant A measure of the intrinsic strength of a force. The coupling constant of a particular force determines how strongly a particle couples to the associated field. For example, $\alpha = e^2/\hbar c = \frac{1}{137}$ (or, equivalently, electric charge e) specifies the strength of the coupling of charged particles to the electromagnetic field.

cross-section (σ) The basic measure of the probability that particles will interact. It corresponds to the effective target area (in, for example, cm^2) seen by the ingoing particles. It can be derived from the quantum-mechanical interaction probability by multiplying by factors such as the flux of particles entering the interaction region. A convenient unit for measuring cross-section is the barn (symbol: b), defined as $1\,b = 10^{-24}\,cm^2$. Typical hadronic cross-sections are measured in milli-barns; $1\,mb = 10^{-27}\,cm^2$. However, neutrino collision cross-sections are typically much smaller, $10^{-39}\,cm^2$.

DESY The German national laboratory for high-energy physics, located near Hamburg. It is the home of the e^+e^- storage rings DORIS and PETRA, and the electron–proton machine, HERA.

deuteron The nucleus of deuterium, an isotope of hydrogen. It consists of one proton and one neutron bound together.

diffraction A property which distinguishes wave-like motions. When a wave is incident upon a barrier which is broken by a narrow slit (of comparable size to the wavelength), then the slit will act as a new isotopic source of secondary waves.

dimensions Physically significant quantities usually have dimensions associated with them. The fundamental dimensions are those of mass M, length L and time T. The dimensions of other quantities can be expressed in terms of these fundamental dimensions. So, for example, momentum has dimensions of mass × velocity (MLT^{-1}) and energy dimensions of force × distance (ML^2T^{-2}). One may define certain quantities in which the dimensions cancel out. These dimensionless quantities are significant in that they are independent of the conventions used to define units of mass, length and time.

eigenstate, eigenvalue   The eigenvalue of a matrix **M** is a number λ which satisfies the equation

$$\mathbf{M}\psi = \lambda\psi, \quad \text{with} \quad \psi \neq 0.$$

In quantum mechanics, the matrix **M** will correspond to a particular dynamical variable (such as position, energy or momentum) and λ will correspond to the value obtained by measuring that dynamical variable if the system is in the state described by ψ. ψ is called an eigenstate of the system.

elastic scattering   Particle reactions in which the same particles emerge from the reaction as entered it (e.g. $\pi^- p \rightarrow \pi^- p$). In *inelastic* scattering, where different and/or new particles emerge, energy is used to create new particles.

electron   A negatively charged spin-$\frac{1}{2}$ particle, which interacts via the electromagnetic, weak and gravitational forces. It has a mass of 0.511 MeV/c^2, some 1800 times lighter than the proton.

Fermilab   The Fermi National Accelerator Laboratory, in Batavia, Illinois, USA. Fermilab is the home of the Tevatron, the world's most powerful accelerator, a $p\bar{p}$ collider with a maximum collision energy of 1.8 TeV ($= 1800$ GeV $= 1.8 \times 10^{12}$ eV).

fermion   Any particle with half-integer spin: $\frac{1}{2}\hbar$, $\frac{3}{2}\hbar$, $\frac{5}{2}\hbar$, etc. All fermions obey Pauli's exclusion principle.

flavour   The term used to describe different quark types. There are six quark flavours: up, down, strange, charm, bottom and top.

gamma (γ) rays   Rays first discovered in radioactive material, and later identified as very high energy photons.

gauge theory   A theory whose dynamics originate from a symmetry. That is, the formulae describing the theory (in particular, the Lagrangian) are unchanged under certain symmetry transformations, called 'gauge' transformations. For example, the equations of classical electrodynamics are invariant under local redefinitions of the electrostatic potential. This symmetry is ultimately responsible for the conservation of electric charge. However, in quantum electrodynamics this gauge symmetry is reinterpreted as invariance under local redefinitions of the phase of the electron wavefunction. The term 'gauge theory' is an archaic one, coming from earlier theories which were based on invariance under transformation of scale (i.e. gauge).

generation   Leptons and quarks come in three related sets, called generations, consisting of two leptons and two quarks. The first generation consists of (e^-, ν_e; u, d). The second and third generations consist of (μ^-, ν_μ; c, s) and (τ^-, ν_τ; t, b) respectively.

gluon, glueball   Gluons are the massless gauge bosons of QCD which mediate the strong colour force between quarks. Because of the non-Abelian structure of the theory, gluons can interact with themselves, and may form particles consisting of gluons bound together. The existence of these 'glueballs' has yet to be confirmed.

Goldstone boson   A massless spin-0 particle which arises whenever a (continuous) global symmetry is spontaneously broken.

graviton   A massless spin-2 particle which is the hypothetical quantum of the gravitational field. It mediates the force of gravity in a similar way to that in which the spin-1 gauge bosons (i.e. the photon, $W^\pm$, Z^0, and gluons) mediate the other forces.

group theory   The branch of mathematics which describes symmetry. A mathematical group **G** is defined as a collection of elements $\{a, b, c, \ldots\}$ with the properties:

(1) if a and b are in group **G**, then the product of the two elements, ab, is also in **G**;
(2) there is a unit element e such that $ae = a$ for all elements a in **G**;
(3) each element a has an inverse a^{-1} such that $aa^{-1} = e$.

So, for instance, the rotations of an (x, y) coordinate system about the z-axis form a group, since the effect of two rotations $\theta_1\theta_2$ is equivalent to the effect of one big rotation θ_3. Such a group is called a continuous group as the angles of rotation can vary continuously.

In general, the elements of a group may be represented by matrices, which form various 'representations' of the group. These representations may be used to determine how a physical system changes under the action of symmetry transformations. Moreover, when a system possesses a symmetry which is described by a group **G** (i.e. when its equations of motion are left invariant under group transformations), then the various representations specify the symmetry properties of the relevant degrees of freedom.

For example, hadrons appear to possess an $SU(3)$ 'flavour' symmetry. The fundamental 3-dimensional representation of $SU(3)$ contains the three flavour degrees of freedom associated with the up, down and strange quarks: $\mathbf{3} = (u, d, s)$. The eight-dimensional representation **8** of $SU(3)$ contains the eight flavour degrees of freedom associated with the meson and baryon octets (see Chapter 10). Furthermore, a given representation completely specifies the flavour quantum numbers of associated particles.

The same is true for local gauge (i.e. dynamical) symmetries. For example, the theory of QCD possesses a local $SU(3)_C$ colour symmetry. The three-dimensional representation of $SU(3)_C$ contains the three colour degrees of freedom associated with the colour charges red, green and blue: $\mathbf{3} = (r, g, b)$. The eight-dimensional representation contains the eight colour degrees of freedom associated with the eight gluons – the gauge bosons of QCD.

Discrete groups having a finite number of elements are associated with discrete symmetry transformations, such as parity. For example, the discrete group corresponding to parity has only two elements: $\mathbf{P}$ and $\mathbf{P}^2 = e$.

hadron The generic name for any particle which experiences the strong nuclear force.

helicity The projection of a particle's spin along its direction of motion. The helicity of a particle is described as being either left- or right-handed depending on whether its spin projection is in the direction of motion or against it. See Figure 13.3.

Higgs boson A hypothetical, spinless particle that plays an important role in the Glashow–Weinberg–Salam electroweak theory (and in other theories involving spontaneous symmetry breaking, e.g. GUTs).

Higgs mechanism A mechanism by which gauge bosons acquire mass through spontaneous symmetry breaking. In the Glashow–Weinberg–Salam electroweak model, for example, Higgs fields are introduced into the theory in a gauge-invariant way. However, the state of minimum energy breaks the local gauge symmetry, generating masses for the $W^\pm$ and Z^0 bosons, and giving rise to a real, observable Higgs boson, ϕ'.

hyperon A baryon with non-zero strangeness.

isotopic spin or isospin A concept introduced by Heisenberg in 1932 to describe the charge independence of the strong nuclear force. Since the strong force cannot distinguish between a proton and a neutron, Heisenberg proposed that these particles were actually different states of a single particle – the nucleon. He argued that just as the electron comes in two different spin states, so the nucleon comes in two different 'isospin' states. So, isospin is a concept analogous to spin which is conserved by the strong interaction. The nucleon is an isospin-$\frac{1}{2}$ particle, and its third component of isospin determines whether we are talking about a proton ($I_3 = +\frac{1}{2}$) or a neutron ($I_3 = -\frac{1}{2}$).

***K* meson or kaon** The name of particular spin-0 mesons with non-zero strangeness quantum numbers.

Kelvin Unit of absolute temperature.

Lagrangian A mathematical expression summarising the properties and interactions of a physical system. It is essentially the difference between the kinetic energy and potential energy of the system. Moreover, one can derive the system's dynamical equations of motion directly from the Lagrangian.

lepton The generic name for any spin-$\frac{1}{2}$ particle which does not feel the strong nuclear force. The six known leptons are the electron, the muon, the tau lepton, and their respective neutrinos. The name was originally coined to refer to light particles.

lifetime The time it takes for a sample of identical particles to decay to $1/e$ of its initial population ($e \approx 2.718$). A related concept is 'half-life', being the time it takes for the number of particles to halve. Half-life, $\tau_{1/2}$, is related to lifetime, τ, by $\tau_{1/2} = (\ln 2)\tau$.

magnetic moment A measure of the extent to which a physical system (e.g. an atom, or nucleus, or particle) behaves like a tiny magnet. It is generally measured in units of magnetons, i.e. $e\hbar/2mc$.

magnetic monopole A hypothetical particle that carries an isolated north or south magnetic pole. This is in contrast to magnets which are north–south pole pairs. If magnetic monopoles exist, they must be very massive.

mass-shell In quantum mechanics, a particle's energy and momentum are essentially independent of each other. A particle is said to be 'on mass-shell' when its energy and momentum satisfy the formula from special relativity:

$$E^2 = p^2 c^2 + m_0^2 c^4$$

which is necessary for it to exist as a real observable particle. Otherwise, the particle is 'virtual'.

meson The generic name for any strongly interacting particle with integer spin in units of $\hbar$ (e.g. the pion and kaon).

muon (μ) A second-generation lepton. It is essentially a more massive electron.

natural units Units of length, time, mass, etc. in which the fundamental constants c (the speed of light), $\hbar$ (Planck's constant) and k_B (Boltzmann's constant) are equal to unity. That is, c, $\hbar$ and k_B have the numerical value 1. (For example, if we measure length in light-years and time in years, then $c = 1$ light-year per year.) The use of natural units allows these constants to be omitted from mathematical equations, leading to less-cluttered calculations. In natural units, $E = mc^2$ becomes $E = m$ and $E = k_B T$ becomes $E = T$, so that both mass and temperature can be expressed in units of energy. (Of course, the correct factors of c, $\hbar$ and k_B must be inserted at the end of a calculation to obtain measurable quantities.)

neutral-current reactions Weak-interaction reactions in which no electric charge is exchanged between the colliding particles. Observation of such reactions in 1973 provided important support for the then-developing gauge theory of the weak interactions. We now know that these reactions are mediated by the exchange of a massive, neutral gauge boson – the Z^0.

neutrino An electrically neutral, massless particle of spin-$\frac{1}{2}$, which interacts only by the weak force and gravity. It was first postulated by Pauli in 1930 to ensure conservation of energy and angular momentum in nuclear β decay. Three different types of neutrinos are known to exist corresponding to the three massive leptons: ν_e, ν_μ and ν_τ.

neutron One of the constituents of the atomic nucleus discovered in 1932. It is bound into atomic nuclei by the strong nuclear force. Free neutrons decay slowly via the weak nuclear force. Despite being electrically neutral, the neutron possesses both an electric dipole moment (as if it were made of positive and negative charges separated a minute distance) and a magnetic moment, indicating some internal substructure.

Noether's theorem A mathematical theorem that states that for every symmetry of the Lagrangian of a physical system (i.e. for every set of transformations under which the Lagrangian is invariant), there will be some quantity which is conserved by the dynamics of the system.

nucleon The generic name for the proton and the neutron.

nucleosynthesis The process by which the light elements (deuterium, helium, lithium) were synthesised in the first few minutes after the Big Bang. See Section 42.3.

parity The operation which reverses the signs of the coordinate axes used to describe a system, i.e. $(x, y, z) \rightarrow (-x, -y, -z)$.

parton A generic term used to describe any particle which may be present inside nucleons. It includes quarks, antiquarks and gluons.

Pauli's exclusion principle Two identical fermions cannot occupy the same quantum state (i.e. cannot have the same charge, spin, momentum, quantum numbers etc. within the same region of space).

phase A number (usually expressed as an angle between $0°$ and $360°$) which characterises a wave. The phase of a wave corresponds to the position in its cycle relative to an arbitrary reference point. It is a measure of how far away a wave crest or trough is.

photon (γ) The quantum of the electromagnetic field. It is the massless spin-1 gauge boson of QED. Virtual photons mediate the electromagnetic force between charged particles. Virtual photons can also adopt a mass for a short period, in accordance with Heisenberg's uncertainty principle.

Planck units Fundamental units of length, time, mass, energy, etc. involving Planck's quantum constant, $\hbar$, Newton's gravitational constant, G, and the speed of light, c. As they incorporate both the quantum and gravitational constants, the Planck units play a key role in theories of quantum gravity. See Table 40.1.

positron The antiparticle of the electron, discovered by Anderson in 1934. It has the same mass and spin as the electron, but opposite charge and magnetic moment.

propagator The mathematical expression used to describe the propagation in space–time of virtual particles.

proton One of the constituents of the atomic nucleus. It is a spin-$\frac{1}{2}$ particle carrying positive electric charge. The proton is the lightest baryon and, as a result, is the particle into which all other baryons eventually decay. It is believed to be absolutely stable, but certain theories (GUTs) predict it will decay very, very slowly.

quantum chromodynamics (QCD) The quantum field theory describing the interactions of quarks through the strong 'colour' field (whose quanta are gluons). QCD is a gauge theory with the non-Abelian gauge symmetry group $SU(3)_C$.

quantum electrodynamics (QED) The quantum field theory describing the interactions between electrically charged particles through the electromagnetic field (whose quantum is the photon). QED is a gauge theory with the Abelian gauge symmetry group $U(1)$.

quantum field theory The theory used to describe the physics of elementary particles. According to this theory, quantum fields are the ultimate reality and particles are merely the localised quanta of these fields.

quantum-mechanical amplitude A mathematical quantity in quantum mechanics whose absolute square determines the probability of a particular process occurring. (Symbol: M or m.)

quantum theory The theory used to describe physical systems which are very small, of atomic dimensions or less. A feature of the theory is that certain quantities (e.g. energy, angular momentum, light) can only exist in certain discrete amounts, called quanta.

quark A spin-$\frac{1}{2}$ particle with fractional electric charge ($+\frac{2}{3}$ or $-\frac{1}{3}$). Baryons are composed of three (valence) quarks which are bound together by the strong colour forces, and mesons consist of a bound quark and antiquark. Quarks come in six flavours (u, d, s, c, b, t) and three colours (r, g, b).

renormalisation The process which ensures that the basic quantities in quantum field theory (e.g. in QED: the photon, electron and electric charge) are well-defined and not infinite.

resonance particles or resonances Hadronic particles which exist for only a very brief time (10^{-23} seconds) before decaying into hadrons.

r.f. (radio-frequency) power Electromagnetic fields alternating at the frequencies of radio waves (up to 10^{10} Hz), which can be used to accelerate charged particles in accelerators.

scaling The phenomenon observed in deep inelastic scattering, and predicted by James Bjørken, whereby the structure functions which describe the shape of the nucleon depend not on the energy or momentum involved in the reaction, but on some dimensionless ratio of the two. The structure functions are hence independent of any dimensional scale.

singularity A point in space–time at which the space–time curvature and other physical quantities become infinite and the laws of physics break down.

SLAC The acryonym for the Stanford Linear Accelerator Center at Stanford University in California, USA. It is distinguished by having a 2-mile-long linear accelerator in which electrons and positrons can be accelerated for subsequent injection into storage rings such as PEP, an e^+e^- collider which was commissioned in 1980. It was in the SPEAR rings at SLAC that the J/ψ (psi) meson and the τ (tau) lepton were first observed in the mid-1970s. However, the most fascinating of SLAC's facilities is the novel SLC (Stanford Linear Collider), consisting of the old linear accelerator together with two new collider arcs.

spin The intrinsic angular momentum possessed by many particles. It can be thought of as resulting from the particles spinning about an axis through their centres. In contrast to orbital angular momentum, spin is quantised in integer and half-integer units of $\hbar$. Fundamentally, spin describes how quantum fields transform under the transformations of special relativity.

spontaneous symmetry breaking Any situation in physics in which the ground state (i.e. the state of minimum energy) of a system has less symmetry than the system itself. For example, the state of minimum energy for an iron magnet is that in which the atomic spins are all aligned in the same direction, giving rise to a net macroscopic magnetic field. By selecting a particular direction in space, the magnetic field has broken the rotational symmetry of the system. However, if the energy of the system is raised, the symmetry may be restored (e.g. the application of heat to an iron magnet destroys the magnetic field and restores rotational symmetry).

standard model This refers collectively to the successful theories of QCD and the Glashow–Weinberg–Salam electroweak model.

strangeness A quantum number associated with the strange quark. Strangeness is conserved by the strong nuclear force.

string theory A theory in which the fundamental constituents of matter are not particles but tiny one-dimensional objects, which we can think of as strings. These strings are so minute (only 10^{-33} cm long) that, even at current experimental energies, they seem to behave just like particles. So, according to string theory, what we call 'elementary particles' are actually tiny strings, each of which is vibrating in a way characteristic of the particular 'elementary particle'.

supergravity A supersymmetric theory of gravity in which the graviton is accompanied by a spin-$\frac{3}{2}$ particle called the 'gravitino'. In supergravity theories, supersymmetry has been promoted to the status of a local gauge symmetry.

supersymmetry A symmetry relating fermions and bosons. If supersymmetry is a true symmetry of nature, then every 'ordinary' particle has a corresponding 'superpartner' which differs in spin by half a unit.

top The sixth, as yet undiscovered, flavour of quark.

vacuum The state of minimum energy (or ground state) of a quantum theory. It is the quantum state in which no real particles are present. However, because of Heisenberg's uncertainty principle, the vacuum is actually seething with *virtual* particles which constantly materialise, propagate a short distance and then disappear. See Section 4.9.

APPENDIX 3

List of symbols

a	acceleration
A	electromagnetic gauge field
b	bottom quark flavour; blue quark colour
B	gauge field of $U(1)^W$
B	baryon number
c	charm quark flavour
c	speed of light
C	charge conjugation operator
d	down quark flavour
D	meson
e	electron
e	electronic charge
E	energy; electric field strength
f	parton probability distribution
F	force
$F_1\,F_2\,F_3$	deep inelastic structure functions
g	green quark colour; gluon
g	g-factor of the electron; general coupling constant
G	Newton's constant

G_F	Fermi's coupling constant	**p**	momentum vector
G	Group	P	probability (of occurrence of quantum-mechanical event); polarisation; fraction of nucleon momentum carried by the quark
h	hadron		
h	Planck's constant; strangeness-conserving weak hadronic current		
		P	parity operator
H	magnetic field strength	q	quark
H^W	weak hadronic current	q^2	momentum transfer squared in deep inelastic scattering
I	isospin		
I_3	third component of isospin	Q	electric charge (in units of e)
J	spin	r	red quark colour
J^W	total weak current	r	magnitude of distance
K	kaon	R	Reggeon
K	constant of proportionality	R	vacuum expectation value of Higgs fields; ratio of cross-sections for e^+e^- into hadrons to e^+e^- into muon–antimuon pair
$l_e\, l_\mu$	electronic or muonic lepton		
L	Lagrangian; length; lepton number		
		s	strange quark flavour
L^W	weak leptonic current	s	space–time interval; spin
$\mathcal{L}$	Lagrangian density		
m	mass; quantum-mechanical amplitude of sub-processes	s_z	third component of spin
		$s^\pm$	strangeness-changing weak hadronic current
M	total quantum-mechanical amplitude of a process	S	strangeness quantum number
M	matrix	$SU(2)$ $SU(3)$ $SU(5)$	special unitary groups of transformations of order 2, 3 and 5 respectively
n	neutron		
n	an integer		
N	nucleon	t	top quark flavour
N*	baryon resonances	t	time
N	numbers of . . .	T	temperature
$N_L\, N_R$	number of left-spinning, (right-spinning)	**T**	time-reversal operator
		u	up quark flavour
p	proton	u	magnitude of velocity
p	magnitude of momentum	v	magnitude of velocity
		v	velocity vector

$W^{\pm}$	W boson	ν	(nu) neutrino; deep inelastic scattering energy transfer; frequency
x	deep inelastic scattering variable; spatial separation		
$\mathbf{x}$	position vector	ξ	(xi)
X	massive gauge bosons predicted by GUTs; unspecified final state of reaction	Ξ	xi hyperon
		o	(omicron)
y	fraction of energy transferred in deep inelastic scattering	π	(pi) 3.141 5927; pion
Y	hypercharge quantum number	ρ	(rho) rho meson; hypothetical hadronic isospin gauge particle
Z^0	Z boson		
α	(alpha) α radiation or particles (helium nuclei); electromagnetic fine structure constant	σ	(sigma) cross-section
		$\sum$	summation over . . .
		Σ	sigma hyperon
β	(beta) β radiation or particles (electrons)	τ	(tau) tau heavy lepton; tau meson (archaic); lifetime, duration
γ	(gamma) γ radiation or particles (photons)		
		υ	(upsilon)
Γ	Fermi's weak interaction couplings	Υ	upsilon meson
δ	(delta) infinitesimal increment in variable in calculus	ϕ	(phi) Higgs particles
		χ	(chi) χ meson
Δ	infinitesimal amount of . . .; delta baryon	ψ	(psi) quantum-mechanical wavefunction; psi meson
ε	(epsilon) small number		
ζ	(zeta)	ω	(omega) omega meson
η	(eta)	Ω^-	omega minus baryon
θ	(theta) theta meson (archaic); an angle	$=$	is equal to
		$\equiv$	is identical with
θ_C	the Cabbibo angle	$\approx$	is approximately equal to
θ_W	the weak angle	$a > b$	a is greater than b
ι	(iota)	$a < b$	a is less than b
κ	(kappa)	$a \leqslant b$	a is equal to, or less than, b
λ	(lambda) wavelength	$a \geqslant b$	a is equal to, or greater than, b
Λ	lambda hyperon	$a \supset b$	a contains b
μ	(mu) muon	$\langle f\|M\|i \rangle$	initial i and final f states connected by a quantum-mechanical amplitude

APPENDIX 4

Bibliography

This bibliography provides two sets of references on most of the subjects dealt with in this book. The first, non-specialist, category includes articles and books accessible to the audience of books such as this one. The specialist category includes material for the professional student of physics and is generally aimed at the level of a third-year undergraduate or first-year postgraduate.

(NS) denotes the non-specialist category.

(S) denotes the specialist category.

Part 0

(NS) *From X-rays to Quarks*, E. Segrè. Freeman, San Francisco, 1980.

(NS) *Relativity*, Albert Einstein. Methuen, London, 1920.

(NS) 'The classical vacuum', T. H. Boyer. *Scientific American*, **253** (2), 56–62, August 1985.

(S) *Special Relativity*, A. P. French. Nelson, London, 1968.

(S) *Simple Quantum Physics*, P. Landshoff & A. Metherell. Cambridge University Press, 1979.

(S) *Relativistic Quantum Mechanics*, I. J. R. Aitchison. Macmillan, London, 1972.

(S) *Quantum Field Theory*, F. Mandl & G. Shaw. Wiley, Chichester, 1984.

Part 1 and general

(NS) *The Forces of Nature*, P. C. W. Davies. Cambridge University Press, 1979.

(NS) *The Particle Play*, J. C. Polkinghorne. Freeman, Oxford, 1979.

(NS) *The Nature of Matter*, J. H. Mulvey (ed.). Clarendon, Oxford, 1981.

(NS) *The Cosmic Onion*, F. Close. Heinemann, London, 1983.

(NS) *The Particle Explosion*, F. Close, M. Marten & C. Sutton. Oxford University Press, 1987.

(S) *An Introduction to High-Energy Physics*, D. H. Perkins. Addison-Wesley, Reading, Mass. 3rd edn), 1987. (An excellent textbook on most of the material mentioned in this book.)

(S) *Symmetry Principles in Elementary Particle Physics*, W. M. Gibson & B. R. Pollard. Cambridge University Press, 1976.

Part 2

(NS) 'Resonance particles', R. D. Hill. *Scientific American*, **208** (1), January 1963.

(NS) 'Strongly interacting particles', G. F. Chew, M. Gell-Mann & A. H. Rosenfeld. *Scientific American*, **210** (2), February 1964.

(NS) 'Dual-resonance models of elementary particles', J. H. Schwarz. *Scientific American*, **232** (5), 61–7, February 1975.

(S) *High Energy Hadron Physics*, M. L. Perl. Wiley, New York, 1974. (A thorough textbook on pre-QCD hadron physics.)

(S) 'High energy hadron collisions: a point of view', J. P. Aurenche & J. E. Paton. *Reports on Progress in Physics*, **39** (2), February 1976.

Parts 3 and 4

(NS) 'The weak interactions', S. B. Treiman. *Scientific American*, March 1959.

(NS) 'The two-neutrino experiment', L. M. Lederman. *Scientific American*, **208** (3), March 1963.

(NS) 'Violations of symmetry in physics', E. P. Wigner. *Scientific American*, **213** (6), 28–36, December 1965.

(NS) 'The detection of weak neutral currents',

D. B. Cline, A. K. Mann & C. Rubbia. *Scientific American*, **231** (6), 108–119, December 1974.

(NS) 'Weak interactions', M. K. Gaillard. *Nature*, **279**, 585–9, June 1979.

(S) *Weak Interactions*, D. Bailin. Adam Hilger, Bristol (2nd edn), 1982.

Part 5

(NS) 'Unified theories of elementary-particle interaction', S. Weinberg. *Scientific American*, **231** (1), 50–9, July 1974.

(NS) 'Gauge theories of the forces between elementary particles', G. 't Hooft. *Scientific American*, **242** (6), 104–138, June 1980.

(NS) 'Elementary particles and forces', C. Quigg. *Scientific American*, **252** (4), 64–75, April 1985.

(NS) 'The Higgs boson', M. J. G. Veltman. *Scientific American*, **255** (5), 88–94, November 1986.

(NS) 'The search for intermediate vector bosons', D. B. Cline, C. Rubbia & S. van der Meer. *Scientific American*, **246** (3), 38–49, March 1982.

(NS) *Story of the W and Z*, P. Watkins. Cambridge University Press, 1986.

(NS) 'The superconducting super-collider', J. D. Jackson, M. Tigner & S. Wojcicki. *Scientific American*, **254** (3), 56–67, March 1986.

(S) *An Informal Introduction to Gauge Theories*, I. J. R. Aitchison. Cambridge University Press, 1982.

(S) *Gauge Theories in Particle Physics*, I. J. R. Aitchison & A. J. G. Hey. Hilger, Bristol, 1982.

(S) *Gauge Theories of Weak Interactions*, J. C. Taylor. Cambridge University Press, 1978.

Part 6

(S) 'Inelastic lepton-nucleon scattering', D. H. Perkins. *Reports on Progress in Physics*, **40**, 409–81, 1977.

(S) *An Introduction to Quarks and Partons*, F. E. Close. Academic Press, London, 1979.

Part 7

(NS) 'Quantum chromodynamics', W. Marciano & H. Pagels. *Nature*, **279**, 479–83, June 1979.

(NS) 'Quark confinement', R. L. Jaffe. *Nature*, **268**, 201–9, July 1977.

(S) 'Quantum chromodynamics', W. Marciano & H. Pagels. *Physics Reports*, **36C**, 137–276, 1978.

(S) *Lectures on Lepton Nuclear Scattering and Quantum Chromodynamics*, W. B. Atwood, J. D. Bjørken, S. J. Brodsky & R. Stroynowski. Birkhauser, Boston, 1982.

Part 8

(NS) 'Electron–positron collisions', A. M. Litke & R. Wilson. *Scientific American*, **229** (4), 104–13, October 1973.

(NS) 'Fundamental particles with charm', R. F. Schwitters. *Scientific American*, **237** (4), 56–70, October 1977.

(NS) 'The upsilon particle', L. M. Lederman. *Scientific American*, **239** (4), 60–8, October 1978.

(NS) 'The tau heavy lepton', M. L. Perl. *Nature*, **275**, 273–7, September 1978.

(NS) 'Particles with naked beauty', N. B. Mistry, R. A. Poling & E. H. Thorndike. *Scientific American*, **249** (1), 98–107, July 1983.

(NS) 'The Stanford linear collider', J. R. Rees. *Scientific American*, **261** (4), 36–43, October 1989.

(NS) 'LEP', Search and discovery, *Physics Today*, **42** (10), 17–20, October 1989.

Part 9

(NS) 'Unified theory of elementary-particle forces', H. Georgi & S. L. Glashow. *Physics Today*, **33** (9), September 1980.

(NS) 'A unified theory of elementary particles and forces', H. Georgi. *Scientific American*, **244** (4), 40–55, April 1981.

(NS) 'The cosmic asymmetry between matter and antimatter', F. Wilczek. *Scientific American*, **243** (6), 60–68, December 1980.

(NS) *A Brief History of Time*, S. W. Hawking. Bantam Press, 1988.

(NS) 'Quantum gravity', B. C. De Witt.

Scientific American, **249** (6), 104–115, December 1983.

(NS) 'The quantum mechanics of black holes', S. W. Hawking. *Scientific American*, **236** (1), 34–40, January 1977.

(NS) 'Black-hole thermodynamics', J. D. Bekenstein. *Physics Today*, **33** (1), 24–31, January 1980.

(NS) 'Is nature supersymmetric?', H. E. Haber & G. L. Kane. *Scientific American*, **254** (6), 42–50, June 1986.

(NS) 'The hidden dimensions of space–time', D. Z. Freedman & P. van Nieuwenhuizen. *Scientific American*, **252** (3), 62–69, March 1985.

(NS) *The First Three Minutes*, S. Weinberg. Basic Books, New York, 1977.

(NS) 'Cosmology and elementary particle physics', M. S. Turner & D. N. Schramm. *Physics Today*, **32** (9), September 1979.

(NS) 'Particle accelerators test cosmological theory', D. N. Schramm & G. Steigman. *Scientific American*, **258** (6), 44–50, June 1988.

(NS) 'Dark matter in the universe', L. M. Krauss. *Scientific American*, **255** (6), 50–60, December 1986.

(NS) 'The search for dark matter in the laboratory', B. Moskowitz. *New Scientist*, 39–42, 15 April 1989.

(NS) 'How a supernova explodes', H. A. Bethe & G. Brown. *Scientific American*, **252** (5), 40–48, May 1985.

(NS) 'The great supernova of 1987', S. Woosley & T. Weaver. *Scientific American*, **261** (2), 24–32, August 1989.

(NS) 'The inflationary universe', A. H. Guth & P. J. Steinhardt. *Scientific American*, **250** (5), May 1984.

(NS) 'Superstrings', M. B. Green. *Scientific American*, **255** (3), 44–56, September 1986.

(NS) 'Solitons', C. Rebbi. *Scientific American*, **240** (2), 76–92, February 1979.

(NS) 'Superheavy magnetic monopoles'. R. A. Carrigan & W. P. Trower. *Scientific American*, **246** (4), 91–99, April 1982.

(NS) *Superstrings: A Theory of Everything?*, P. C. W. Davies & J. Brown. Cambridge University Press, 1988.

(S) *Grand Unified Theories*, G. G. Ross. Benjamin/Cummings, 1985.

(S) 'Why superstrings?', D. Bailin. *Contemporary Physics*, **30**, 237–250 (1989).

General sources of information

For the non-specialist interested in keeping abreast of the modern developments in particle physics, the following magazines and journals combine to form a good coverage: *New Scientist*, *Scientific American*, *Nature*, *Physics Today* and *CERN Courier*.

For the specialist, the most up-to-date letters appear in *Physical Review Letters* and *Physics Letters*. The mainstream journals on the subject are *The Physical Review* (*D*), *Nuclear Physics* (*B*) and, to a lesser extent, *Nuovo Cimento* and *Journal of Physics* (*G*). Review articles on the subject generally appear in *Physics Reports*, *Reviews of Modern Physics* and *Reports on Progress in Physics*. For up-to-date, comprehensive reviews of the entire field the best sources are the summer school lecture note compilations and the conference proceedings on the subject published periodically; generally by CERN or SLAC.

APPENDIX 5

Abridged particle table

The following list is a complete account of the particles known to date. Only the main features of the particle decays have been listed. No long-standing unconfirmed particles have been included, nor have the various dibaryon states identified to date. For full details see the *Review of Particle Properties*, published by the Particle Data Group at CERN (*Physics Letters*, **B239**, April 1990.)

Table A.5. *Abridged particle table*

Particle	Parity P Isospin I Spin J $I(J^P)$	Mass (MeV)	Lifetime τ (seconds) Width Γ (MeV)	Decays Main mode	Fraction (%)
GAUGE BOSONS					
γ	$0(1^-)$	$0\,(<3\times10^{-33})$	Stable		
$W^\pm$	$1(1^-)$	$80.6\pm0.4\,\text{GeV}$	$\Gamma=2.25\pm0.14\,\text{GeV}$	$e\nu,\mu\nu,\tau\nu$; hadrons	100
Z^0	$1(1^-)$	$91.177\pm0.031\,\text{GeV}$	$\Gamma=2.496\pm0.016\,\text{GeV}$	$e^+e^-,\mu^+\mu^-,\tau^+\tau^-,\nu\bar\nu$; hadrons	30 70
LEPTONS					
ν_e	$J=\tfrac12$	$0\,(<0.000\,010)$	Stable		
e	$J=\tfrac12$	$0.510\,999\,06\pm0.000\,000\,15$	Stable		

Table A.5. (*cont.*)

Particle	Parity P Isospin I Spin J $I(J^P)$	Mass (MeV)	Lifetime τ (seconds) Width Γ (MeV)	Decays Main mode	Fraction (%)
ν_μ	$J = \frac{1}{2}$	0 (<0.25)	Stable		
μ	$J = \frac{1}{2}$	$105.658\,39 \pm 0.000\,06$	$\tau = (2.197\,03 \pm 0.000\,04)$ $\times 10^{-6}$	$\bar{\mu} \to e^- \bar{\nu}\nu$ $e^- \bar{\nu}\nu\gamma$	98.6 1.4
ν_τ	$J = \frac{1}{2}$	0 (<35)	?		
τ	$J = \frac{1}{2}$	1784.2 ± 3.2	$\tau = (3.04 \pm 0.09) \times 10^{-13}$	$\tau^- \to \mu^- \bar{\nu}\nu$ $\to e^- \bar{\nu}\nu$ $\to$ hadrons	18 18 64

NON-STRANGE MESONS

Particle	$I(J^P)$	Mass (MeV)	Lifetime/Width	Main mode	Fraction (%)
$\pi^\pm$	$1(0^-)$	139.5675 ± 0.0004	$\tau = (2.6030 \pm 0.0023) \times 10^{-8}$	$\pi^+ \to \mu^+ \nu$	99.99
π^0	$1(0^-)$	134.9739 ± 0.0006	$\tau = (0.83 \pm 0.06) \times 10^{-16}$	$\gamma\gamma$ $\gamma e^+ e^-$	98.8 1.2
η	$0(0^-)$	548.8 ± 0.6	$\Gamma = (1.19 \pm 0.12)\,\text{keV}$	$\gamma\gamma$ $3\pi^0$ $\pi^+\pi^-\pi^0$	39.1 31.8 23.7
$\rho(770)$	$1(1^-)$	768.3 ± 0.5	$\Gamma = 149 \pm 2.9$	$\pi\pi$	≈ 100
$\omega(783)$	$0(1^-)$	782.0 ± 0.1	$\Gamma = 8.5 \pm 0.1$	$\pi^+\pi^-\pi^0$ $\pi^0\gamma$	89.3 8.0
$\eta'(958)$	$0(0^-)$	957.50 ± 0.24	$\Gamma = 0.21 \pm 0.02$	$\eta\pi\pi$ $\rho^0\gamma$	44.1 30.1
$f_0(975)$	$0(0^+)$	976 ± 3	$\Gamma = 34 \pm 6$	$\pi\pi$ $K\bar{K}$	78 22
$a_0(980)$	$1(0^+)$	983 ± 3	$\Gamma = 57 \pm 11$	$\eta\pi$ $K\bar{K}$	? ?
$\phi(1020)$	$0(1^-)$	1019.41 ± 0.01	$\Gamma = 4.41 \pm 0.05$	K^+K^- $K_{\text{LONG}}K_{\text{SHORT}}$ $\pi^+\pi^-\pi^0$	49.1 34.6 14.8
$h_1(1170)$	$0(1^+)$	1170 ± 21	$\Gamma = 311 \pm 33$	$\rho\pi$	?
$b_1(1235)$	$1(1^+)$	1233 ± 10	$\Gamma = 150 \pm 10$	$\omega\pi$	most
$a_1(1260)$	$1(1^+)$	1260 ± 30	$\Gamma = 300\text{--}600$	$\rho\pi$	most

Table A.5. (*cont.*)

Particle	Parity P Isospin I Spin J $I(J^P)$	Mass (MeV)	Lifetime τ (seconds) Width Γ (MeV)	Decays Main mode	Fraction (%)
$f_2(1270)$	$0(2^+)$	1274 ± 5	$\Gamma = 185 \pm 20$	$\pi\pi$ $\pi^+\pi^-2\pi^0$ $K\bar{K}$	86 7 4
$f_1(1285)$	$0(1^+)$	1283 ± 5	$\Gamma = 25 \pm 3$	$K\bar{K}\pi$ $\eta\pi\pi$ $a_0\pi$	11 49 36
$\pi(1300)$	$1(0^-)$	1300 ± 100	$\Gamma = 200\text{--}600$	Tentative assignment	
$a_2(1320)$	$1(2^+)$	1318 ± 1	$\Gamma = 110 \pm 5$	$\rho\pi$ $\eta\pi$ $\omega\pi\pi$	70.1 14.5 10.6
$f_0(1400)$	$0(0^+)$	≈ 1400	$\Gamma = 150\text{--}400$	$\pi\pi$	most
$f_1(1420)$	$0(1^+)$	1425 ± 15	$\Gamma = 55 \pm 3$	$K\bar{K}\pi$	most
$f_2'(1525)$	$0(2^+)$	1525 ± 5	$\Gamma = 76 \pm 10$	$K\bar{K}$	most
$\omega_3(1670)$	$0(3^-)$	1668 ± 5	$\Gamma = 166 \pm 15$	$3\pi, 5\pi$	most
$\pi_2(1670)$	$1(2^-)$	1665 ± 20	$\Gamma = 250 \pm 20$	$f_2(1270)\pi$ $\rho\pi$	55 36
$\phi(1680)$	$0(1^-)$	1680 ± 50	$\Gamma = 150 \pm 50$	$K^*\bar{K} + \bar{K}^*K$	most
$\rho_3(1690)$	$1(3^-)$	1691 ± 5	$\Gamma = 215 \pm 20$	$2\pi, 4\pi$	most
$\rho'(1700)$	$1(1^-)$	1700 ± 20	$\Gamma = 235 \pm 50$	4π $\pi\pi\rho$	
$f_4(2050)$	$0(4^+)$	2049 ± 11	$\Gamma = 203 \pm 13$	$\pi\pi$ $\omega\omega$	17 25
$\eta_c(2980)$	$0(\ )$	2979.6 ± 1.7	$\Gamma = 10.3 \pm 3.4$	$\eta\pi^+\pi^-$ others	4 96
$J/\psi(3097)$	$0(1^-)$	3096.9 ± 0.1	$\Gamma = 0.068 \pm 0.009$	e^+e^- $\mu^+\mu^-$ hadrons	6.9 6.9 86.2
$\chi_{C0}(3415)$	$0(0^+)$	3415.0 ± 1.0	$\Gamma = 13.5 \pm 5.3$	$2(\pi^+\pi^-)$ $\pi^+\pi^-K^+K^-$	3.7 3.0
$\chi_{C1}(3510)$	$0(1^+)$	3510.6 ± 0.5	$\Gamma < 1.3$	$\gamma\psi$ $3(\pi^+\pi^-)$	27.3 2.2

Table A.5. (*cont.*)

Particle	Parity P Isospin I Spin J $I(J^P)$	Mass (MeV)	Lifetime τ (seconds) Width Γ (MeV)	Decays Main mode	Fraction (%)
$\chi_{C2}(3555)$	$0(2^+)$	3556.3 ± 0.4	$\Gamma = 2.6 \pm 1$	$\gamma\psi$ $2(\pi^+\pi^-)$	13.5 2.3
$\psi(3685)$ or ψ'	$0(1^-)$	3686.0 ± 0.1	$\Gamma = 0.243 \pm 0.043$	e^+e^- $\mu^+\mu^-$ hadrons	0.09 0.08 98.2
$\psi(3770)$ or ψ''	(1^-)	3769.9 ± 2.5	$\Gamma = 24 \pm 3$	$D\bar{D}$	most
$\psi(4040)$	(1^-)	4040 ± 10	$\Gamma = 52 \pm 10$	hadrons	most
$\psi(4160)$	(1^-)	4159 ± 20	$\Gamma = 78 \pm 20$	hadrons	most
$\psi(4415)$	(1^-)	4415 ± 6	$\Gamma = 43 \pm 15$	hadrons	most
$\Upsilon(9460)$	(1^-)	9460.3 ± 0.2	$\Gamma = 0.052 \pm 0.003$	$\tau^+\tau^-$ $\mu^+\mu^-$ e^+e^-	3.0 2.6 2.5
χ_{b0}	$?(0^+)$	9859.8 ± 1.3	?	$\Upsilon\gamma$	
χ_{b1}	$?(1^+)$	9891.9 ± 0.7	?	$\Upsilon\gamma$	
χ_{b2}	$?(2^+)$	9913.2 ± 0.6	?	$\Upsilon\gamma$	
$\Upsilon(10023)$	$?(1^-)$	$10\,023 \pm 0.3$	$\Gamma = 0.044 \pm 0.009$	$\Upsilon\pi\pi$ e^+e^- $\mu^+\mu^-$	27.3 1.7 1.4
$\Upsilon(10355)$	$?(1^-)$	$10\,335 \pm 0.5$	$\Gamma = 0.024 \pm 0.003$	e^+e^-	seen
$\Upsilon(10580)$	$?(1^-)$	$10\,580 \pm 3.5$	$\Gamma = 24 \pm 2$	e^+e^-	seen
STRANGE MESONS					
$K^\pm$	$\frac{1}{2}(0^-)$	493.646 ± 0.009	$\tau = (1.2371 \pm 0.0028) \times 10^{-8}$	$K^+ \to \mu^+\nu$ $\pi^+\pi^0$	63.5 21.2
K^0 $\bar{K}^0$	$\frac{1}{2}(0^-)$	497.67 ± 0.03		50% K_{SHORT}	50% K_{LONG}
K^0_{SHORT}	$\frac{1}{2}(0^-)$		$\tau = (0.8923 \pm 0.0022) \times 10^{-10}$	$\pi^+\pi^-$ $\pi^0\pi^0$	68.4 31.4

Table A.5. (*cont.*)

Particle	Parity P Isospin I Spin J $I(J^P)$	Mass (MeV)	Lifetime τ (seconds) Width Γ (MeV)	Decays Main mode	Fraction (%)
K^0_{LONG}	$\frac{1}{2}(0^-)$		$\tau = (5.183 \pm 0.040) \times 10^{-8}$	$\pi^0\pi^0\pi^0$	21.5
				$\pi^+\pi^-\pi^0$	12.4
				$\pi^\pm\mu^\mp\nu$	27.1
				$\pi^\pm e^\mp\nu$	38.7
$K^*(892)$	$\frac{1}{2}(1^-)$	891.83 ± 0.24	$\Gamma = 49.8 \pm 0.8$	$K\pi$	most
$K_1(1270)$	$\frac{1}{2}(1^+)$	1270 ± 10	$\Gamma = 90 \pm 20$	$K\rho$	42
				$K\pi$	28
				$K^*\pi$	16
$K_1(1400)$	$\frac{1}{2}(1^+)$	1402 ± 7	$\Gamma = 174 \pm 13$	$K\pi$	most

CHARMED, NON-STRANGE MESONS

Particle	$I(J^P)$	Mass (MeV)	Lifetime / Width	Main mode	Fraction (%)
$D^\pm$	$\frac{1}{2}(0^-)$	1869.4 ± 0.6	$\tau = (10.62 \pm 0.28) \times 10^{-13}$	$K^0\overline{K^0}X$	48
				eX	19
				KX	22
D^0 $\overline{D^0}$	$\frac{1}{2}(0^-)$	1864.7 ± 0.6	$\tau = (4.21 \pm 0.10) \times 10^{-13}$	$K\overline{K^0}X$	33
				KX	49
				eX	7.7

CHARMED, STRANGE MESON

Particle	$I(J^P)$	Mass (MeV)	Lifetime / Width	Main mode	Fraction (%)
D_S^+ (or F^+)	$0(0^-)$	1968.8 ± 0.7	$\tau = (4.4 \pm 0.3) \times 10^{-13}$	$\phi\pi$	2.7
				$\phi\pi\pi$	1.3
				ρK^*K	5.8

BOTTOM, NON-STRANGE MESON

Particle	$I(J^P)$	Mass (MeV)	Lifetime / Width	Main mode	Fraction (%)
$B^\pm$	$\frac{1}{2}(0^-)$	5277.6 ± 1.4	$(11.8 \pm 1.1) \times 10^{-13}$	$J/\psi K^+$	
				$Dn\pi$	
B^0 $\overline{B^0}$	$\frac{1}{2}(0^-)$	5279.4 ± 1.5	?	$D^*l\nu$	10

NON-STRANGE BARYONS

Particle	$I(J^P)$	Mass (MeV)	Lifetime / Width	Main mode	Fraction (%)
p	$\frac{1}{2}(\frac{1}{2}^+)$	938.27231 ± 0.00028	Stable? $\tau > 10^{32}$ (yr)		
n	$\frac{1}{2}(\frac{1}{2}^+)$	939.56563 ± 0.00028	$\tau = 888.6 \pm 3.5$	$pe^-\bar{\nu}_e$	≈ 100
$N(1440)$	$\frac{1}{2}(\frac{1}{2}^+)$	1400–1480	$\Gamma = 120$–350	$N\pi$	50–70
				$\Delta\pi$	12–28

Table A.5. (*cont.*)

Particle	Parity P Isospin I Spin J $I(J^P)$	Mass (MeV)	Lifetime τ (seconds) Width Γ (MeV)	Decays Main mode	Fraction (%)
N(1520)	$\frac{1}{2}(\frac{3}{2}^-)$	1510–1530	$\Gamma = 100$–140	$N\pi$ $\Delta\pi$ $N\rho$	50–60 15–25 12–25
N(1535)	$\frac{1}{2}(\frac{1}{2}^-)$	1520–1560	$\Gamma = 100$–250	$N\pi$ $N\eta$	35–50 40–65
N(1650)	$\frac{1}{2}(\frac{1}{2}^-)$	1620–1680	$\Gamma = 100$–200	$N\pi$ $N\pi\pi$	55–65 ≈ 30
N(1675)	$\frac{1}{2}(\frac{5}{2}^-)$	1660–1690	$\Gamma = 120$–180	$N\pi$ $N\pi\pi$	≈ 40 ≈ 60
N(1680)	$\frac{1}{2}(\frac{5}{2}^+)$	1670–1690	$\Gamma = 110$–140	$N\pi$ $N\pi\pi$	≈ 60 ≈ 40

NON-STRANGE DELTA RESONANCES (Δ)

Particle	$I(J^P)$	Mass (MeV)	Width Γ (MeV)	Main mode	Fraction (%)
$\Delta(1232)$	$\frac{3}{2}(\frac{3}{2}^+)$	1230–1234	$\Gamma = 110$–120	$N\pi$	99
$\Delta(1620)$	$\frac{3}{2}(\frac{1}{2}^-)$	1600–1650	$\Gamma = 120$–160	$N\pi$ $N\pi\pi$	≈ 30 ≈ 70
$\Delta(1700)$	$\frac{3}{2}(\frac{3}{2}^-)$	1630–1740	$\Gamma = 190$–300	$N\pi$ $N\pi\pi$	≈ 15 ≈ 85
$\Delta(1900)$	$\frac{3}{2}(\frac{1}{2}^-)$	1850–2000	$\Gamma = 130$–300	$N\pi$ ΣK	≈ 10 ?
$\Delta(1905)$	$\frac{3}{2}(\frac{5}{2}^+)$	1890–1920	$\Gamma = 250$–400	$N\pi$ $N\pi\pi$	≈ 10 < 75
$\Delta(1910)$	$\frac{3}{2}(\frac{1}{2}^+)$	1850–1950	$\Gamma = 200$–330	$N\pi$ ΣK $N\pi\pi$	≈ 20 ? < 75

STRANGENESS-1, LAMBDA RESONANCES (Λ)

Particle	$I(J^P)$	Mass (MeV)	Lifetime τ / Width Γ	Main mode	Fraction (%)
Λ	$0(\frac{1}{2}^+)$	1115.63 ± 0.05	$\tau = (2.632 \pm 0.020) \times 10^{-10}$	$p\pi^-$ $n\pi^0$	64.1 35.7
$\Lambda(1405)$	$0(\frac{1}{2}^-)$	1405 ± 5	$\Gamma = 55 \pm 10$	$\Sigma\pi$	100
$\Lambda(1520)$	$0(\frac{3}{2}^-)$	1519.4 ± 1.0	$\Gamma = 15.6 \pm 1.0$	$N\bar{K}$ $\Sigma\pi$	45 42

Table A.5. (*cont.*)

Particle	Parity P Isospin I Spin J $I(J^P)$	Mass (MeV)	Lifetime τ (seconds) Width Γ (MeV)	Decays Main mode	Fraction (%)
$\Lambda(1600)$	$0(\frac{1}{2}^+)$	1560–1700	$\Gamma = 50$–250	$N\bar{K}$ $\Sigma\pi$	≈ 25 ≈ 50
$\Lambda(1670)$	$0(\frac{1}{2}^-)$	1660–1680	$\Gamma = 25$–50	$N\bar{K}$ $\Sigma\pi$ $\Delta\eta$	≈ 20 ≈ 40 ≈ 25
$\Lambda(1690)$	$0(\frac{3}{2}^-)$	1685–1695	$\Gamma = 50$–70	$N\bar{K}$ $\Sigma\pi$	≈ 25 ≈ 30
$\Lambda(1800)$	$0(\frac{1}{2}^-)$	1720–1850	$\Gamma = 200$–400	$N\bar{K}$ $\Sigma\pi$	≈ 35 seen

STRANGENESS-1, SIGMA RESONANCES (Σ)

Particle	$I(J^P)$	Mass (MeV)	Lifetime τ / Width Γ	Main mode	Fraction (%)
Σ^+	$1(\frac{1}{2}^+)$	1189.36 ± 0.06	$\tau = (0.800 \pm 0.004) \times 10^{-10}$	$p\pi^0$ $n\pi^+$	52 48
Σ^0	$1(\frac{1}{2}^+)$	1192.55 ± 0.08	$\tau = (7.4 \pm 0.7) \times 10^{-20}$	$\Lambda\gamma$	≈ 100
Σ^-	$1(\frac{1}{2}^+)$	1197.43 ± 0.06	$\tau = (1.482 \pm 0.011) \times 10^{-10}$	$n\pi^-$	≈ 100
$\Sigma(1385)$	$1(\frac{3}{2}^+)$	$(+)1382.8 \pm 0.4$ $(0)\ 1383.7 \pm 1.0$ $(-)1387.4 \pm 0.6$	$(+)\Gamma = 36 \pm 1$ $(0)\ \Gamma = 36 \pm 5$ $(-)\Gamma = 39 \pm 2$	$\Lambda\pi$ $\Sigma\pi$	88 12
$\Sigma(1660)$	$1(\frac{1}{2}^+)$	1630–1690	$\Gamma = 40$–200	$N\bar{K}$	≈ 20
$\Sigma(1670)$	$1(\frac{3}{2}^-)$	1665–1685	$\Gamma = 40$–80	$N\bar{K}$ $\Lambda\pi$ $\Sigma\pi$	≈ 10 ≈ 10 ≈ 50
$\Sigma(1750)$	$1(\frac{1}{2}^-)$	1730–1800	$\Gamma = 60$–160	$N\bar{K}$ $\Sigma\eta$	≈ 30 ≈ 40
$\Sigma(1775)$	$1(\frac{5}{2}^-)$	1770–1780	$\Gamma = 105$–135	$N\bar{K}$ $\Lambda\pi$ $\Lambda(1520)\pi$	≈ 40 ≈ 20 ≈ 20
$\Sigma(1915)$	$1(\frac{5}{2}^+)$	1900–1935	$\Gamma = 80$–160	$N\bar{K}$ $\Lambda\pi$	≈ 10 seen

STRANGENESS-2, CASCADE RESONANCES

Particle	$I(J^P)$	Mass (MeV)	Lifetime τ / Width Γ	Main mode	Fraction (%)
Ξ^0	$\frac{1}{2}(\frac{1}{2}^+)$	1314.9 ± 0.6	$\tau = (2.9 \pm 0.1) \times 10^{-10}$	$\Lambda\pi^0$	≈ 100
Ξ^-	$\frac{1}{2}(\frac{1}{2}^+)$	1321.32 ± 0.13	$\tau = (1.641 \pm 0.016) \times 10^{-10}$	$\Lambda\pi^-$	≈ 100

Table A.5. (*cont.*)

Particle	Parity P Isospin I Spin J $I(J^P)$	Mass (MeV)	Lifetime τ (seconds) Width Γ (MeV)	Decays Main mode	Fraction (%)
$\Xi(1530)$	$\frac{1}{2}(\frac{3}{2}^+)$	(0) 1531.8 ± 0.3 $(-)1535.0 \pm 0.6$	$\Gamma = 9.1 \pm 0.5$ $\Gamma = 9.9 \pm 1.9$	$\Xi\pi$	≈ 100
$\Xi(1820)$	$\frac{1}{2}(\frac{3}{2})$	1823 ± 5	$\Gamma \approx 24$	$\Lambda\bar{K}$ $\Sigma\bar{K}$ $\Xi(1530)\pi$	dominant
$\Xi(2030)$	$\frac{1}{2}(?)$	2025 ± 5	$\Gamma \approx 20$	$\Lambda\bar{K}$ $\Sigma\bar{K}$	≈ 20 ≈ 80

STRANGENESS-3, OMEGA BARYON

Ω^-	$0(\frac{3}{2}^+)$	1672.43 ± 0.32	$\tau = (0.822 \pm 0.012) \times 10^{-10}$	ΛK^- $\Xi^0\pi^-$ $\Xi^-\pi^0$	68 24 9

NON-STRANGE, CHARMED BARYON

Λ_c^+	$0(\frac{1}{2}^+)$	2285.2 ± 1.5	$\tau = (1.91 \pm 0.2) \times 10^{-13}$	$pK^-\pi^+$ $p\bar{K}^0$ e^+X pe^+X	2.6 1.5 4.5 1.8

Name index

Subject index

Abelian group, 220
absolute temperature, 220
alpha-rays, particles, 4, 220
amplitude, quantum mechanical, 30
angular momentum, 16, 45, 63, 220
antimatter, antiparticles, 26–8, 220
associated production, 54
asymptotic freedom
  in QCD, 120, 137–40, 220
  in GUTs, 173–4
atoms
  in matter, 1–2
  Bohr's model, 16–17, 18
  Rutherford's model, 5
  Thomson's model, 3
  atomic spectra, 16–17
axial vector, 71
axion, 209, 220

baryogenesis, 176–7, 220
baryon
  definition of, 51, 220
  baryon number, 51
beta-rays, particles, 4, 220
Bhabba scattering, 149
Big Bang theory, 178, 193, 195–6, 220
black hole
  properties, 181–3
  event horizon, 181
  Schwartzchild radius, 181
  Hawking temperature, 182
bosons, 24, 220
bottom quark, 159–60

Brookhaven National Accelerator Laboratory (USA), 54, 60, 76
bubble chamber, 4, 121

Casimir effect, 33–4
CERN, Switzerland, 78, 88, 101, 106–11, 121, 146, 148, 162, 167, 221
CP symmetry, 47, 70, 77–9, 176
CPT theorem, 47–8
Cabbibo angle, 85, 86, 221
charge conjugation symmetry, 47
charm
  hypothesis of, 104, 221
  discovery of, 153–8
charmonium, 155
chromostatic force, 144
cloud chamber, 4
colour
  introduction of, 129, 221
  evidence for, 130
  multiplets of, 131–2
  confinement of, 133
coordinates, 7
cosmological parameters
  cosmological constant, 221
  critical density, 194
  Hubble constant, 193
  ratio baryons to photons, 196
  scale factor, 195
cosmology and particle physics
  general, 193–201
  neutrinos, 176, 196–7

baryogenesis, 176–7, 220
nucleosynthesis, 193, 196
cosmic microwave background ratiation, 193
dark matter, 197–8
Coulomb's law, 39, 138, 157
coupling constant, 221
cross-section, definition, 57, 221
crystal ball detector, 155–6
current–current theory
  of the weak force, 80–2, 84–6
  leptonic current, 81
  hadronic current, 84
  current algebra, 85

DESY, 148, 159, 162, 221
D-mesons, 154
deep inelastic scattering
  introduction to, 112–14, 118–20
  electron–nucleon, 114–17
  neutrino–nucleon, 120–3
  structure functions in, 114, 116–17, 118
deuteron, 51, 221
diffraction, 3, 18, 221
dimensionality, 221
Dirac equation, 25
Dirac sea, 26

eigenstate, 77, 222
Einstein's field equations of gravity, 38, 180
elastic scattering, 222
electromagnetism, 39–40
electron
  definition, 222

discovery, 2–3
spin, 22–3
magnetic moment, 22
gyromagnetic ratio, 26
polarisation in β decay, 72
electron number, 75
scattering off nucleons, 114–17
scattering with neutrinos, 83–4
collisions with positrons, 147–70
electronvolt, 219
electroweak theory, 98–105
entropy, 182, 186
ether, 8
exclusion principle
  Pauli's formulation of, 23, 225
  role in the nucleus, 43
  role in QCD, 129
extra, hidden dimensions, 191, 204, 207–8

Fermilab, 153, 159, 222
fermion, 24, 222
Fermi's theory of β-decay, 71–4
Feynman diagrams, rules, 30–1
Feynman's sum over histories, 184, 205
field theory, quantum, 28–31
fine structure constant, 39, 219
flavours of quarks, 61, 222
four-vectors, 13
Fourier sum, 28
frame (of reference), 7

BL 44 Jog 29

KU-621-754

NOTICE

It is essential to remember that all the libraries mentioned in this work, whether connected with the University of Oxford or not, are independently owned collections maintained primarily for the use of their own members. It is imperative, therefore, that others who may wish to use these libraries should obtain permission beforehand, preferably by writing, from the librarians concerned. Those who plan to visit Oxford during the Long Vacation in the summer and who wish to work in any of these libraries, are advised to write well in advance.

There are no standard times of opening; details of many, with other information, are given in the leaflet, *Oxford Libraries*, issued by the University Registry and available in the Bodleian Library; and in the *Vade Mecum*, a general guide to University activities published by and for the use of Oxford students at the beginning of each term. The *Union List of Serials in the Science Area, Oxford, Stage II* (1970), includes particulars of opening times, borrowing regulations, etc., of some fifty scientific and medical libraries connected with the University.

OXFORD LIBRARIES OUTSIDE THE BODLEIAN

A GUIDE

Compiled by

Paul Morgan

Published by the Oxford Bibliographical Society
and the
Bodleian Library,
OXFORD
1974

© 1972 Oxford Bibliographical Society
ISBN 0 901420 08 5

© 1972 Bodleian Library, Oxford
ISBN 0 900177 17 9

Republished 1974

Printed in Great Britain by
Scolar Press Limited, Menston, Yorkshire

M. S.
DOCTISSIMORVM VIRORVM
STRICKLAND GIBSON
L. W. HANSON
QVI MIHI VIAM
FELICITER OSTENDERVNT

CONTENTS

Note: The following are either dependent libraries or sections of the Bodleian and do not come within the scope of this guide:

Bodleian Law Library
Hooke Library
Indian Institute Library
John Johnson Collection
Radcliffe Camera
Radcliffe Science Library
Rhodes House Library

FOREWORD

The high quality of Oxford as a library centre results not only from the Bodleian but also from the wide range of other libraries which complement its resources. These libraries present a great variety, from the richness of a college like Merton, antedating both the Bodleian and the invention of printing, to a small but precious collection like that of the Department of the History of Art.

The need to enable scholars to learn easily of these resources has long been recognized. It was realized in the sixteenth century by Thomas James, Bodley's first Librarian, who surveyed the toil needed to supply the need, groaned silently to himself (*tacitus apud me gemebam*), and set to work to prepare his *Ecloga Oxonio-Cantabrigiensis*, which was published in 1600. This is the first union catalogue for Oxford libraries: it lists the manuscript holdings of the Colleges and includes Cambridge also. The second union catalogue for Oxford libraries, one may say, is still awaited.

The creation of a union list of pre-1801 books in Oxford, Cambridge, and the British Museum is at present being actively studied. An experimental scheme, Project LOC, sponsored by the Andrew W. Mellon Foundation, has tested techniques and estimated the cost. Among other statistical conclusions, it has suggested on the basis of sampling that the proportion of pre-1801 works held by College and other Oxford libraries to those in the Bodleian is as five to three. This figure, however tentative, emphasizes the importance of the non-Bodleian Oxford holdings, and suggests that a great bulk of early library material, much of it unexploited, is to be found within three miles of Carfax. All the greater is the need for a description and a qualitative appraisal of the collections of 'other Oxford libraries'.

This is what Mr. Paul Morgan has undertaken in *Oxford Libraries Outside the Bodleian*. His duties have for some years included the revision of entries from Oxford libraries for the new edition of Pollard and Redgrave's *Short-title Catalogue* of English books published before 1641. In the course of this activity he has acquired a deep and unique knowledge of the contents of Oxford libraries. In the preparation of this book, he has extended his range to cover the collections in general, both printed and manuscript. There can be no doubt that his survey, thorough, balanced, and indexed, will be of the utmost value to the scholar and the research worker.

It is a pleasure to express thanks for their co-operation to the librarians and governing bodies of the libraries whose collections are described and to the Oxford Bibliographical Society for encouraging and sponsoring this work, which appears with the joint imprint of that Society and of the Bodleian Library.

ROBERT SHACKLETON
Bodley's Librarian

19 JUNE 1972

INTRODUCTION

General

The inspiration for this guide to Oxford libraries outside the Bodleian complex[1] came from Dr. A. N. L. Munby's *Cambridge College Libraries*, first published in 1960; in the following year I was appointed to check the British entries in the Oxford Inter-Collegiate Catalogue of Books printed before 1641, mentioned below. I have, then, had an opportunity to work in Oxford libraries, and the material that I gathered in the course of doing so has led me to a more detailed and historical treatment than was possible for Dr. Munby, but what he wrote about Cambridge college libraries is equally applicable to those in Oxford:

Each College to the best of its ability provides a working library for its own members, but it is not to that aspect of the College Libraries that this guide refers. Nearly all the older Colleges have substantial collections of books from the fifteenth century onwards, covering the ancient academic curriculum, the Greek and Latin classics, mathematics, law, medicine and, of course, divinity. I have made no entries in these fields unless a College's holding of books in one of these subjects is particularly strong. Most College libraries in addition have received in the course of five centuries special collections by gift or bequest, some of the highest value and importance.

The same principles have been adopted here for all the older libraries. It is perhaps necessary to point out to those unfamiliar with Oxford (or Cambridge) that colleges and halls, as distinct from the University, have usually been the first objects of loyalty and recipients of the generosity of their members.

Besides those in colleges and halls, there are over a hundred other libraries in Oxford attached to faculties, institutes, departments, and museums of the University, and in addition a few belonging to non-university institutions, such as the City Library, or the Oxford Polytechnic. These special libraries generally reveal their nature in their titles and it is presumed that research workers will naturally gravitate towards those serving their particular subjects. This type varies from small working collections of a few hundred volumes to large research libraries like the Taylor Institution. Descriptions of the minority containing unusual or older material are given in a separate sequence following the college and hall libraries. Those of all types not described at length are listed in Appendix I.

[1] *i.e.* the main Bodleian Library (which includes the Radcliffe Camera and the John Johnson Collection), and its dependents, the Radcliffe Science Library (with the Hooke Library), the Bodleian Law Library, Indian Institute Library and Rhodes House Library.

Previous surveys of Oxford libraries have generally been rather selective. The earliest was probably Louis Jacob, *Traicté des plus belles Biblio-
thèques dans le Monde*, published in Paris in 1644[2], based chiefly on the lists
of manuscripts in Thomas James, *Ecloga Oxonio-Cantabrigiensis* of 1600.
Then there was nothing of note until William Clarke described All Souls,
the old Ashmolean Museum, Christ Church, Corpus Christi, St. John's
and the Radcliffe Library on Dibdinian lines in his *Repertorium Biblio-
graphicum* of 1819. Ernest C. Thomas, sometime Librarian of the Oxford
Union Society, gave a general outline description of Oxford libraries at
the first annual meeting of the Library Association in 1878.[3] In the present
century, Strickland Gibson gave a general historical description of most of
the older college libraries in his *Some Oxford Libraries*, published in 1914.
Many Oxford libraries have necessarily brief entries in current reference
works such as the *Aslib Directory*, *The Libraries, Museums and Art Galleries
Year Book*, and the *European Library Directory*, edited by Richard C. Lewan-
ski in 1968. A good general survey of the main libraries is given in the
Handbook to the University of Oxford, published by the Clarendon Press and
frequently revised. A leaflet on Oxford libraries, designed to help undergraduates and giving useful practical information, is available free at the
Bodleian Library.

Summary histories of most college libraries are given in the *Victoria
History of the County of Oxford*, vol. iii, published in 1954, while the architectural history of library buildings erected before 1714 can be found in
the City of Oxford volume of the Royal Commission on Historical
Monuments, published in 1939.

From their earliest days in the thirteenth century each college acquired
books, principally by gift or bequest, and these were kept in two distinct
categories: a chained library, and a collection of books available for loan
to members. Chains were not removed till the second half of the eighteenth
century; the first library to do so was apparently All Souls about 1756
(about the same time as the Bodleian), and the last Magdalen College in
1799, while the loan system continued into the sixteenth century and as
late as 1596 at Lincoln College. Details of books in Oxford libraries in
pre-Reformation times and no longer in their original locations can be
found in N. R. Ker, *Medieval Libraries of Great Britain; a List of Surviving
Books*, (2nd edition, 1964), which, in addition, gives references to works in
which details of what has not strayed are printed. What happened in the
sixteenth century has also been described by Dr. Ker in his 'Oxford
College Libraries in the Sixteenth Century',[4] while descriptions of many
of the actual books are given in the Bodleian Library exhibition catalogue,
Oxford College Libraries in 1556 (1956). Later developments, particularly
in regard to fittings and architecture, are outlined by Dr. J. N. L. Myres
in his chapter 'Oxford Libraries in the Seventeenth and Eighteenth

[2]   Pp. 252–68.
[3]   Printed in its *Transactions*, 24–8.
[4]   *BLR*, vi (1959), 459–515.

Centuries' in *The English Library before 1700*, edited by Francis Wormald and C. E. Wright (1958). Eighteenth-century history is also well covered in Strickland Gibson's *Some Oxford Libraries*, mentioned above.

The custom of relying mainly on benefactions for acquisitions, a duty insisted upon at Merton as early as 1276, continued after the Reformation, but more money began to be spent on purchases, first from college income (as at Magdalen College, 1535; Queen's, 1560; Balliol, 1572) and later, in the seventeenth century, from fees (as at Oriel, 1662, or St. Edmund Hall, 1685) or bequests (as at Trinity in 1640). Sir Thomas Bodley's introduction of a Benefactors' Register into his library in 1604—a finely bound volume in which names of donors with details of their gifts were handsomely inscribed, kept in a prominent place to catch the eyes of visitors and possible donors—was gradually imitated by the colleges, not necessarily only for gifts of books. All Souls began one about the same time as the Bodleian and Christ Church followed in 1614, then New College in 1617, St. John's in 1620 and Queen's in 1621. During the Commonwealth, Brasenose and Magdalen Hall started similar volumes, but the practice really spread after the Restoration beginning with an ordinarily bound volume at Trinity in 1665 and then with extraordinarily finely bound ones such as those at University College in 1674, Magdalen College, Oriel and a new one at Queen's in 1680, the Ashmolean Museum in 1683, St. Edmund Hall in 1685, Corpus in 1690, Pembroke in 1692, Exeter in 1703, and lastly Jesus in the early eighteenth century. Balliol, Merton and St. Mary Hall also acquired such registers in the late seventeenth century; that at Lincoln, purchased in 1676, is now lost. It was the practice usually to enter earlier benefactions, with varying degrees of thoroughness, but a few colleges, among them Brasenose and Christ Church, did not. In the post-Restoration period also, Brasenose, University College, St. Edmund Hall and Queen's all built, or adapted, libraries.

The statutes of 1438 at All Souls laid down that all books given must be marked both with the donor's name and that of the College, possibly giving formal expression to a practice adopted in the older colleges, though not at New College nor at University College. Later foundations followed the custom; some, such as Balliol, were particularly thorough; others, such as Brasenose, added few donors' names to books before the mid-seventeenth century. It was also the custom from early times for members to give a present of either one or more books, or money for book purchases, when leaving college; a study of these donors can reveal names not otherwise known. At Brasenose, the new graduates each year between 1650 and 1764 combined to purchase volumes, suitably in-scribed with all their names. This custom of individual gifts continues today, including the more modern foundations such as Lady Margaret Hall, while the recently-instituted visiting fellows at All Souls each year join together in making a presentation of a major work to the library.

Whatever the general condition of the University may have been in Oxford in the eighteenth century, it was certainly not somnolent so far as its college libraries were concerned. With the exception of Magdalen, Merton and Trinity colleges, there was great activity. All Souls, Christ Church, Lincoln, New College and Oriel received both important benefactions and erected new library buildings, while notable gifts of books came to all the other colleges then in existence. The removal of chains during the second half of the century, already mentioned, betokens some activity also. The earlier part of the nineteenth century, before the influence of the Tractarian Movement and the reforms of the 1854 and 1877 Acts were felt, was much quieter for Oxford college libraries.

The second half of the nineteenth century saw two noteworthy developments in Oxford. The first was the gradual growth of library facilities for undergraduates, consequent upon the increase in numbers sitting for examinations. Until this period, only senior members were allowed in most college libraries, though at some, such as Christ Church, tutors were permitted to borrow books for their pupils' use. There had been earlier, but short-lived, experiments; Christ Church undergraduates formed a library in a coffee-house as early as 1668, while Trinity had a separate junior library for a time in the 1680s; the taberdars at Queen's had their own library from the late seventeenth century until 1840. According to Hearne, an undergraduate library had been mooted at Jesus in 1725, but nothing more is known about it; books for 'the young gentlemen' were bought there in 1795. University College had a separate undergraduate library in the later eighteenth century. The Oxford Union Society, founded in 1823, started a library about 1830 which has since been used for both work and recreation and is very likely to have been the only one familiar to the majority of students for a large part of the nineteenth century. Merton, in contrast, admitted undergraduates for one hour a week only to its library in 1827, extended to three hours daily in 1899, while until 1902 Oriel kept one cupboard of books for its junior members. Other colleges provided separate rooms for undergraduate libraries; Jesus did so in 1865, Balliol in 1871 and Trinity in 1877, followed by Pembroke and Christ Church in the 1880s; Corpus and Lincoln had such rooms in the later nineteenth century.

The second development was an attempt about 1871 to get particular college libraries to specialize in a particular subject, and make those books available to all members of the University. Thus Oriel opted for comparative philology; Balliol for philosophy; Merton, Trinity and Brasenose for certain periods of history; Lincoln for theology; and Worcester for classical archaeology. But only Oriel, Balliol and Worcester really made a serious attempt to carry out the scheme, and it lapsed.[5] All Souls, however, independently of this scheme, specialised in law and history from 1868 and permitted other members of the University to use its library. College libraries, in general, have continued to consider that their main function

[5]   Thomas, 26–7, also commented on by F. Madan in *Notes and Queries*, series 6, ii (1880), 322.

is to supply whatever is needed by members of the college, even if this means spreading their limited resources thinly over many subjects.

Special subject libraries in Oxford go back to 1619, when small collections of books were attached to the Savilian chairs of geometry and astronomy. The Botanic Garden, established in 1621, and the Ashmolean Museum, founded in 1677, also built up libraries, but all these early acquisitions have now been presented to or deposited in the Bodleian. During the nineteenth century a number of special University departments were started and most have developed libraries for the use of their members, such as Geology (1813) and the Hope Department of Entomology (1840). The Taylor Institution, whose effective history began in 1845, has grown into the major research library in Oxford for modern European languages and literatures.

Another parallel development dating from the later nineteenth century has been the increase in the number of specialist theological libraries. The first was that at Pusey House, founded in 1884, with a Tractarian emphasis. Then came the transfer to Oxford of several theological colleges connected with nonconformist denominations, and containing interesting libraries derived from the old dissenting academies, beginning with Mansfield, which arrived in 1886, Manchester following in 1889. Rather similarly, in this century certain Roman Catholic orders have re-established houses dissolved at the Reformation, some of which, like Blackfriars, Greyfriars and Campion Hall, possess good collections of books.

The growth in the number of libraries in Oxford has accelerated during the twentieth century; as new subjects are introduced, so departmental libraries tend to be built up as well, such as that in the Computing Laboratory, or the numerous small libraries in the Radcliffe Infirmary and other hospitals. In addition, independent faculty libraries have been set up, such as History in 1908, and English in 1914, though Law and Natural Science continue to form part of the Bodleian complex and the library of the Ashmolean Museum has become the faculty library for Near Eastern and European archaeology, classical archaeology and literature. The new colleges established since 1945 are also beginning to acquire books. Details of the present situation regarding size of holdings, finance and administration in faculty, departmental, institute and other libraries (but not those in colleges) can be found in the report made in 1966 of the Committee on University Libraries[6] under the chairmanship of Dr. Robert Shackleton.

It is, of course, inevitable that active libraries have to dispose of book stock from time to time in order to make shelf-space for current intake, and in libraries that have flourished over several centuries, as so many have in Oxford, none can be said to have discarded nothing. Details are given in the separate accounts of known instances. Libraries which do not appear to have got rid of very much are: Brasenose, Christ Church, Jesus, Magdalen

[6]   Published as Supplement no. 1 to the *Gazette*, xcvii no. 3295, Nov. 1966.

College, Manchester, New College, Trinity and Worcester; the Union Society also preserves a high proportion. On the other hand, appreciable numbers have been sold or given away by: All Souls, Balliol, Hertford (in its Magdalen Hall period), Mansfield, Queen's, and University. St. John's disposed of a certain amount in the seventeenth century and Wadham in the eighteenth, but they still have substantial collections of their older books.

Method

A summary of the history of each library, followed by outline descriptions of its manuscripts, archives, printed books and catalogues have been given, with a tendency to treat each part chronologically. Attention has been concentrated on the benefactors, with general descriptions of what they gave, rather than on individual manuscripts and books since it was thought that, apart from the intrusion of personal predilections, the latter course would give a misleading emphasis. References are made as far as possible to published work dealing with particular groups of books or manuscripts, or aspects of a library, but no attempt has been made to mention the many publications on individual manuscripts.

Statistics

The approximate total number of books and pamphlets has been given wherever possible, together with the number of current periodicals taken; these figures are based on various sources, such as information provided by the Secretary of the Libraries Board, the Shackleton Report, and the current directories mentioned above. They are put here in order to provide some idea of the size of a library, but must be regarded only as estimates in many cases, especially the older institutions. Also given are the approximate number of incunabula (kindly contributed by Dr. D. E. Rhodes), and of English books printed or published in Great Britain or printed in English abroad up to 1640 (here designated *STC*), based on the Inter-Collegiate Catalogue. For some of the smaller libraries, the number of foreign books up to 1640, and also the number of items printed or published in Britain between 1641 and 1700 (here designated 'Wing'), are known and have, therefore, been included.

Manuscripts

Thomas James, later Sir Thomas Bodley's choice as his first librarian, attempted a union catalogue of manuscripts in Oxford and Cambridge libraries in 1600, entitled *Ecloga Oxonio-Cantabrigiensis*. This was superseded in 1697 by *Catalogi Librorum Manuscriptorum Angliæ et Hiberniæ in unum Collecti*, in turn made out of date so far as Oxford was concerned in 1852 by H. O. Coxe with his *Catalogus Codicum MSS. qui in Collegiis Aulisque Oxoniensibus hodie adservantur*; this does not include Christ Church, Pembroke, St. Edmund Hall or modern foundations. Since Coxe, a few colleges have published separate catalogues, such as Christ Church (1867) and

Balliol (1963), or have one in an advanced stage of preparation, as Keble.

Oriental manuscripts in All Souls, Balliol, Brasenose, Corpus Christi, Exeter, Jesus, Magdalen, Merton, New College, Pusey House, Queen's, St. John's and Wadham are surveyed by J. D. Pearson in his *Oriental Manuscript Collections in the Libraries of Great Britain and Ireland* (1954). Hebrew manuscripts in Balliol, Christ Church, Corpus Christi, Jesus, Merton, Oriel, St. John's and Worcester are described in A. Neubauer, *Catalogue of the Hebrew Manuscripts in the Bodleian Library and in the College Libraries of Oxford* (1886).

There are also typescript lists of certain later acquisitions, copies of which are available in Duke Humfrey Reading Room in the Bodleian; such lists and where they can be seen are indicated in the separate descriptions. There are also collections of papers or correspondence of individuals, many of comparatively modern date, scattered round Oxford libraries; a select list, compiled mainly from personal observation, is given in Appendix II.

Some colleges and other libraries have deposited their manuscript collections in the Bodleian, where they can be consulted in Duke Humfrey Reading Room without prior permission; these are indicated in the separate accounts. Manuscripts presented to or deposited with the Oxford Colonial Records Project are kept in Rhodes House Library and, therefore, do not come within the scope of this work.

Archives

No union calendar or catalogue of Oxford archives exists. The selective descriptions of several college archives in the early volumes of the Historical Manuscripts Commission are generally unsatisfactory. While most institutions now have some sort of calendar or guide, standards vary greatly. Those which have been reproduced for limited distribution through the National Register of Archives or are available in another form in Duke Humfrey Reading Room are indicated. In connection with the History of the University of Oxford Project currently being directed by Mr. T. H. Aston of Corpus Christi College, summary handlists of the archives of each institution within the University are being compiled and I have been privileged to use them in preparing this *Guide*. It is hoped that they will in due course be edited and made generally available.

The official custodian of archives varies with each institution; it is hoped that the correct officer has been named in each case.

Printed Books

(i) *Incunabula*

Brief entries with author or entry-word, short title, place of printing, date and format are included in the Inter-Collegiate Catalogue, available for consultation on application at the Bodleian. More detailed descriptions are currently being written by Dr. D. E. Rhodes, of the Department of Printed Books of the British Museum, but are not yet quite complete.

There are approximately 2,500 items in all. Some (but by no means all) of the holdings of the following are included in the *Gesamtkatalog der Wiegendrucke*: All Souls, Balliol, Brasenose, Christ Church, Corpus Christi, Exeter, Hertford, Jesus, Keble, Lincoln, Magdalen, Merton, New College, Oriel, Pusey House, St. John's, Taylor Institution, Trinity, Wadham and Worcester.

(ii) *Early printed books to 1640*

The Inter-Collegiate Catalogue, available for consultation on application at the Bodleian, was originally compiled by conflating the entries in the separate catalogues for all items printed before 1641. The cards have now been divided into two sequences: (a) Books printed or published outside Britain and not in English; (b) Books printed or published in Britain, and books in English printed abroad. This second section follows the numerical arrangement of the Bibliographical Society's *Short-title Catalogue of Books Printed in England, Scotland and Ireland, and of English Books Printed Abroad, 1475–1640*, first published in 1926 (new edition in preparation). It is perhaps worth noting that whereas the *Short-title Catalogue* gives Oxford locations outside the Bodleian for rare items only, the Inter-Collegiate Catalogue records numerous copies of a large number of works, including those listed in E. G. Duff's *Fifteenth Century English Books* (1917) and A. F. Allison and D. M. Rogers, *A Catalogue of Catholic Books in English Printed Abroad or Secretly in England, 1558–1640* (1956). This second section covers over 21,600 items, not including over 1,000 identified fragments in bindings.

(iii) *English Books 1641–1700*

Some locations in Balliol, Brasenose, Christ Church, Corpus Christi, Exeter, Magdalen, Merton, New College, Pembroke, Wadham, and Worcester, and the Oxford University Press, are given by Donald Wing in his *Short-title Catalogue of Books Printed in England . . . 1641–1700* (1945–51; new edition in preparation); they must be treated with extreme caution since they were based to a large extent on obsolete printed catalogues.

(iv) *Project LOC*

This scheme to get all books printed before 1801 in the British Museum and all Oxford and Cambridge libraries entered in a catalogue in machine-readable form is in an early and experimental stage. Application to consult what has so far been done must be made in the first instance to the Keeper of Catalogues, Bodleian Library.

(v) *Periodicals*

The two most up-to-date guides to periodical holdings in Oxford libraries are: (a) *Current [i.e. 1925 or later] Foreign and Commonwealth Periodicals in the Bodleian Library and other Oxford Libraries*, published in 1953 with Supplements in 1956 and 1962. All entries from these, with more recent accessions, are recorded in a card catalogue kept in the Lower Reading Room of the Bodleian.

(b) *Union List of Serials in the Science Area, Oxford; Stage II,* published in 1970, with its *First Supplement* (1971) records part of the periodical holdings of the Radcliffe Science Library and all those in some fifty scientific and medical libraries. A wider range of medical libraries is covered by the *List of Medical Periodicals Taken in the United Oxford Hospitals* (1964), supplemented, but not entirely superseded, by the Oxford Joint Medical Libraries Committee's *Union List of Periodicals Held in Medical Libraries of the Oxford Hospital Region* (1971); the latter, modestly described as a pilot survey, includes libraries in a wide area around the City of Oxford.

A smaller group is covered in *Arabic Periodicals in Oxford; a Union List Compiled by D. Hopwood* [1968], which records the holdings of the Bodleian, the Oriental Institute and the Middle East Centre of St. Antony's College.

(vi) *Music*

Holdings in the following libraries are recorded in Edith B. Schnapper, *British Union–Catalogue of Early Music Printed before the Year 1801* (1957): Brasenose, Christ Church, Corpus Christi, Lincoln, Magdalen, New College, Oriel, Queen's and St. John's, the Music Faculty Library and the University Music Club. [Meredith Moon], *A Checklist of Books on Music Published before 1800 in the Bodleian* [1965], gives a few locations in other Oxford libraries, but does not aim at full coverage. Brief details of the musical holdings and numbers of scores in Brasenose, Christ Church, New College, Oriel, Queen's, St. John's, and the Music Faculty Library can be found in Maureen W. Long, *Music in British Libraries: a Directory of Resources,* published by the Library Association in 1971.

(vii) *Theology*

A brief general survey is given in R. G. Chapman, 'Some Theological Holdings in Oxford; a preliminary Survey' in *Bulletin of the Association of British Theological and Philosophical Libraries,* no. xiii (1960), 3–7.

(viii) *Scientific Books*

Some of the collections of older scientific books in college libraries were surveyed in general terms by R. T. Gunther in *Oxford and the History of Science. With an Appendix on Scientific Collections in College Libraries. Inaugural lecture . . . 1934* (1934), 38–49; this was reprinted in his *Early Science in Oxford,* xi (1937), 325–36.

(ix) *Middle East*

The Middle East Centre of St. Antony's College maintains a union catalogue of relevant holdings in its own library and those of Nuffield College; Department of Eastern Art, Ashmolean Museum; Rhodes House Library; School of Geography; Institutes of Agricultural Economics, of Commonwealth Studies, of Economics and Statistics, of Social Anthropology, and the Oriental Institute. Copies of this catalogue are deposited in the Oriental Reading Room of the Bodleian and in the Oriental Institute. For a list of Arabic periodicals, see Section (v) above.

(x) *Latin America*

The Latin American Centre of St. Antony's College maintains a union catalogue of relevant modern books in its own library, the Bodleian, the Taylor Institution, the Pitt Rivers Museum (Balfour Library), and the Institutes of Economics and Statistics, and of Social Anthropology.

Bindings

Oxford awaits a survey such as G. D. Hobson made for Cambridge. A few of the more outstanding specimens are described in the Bodleian exhibition catalogue, *Fine Bindings 1500–1700 from Oxford Libraries* (1968). Numerous early Oxford bindings in college libraries are listed in N. R. Ker, *Pastedowns in Oxford Bindings, with a survey of Oxford Binding, c. 1515–1620* (Oxford Bibliographical Society Publications, new series v, 1954). Strickland Gibson in his *Early Oxford Bindings* (1903), which surveys the history of bookbinding in Oxford down to the Civil War, also quotes many college examples.

Catalogues

Under this heading, surviving catalogues known to the compiler are mentioned, together with the type currently used. The survey of older catalogues does not pretend to be complete and should be treated as a preliminary outline only. In this connection, it should be pointed out that many of the Oxford lists entered in Sears Jayne, *Library Catalogues of the English Renaissance* (1956) are not mentioned here, since they are really parts of larger works such as Benefactors' Registers.

Acknowledgements

It must be self-evident that I am deeply indebted to the large number of Oxford librarians and their assistants who so readily granted me both the freedom of their shelves and the benefit of their knowledge and advice; to them all I wish to express my sincerest thanks. With so many willing helpers, it is perhaps invidious to mention names, but I am particularly grateful to Mr. T. H. Aston and Mr. David Vaisey for contributing an account of the University Archives. Mr. Aston also kindly let me make use of the handlists of college archives prepared under his supervision for the History of the University Project, thereby considerably enlarging these sections.

Dr. R. W. Hunt, Mr. Harry Carter and Miss G. M. Briggs read through the whole in typescript and made many valuable suggestions, but the no doubt numerous errors that remain are my own responsibility. Mr. H. F. Alexander, with his unique knowledge of Oxford scientific and medical libraries, willingly guided me around the Science Area and the hospitals. Miss M. E. Grinyer, Secretary of the Libraries Board, kindly checked the lists of libraries and provided up-to-date statistics for those in University institutes, faculties and departments. Dr. A. B. Emden, Dr. N. R. Ker and Mr. J. S. G. Simmons gave freely of their special knowledge of Oxford libraries, while many colleagues on the staff of the Bodleian Library were always ready to assist.

ABBREVIATIONS

BLR	*Bodleian Library Record*
BQR	*Bodleian Quarterly Record*
Coxe	H. O. Coxe, *Catalogus Codicum MSS. qui in Collegiis Aulisque Oxoniensibus hodie adservantur.* 2 pts. 1852.
DNB	*Dictionary of National Biography.*
Duke Humfrey	Manuscripts and early printed books reading room in the Bodleian Library.
Fine Bindings 1500–1700	Bodleian Library, *Fine Bindings 1500–1700 from Oxford Libraries; Catalogue of an Exhibition.* 1968.
Gazette	*Oxford University Gazette*
Gunther	R. T. Gunther, *Oxford and the History of Science. With an Appendix on Scientific Collections in College Libraries. Inaugural Lecture . . . 1934.* 1934. Reprinted in his *Early Science in Oxford* xi (1937), 325–36.
Ker	N. R. Ker, 'Oxford College Libraries in the Sixteenth Century', *BLR*, vi (1959), 459–515.
Oxford College Libraries in 1556	Bodleian Library, *Oxford College Libraries in 1556; Guide to an Exhibition held in 1956.* 1956.
STC	*A Short-Title Catalogue of Books Printed in England, Scotland and Ireland, and of English Books Printed Abroad, 1475–1640.* Compiled by A. W. Pollard and G. R. Redgrave [and others], 1926.
Streeter	B. H. Streeter, *The Chained Library.* 1931.
Union List	*Union List of Serials in the Science Area, Oxford, Stage II.* 1970, and *First Supplement,* 1971.
VCH Oxon. [or] *Berks.*	*Victoria History of the County of Oxford* [or] *Berkshire.*
Wing	D. Wing, *Short-title Catalogue of Books Printed in England, Scotland, Ireland, Wales and British America and of English Books Printed in Other Countries, 1641–1700.* 3 vols. 1945–51.

In each section, full details of a work are only given when it is first mentioned; subsequent references within that section are abbreviated to authors' surnames, or to short titles.

I

COLLEGE AND HALL LIBRARIES

ALL SOULS COLLEGE

THE CODRINGTON LIBRARY

1438

Stock: 110,000   Current periodicals: 82   Incunabula: 338   *STC*: 949

All Souls College, founded titularly by Henry VI and really by Henry Chichele, Archbishop of Canterbury from 1414 to his death in 1443, has a unique reputation as a scholarly college without undergraduates; its library is equally remarkable.[1]

The original statutes laid down that any books given should be marked both with the donor's name and that of the College, and it is interesting that all colleges founded afterwards copy this regulation.[2] Actually, inscriptions of this nature are commonly found in Oxford college libraries whatever their date of foundation may be. As elsewhere, there were originally two distinct collections of books: those chained in the library, and those given out yearly to members on loan at *electiones*; the distinction between these two classes is expressly made in the Founder's statutes.[3] All Souls is also fortunate that more sixteenth-century lists and inventories of its books have survived than in most other colleges,[4] showing that a good, all-round library had been assembled, unlike the specialist theological one at Merton.

Books were purchased in the 1540s, but not so many as at Merton, with a bias towards the Fathers, the classics and law.[5] Rather prophetically,

---

[1]   This account was written before the publication of Sir Edmund Craster's *The History of All Souls College Library* (1971), the first full history of an Oxford college library; some amendments have subsequently been made; detailed references are not given, since the facts in this summary account can mostly be found in the longer work. Of shorter descriptions, the best is in *VCH Oxon.* iii, 173–6, 180–2; other general accounts are by R. E. P[rothero, 1st Baron Ernle] in *Notes and Queries*, series 6, ii (1880), 421–2, and by Sir Charles Grant Robertson, *All Souls College* (College Histories, 1899), 214–22. Sir Charles Oman, *Report on the History of the Codrington Library*. (Published for College use only, 1929), is solely concerned with the eighteenth and nineteenth centuries.

[2]   Ker, 460; and also his *Records of All Souls College Library, 1437–1600* (Oxford Bibliographical Society Publications, new series xvi, 1971), 170.

[3]   E. F. Jacob, 'The Chained Books of All Souls, *Times Literary Supplement* 21 Jan. 1932, 44, correcting Streeter, 181–2. A list of *libri distribuendi, c.* 1440, was printed as an appendix to E. F. Jacob, 'Two Lives of Archbishop Chichele', *Bulletin of the John Rylands Library*, xvi (1932), 469–81; it is also in Ker, *Records*, 14–5; *cf.* Ker, 477.

[4]   Ker, 461–2, 477, 493; published in full in his *Records*.

[5]   Ker, 482, 493–97.

Edward VI's Visitors in 1549 considered assigning the specialist study of law to All Souls,[6] something that really came about two hundred years later. Nevertheless, the bulk of accessions came from gifts and bequests of members from its foundation to the present century. A Benefactors' Register, recording earlier donations, was started in 1604–5, the same time that the Bodleian adopted this practice, but before other Oxford colleges did so. Until 1641 only gifts of books were entered in it, but after the Restoration other types of donations appear. A regular book-fund of £17 a year was started in 1674, augmented by certain degree fees from 1713; library accounts survive from 1687.

The eighteenth century was a period of great activity. In 1710, Christopher Codrington died, bequeathing £10,000 and his own books, estimated by some to number 12,000, to All Souls, immediately making it one of the leading college libraries. The money was used to build the present magnificent library, the longest room in Oxford, named after the benefactor and completed by 1751. The organisation and arrangements were in the capable hands of Sir William Blackstone (1723–1780), the lawyer, whose object was to make it 'as fine an instrument as possible for legal and humane studies'.[7] Perhaps it was his influence that caused a Library Committee to be established in 1751, which still continues in an altered form.[8] The earliest borrowers' register dates from the opening of the new building. Chains were still being added in 1697, but since Codrington's own books were housed in a separate locked room until the new building was finished, they were not chained; presumably the older stock had their chains removed when they were transferred.

Benefactions have continued steadily since then, both of manuscripts and printed books, while works on theology, classics, law, literature, fine arts, mathematics, botany and medicine were purchased. By the time of the University Commission of 1852, however, the library had become neglected,[9] but as a consequence of the reforming spirit then abroad, it was decided to open it from 1867 to graduates and undergraduates connected with the new School of Law and Modern History. Since this date, library purchases have been largely restricted to those subjects together with political science and economics. Donations continued, especially from fellows of their own publications,[10] so although there is no *alumnus* collection as such, comparable to those at Magdalen or Oriel, the library is rich in the writings of members of the College. A fresh development is for the recently instituted visiting fellows to join together each year to present a major work to the library.

[6] *Oxford College Libraries in 1556*, 5.
[7] *VCH Oxon.* iii, 181.
[8] Oman, 1. Its *Annual Reports* for 1869 onwards are printed; the principal benefactions are listed each year.
[9] Oman, 2; *cf.* C. H. Robarts, *Some Remarks Concerning the Library of All Souls College* (1867).
[10] Listed under a separate heading in each *Annual Report*.

The problem of shelf-space, common to all libraries, had become particularly acute by the 1920s, and was only eased by the drastic solution of disposing of several thousand volumes from little used sections to other institutions in 1926. Indeed, a fluid policy over accessions has always been adopted; it has rarely been the practice to accept an entire private library, but only a selection, from the Digges bequest of 1643 onwards. Duplicates were sold at least as early as 1751.

Manuscripts

Coxe[11] in the mid-nineteenth century described just under three hundred manuscripts; a typescript calendar of a further ninety-four has been compiled.[12] The founders both made gifts, Archbishop Chichele presenting forty-nine manuscripts in 1438 and Henry VI twenty-seven in 1440.[13] The medieval manuscripts include some Psalters (of which the illuminated thirteenth-century *Amesbury Psalter* is the most notable),[14] Biblical and other codices, as well as a good collection of medicine, canon and civil law.[15] There was no further large accession until the end of the fifteenth century when James Goldwell, a former fellow, later Bishop of Norwich from 1472 to his death in 1499, left some thirty manuscripts.[16] During the next century, in 1575, the library of David Pole (d. 1568), sometime Bishop of Peterborough, was received; it consisted of 250 manuscripts and printed books, the most extensive bequest to any Oxford library before 1600.[17]

The great increase between the seventeenth and nineteenth centuries was mainly due to the presentation in 1786 by Luttrell Wynne (1740–1814) of the papers of his ancestors, Narcissus Luttrell (1657–1732) and Owen Wynne (*fl.* 1650–90). This group is most important for it includes not only the manuscript of Luttrell's *Brief Historical Relation of State Affairs from September 1678 to April 1714*, but many volumes of Parliamentary journals and state papers, as well as documents concerned with the East India Company between 1619 and 1685.[18] Since Owen Wynne had been secretary to Sir Leoline Jenkins (1625–1685), sometime Principal of and a benefactor to Jesus College, Oxford, there are many papers connected with maritime law. Of more literary than historical interest among the

---

[11]   Pt. II, section 1 (a revised edition of his separately published catalogue of 1842).
[12]   Copy inserted in the Duke Humfrey copy of Coxe. A selective list of these additions is in *Bulletin of the Institute of Historical Research*, ii, no. 6 (1925), 99.
[13]   Only three of the latter are now in All Souls; R. Weiss, 'Henry VI and the Library of All Souls College', *English Historical Review*, lvii (1942), 102–5.
[14]   Presented by Daniel Lysons (1727–1800).
[15]   Some of the more notable are listed in *VCH Oxon.* iii, 182, and Grant Robertson, 217–9.
[16]   Listed in A. B. Emden, *A Biographical Register of the University of Oxford to* A.D. *1500*, ii (1958), 785.
[17]   Ker, 501–2.
[18]   Calendared in S. A. Khan, *Sources for the History of British India in the Seventeenth Century* (1926), 191–204; summarised in M. D. Wainwright and Noel Matthews, *A Guide to Western Manuscripts and Documents in the British Isles Relating to South and South-East Asia* (1965), 337–8.

Luttrell manuscripts are six notebooks of Humphrey Dyson (d. 1632), a book-collecting lawyer.[19] Other groups of seventeenth-century interest are nearly 400 drawings and plans by Sir Christopher Wren (1632–1723),[20] letters from Sir George Downing to Sir William Temple, 1664–67, and various heraldic documents chiefly connected with the younger Sir Henry St. George (1625–1715).

Eighteenth- and nineteenth-century manuscripts are not so numerous. There are some thirty-five volumes of notes of Sir William Blackstone's lectures, possibly made by an amanuensis or a student. Some interesting documents relating to the Madras Presidency and the southern division of India, *c.* 1767–1807, supplement the East India Company papers already mentioned. All Souls also houses the papers of Sir Charles Vaughan (1774–1849), traveller and diplomat at Madrid 1810–20, and in Washington 1825–35[21], and the correspondence from 1835 to 1859 between Alexis de Tocqueville and Henry Reeve.

There are also about fifty Oriental manuscripts.

Archives

The earlier archives of All Souls were deposited for the most part in the Bodleian in 1966. They form a large collection,[22] some going back to the thirteenth century. Besides the original charter, fifteenth-century building accounts,[23] administrative records and accounts from 1446 and estate rolls from 1443, there are also numerous letters and injunctions from successive archbishops, visitors of the College. The accounts were used by J. E. Thorold Rogers in his *History of Agriculture and Prices in England.*[24]

A calendar, arranged by place, compiled by C. Trice Martin, was published in 1877[25]; it lists documents relating to about fifty College properties in various parts of Britain, and also the College's own administrative records.

The College Librarian is responsible for all matters connected with the archives.

19   W. A. Jackson, 'Humphrey Dyson's Library', *Papers of the Bibliographical Society of America*, xliii (1949), 279–87; revised in his *Records of a Bibliographer* (1967), 135–41.
20   See *Catalogue of Sir Chr. Wren's Drawings at All Souls* [*etc.*] (Wren Society, xx, 1943), 1–33.
21   A calendar by J. A. Doyle has been reproduced by the National Register of Archives and a copy is in Duke Humfrey. See also J. A. Doyle, 'American Papers of Sir Charles Richard Vaughan', *American Historical Review*, vii (1902–3), 304–29, 500–33; and B. R. Crick and M. Alman, *A Guide to Manuscripts Relating to America in Great Britain* (1961), 390–4.
22   An outline description is in *VCH Oxon.* iii, 182; see also E. F. Jacob, 'All Souls College Archives', *Oxoniensia*, xxxiii (1969), 89–91.
23   E. F. Jacob, 'The Building of All Souls College', in *Historical Essays in Honour of James Tait* (1923), 121–35.
24   7 vols., 1866–1902; see Preface to vol. 6, p. 5.
25   C. T. Martin, *Catalogue of the Archives in the Muniment Room of All Souls College* (1877). The master copy, kept in the College, has been heavily annotated; a xerox copy is available in Duke Humfrey.

Printed Books

Books on theology, classics, law and medicine were bought as early as the 1540s, of which a fair proportion survive,[26] while the bequest of David Pole, received in 1575, contained printed books as well as manuscripts. There were no other large bequests for nearly a century; John Morris (d. 1648), Regius Professor of Hebrew, left an annuity of £5 for the purchase of Hebrew books both to All Souls and to Christ Church; this fund did not become operative until after the widow's death in 1681. Hebrew books were regularly acquired with the money, but in 1927 All Souls agreed to let Christ Church have the use of the whole benefaction. Dudley Digges (1613–43) left a general library including English and European literary works, unusual in a college library at that period; the books were stored until incorporated in the library in 1665.[27]

Besides money for a new building, Christopher Codrington left several thousand volumes which at once made All Souls one of the more important college libraries in Oxford. After the completion of the library in 1751, more bequests were received from fellows: a general collection, for instance, from Anthony Jones in 1769 and another larger, general library from Ralph Freman in 1774; the latter, comprising books owned by his father, another Ralph (d. 1746), Sir Thomas Brograve and Thomas Leigh (d. 1686), of Bishop's Stortford, is particularly interesting for English and Italian literature of the seventeenth and eighteenth centuries and six-teenth- and seventeenth-century pamphlets.

Although there were changes in the organisation of the library during the nineteenth century, there do not appear to have been many significant gifts, except for some Spanish literature of the seventeenth and eighteenth centuries from Peter Frye Hony (d. 1876), supplementing the older books in Dudley Digges's bequest.

During the present century, there has been a series of interesting bene-factions, mainly associated with the specialization in law and history after 1868. J. A. Doyle (1844–1907), a former fellow and librarian, bequeathed an excellent collection especially strong in Americana and British colonial history, as well as English literature; his sporting books were divided in 1945 between the Bodleian Library, the Kennel Club and the Shire Horse Society. Between 1907 and 1913, Miss Katherine Stiles gave numerous books on the American Civil War, thereby adding to the usefulness of Doyle's bequest.

As befits the College of the Chichele Professor of the History of War (styled Military History until 1946), there have been several donations dealing with that subject: some 300 volumes from his own collection in memory of Sir Foster H. E. Cunliffe, Bart., killed in action in 1916; a series of gifts from Colonel John Leslie between 1920 and his death in 1943, mainly concerned with artillery; and over 2,000 volumes (of which about 1,600 were retained) on the Napoleonic period, given by Colonel Ramsay

[26] Ker, 482, 493–7; *Oxford College Libraries in 1556, passim.*
[27] Streeter, 172.

Phipps in 1920. These, together with books purchased (especially in the field of strategic studies), make the Codrington Library one of the best military libraries outside London.

A very different theme dominated the next bequest, that of W. P. Ker in 1923, who left money and his collection of Scandinavian literature; about 400 items were passed on to University College, London;[28] he had been a regular benefactor during his lifetime.

Another consequence of the re-organisation of historical teaching in Oxford which had led to the establishment of the chair of military history in 1909, was that All Souls also began to specialize in political science and economy. These subjects were greatly helped by financial assistance from R. H. Brand, the first Baron, in 1923 and the gift in 1930 of some volumes from the library of Nassau Senior (1790–1864) from his grand-daughter, Mrs. St. Loe Strachey.

In the following year, 1931, F. F. Urquhart (1868–1934), a fellow of Balliol, presented a collection of about 150 books and pamphlets dealing with the question of the authorship of the *Letters of Junius*, many formerly owned by his mother's relative, Chichester Fortescue, first Baron Carling-ford (1823–98); further papers and books from this collection were received in 1944 from an executor. Sir Charles Oman, librarian for over fifty years, also gave in 1931 about sixty volumes of early editions of the works of the late eighteenth-century writers Richard Graves (sometime fellow), William Shenstone and William Somervile, bequeathed to him by W. H. Hutton (1860–1930) to dispose of as he wished.

An interesting group of peerage cases heard in the House of Lords was presented in 1932 by Granville Proby,[29] a former Clerk in that House. Since that date, there has been no particularly large donation, though fellows have continued the custom of giving copies of their own publications while some have bequeathed small collections, such as Spenser Wilkinson in 1936, Sir Charles Oman in 1946 and G. M. Young in 1959. The widow of T. R. Buchanan (1846–1911), a former librarian of the College and Member of Parliament, in 1941 gave forty books selected from his choice library.

As already mentioned, problems of space during the last fifty years have led to a certain amount of discarding, though works by former fellows and 'valuable books' [sic] have been retained.[30] After the concentration on law and history began in 1868, no books on the natural sciences, mathematics or medicine were acquired, and in 1926 1,200 volumes on these subjects went to the Brotherton Library of the University of Leeds; 400 more volumes together with Oriental books and the Bible Clerk's small library were transferred to the Bodleian. In the next year, many items of eighteenth-century theology were also divided between these two institutions. Some

---

[28]   In 1945 the etymological and philological sections were placed by All Souls on permanent loan in the Taylor Institution.

[29]   *Cf.* the rather similar collection presented by Lord Donoughmore to New College.

[30]   See especially Oman, 6–7, and *Annual Reports* from 1919.

Hebrew and Arabic works were passed over to the Regius Professor of Hebrew and to the London School of Oriental Studies in 1934. Various sets of little-used law reports together with the section on Roman-Dutch law have been given to the Bodleian at various times, while Indian law reports were sent both to the Indian Institute Library in Oxford and to Manchester University in 1939. Other British and foreign universities have benefited from this policy as well, such as the University of Belgrade in 1919 and the University of Hull in 1950.

The Codrington Library is particularly rich in fine bindings, especially French and Italian examples of the sixteenth and seventeenth centuries. Their interest was first appreciated by T. R. Buchanan, already mentioned,[31] when he was librarian between 1878 and 1882, and it was he who inspired the publication in 1880 of *Bookbindings in the Library of All Souls College; Twelve Plates drawn by J. J. Wild.*

This library has strong holdings in several subjects; law and history—especially military, imperial and French history—are very well represented. Unlike most Oxford college libraries, older English, French, Italian and Spanish literary works are present in force, including some English plays of the seventeenth and eighteenth centuries. Besides the various specialist bequests, the Codrington Library possesses a quantity of English newspapers published in the first half of the eighteenth century, mainly printed in London, but including runs of the *Bristol Post Boy*, the *Manchester Newsletter*, and the *Northampton Mercury*, as well as many early printed books.

Catalogues

The various lists and inventories of the sixteenth and seventeenth centuries have been fully described elsewhere.[32] The main ones are: a list of books included in the inventory presented to the Marian Commissioners in 1556; an author catalogue of 1631 with only ninety-six books, and another of *c.* 1635, showing 1,225 manuscripts and printed books. This last was in use until 1665 when a new one was compiled, superseded in its turn in 1689. Calendars of manuscripts have been mentioned above. Interleaved copies of the Bodleian catalogues of printed books of 1738 and 1843, annotated with All Souls' holdings, have survived. Printing of a catalogue was started in 1911 but stopped in 1917 when A–C had been completed; this was published in 1923 and the project abandoned in favour of the present name catalogue on cards, originally compiled between 1925 and 1934.

A manuscript short-title catalogue of incunabula was made in 1888 and is available in Duke Humfrey; another specialist catalogue, of canon and civil law, made by G. Hawke of Wadham in 1893, is kept in the Codrington Library.

---

[31] S. Gibson, 'Bookbindings in the Buchanan Collection', *BLR*, ii (1941), 6–12.
[32] Ker, *Records, passim; Oxford College Libraries in 1556*, nos. 1, 4; Sears Jayne, *Library Catalogues of the English Renaissance* (1956), 63, 74–5, *etc.*

BALLIOL COLLEGE

c. 1263

Stock: 63,000   Current periodicals: 90   Incunabula: 58   *STC*: 1,223

The exact date of the foundation of this College by John Balliol and his wife Dervorguilla, of Barnard Castle, is unknown. Books were presented, it is believed, in its earliest period but information about acquisitions really begins in the fifteenth century, and a considerable number of manuscripts given in that century still survive.

The growth and history of this library have been well described elsewhere so it is unnecessary to go into great detail here.[1] The Edwardian reformers removed and burnt some manuscripts in 1553 and more were sold by Robert Parsons while a fellow between 1568 and 1574, but this was during his Calvinist phase before he became a Jesuit missionary and so the proceeds were used to buy Protestant theological works, still on the shelves.[2] Library expenditure began in 1572; money was spent on patristics, though the books are no longer extant.[3] From the latter part of the sixteenth century onwards it was the custom, as elsewhere in Oxford, for members of the College to present one or more volumes on admission or graduation and many survive, appropriately inscribed by hand or with a printed label.

The first large benefaction of printed books came during the late seventeenth century, when a Donors' Register, recording gifts from about 1620 and used until 1841, was started, since when there has been a steady series. An unusual documentary survival is a record of books borrowed in 1679–80 and 1693–1703, mixed up with rough notes of College business probably made by Roger Mander, Master 1687–1704. Chaining continued until 1767–8 and the old chains were possibly sold in 1791–2.[4]

A separate library for undergraduates was established in 1871,[5] an early consequence of Jowett's mastership. New Inn Hall together with its few

---

[1] For the history of the manuscript collection see Sir Roger A. B. Mynors, *Catalogue of the Manuscripts of Balliol College, Oxford* (1963), xi–liii (probably the best account of the growth of a medieval college library). The main events in the library's history are to be found in H. W. Carless Davis, *A History of Balliol College*, Revised by R. H. C. Davis and R. Hunt (1963), while a useful summary is given by R. W. Hunt in *VCH Oxon.* iii, 88. All these supersede T. K. Cheyne's description in *Notes and Queries* series 6, iii (1881), 61–2.

[2] Davis, 78.

[3] Ker, 460, 484.

[4] Mynors, lii.

[5] *VCH Oxon.* iii, 87.

manuscripts and printed books was absorbed by Balliol in 1887. More recently the College has combined with its neighbour, Trinity, to form a joint scientific library.

There are now more than 450 manuscripts and over 60,000 printed books in the Library. More than 3,000 theological works, printed after 1600, were discarded in 1928, of which about 1,000 went to the Bodleian and the remainder were sold.[6] Since 1921, the more important accessions and gifts, together with books, etc., written by Balliol men, have been listed each year in the *Balliol College Record*.

Manuscripts

Balliol is the first Oxford college to publish a modern catalogue of its manuscripts, leading here as in other fields of academic work. Coxe described 363,[7] but Sir Roger Mynors' *Catalogue* of 1963, already mentioned, covers 450. The College is still rich in medieval manuscripts, in spite of the depredations, besides which there is a small number of Oriental ones. The collection was started by William Gray (*c.* 1416–1478), Bishop of Ely, who gave more than half of the surviving medieval library, reflecting his scholarly interests. There were several smaller donations during the fifteenth century but no further large gifts.

Besides the medieval manuscripts, there are also groups from later centuries. Some seventeenth-century legal manuscripts come from New Inn Hall, one or two formerly owned by Sir William Blackstone (1723–1780), sometime Principal. The eighteenth century is represented by correspondence between Theophilus Leigh, Master 1726–85, and the first Duchess of Chandos, and the papers of Henry Fisher (1685–1773), a West Country clergyman of Jacobite sympathies.

The nineteenth-century manuscripts, of considerable literary and historical interest, are much more numerous. As for former members, it is not surprising that the College has papers or letters of Matthew Arnold, A. C. Swinburne, A. H. Clough, Benjamin Jowett, A. P. Stanley and J. W. Mackail, but a group of letters to George Eliot, the manuscript of Byron's *Monk of Athos*, letters from John Ruskin and P. B. Shelley, the printer's copy of George MacDonald's *Diary of an Old Soul*, with poems by and papers relating to Robert Browning can also be found here.[8] Less famous members of the College are also represented by various manuscripts, such as the autobiography and papers of Frederick Oakeley (1802–1880), or a diary and some papers of Ernest Walker (1870–1949), Balliol's Director of Music for many years.

There are also several groups either omitted from Sir Roger Mynors' *Catalogue*, or received since its publication in 1963. Briefly, they are:

[6] Ibid., 88.
[7] Pt. I, section 2.
[8] The Browning, Byron, and Mackail manuscripts are described in the *Literary Supplement*, [pp. 4–5] to the *Oxford Magazine* of 14 May 1925.

The papers of Richard Jenkyns, Master from 1819 to 1854.

The papers of David Urquhart (1805–1877), of particular importance for Near Eastern and diplomatic history, presented by his son, F. F. Urquhart, fellow 1896–1934.[9]

Such papers as survive of T. H. Green (1836–1882), the philosopher.

The surviving papers of Benjamin Jowett (1817–1893), fellow from 1838 and Master from 1870 to his death; this is a small group as Jowett left instructions that his letters should be destroyed; an important batch to escape is his long correspondence with Florence Nightingale.

The papers of Sir Robert Morier (1826–1893), the diplomatist, presented in 1965.

The papers of Arthur Lionel Smith (1850–1924), Master 1916 to 1924.

The surviving papers of Andrew Cecil Bradley (1851–1935), the Shakespearean scholar.

The papers of John Alexander Smith (1864–1939), the philosopher.

The typescript of the journal of Sir Harold Nicolson (1886–1968), of both historical and literary interest.[10]

Archives

Except for some records concerned with statutes and property, the bulk of the archives date from 1514, when the registers of college business start. A haphazard selection was described by H. T. Riley for the Historical Manuscripts Commission,[11] while H. E. Salter edited the deeds concerned with land and buildings in Oxford.[12] There are also archives relating to land in several parts of Britain, such as London, Essex, the West Country and the Midlands; five volumes of transcripts, arranged by place, made in the later nineteenth century, are now in the College library.

Bursarial accounts begin in 1568, and battels in 1576. Admission and benefactors' registers survive from 1636. There are also some records of various undergraduate societies; to select a few, mention can be made of the Boat Club journal and accounts from 1835–1940; the Shakespearean Society cast lists from c. 1840–50; the Debating Society minutes 1889–1960, and several athletic, musical and social societies from the 1880s to the 1930s. A small amount of material relating to New Inn Hall is kept in the College library, including general accounts 1831–68; buttery and battel books 1833–66 and an entrance book 1831–80.

9   A typescript calendar has been compiled. Relevant letters are listed in M. D. Wainwright and Noel Matthews, *A Guide to Western Manuscripts and Documents in the British Isles relating to South and South East Asia* (1965), 338.

10   Access is restricted; three volumes of selections from 1930 to 1962 have been published.

11   *4th Report* (1874), 442–51.

12   *Oxford Deeds of Balliol College.* (Oxford Historical Society, lxxiv, 1913.) Also published separately.

Handlists, compiled by George Parker in 1889, cover the whole collection and are well indexed. There is a further handlist of documents concerned with internal administration from the sixteenth to the nineteenth centuries. A summary handlist has been made for the History of the University Project.

The archives are in the custody of the Bursar, to whom application must be made for permission to consult.

Printed Books

The custom of individuals presenting a single or a few volumes has already been mentioned; such gifts were nearly always concerned with the subjects currently taught. The earliest benefaction that brought in books outside the ordinary run was that of Sir Thomas Wendy (1613–1673), whose library of over 2,000 volumes reached the College in 1677,[13] immediately increasing its size by a third[14] and, as Wood wrote, 'by which addition, consisting chiefly of choice books, this Library is accounted one of the best in Oxon.'[15] Nicholas Crouch, fellow 1641–89, bequeathed such books as the College cared to select from his library; it would seem that most were taken. This working library of a scholar with an appendage of tract volumes on all kinds of topics contains many scarce works. Crouch had a habit of noting on fly-leaves the price paid and the cost of binding, occasionally giving names.

Another academic collection was received after the death of Roger Mander, Master 1687–1704. Later in the eighteenth century a number of books formerly owned by Nathaniel Crynes (d. 1745) were acquired, probably duplicates from the bequests to the Bodleian and St. John's. George Coningesby (1693–1766), a Herefordshire clergyman with bibliophilic tastes, bequeathed an extensive library containing many early printed books on a wide range of subjects.

There were no large benefactions for nearly a century, but in 1863 the College was given by H. H. Norris the large theological library formed by Henry Norris (d. 1853), mainly English books. This gift was not appreciated at the time and kept separately; T. K. Cheyne in 1881 was very scathing about this 'lower Theology',[16] so it is not surprising that many of Norris's books were among those sold in 1928.

During the mastership of Benjamin Jowett between 1870 and 1893, when Balliol's reputation grew enormously, few important benefactions were

13 Sir Thomas Wendy's will does not specifically mention books; Balliol was named as the residual legatee, so the College probably accepted them in lieu of land. See L. K. Hindmarsh, 'The Benefactors of Balliol', unpublished MS. [c. 1950] in the College library.
14 An inscription formerly on Wendy's portrait included 'e cujus musaeo instructiss: haec Biblioth: tertia parte auctior evasit', quoted in Anthony Wood, *History and Antiquities of the University of Oxford.* [Edited] by J. Gutch, ii (1796), 98.
15 Ibid., 90.
16 Cheyne, 61.

received. The remnant of the library of New Inn Hall, mainly eighteenth-century English law books, came when the Hall was absorbed by the College in 1887, and the books of Arnold Toynbee (1852–1883), the pioneer economic historian, were given on his death. Jowett's influence, however, was felt later when his former students made gifts or bequests, such as Paget Toynbee (1855–1932), whose library of Italian literature and history was divided between the Bodleian and Balliol. Jowett himself bequeathed his books; his gift does not contain many rare items, but is interesting both as the working collection of a classical scholar and teacher, and for the numerous presentation copies from friends and students.

Another famous Balliol teacher, F. F. Urquhart (1868–1934), left his library, mainly on history or art, to the College. Percy Hide (1874–1938), who bequeathed his books on British India, and Sir Robert C. K. Ensor (1877–1958), who left part of his historical library, were both former members of the College. The library of Sir Maurice Powicke (1879–1963), the medieval historian, was mainly a working collection but resembles that of Jowett in the large number of presentations from former students.

Two small groups not usually found in a college library are a small collection of books connected with the Salvation Army and another of editions of Edward Fitzgerald's translation of the *Rubaiyat of Omar Khayyam*; the provenance of the former is unknown, but the latter was once owned by F. York Powell (1850–1904). There are also some scarce early editions of Bibles and Prayer Books.

The number of early scientific and medical works is small, but interesting; the majority arrived as gifts from individuals,[17] though the medical section was notably enlarged by a John Harris in 1666, while the presence of a large number of bound volumes of medical tracts of the late seventeenth century indicate an interest in this subject at that period.

Catalogues

Balliol has an interesting series of catalogues of printed books. The earliest is a ledger made about 1697, superseded by another ledger in 1709. Another ledger contains the fair copy of a catalogue with entries for A–M only, with the rest of the book used for bursar's notes and college business during the 1670s; the relation of this list to the College library is obscure. A new catalogue was made in 1721; a marked copy of the 1738 Bodleian catalogue was superseded by a further one in the early nineteenth century. Coincidentally with the opening of a separate undergraduate library in 1871, a *Catalogue of the Printed Books in Balliol College Library* was published, which can be misleading since numerous items recorded in it were included in the 1928 sale.[18] It was cut up and the entries pasted in guard-books, to which new accessions were added until 1963, since when a card catalogue

[17]   A selection of over thirty names of donors is given in Gunther, 39–40.
[18]   This *Catalogue* was used for Balliol locations in Donald Wing's *Short-title Catalogue. . . 1641–1700* (1945–51), thereby including many items discarded in 1928.

has been employed. There is a separate guard-book catalogue of tracts and pamphlets, superseding a ledger made in the late eighteenth century.

A manuscript catalogue of books in Balliol, but not in the Bodleian printed catalogue of 1738, was made about 1800 and is kept in Duke Humfrey.

BLACKFRIARS

PRIORY OF THE HOLY GHOST ORDER OF PREACHERS

ST. GILES

1921

Incunabula: 26   *STC*: 35   Pre-1641 foreign books: 230   Wing: 54

Note:—Since the main library is within the conventual enclosure, women readers cannot have direct access to the catalogue and shelves; alternative accommodation can usually be arranged.

The Dominicans first established a house in Oxford in 1221, but were dissolved at the Reformation and not restored until seven hundred years after their original foundation. Blackfriars is not officially part of the University, though its members make such use of the teaching and research facilities as they wish.

The library specialises in works on theology and philosophy rather than ecclesiastical history, with naturally an emphasis on the Dominican order. There is a collection of nearly 350 books printed before 1700. St. Thomas Aquinas, and members of the order like Leonardus de Utino, Luis of Granada and Robert Holcot are well represented, often by editions not found elsewhere in Oxford. Many of these older volumes were formerly at smaller Dominican houses, such as those at Haverstock Hill, London; Hawkesyard, Staffs.; Manchester; Newcastle; Stone, Staffs.; and Woodchester, Glos.

Besides these early printed books, the library also has some groups of literary interest. André Raffalovich (1864–1934)[1] bequeathed a small collection that included both presentation copies to himself from other contemporary writers and some editions of William Beckford. A large number of items printed at the St. Dominic's Press, Ditchling, Sussex, together with a few specimens from other private presses and of English literature of the 1920s, were given by Michael Gerveys Sewell.

The catalogue is on cards and the Library of Congress classification is used.

---

[1]   Brocard Sewell, ed., *Two friends: John Gray and André Raffalovich: essays biographical and critical* (1963), 138, 149.

BRASENOSE COLLEGE

1509

Stock: 37,000   Current periodicals: 80   Incunabula: 83   *STC*: 753

Brasenose College was founded in 1509 by William Smyth (d. 1514), Bishop of Lincoln, and Sir Richard Sutton (d. 1524). The sites of several medieval halls were acquired in the early sixteenth century, including one named Brasenose, but except for one manuscript now in the Bodleian formerly belonging to Staple Hall,[1] no books from these foundations are known to have survived.[2] Books and manuscripts were received from the beginning, starting with a bequest from Edmund Croston (d. 1508), Principal of Brasenose Hall, while a library was made part of the first College buildings; Hugh Oldham, Bishop of Exeter, made a donation towards its furnishings about 1511[3] and glass, desks and chains were bought about 1520.[4] The revised statutes of 1520–21 laid down regulations about books, showing that there were a chained and a circulating collection as in other then existing colleges.[5]

Donations were received from the Founders and others in the early years, but none was particularly large or remarkable, unlike the collections received by All Souls, Lincoln and Magdalen during the previous century and Corpus in the first half of the sixteenth.[6] Money was spent on the library at this period, but whether on the purchase of books or other purposes is unknown.[7] An inventory made in 1556 to satisfy Cardinal Pole's Commissioners shows that Brasenose then owned 102 volumes (if all are recorded), mostly gifts but including new editions of the Fathers, all of which are still present.[8] Money for the library was given or bequeathed on several occasions during the sixteenth and seventeenth

---

[1]   Now Ashmole MS. 748.
[2]   For general accounts of the library see F. Madan, 'The Library of Brasenose College, Oxford', *Notes and Queries*, series 6, ii (1880), 321–2; John Buchan, 1st Baron Tweedsmuir, *Brasenose College*. (College Histories, 1898), 89–91; *VCH Oxon.* iii, 213. Numerous references to the library are scattered through *Brasenose College Quatercentenary Monographs*, i–xiv, (1909–10), published separately and as Oxford Historical Society, lii–liv, referred to here only by *Monograph* number and page.
[3]   *Monograph* iv, 8.
[4]   *Monograph* ix, 25, note 2.
[5]   Ibid., 25–6.
[6]   Ker, 475.
[7]   Ker, 485, says there is no direct evidence for the purchase of books between 1535 and 1558, but I. S. Leadam in *Monograph* ix, 152, using the Bursars' Rolls, records about 26s. 8d. a year on average spent on the library.
[8]   *Oxford College Libraries in 1556*, 6, 15; Ker, 485; the Ambrose stated there as missing is probably the Basle, 1538, edition still in the library.

centuries. The practice of writing the names of donors inside their presents was not often observed before the mid-seventeenth century.

A new library was built after the Restoration, the books being put on the shelves in 1664; this room was considerably altered and remodelled in 1779–80,[9] but it has been restored more nearly to its original appearance in recent years.[10]

A Benefactors' Register was begun in 1657, when the rebuilding was first planned, but unlike those at other colleges dating from this period, earlier gifts were not recorded;[11] library entries cease in the early eighteenth century, though continued to the twentieth for plate. A regular source of new books between 1650 and 1764 was the custom of each year's batch of new graduates—'determining bachelors'—combining to present a volume to the library, presumably some work suggested by the fellows.[12] It is, of course, not uncommon for men going down from Oxford to give books as individuals; the combination of several adopted at Brasenose is not found elsewhere in Oxford, though known at Edinburgh University as early as 1596.[13]

Besides relying on gifts, the College made a small fixed contribution to the Library from 1669.[14] The office of librarian was joined to that of *Custos Jocalium*, a post held by a fellow for about a year at a time, until 1826, when the duties were separated; names from 1614 are recorded.[15]

Throughout the seventeenth and eighteenth centuries, besides actual books, monetary gifts were also made; to take a few examples, Thomas Allen left £20 in 1637, William Grimbaldson £1,000 in 1725 and Samuel Malbon £50 in 1788 for book purchases.[16] The large collection bequeathed by Francis Yarborough, Principal from 1745 to his death in 1770, caused considerable internal re-arrangement; in 1780 James Wyatt refitted the library, blocking up the large west window as well as some smaller ones; the latter were re-opened during the re-organisation of 1950–1956.[17] Chaining continued until the 1780 re-fitting; purchases of chains are recorded at intervals between 1520 and 1673,[18] but Falconer Madan's

9   *Monograph* iii, 23, 24.
10   *The Brazen Nose*, x (1955–6), 98; [xii] (1960), 7.
11   *Monograph* iv lists benefactions; one cover of an original volume, rebound by Douglas Cockerell in 1902, and two pages are illustrated on Plates III and VI. Ker, 462, implies there is no Register at Brasenose, but presumably only means that the sixteenth-century gifts are not recorded.
12   A card index, arranged by date, has been compiled.
13   Edinburgh University Library, *Benefactors of the Library in Five Centuries; an Exhibition*. . . . (1963), no. 12.
14   The library account book, covering 1668/9 to 1872, is kept in the library, not with the Archives.
15   *Brasenose College Register, 1509–1909*, ii (1909), 92–4, 110; see also *Monograph* iv, 23 and vi, 22.
16   *Monograph* iv, 22, 31, 39.
17   *Monograph* iii, 34–5; *The Brazen Nose*, x (1955–6), 98.
18   R. W. Jeffery, 'The Bursars' Account Books', *The Brazen Nose*, iv (1924), 25; *Monograph* iii, 23 and ix, 25.

claim that Brasenose 'was the last library in Oxford to retain the custom', though frequently repeated,[19] is incorrect; that honour belongs to Magdalen College where they were not removed until 1799.[20]

There was no reading room provision for undergraduates until the 1870s, when a separate library was established,[21] transferred to part of the new buildings in 1887.[22] A decade later, the main library was converted into an undergraduate reading room and older books moved into the Tower.[23]

During the present century the library has continued to grow, through the generosity of members as well as by purchases. For instance, Albert Watson, Principal 1886–89, gave large endowments both in 1889 and by bequest in 1904.[24] Expansion has naturally caused problems of space, somewhat alleviated by making an underground bookstore about 1930,[25] new rooms for history and law in recent years,[26] as well as more stack rooms.[27]

Manuscripts

Coxe[28] lists twenty-two manuscripts, much fewer than its near contemporary, Corpus, all single gifts except for two from William Smyth, the Founder. They consist mainly of the usual theological and classical texts, including a tenth-century Terence once Cardinal Bembo's, a thirteenth-century Vulgate and a fifteenth-century Cicero.[29] Since the publication of Coxe's list, more have been acquired, of which a calendar is available in Duke Humfrey. Among these additions are several manuscripts of Alexander Nowell (d. 1602), Dean of St. Paul's and Principal in 1595, and of other *alumni* such as Reginald Heber (1783–1826), Bishop of Calcutta, and Michael Wodhull (1740–1816). A collection of twenty-one legal manuscripts of the seventeenth and eighteenth centuries formed part of the bequest of W. T. S. Stallybrass (1883–1948), Principal 1936–48.

The groups so far mentioned are all deposited in the Bodleian,[30] where they may be consulted. There are also a few, of a miscellaneous nature, kept in the College library, such as the commonplace book of Thomas Sixesmith (d. 1651), sometime Vice-Principal, and two poems by R. H. Barham (1789–1845), including one of the *Ingoldsby Legends*. The papers

---

19   Madan, 321; repeated, *e.g.* in Jeffery quoted above, where Brasenose is claimed to be 'the last in England' to unchain.
20   Streeter, xiv.
21   Madan, 321.
22   *Monograph* iii, 60.
23   *The Brazen Nose*, i (1911), 154.
24   *Monograph* iv, 41; *Brasenose College [Report] 1904–5*, 1.
25   *The Brazen Nose*, v (1933), 314–27.
26   Ibid., ix (1949–54), 1, 90, 159.
27   Ibid., x (1955–6), 98.
28   Pt. 2, section 3.
29   Some are listed in Madan, 322 and Buchan, 91.
30   Sir E. Craster, *History of the Bodleian Library, 1845–1945* (1952), 190.

of R. W. Jeffery (1876–1956), a former fellow, are also deposited in the library, but are not to be opened until 2006 A.D.

Archives

The archives of Brasenose are extensive; indeed, it has been remarked that their wealth is so great that much was left untouched by the contributors to the *Quatercentenary Monographs* of 1909.[31] A full calendar, arranged by place, schools, bursarial and College business, was made between 1900 and 1904, while a supplement, mainly of bursarial documents, with subject and chronological indices, has been compiled in recent years; both have been reproduced for the National Register of Archives. A card index is available of items not contained in either calendar or supplement. A partial list of estates owned by the College is given in *Monograph* vi, and a survey of the classes of documents in the Muniment Room was made by R. W. Jeffery.[32] A handlist of muniments has been made also for the History of the University Project. Land is, or has been, owned in nearly twenty English counties stretching from Durham to Kent. The oldest deed is probably one relating to Ivington, Herefordshire, of about 1140;[33] there are also a number of interesting early maps.[34]

An excellent series of records concerned with College business exists, from the earliest days to the present; Bursars' rolls, for example, begin in 1516 and the registers of the Vice-Principal in the reign of Edward VI. These records were widely used for the *Quatercentenary Monographs*, and R. W. Jeffery wrote several papers based on them, such as the Bursars' account books (from 1582),[35] Common Room accounts (1773–1810),[36] charities between 1652 and 1659[37] and an account of land owned in Kensington.[38] An unusual group of records is concerned with schools in the towns or villages of Charlbury (1670–1896); Farnworth (1507–41, 1833); Middleton (1565–1858); and Steeple Aston (1650–1890); they mainly deal with finance, with a few letters and memoranda as well. As already mentioned, the library account book for 1668/9 to 1872 is kept in the College library.

The archives are in the custody of the College Archivist.

---

31   *VCH Oxon.* iii, 213.
32   R. W. Jeffery, 'The Brasenose College Muniment Room', *The Brazen Nose*, v (1933), 290–3.
33   Ibid., iii (1921), 107–8.
34   Those of Burrough, Leics., 1607, and of Folly Bridge, Oxford, *c.* 1500 are reproduced as Plates III–IV of *Monograph* vi; another of Folly Bridge, 1726, is in J. L. G. Mowat, *Sixteen Old Maps of Properties in Oxfordshire in the Possession of Some of the Colleges*, (1881).
35   *The Brazen Nose*, iv (1924), 19–30.
36   Ibid., ii (1915), 55–9.
37   Ibid., iv (1925), 118–29.
38   Ibid., iv (1928), 404–22.

Printed Books

The printed books at Brasenose form an interesting collection built up over the centuries by numerous gifts and bequests from members[39] and a certain amount of steady purchases from the seventeenth century onwards. Little appears to have been discarded, except for a sale of duplicates in 1681, presumably since the regularly occurring problems of space, common to all libraries, have fortunately been solved by the acquisition of more rooms.

As with the manuscripts, the printed books given by the founders and others in the first half of the sixteenth century were neither so numerous nor impressive as those acquired by colleges such as Corpus. It should be noted that more printed books than manuscripts were given at this time. Since presentation inscriptions were not always added, the definite identification of books now on the shelves with recorded donors can only be assumed in many cases. Twelve of those given by William Smyth, the Founder, eight from John Hafter, one of the original fellows, and several presented by John Longland, Bishop of Lincoln, some time before his death in 1547, are still in the library; all are incunabula. This original collection was augmented by gifts from *alumni* once the College was established; John Booth, a Canon of Hereford, for example, left books in 1542, but only two survive.[40] Nearly thirty volumes of the latest editions of the Fathers, acquired possibly by purchase in the mid-sixteenth century, are still on the shelves. Alexander Nowell (d. 1602) left his books to his old College of which some standard works in folio, such as the Magdeburg Centuriators, survive. An early example of the pleasant custom of authors presenting their own books to their college is a copy of the third edition of the *Anatomy of Melancholy* of 1628, given by Robert Burton.

The first really substantial gift, however, was not received till the reign of James I, when Henry Mason (d. 1647) presented books then valued at £1,000; they are nearly all sixteenth- and seventeenth-century theological works, now numbering about 500 volumes. Three smaller collections of rather similar content were received from Edmund Leigh, fellow 1611–41; William Hutchings, Vice-Principal from 1642 to his death in 1647; and Samuel Radcliffe (d. 1648), Principal from 1614 until he died.

No more large accessions were received for over a century, though smaller ones continued, including those from determining bachelors each year until 1764. With the death of Francis Yarborough, the Principal, in 1770 Brasenose received a remarkable library, rich in eighteenth-century theological, philosophical and classical works. Much of the classical literature is annotated by a former owner, Christopher Wasse (d. 1690),

---

[39] A full list of benefactions to 1908 is in *Monograph* iv; select lists of library benefactions are in Madan, 321; Buchan, 90; and *VCH Oxon*. iii, 213. A survey of the rarer early printed books, by R. W. Jeffery, is in *The Brazen Nose*, v (1933), 315–27.

[40] *The Brazen Nose*, iii (1922), 173–4.

of Queens' College, Cambridge, the rival of Bentley. Yarborough's successor as Principal, Ralph Cawley (d. 1777), also bequeathed his books,
as did John Holmes (d. 1795), a former fellow.

There were no similar large benefactions during the nineteenth century,
though there have been some since 1900. A good collection of nineteenth-
century theology, classics and pamphlets was bequeathed by Albert
Watson (1829–1904), Principal 1886–89; another, strong in Roman
history, was given by the widow of H. F. Pelham (1847–1907), sometime
Camden Professor of Ancient History. An important collection of law
books was left by W. T. S. Stallybrass in 1948, with some manuscripts
already mentioned. Smaller groups have also been received, such as some
eighteenth-century natural history from S. P. Duval in 1942, while there
have been several bequests of working libraries from fellows, like K. J.
Spalding (1879–1963) on philosophy and E. S. Cohn (1899–1963) on
history. The Law Library was augmented also with the books of Charles
d'Olivier Farran (1924–1967), presented by his widow, and strong on the
international aspect.

Brasenose College library is particularly rich in copies of the published
works of members. Burton's example in 1628 has been mentioned, but the
present collection is based on that made by W. E. Buckley (1818–1892), and
purchased after his death; it is actively maintained and grows steadily.[41]

Catalogues

The inventory prepared for the Commissioners in 1556 has already been
discussed. In 1635 10s. was paid 'to D. Birch' for 'transcribing the Catalogue', and the same amount was again paid in 1663 to 'Mr. Frankland'
(presumably Thomas Frankland, a fellow then Proctor) for a new
catalogue;[42] both of these have survived. The latter was used until a copy
of the 1738 printed Bodleian catalogue was annotated with the College
holdings. The 1843 printed Bodleian catalogue, however, was not annotated, as it was in the majority of colleges, but a separate ledger, begun
in the early part of the century was used until superseded by another
ledger about 1890; this lasted until about 1930, when the present card
catalogue was started, covering the whole library.

There is also a separate card catalogue of books printed before 1641.
W. E. Buckley's own catalogue of Brasenose authors continued to be
annotated until about 1958, when the practice ceased. Robert Proctor
(1868–1903), the incunabulist, wrote slips for all books and a good many
fragments printed before 1601; they are arranged by place and printer
in three sequences: (i) incunabula; (ii) those printed between 1500 and
1520, except for (iii), those printed in Britain between 1500 and 1600.

---

41   Works mentioned in *Brasenose College Register, 1509–1909* and then in the College
      library are marked with a cross; there is also an alphabetical list of authors in ii,
      153–60. Publications by Brasenose authors have been listed in the annual reports
      between 1904–5 and 1909–10, and since 1909 in *The Brazen Nose*.
42   *The Brazen Nose*, iv (1924), 24.

CAMPION HALL

BREWER STREET

1918

Incunabula: 3   *STC*: 7   Pre-1641 foreign books: 39   Wing: 29

Note: The library is ordinarily closed for the whole of August; during the rest of the vacation there are only very limited facilities.

A private hall for members of the Society of Jesus was established in Oxford in 1896, becoming recognised by the University as a permanent private hall in 1918.

Besides a good working library there is a small collection of manuscripts and older printed books. Many of the papers (but *no* manuscript poetry) of Gerard Manley Hopkins (1844–1889) are housed here;[1] there are also two medieval illuminated manuscripts. There are about 200 books printed before 1800, mainly of a theological or a classical nature, together with Catholic polemics. St. Edmund Campion is represented in several early and scarce editions, while Merton College in 1936 gave a volume once owned by him, which had been in the gift of Robert Barnes in 1594.

---

[1] A catalogue is published as Appendix IV to *The Journals and Papers of Gerard Manley Hopkins*. Edited by Humphry House, completed by G. Storey (1959).

CHRIST CHURCH
1546

Stock: 100,000 plus   Current periodicals: 120   Incunabula: 96
STC: 2,749

Whether Cardinal Wolsey intended to build a library when his College was first planned in 1525 is unknown;[1] there appears to have been no provision for books until the early 1560s,[2] after its re-foundation by Henry VIII in 1546. Some gifts from individuals were made in this decade[3] and many more in the 1580s. Donors' names appear in the Disbursement Books from 1581, and chains were purchased, first in London but from 1592 they were forged by the College smith. As in other colleges, accessions mainly came through donations from members. The practice began in 1599, and continued for many years, of electing a Bachelor of Arts as Library Keeper, who usually handed over to another after taking his master's degree.[4]

The next development was in 1613 when Otho Nicholson, known more widely as the donor of the Carfax Conduit, provided £800 for the restoration of the building; in the following year he gave £100 towards the purchase of books. This year, 1614, was notable for two other library matters: statutes were drafted which laid down that a gift of money or books had to be made when graduating, the value varying with the degree, (which was only legalising an established custom,) and a Donors' Book was started.[5] Liberally decorated with Nicholson's arms, earlier benefactions were not entered in it. It was used rather intermittently until 1841—there are no entries between 1635 and 1653 for example.[6] A card index of names has been compiled in recent years. A final significant decision of the later seventeenth century was the reduction of library fees in 1679.[7]

[1]   W. G. Hiscock, *A Christ Church Miscellany* (1946), 1; possibly the most detailed account of an Oxford college library from the sixteenth to the twentieth centuries (though closely approached by Sir Edmund Craster's account of All Souls Library), hereafter referred to as: Hiscock.

[2]   Hiscock, 3, suggests that the Library 'probably dates from 1563'; Dr. N. R. Ker, however, in *Oxford College Libraries in 1556*, 7, says the 'Library was not begun until 1561'.

[3]   *E.g.* Cicero, *Rhetorica* (Basle, 1541), presented by C. Worsop in 1563 (Shelfmark: A.C. 2. 14).

[4]   Hiscock, 3–4.

[5]   Ibid., 6–7; *Fine Bindings 1500–1700*, no. 128.

[6]   Hiscock, 10.

[7]   Ibid., 14.

During the first half of the eighteenth century several important benefactions were received, culminating in that of William Wake, Archbishop
of Canterbury (1657–1737), for which the old library was quite inadequate
and so the present fine building was erected. Begun in 1717, its decoration
was not completed until 1772, but in 1763 the contents of the old library
with about 18,000 additional volumes were put on the wall bookcases,[8] an
early example of this method. Presumably the chains on books transferred from the old Library were removed beforehand. New regulations
were brought into force in 1776, covering the extra post of Wake Librarian
(established 1754), to be in charge of the Wake manuscripts, in addition
to that of College Librarian; each was helped by an Under-Librarian; the
posts were combined in 1869.[9] The rules governing persons entitled to use
the library remained the same: namely, members who were graduates,
noblemen or gentlemen commoners, thereby excluding commoners,
battlers and servitors.[10]

During the eighteenth and nineteenth centuries, the annual expenditure
on books steadily rose from about £30 to about £100 in the 1780s, and
£200 between 1820 and 1870. Simultaneously a large balance accumulated in the library account, reaching over £600 in 1869 which, in the
following year, was transferred to the College account and the practice
of a separate annual grant was instituted.[11] In 1904 this was about £200,
considered inadequate by the Librarian; Balliol, for example, spent
over £300 then.[12]

Duplicates and surplus volumes have been discarded on several occasions, notably in 1793, 1813 and 1814, bringing in respectively £277,
£505 and £422.[13] Besides books disposed of officially, others have strayed
elsewhere, as has so often happened in college libraries, including some
scarce items.[14]

The relationship between this college library and its undergraduates has
followed the usual Oxford pattern, but it is interesting that in 1668
Anthony Wood records: 'A little before X^tmas, the Xt. Ch. men, yong
men, set a library in Short's coffee hous in the study ther, viz., Rablais,
poems, plaies, etc. One scholar gave a booke of 1s. and chaine 10d..'[15]

The 1776 regulations, restricting the use of the library to noblemen and
gentlemen commoners among the undergraduates, mentioned above, were
amended in 1813 when, in consequence of damage to books taken out,
borrowing was forbidden to junior members, while the library was
opened to those previously excluded provided a suitable recommendation

8 Ibid., 72.
9 Ibid., 103, n. 1.
10 Ibid., 88.
11 Ibid., 97.
12 *The Library 1904.* [A special report by F. Haverfield.]
13 Hiscock, 13, 89.
14 *Cf.* Hiscock, 118–33, 222–31.
15 *Life and Times.* Collected by A. Clark, ii, 147. (Oxford Historical Society, xxi, 1892.)

was given by tutors.[16] Some years later, in 1827, college tutors were allowed to borrow books for their pupils, possibly to regularise an existing situation,[17] but according to the rules[18] undergraduates were forbidden to take down books from the shelves. A separate undergraduate reading room was opened in 1884 for those reading for an honour school, at first under the control of the Librarian, up to 1893, and then separately administered until brought back under the Librarian in 1928 and the whole structure re-organised into its present state.[19]

Christ Church was fortunate to have the services of the late W. G. Hiscock first as Assistant, and later Deputy, Librarian from 1928 till 1962. His scholarly researches into the Library's history and careful examination of its contents have made Christ Church Library more widely known than possibly that of any other Oxford college. This present account is deeply indebted to his works, as the notes reveal.

Manuscripts

Christ Church has an extensive collection of manuscripts, ranging from early Greek through medieval to modern examples. A catalogue was published in 1867;[20] an annotated copy, kept in the College library, has *addenda* in card-index form. C. H. Hoole, in his *Account of some Greek Manuscripts Contained in the Library of Christ Church, Oxford* (1895), called attention to the outstanding Latin, English, Hebrew, Arabic and Welsh manuscripts as well.[21]

The Greek manuscripts[22] came with the bequest of Archbishop Wake in 1737; they were probably brought from Mount Athos about 1724, but how the Archbishop acquired them is unknown. In all there are eighty-six, dating from the eleventh to the eighteenth century, and include patristic writings and Biblical texts.[23]

There are about sixty Latin manuscripts,[24] the earliest of the twelfth century, of varying provenances, mainly of theological, liturgical and legal interest. They include two fifteenth-century ones of Virgil, presented by John Moore in 1766; seven vellum manuscripts of the Vulgate (five dating from the fourteenth century); two French fifteenth-century Books of Hours; an interesting fifteenth-century *Corpus Juris Angliæ*, beginning with Magna Carta and ending with the Statute De Wardis. A lectionary,

---

[16]   Hiscock, 92.
[17]   Ibid., 93.
[18]   Ibid., 123.
[19]   Ibid., 107–11.
[20]   G. W. Kitchin, *Catalogus Codicum Manuscriptorum qui in Bibliotheca Ædis Christi apud Oxonienses Adservantur* (1867). The style follows that used by H. O. Coxe, but there is no list of donors, and the index is not exhaustive.
[21]   On pp. 19–22.
[22]   Kitchin, nos. 1–86.
[23]   *Cf.* C. H. Hoole, *An Account of some Manuscripts of the New Testament hitherto Unedited Contained in the Library of Christ Church, Oxford* (1892).
[24]   Kitchin, nos. 87–144, 231–33.

finely illuminated, made for Thomas Wolsey and dated 1528 was possibly presented by him; it is the companion of the Gospels at Magdalen College. A fourteenth-fifteenth century copy of Ralph Higden's *Polychronicon* was given by Samuel Burton in 1595. In the separate library bequeathed by Richard Allestree in 1681, is a thirteenth-century manuscript of the sermons of St. Bernard. A later group of some interest includes some notes on Isaac Newton's *Principia*, mathematical tables and *Institutiones Astronomicae* by David Gregory (1661–1708), presented by his son, also David, Dean from 1756 to 1767.[25]

The Oriental manuscripts include seventeen in Hebrew[26] (though Dean Fell bequeathed twenty-eight in 1686[27]); there is one on vellum containing the *Halacoth*, dated 1410, and two seventeenth-century works on the Cabala. Among the twenty-nine Arabic manuscripts[28] are four Korans, the oldest of 1441, two fragmentary portions of the *Arabian Nights* and a version of the Gospels. There are three French manuscripts,[29] one a fourteenth-century New Testament, and two in Italian,[30] one a Palladio, *Antichità di Roma*, 1596. The two Welsh manuscripts[31] contain for the most part poems in praise of members of the Salusbury family by William Cynwal (d. 1587 or 1588).

Manuscripts in English[32] probably form a more important group. Some, such as Chaucer's *Canterbury Tales* of *c.* 1420, Lydgate's translation of Dares Phrygius, and three early versions of Wiclif's translation of the Bible are of literary interest, but the majority have historical value. There is the sixteenth-century manuscript of George Cavendish's *Life of Wolsey*, probably used as copy for the first published edition, but pride of place historically must go to the extensive papers of William Wake (1657–1737),[33] covering not only the English ecclesiastical and political world of his time but also a wide range of contacts with Europe and missionary enterprises overseas. Permission to consult the Wake manuscripts must first be obtained from the Wake Trustees.

Another important group is the collection of manuscript music, mainly

25 Hiscock, 116; *David Gregory, Isaac Newton and their Circle. Extracts from David Gregory's Memoranda 1677–1708.* Edited by W. G. Hiscock (1937).
26 Kitchin, nos. 185–201; A. Neubauer, *Catalogue of the Hebrew Manuscripts in the Bodleian Library and in the College Libraries of Oxford*, i (1896), nos. 2444–58; Hoole, *An Account of some Greek Manuscripts*, 21, gives the number as sixteen.
27 Hiscock, 17.
28 Kitchin, nos. 202–30.
29 Ibid., nos. 178–80.
30 Ibid., nos. 181–2.
31 Ibid., nos. 183–4.
32 Ibid., nos. 145–77.
33 Ibid., nos. 234–334; the papers were used in Norman Sykes, *William Wake, Archbishop of Canterbury, 1657–1737.* 2 vols. (1957), who gave an outline list of the correspondence at Christ Church in vol. 2, pp. 272–3; papers concerned with Wake's missionary interests in Asia are listed in M. D. Wainwright and N. Matthews, *A Guide to Western Manuscripts and Documents in the British Isles Relating to South and South East Asia* (1965), 338.

of the later sixteenth and earlier seventeenth century; a partial catalogue has been published.[34]

Two smaller groups worthy of mention are, firstly, the papers of Thomas Gaisford (1779–1855), Dean from 1831 until his death, of interest not only for the history of Greek scholarship but also for his letters to a pupil, Fynes Clinton, which reveal his business activities.[35] Secondly, the manuscripts of H. J. White (1859–1934), Dean 1920–34, concerned with his work on the Vulgate.

Christ Church also has two collections of papers on deposit, which require special permission from the Librarian before being consulted. The earlier is that of John Evelyn (1620–1706),[36] which includes the well-known diary and other writings as well as family letters down to the nineteenth century; letters between Sir Richard Browne (1605–1683), Evelyn's father-in-law, and the first Earl of Clarendon; and a collection of original letters of famous people from the sixteenth century onwards, begun by the diarist and continued by successive generations. A handlist is available. The second consists of the papers and correspondence of Robert, third Marquess of Salisbury (1830–1903), sometime Prime Minister; a calendar of the private Foreign Office correspondence has been prepared and made available through the National Register of Archives; a card index to his domestic correspondence and papers is being made.[37]

Archives

The medieval archives, concerned with monastic houses suppressed to endow Wolsey's Cardinal College, such as the Priory of St. Frideswide, Oseney Abbey,[38] and Daventry Priory, with nine other houses, were deposited in the Bodleian in 1927,[39] where they may be consulted, joining deeds relating to the same places pilfered by Anthony Wood in the late seventeenth century. A calendar has been published.[40]

The remainder of the Christ Church archives are kept in the College

---

[34]   G. E. P. Arkwright, *Catalogue of Music in the Library of Christ Church, Oxford, Pt. 1: Works of ascertained Authorship* (1915); Pt. 2, *Works of unknown Authorship* (1923). Pt. 3, 'A thematic catalogue of anonymous instrumental music' is unpublished; a photocopy is kept in the library. Pts. 1–2 were reprinted in 1971 with a list added of those manuscripts from which negative microfilms have been prepared.

[35]   Hiscock, 103.

[36]   Many references to this collection can be found in W. G. Hiscock, *John Evelyn and Mrs. Godolphin* (1951), and his *John Evelyn and his Family Circle* (1955).

[37]   Items relating to South and South East Asia are listed in Wainwright and Matthews, 338–9; those concerned with America are listed in B. R. Crick and M. Alman, *A Guide to Manuscripts relating to America in Great Britain* (1961), 395–8.

[38]   H. E. Salter, *Cartulary of Oseney Abbey*, 6 vols. (Oxford Historical Society, lxxxix–xci, xcvii–xcviii, ci, 1929–36.)

[39]   Sir Edmund Craster, *History of the Bodleian Library, 1845–1945* (1952), 69, 100, 304–5.

[40]   N. Denholm-Young, *Cartulary of the Medieval Archives of Christ Church*. (Oxford Historical Society, xcii, 1931.)

Muniment Room (though some remain in the College Treasury); they are in the custody of the Archivist to whom application must be made. Battelling books from 1547 have survived, but the Disbursement Books, apart from a mutilated one of 1548, do not begin until 1577.[41] Accounts start in 1561–2. There is a large quantity of deeds and papers concerning estate management relating to land in several parts of the country, the largest collection belonging to an Oxford college. Three typescript catalogues have been prepared:

   i. A catalogue of manorial records.

   ii. A catalogue of maps, plans and drawings. (This covers drawings kept in the Archives and does not refer to the large collections of drawings kept elsewhere in Christ Church.)

   iii. A catalogue of Treasury Books, including miscellaneous items.

The first two are arranged by county and then place and copies of them are kept in Duke Humfrey. A typescript calendar of estate papers in numerous volumes is available in the Muniment Room.

Besides the classes so far mentioned, there are also the S.C.R. papers of C. L. Dodgson (1832–1898), more widely known as 'Lewis Carroll', and some papers of Claude Jenkins (1877–1959), the ecclesiastical historian.

The Evelyn deposit also contains archival material in papers and deeds concerned with family properties in Wotton (Surrey), Deptford and elsewhere.

The library records, in the custody of the Librarian, are more extensive than in any other Oxford college; a handlist is available. The Donors' Book, already mentioned, used c. 1614–1841, is the earliest; there is an Admissions Register covering 1698–1726, and separate accounts from 1712; accounts from 1900 are kept in the Treasury. Lists of books sent to be bound begin in 1822, while there is an order book for 1824–61 and an accession register from 1814 onwards. The minutes of the Library Committee for 1869 and 1878–99 have survived, as has the minute book for the undergraduate Reading Room for 1884–1901 and its suggestion book for 1884–1928.[42] The long series of catalogues is outlined below.

It should be noted that the records of the Dean and Chapter of the Cathedral, so closely intertwined with the College, are in the custody of the Chapter Archivist, to whom application must be made. They are not extensive and mainly take the form of Chapter Act Books and minutes.

Printed Books

Now numbering well over 100,000 items, the printed books and pamphlets make Christ Church the largest and richest library for research material in Oxford outside the Bodleian. Its growth will be outlined and some of the stronger subjects mentioned, but it must be emphasised that this library should be searched by anyone seeking older works on practically any topic.

41   Hiscock, 2.
42   Ibid., 107–10.

There has been a steady flow of small donations from new graduates and old members from the 1560s onwards. The first substantial benefaction was that of Otho Nicholson in 1613–14 which included £100 for the purchase of books, mainly spent on theology; others, like John King, Bishop of London (d. 1621), also gave money for books.[43] Robert Burton (1577–1640), who had given a copy of the first edition of his *Anatomy of Melancholy* in 1621, and had been Librarian from 1626,[44] bequeathed the residue of his books to Christ Church after the Bodleian had had first choice. About 500 volumes were received which have now been re-assembled, a process begun by Sir William Osler in 1907–8; as might be expected from the *Anatomy*, many subjects are represented including rare ephemeral publications.[45] John Morris, Regius Professor of Hebrew from 1626 to his death in 1648, left an annuity of £5 for the purchase of books in Hebrew and a good collection strong in grammars and dictionaries has been built up since this still current fund became operative in 1682.[46]

The eighteenth century was not a period of stagnation, for during the first half some magnificent benefactions were received that are the found-ations of this library's riches today. Of these, the first was the bequest of Henry Aldrich,[47] Dean from 1689 until his death in 1710, containing a remarkable assemblage of some 3,000 theological, classical, mathematical and architectural books and pamphlets; the last includes many Quaker, Civil War and Commonwealth tracts bought from Francis Bugg. In addition the bequest contained about 8,000 pieces of sixteenth- and seventeenth-century music by English and foreign composers, many extra-ordinarily rare,[48] and about 2,000 engravings, European and English.[49] Books owned by Aldrich are notable for their fine condition and binding. He expressed the wish that his nephew, Charles, should be given any 'duplicates' from his library, a term rather loosely interpreted at the time. Charles was Rector of Henley-on-Thames from 1709 until he died in 1737, leaving his books to found a parochial library in that town; a catalogue was published in 1852,[50] but by 1909 the books were in a deplorable condition so a number were brought to Christ Church on permanent loan; a few more were similarly transferred in 1942;[51] they are not

---

43   Ibid., 7.

44   Ibid., 7–9.

45   Ibid., 112; the books are listed in Oxford Bibliographical Society, *Proceedings and Papers*, i (1922–7), 236–46.

46   Hiscock, 16. A similar sum was left to All Souls, but since 1927 the whole benefaction has been allocated to Christ Church.

47   Ibid., 17–30.

48   See Aloys Hiff, *Catalogue of the Printed Music Prior to 1801 now in the Library of Christ Church, Oxford* (1919).

49   Hiscock, 31–7; there is an eighteenth-century catalogue, but a new one is in the course of preparation. The engravings are now in the College Picture Gallery in the custody of the Curator of Pictures.

50   *The Catalogue of the Old Library, at Henley-on-Thames. . . .* 1852.

51   Hiscock, 63–4.

shelved together as a collection. The residue was deposited in Reading University Library in 1957.[52]

In 1722 Lewis Atterbury, brother of Francis, Dean 1711–13, gave between three and four thousand pamphlets, a remarkable collection extending from the early seventeenth century to contemporary publications.[53]

Next came the bequest of William Stratford (1672–1729), a canon, of nearly 5,000 volumes; it resembled that of his colleague, Henry Aldrich, but was of a more general character, including books on natural science, law, history, and literature not owned by the Dean, as well as some sixteenth-century English books. Charles Boyle, fourth Earl of Orrery (1676–1731), editor of the ill-fated *Letters of Phalaris*, had been a favourite pupil of Dean Aldrich, and built up a rather similar general library, though with more medical and scientific works; he bequeathed it to his old college, whither it was brought from London in 1733.[54]

The fourth great benefaction in the first half of the eighteenth century was the bequest of William Wake, Archbishop of Canterbury, in 1737, whose printed books came with the papers and manuscripts already mentioned. This collection of some 5,000 volumes, naturally strong in theology, includes many early-printed books, finely bound large-paper and presentation copies from authors, European as well as English. This was the last really large gift, but smaller ones from this century are also worthy of note. Richard Goodson, organist 1718–41, left his father's small library of music,[55] which augmented that of Aldrich, and David Gregory, Dean 1756–67, left a good general collection. Richard Trevor (1707–1771) gave some forty rare early books, including incunabula, on classics, mathematics, and medicine[56] before leaving Oxford for Durham in 1753, while Cyril Jackson, Dean 1783–1809, gave some books in 1795.

During the nineteenth century no benefactions of note were received, though the flow of small gifts from members continued, and the money spent on purchases gradually increased. Typical examples are a bequest of 144 volumes, mainly on Italian history, from Henry Auber Harvey in 1884, and W. E. Gladstone's numerous presentations, often of his own writings, between 1841 and 1897.[57]

The present century, however, has seen several interesting donations. Frederick York Powell (1850–1904) bequeathed about 800 volumes of Icelandic and Scandinavian literature, many given to him by Guðbrandr Vigfússon (1828–1889);[58] all, except for a set of Icelandic Bibles, have been

---

52 *Library History*, i (1969), 216–7; a catalogue made 1777–8 and a list of 'Books retained by Christ Church, 1909' are also in Reading University Library.
53 Hiscock, 64.
54 Ibid., 64–6, 72–4.
55 Ibid., 23, n. 1.
56 Ibid., 83–4.
57 Ibid., 102.
58 Ibid., 105.

deposited in the English Library. A collection of about 500 volumes on classical literature and strong in editions of and works on Aristophanes was presented in 1908 by the widow of W. G. Rutherford (1853–1907), sometime Headmaster of Westminster School.[59] In the same year a bequest was received of a small group of books mainly on Oxford and Christ Church from T. V. Bayne, a former Librarian.[60] Theological books were bequeathed by Francis Paget (1851–1911), Bishop of Oxford and a former Dean, and Henry Scott Holland (1847–1918), Regius Professor of Divinity. Eighty-four volumes, mostly in the Thai language, presented by Rama VI (1881–1925), King of Siam and a commoner of the House 1900–1, were deposited in the Bodleian in 1967. Besides his papers, H. J. White (1859–1934) left his working library concerned with his edition of the text of the Vulgate, including a number of early books; many had been bequeathed to White by John Wordsworth, Bishop of Salisbury, in 1911. The travelling library of Edward Gibbon, containing twenty-four miniature volumes of Latin authors, was presented by J. A. Stewart in 1933.[61]

A good collection of liturgical texts, especially English Books of Common Prayer, with many drafts and papers connected with Prayer Book revision in the 1920s, formed by Kenneth Gibbs (1856–1935), sometime Archdeacon of St. Albans, was presented by his widow in 1946; there are many early editions, though some have suffered from the old habit of 'making-up' imperfect copies.

In 1946 also Christ Church library took over the administration of the separate library bequeathed by Richard Allestree in 1681 for the use of the Regius Professor of Divinity and his successors, housed over the south Cloister.[62] Although it naturally has a bias towards theology, there are books on many other subjects—classics, science, medicine, mathematics and patristics, for example; 138 were formerly owned by Henry Hammond (1605–1660).

Kept with the Allestree library is the parochial library of Wotton-under-Edge, Gloucestershire (a Christ Church living) deposited for safe-keeping. John Okes (d. 1710), a St. Edmund Hall man, left it to his native place; there are now about 300 volumes,[63] chiefly seventeenth-century theology, with a bias towards Oriental studies. Many of the books bear the names of members of the Cholmondely family, patron of Okes' living at Whitegate, Cheshire, which he held from 1665 till deprived as a non-juror in 1689.

The most important deposited collection is the library of the Evelyn

---

[59]   Ibid., 105.
[60]   Ibid., 102.
[61]   Ibid., 106.
[62]   Ibid., 14–5, where rather too rosy a picture of the physical condition of the books is given.
[63]   Central Council for the Care of Churches, *The Parochial Libraries of the Church of England* (1959), 107.

family of Wotton, Surrey. It contains the majority of books formerly owned by John Evelyn (1620–1706), the diarist, in their distinctive bindings, reflecting his wide interests, such as silviculture, horticulture and architecture, as well as copies of his own published works.[64] There are numerous sixteenth- and seventeenth-century books, with some owned by his father-in-law, Sir Richard Browne (1605–1683), sometime English Ambassador in Paris. Special application must be made to the Librarian before this collection can be used.

An attempt has been made to indicate the strong points of individual benefactions. Taken as a whole, Christ Church library is particularly rich in music, theology, classics, travel books, numismatics, early science and medicine,[65] and Hebrew studies. The large pamphlet collection has many rare items, especially of the seventeenth and eighteenth centuries, but there is also a large number from the sixteenth century. The number of printed plays and poetry is not negligible either. But it is invidious to emphasise subjects, for it would be unwise for anyone seeking scarce material from the sixteenth to the nineteenth centuries to neglect this library.

Catalogues

There is a full series of catalogues kept with the Library records, more than can be mentioned here. There was none at all during the first century of the library's existence; the earliest, compiled in 1665, shows about 2,000 volumes; this was used until the 1674 printed Bodleian catalogue was published, when a copy was interleaved and annotated.[66] Separate contemporary catalogues have survived of the great benefactions of the eighteenth century—Aldrich, Stratford, Orrery and Wake. The duplicates sale of 1813 financed the compilation of a new catalogue, supplanting the existing one in three sequences.[67] Next the pamphlets were re-catalogued during the 1820s,[68] and between 1844 and 1847 a catalogue of music was made by H. E. Havergal;[69] lists of music in the Aldrich bequest made by Charles Burney and William Boyce, about 1778, also survive. Between 1928 and 1935 the library was re-organised, the current catalogue, first written in the 1890s, thoroughly checked, and a separate card catalogue of pamphlets made; at the same time names of former owners written in books were recorded.[70] A nineteenth-century list of autographs in the Allestree Library is also among the library records.

The current catalogues are:

---

[64] Works in this collection are included in the locations given in Sir Geoffrey Keynes, *John Evelyn: a Study in Bibliophily with a Bibliography of his Writings*, 2nd ed. (1968); there is a general description of the Library on pp. 6–30.
[65] *Cf.* Gunther, 46.
[66] Hiscock, 11–4.
[67] Ibid., 89–90.
[68] Ibid., 96, 118.
[69] Ibid., 96, 127.
[70] Ibid., 111–7.

   i. A copy of Kitchin's catalogue of manuscripts, with *addenda* on cards.

   ii. A copy of Arkwright's catalogue of manuscript music, annotated.

  iii. Author catalogue of the printed books, written in six volumes.

  iv. Card catalogue of pamphlets.

   v.   ,,   ,,   ,, the Allestree Library.

  vi.   ,,   ,,   ,,   Evelyn deposit.

 vii.   ,,   ,,   ,,   library from Wotton-under-Edge.

viii. Annotated copy of the published catalogue of printed music by Hiff; (these items are also included in Edith B. Schnapper, *The British Union-Catalogue of early Music Printed before the Year 1801.* (1957).)

  ix. A copy of Sir A. E. Cowley, *A Concise Catalogue of the Hebrew Printed Books in the Bodleian Library* (1929), annotated with relevant holdings.

Until the new edition now in preparation is published, the Christ Church locations in Donald Wing's *Short-title Catalogue 1641–1700*, like those of Balliol and Merton, must be used with caution and only after checking W. G. Hiscock's *The Christ Church Supplement to Wing's* Short-title Catalogue 1641–1700 (1956); the same author also published in duplicated form in a limited edition a very useful list, *The Christ Church Holdings in Wing's Short-title Catalogue 1641–1700 of which less than five Copies are Recorded in the United Kingdom* (1956).

CORPUS CHRISTI COLLEGE

1517

Stock: 43,000   Current periodicals: 112   Incunabula: 258   *STC*: 1,931

Corpus Christi College was founded by Richard Foxe (1448?–1528), a statesman-cleric, to train a supply of educated clergy and the original statutes made provision for a library and its government. These incorporated the idea of a circulating library besides laying down that books were not to be chained unless worthy in form and contents.[1] Erasmus foretold that the *trilinguis bibliotheca* would become one of the chief scholarly attractions of Oxford,[2] and, indeed a good collection of theological and classical works was soon formed. Though funds were used to acquire books in the sixteenth century, as in few other colleges, the magnificent gifts of the Founder and others made few purchases necessary.[3] In 1597 Richard Cobbe bequeathed £10 a year for the library and this was the only income until 1791, when graduate members began to be charged quarterly, and gentlemen-commoners had to pay entrance fees, towards upkeep.[4] Entries for book purchases appear from 1602 onwards in the *Libri Magni*, the annual College accounts; 265 printed books were bought between 1602 and 1617, for example.[5] Separate library accounts begin in 1769. The statutes of 1856 made provision for a more regular financial basis. A separate library for undergraduates was in existence in the late nineteenth century, and a College Library Committee has been functioning since 1889.

A *Liber Benefactorum* was started *c*. 1690, which records earlier gifts, and is complete up to 1803; fuller particulars are found in a list made about 1700.[6] J. G. Milne (1867–1951), fellow and numismatist, who was Librarian between 1933 and 1946, made some useful indices of former owners and bindings, while many donors are noted in his *Early History of Corpus Christi College*.[7]

The original library, built in the early sixteenth century and situated in the south range of the front quadrangle, still has its contemporary

1 Ker, 464; J. G. Milne, *The Early History of Corpus Christi College, Oxford* (1946), 37.
2 *Epistolae*, ed. P. S. and H. M. Allen, iii (1913), 619.
3 *Oxford College Libraries in 1556*, 7.
4 Milne, 38; J. R. Liddell, 'The Library of Corpus Christi College, Oxford, in the Sixteenth Century', *The Library*, series IV, xviii (1938), 392.
5 J. R. Liddell, *The Library of Corpus Christi College, 1517–1617* (B.Litt. thesis, Oxford, 1933), ff. 93–110.
6 Ker, 462; J. G. Milne, 'The Muniments of Corpus Christi College', *Oxoniensia*, ii (1937), 130; *Fine Bindings 1500–1700*, no. 215.
7 Pp. 45–53.

bookcases and it is refreshing to see readers working in this attractive setting to-day, although books have naturally spread into adjoining rooms. Though Corpus has been small in numbers, it has had a high proportion of distinguished scholars as members down the centuries, and this quality is reflected in the richness of the library's holdings, both of manuscripts and of printed books.

Manuscripts

Coxe[8] describes 393 manuscripts, while a supplement, compiled by Charles Plummer, listing forty-four more, was printed in 1900. Subsequent additions have been added by hand or in typescript to the copy of Coxe kept in Duke Humfrey; many of the older manuscripts were deposited in the Bodleian in 1935.[9]

The Founder, Richard Foxe, gave about 150 manuscripts and books, while John Claimond (1457?–1537), the first President, gave about the same number, and some more came from Thomas Walshe (d. 1528). These were mainly concerned with theological, liturgical and classical matters. Foxe's gift included many Latin (but no Greek) manuscripts acquired from John Shirwood, Bishop of Durham, which have been fully described elsewhere.[10] A catalogue of 1589[11] lists 371 titles, of which 310 can still be identified as present in the library.

The bequest of Brian Twyne, a fellow, in 1644, included at least sixty-three manuscripts formerly owned by Dr. John Dee (1527–1608), some interesting not only for medieval science, medicine and mathematics but also for the monastic provenance of others;[12] the College unfortunately did not get Twyne's own papers.[13] Another seventeenth-century collection is that of Christopher Wase (1625?–1690), which probably came to the College through his son of the same name, who was a fellow. It is important for contemporary English history, and especially for details of grammar schools.[14]

The papers of Thomas Hornsby (1733–1810), covering mainly 1768 to 1774 and concerned with his astronomical studies, represent the eighteenth century, but later accretions are more numerous. There are letters to William Little, one of the editors of the *Shorter Oxford English Dictionary* chiefly for the years 1867–71, and letters and papers of Thomas Fowler (1832–1904), President 1881–1904, dealing with his researches into the College's history as well as his philosophical writings. There are also some

8   Pt. 2, section 4.
9   Sir E. Craster, *History of the Bodleian Library, 1845–1945* (1952), 289–93.
10   P. S. Allen, 'Bishop Shirwood of Durham and his Library', *English Historical Review*, xxv (1910), 445–56.
11   Printed in Liddell.
12   Craster, 291–2.
13   R. F. Ovenell, 'Brian Twyne's Library', *Oxford Bibliographical Society Publications*, new series iv (1952), 5.
14   An index by place is included in P. J. Wallis, 'The Wase School Collection', *BLR* iv (1952), 78–104.

of the papers of Percy Stafford Allen (1869–1933), the editor of the letters of Erasmus. More recently, the College has received the correspondence, papers and diaries of Sir Robert C. K. Ensor (1877–1958), which are important for the study of modern British foreign policy and the history of socialism, to say nothing of his hobby of ornithology.[15] The papers of A. Francis Hemming (1893–1964), an *alumnus*, including many concerned with the International Committee for Non-Intervention in Spain, 1936–1939, have also been deposited.

Numerous fragments of early manuscripts from the bindings of books in the College library were extracted, arranged and classified in albums by J. G. Milne and R. G. C. Proctor (1868–1903), the incunabulist, during the 1890s; these are deposited in the Bodleian.

Archives

The Founder laid down strict regulations for preserving the muniments of College property,[16] and a classified account has been published.[17] Hand-lists have been prepared for the National Register of Archives (available in Duke Humfrey) and also for the History of the University Project. A brief description by H. T. Riley was written for the Historical Manuscripts Commission.[18] Brian Twyne's transcripts of evidences connected with the College, made in the early seventeenth century, fill thirty volumes and can be used in some ways as a guide.

Besides the original charter and statutes, the Founder's correspondence with the first President, John Claimond, has survived. Admission registers are complete. Bursarial accounts begin in 1521, battels and buttery records in 1648–9, but there are gaps. The College archives also include some records of undergraduate life, such as J. C. R. resolutions and minute books from 1811, Boat Club annals and accounts from 1858, besides some later nineteenth-century essay and debating club minutes. There are also documents concerned with schools at Cheltenham, Grantham and Manchester, from the seventeenth to the nineteenth centuries.

Land was owned in several parts of the country and the records of the manors of Lower Heyford, Oxon., West Hendred, Berks., and Pertenhall, Beds., are especially noteworthy. The Berkshire deeds were described in detail by J. G. Milne.[19] A handlist of the estate and central records is available. Maps of Oxfordshire properties at Lower Heyford, Whitehill near Tackley, and Cowley have been published.[20] An unusual type of archive in Oxford is the extensive collection of the papers of Robert

---

15   A typescript calendar is available in Duke Humfrey.
16   J. G. Milne, *Early History*, 54–72.
17   J. G. Milne, 'The Muniments of Corpus Christi College', *Bulletin of the Institute of Historical Research*, x (1933) 105–8, and also in *Oxoniensia*, ii (1937), 129–33.
18   *2nd Report*, (1871), 126.
19   'The Berkshire Muniments of Corpus Christi College, Oxford', *Berkshire Archaeological Journal*, xlvi (1942), 35–44, 78–87.
20   J. L. G. Mowat, *Sixteen Old Maps of Properties in Oxfordshire in the Possession of Some of the Colleges* (1888), Maps 1–7.

Newlin, steward of the College between *c.* 1645 and 1700, except for during the Commonwealth; he never threw anything away and so left intact all his correspondence and material connected with College business.

The archives are in the custody of the Librarian.

Printed Books

The gifts of the Founder and John Claimond, already mentioned, included many printed books dealing with contemporary classical and theological studies. As in other colleges, there was a fair number of donations of one or two volumes from individuals, but a substantial quantity came in the bequest of Thomas Greneway, the fifth President, who died in 1571, which reflects his interest in the Protestant theology of the time. As with the manuscripts, these early gifts contained practically no scientific or medical works.

By his will, John Rainolds (1549–1607), one of the leading scholars of his day and President 1598–1607, left to the College one hundred books, mostly Biblical, but in actual fact more were received, including a few botanical, scientific and geographical works.

Later in this century, one of the most varied collections of books to be found in any Oxford college was bequeathed by Brian Twyne (*c.* 1580–1644), sometime Reader in Greek at Corpus and the first Keeper of the University Archives. A full description with a catalogue has been published,[21] and it is sufficient to say here that all subjects are represented while many of the ephemeral English printed works, especially those concerned with astrology, farriery, medicine and the light reading of the time, are extremely rare. Books on antiquarian matters and Greek studies are fairly numerous also; in all there are about 750 items.

In the latter part of the century, two further bequests, nearly as miscellaneous as that of Brian Twyne, came from Richard Samwayes (*c.* 1625–1669), fellow, and John Rosewell (d. 1684), Headmaster of Eton. In consequence, the library is particularly strong in Civil War and other seventeenth-century tracts.

More collections arrived during the first half of the eighteenth century. The medical section was strengthened by the bequest of William Creed (d. 1711), while that of Thomas Turner (1645–1714), President 1688–1714, was again strong in ephemeral pamphlets and theology. A benefaction from Cuthbert Ellison in 1719 increased the number of scientific books.

Italian history and topography are not often found in college libraries, but the collection on these subjects formed by Henry Hare, third Baron Coleraine (1693–1749) and his father was bequeathed to Corpus; it included thirty volumes of sketches and drawings.

The rest of the eighteenth and all the nineteenth century saw no large

21   Ovenell, 3–42.

benefactions or bequests; several have been received since 1900. Part of the historical library formed by Robert Laing (later Cuthbert Shields), fellow, came to Corpus after his death in 1908. Shadworth Hodgson (1832–1912), fellow, left his important philosophical collection to his College, while many of the books of Charles Plummer (1851–1928), fellow, chiefly of historical interest, were received after his death. In 1919 a select collection of works on the Peninsular War came in a bequest from Askell Benton (1880–1918), an *alumnus*. There is also a group of early editions of works by Erasmus formerly owned by the editor of his letters, P. S. Allen (1869–1933), President 1924–33. In 1957 the library bequeathed by John Conington (1825–1869) for the special use of holders of the Corpus Professorship of Latin was deposited here.

R. G. C. Proctor and J. G. Milne removed many of the printed fragments from bindings in the 1890s simultaneously with the manuscript pieces, and arranged them in albums, kept in the library. Proctor's unique knowledge of incunabula enabled him to identify the majority of the earlier fragments.

There is also a small collection of works by *alumni*, not so complete as in other colleges, but it is being expanded.

The riches of this distinguished library make a summary impossible; some of the outstanding books are mentioned in J. G. Milne's *Early History*,[22] and some of the early scientific books were described by R. T. Gunther,[23] who stresses the 'strange absence of the ordinary mathematical books' found in other colleges.

Catalogues

The earliest catalogues of 1589, the special lists of Bishop Shirwood's manuscripts, of the archives and of Twyne's books, have already been mentioned. Brian Twyne himself compiled a shelf-list about 1605, but it is incomplete, recording only about ninety-three manuscripts and printed books.[24] There is an interleaved Bodleian catalogue of 1843 annotated with Corpus locations, while a modern card catalogue is maintained; the latter does not yet include all the tracts or printed fragments. R. G. C. Proctor compiled a catalogue of the incunabula, which is now kept in Duke Humfrey.

Edward Edwards (1812–1886), the pioneer of public libraries, was a friend of Charles Plummer and worked in Corpus between 1877 and 1880 after completing the catalogue of Queen's; he made several shelf-lists which still survive while annotations in ink in his distinctive hand on title-pages are also noticeable.[25]

[22] Pp. 39–45.
[23] Pp. 45–6.
[24] Jesus College, Oxford, MS. 30, pp. 156–86.
[25] W. A. Munford, *Edward Edwards* (1963), 187.

EXETER COLLEGE

1314

Stock: 61,000   Current periodicals: 38   Incunabula: 74   *STC*: 759

Exeter College was founded by Walter de Stapledon for men from the West Country and the Channel Islands, connections which are still maintained. A library was built in 1383, the first of a series of buildings for housing books in this College.[1] A former chapel converted to a library was burnt down in December 1709, endangering the neighbouring Bodleian and alarming Thomas Hearne;[2] some of the surviving books still show signs of this ordeal. A new library building was completed in 1788. The present building, together with the chapel, was designed by Sir Gilbert Scott and opened in 1855.[3]

Books were purchased for the College from its foundation and a list of those chained in the fourteenth century survives,[4] while the titles of various liturgical works acquired for the chapel between 1547 and 1566 are known.[5]

Beginning with Sir William Petre's gifts in 1567, there has been a stream of gifts and bequests to the library up to the present day. Library fees are recorded with the Dean's accounts from 1760. A Benefactors' Book, recording earlier gifts as in other colleges, was begun in 1703.[6] Some seventeenth- and eighteenth-century theological works were discarded in 1971. Borrowers' registers begin in 1785.

Taken as a whole, this college library is perhaps neither so large nor quite so full of treasures as its antiquity might suggest, but hardly surprising in view of the 1709 fire; nevertheless, it does have many rarities.

Manuscripts

Coxe[7] describes 184 manuscripts, chiefly individual gifts, though Roger Keys (d. 1477), Archdeacon of Barnstaple and Warden of All Souls, presented several. In Joseph Sandford's bequest of 1774 there is a group of manuscripts formerly owned by Sir William Glynne, Bart., of interest to historians of sixteenth- and seventeenth-century England. A few

---

1   C. W. Boase, *Registrum Collegii Exoniensis. Register of Exeter College, Oxford*. New ed. (Oxford Historical Society, xxvii, 1894), pp. xxviii–ix, xlviii–ix.
2   Ibid., cxxxv–vi, 316.
3   Ibid., 359.
4   Ibid., xxxv–vii, xlviii–ix.
5   Ibid., lxxvii–viii.
6   Benefactions up to 1800 are summarised in Boase, 268–74.
7   Pt. 1, section 4.

of the liturgical manuscripts are described by C. W. Boase.[8] The R. M. Dawkins bequest of 1955, now deposited in the Taylor Institution, includes some manuscripts of Modern Greek and Byzantine interest.

Archives

The College archives are separately housed. H. T. Riley gave a short description of them to the Historical Manuscripts Commission[9] which includes an account of the older records and charters, together with deeds relating to property in Oxford. Many go back to the foundation in 1314; accounts begin in 1324 and registers of College business in 1540. Admission records date from 1574, buttery books from 1593 and bursary books from 1596. Junior Common Room minutes start in 1887, and there are records of various undergraduate, athletic, sporting, debating and literary societies from 1854 onwards. Cellar books with details of vintages ordered, in stock, amount consumed and battel payments begin in 1785.[10]

An index was made in the eighteenth century, and there is a modern handlist with brief entries. A summary handlist was compiled in 1970 for the History of the University Project.

For permission to consult, application must be made to the Keeper of the Archives, Exeter College.

Printed Books

It seems probable that Exeter remained without modern printed books, except for the liturgical works already mentioned, until after the accession of Elizabeth I,[11] and the virtual refounding of the College by Sir William Petre. John Dotyn, a Somerset clergyman and *alumnus*, in 1561 left twenty-three books, nearly all medical, to the College,[12] the only recorded accession of works on this subject before Sir William Paddy's gifts to St. John's early in the reign of James I. None of these has survived.

Besides his endowments of property, Sir William Petre presented a set of the Latin Fathers in 1567. Sir William, an *alumnus* who achieved success as a statesman and ambassador, was a friend of Thomas Wotton, the bibliophile, who bought books in France for Sir William occasionally.[13] Wotton had most of his own library bound very distinctively and sumptuously in Paris, so it is not surprising that this set of the Fathers is similarly bound, and indeed two volumes bear Wotton's armorial stamp.[14]

As in most other college libraries, there were many gifts of particular volumes by individuals, beginning in the latter part of the sixteenth

8   x, clxv–vii.
9   *2nd Report* (1871), 127–30.
10 J. Haslam, 'College Wine Cellars', *Oxoniensia* [forthcoming].
11 *Oxford College Libraries in 1556*, 7.
12 Boase, 58.
13 Boase, lxxxi–ii.
14 *Fine Bindings, 1500–1700*, nos. 46–7, 49, 54, 56.

century.[15] Bequests for money for the purchase of books were also made, such as £120 from Samuel Cosens in 1668.[16] Items bought with these funds and still in the library are preponderantly of theological and classical interest.

How many works were destroyed in the fire of 1709 is unknown; Hearne recorded that 'only one stall of books, or thereabouts' was saved;[17] several volumes have charred edges still. The library was soon refurnished and gifts of money and books continued to flow in: £5 from Thomas Mathews in 1712 and £20 from Richard Eastchurch shortly afterwards, for example.[18] Richard Hutchins left his library to the College in 1718 and Samuel Conant did the same the following year; the latter collection was large, revealing a wide range of interests beyond theology. The bequest of Edward Richards in 1729 was strong in classical authors, while that of Joseph Sandford in 1774 was both extensive and miscellaneous in character.

During the nineteenth century there was no large benefaction before 1889, when the widow of Alfred Edersheim (1825–1889) presented that scholar's notable collection of Hebrew, Talmudic and early Christian literature with relevant secondary works on condition that books should be lent freely to all members of the University.[19] In order to implement this provision, the Hebrew section has been placed on permanent loan in the library of the Oriental Institute.

More recently, R. M. Dawkins (1871–1955) bequeathed his large library of modern Greek and Byzantine history and literature to the College, but this has been deposited in the Taylor Institution.

Catalogues

The current catalogue is on cards, kept in two sequences: one for books on open shelves, the other for those in the stacks. Besides a hand-written eighteenth-century catalogue, Bodleian catalogues of 1738 and 1843 annotated with Exeter holdings survive. There is a separate catalogue of the Edersheim collection, with the items deposited in the Oriental Institute suitably marked.

A manuscript catalogue of books in Exeter but not in the Bodleian was compiled in 1800 and is kept in Duke Humfrey.

Books printed or published in Britain before 1641 are shelved in a separate sequence arranged by the *STC* number of the first item in each volume. Books printed abroad before 1601 are arranged by place of printing.

---

15   Boase, xcvi.
16   Ibid., 269.
17   Ibid., cxxxv.
18   Ibid., 269.
19   Ibid., clxviii.

GREYFRIARS

IFFLEY ROAD

1910

STC: 11   Pre-1641 foreign books: 57   Wing: 12

The Franciscans first came to Oxford in 1224 and stayed until the Reformation. The Capuchin branch returned at the beginning of this century, later opening a hostel in 1910. The Priory, built 1930–31, was recognised by the University as a permanent private hall in 1957.[1]

There is a small specialist library, chiefly concerned with Franciscan history and thought; its backbone is the collection of A. G. Little (1863–1945), the historian of the Order in Britain, including a certain number of his manuscript notes and letters;[2] it was given by his widow in 1949. Correspondence and papers connected with the Roger Bacon celebrations of 1914, collected by F. A. Dixey, are also here. There are nearly a hundred books or pamphlets printed before 1700, among which are several rare editions of Franciscan devotional works.

Two other groups are worthy of notice: an extensive range of Catholic pamphlets mainly concerned with educational and sociological questions in the early years of this century, presented by Leslie A. St. L. Toke (1871–1944), a Balliol man; and a good collection of books on Egyptian, Syrian and Coptic thought and belief, presented by Samuel Webster Allen (1844–1908), sometime Bishop of Shrewsbury.

[1]   *Oxford Magazine*, lxxvi (1957), 150–2.
[2]   The bulk of his papers are now in the Bodleian.

HERTFORD COLLEGE
1874

Stock: 12,000 Incunabula: 5 *STC*: 192

Note: The Librarian stresses that this library is not open to the general public; those who wish to consult books are requested in general not to apply between 31 July and 15 September, nor within seven days before or after Christmas and Easter. If the librarian is then available he will readily entertain applications between the dates mentioned, but no service can normally be guaranteed.

Although the present College is a nineteenth-century creation, it is the heir of older foundations: Hart Hall, later the first Hertford College, and Magdalen Hall. Its rather chequered history can be found elsewhere,[1] so it is sufficient to say here that Hart Hall, of medieval origin and connected with Exeter College, became the first Hertford College in 1740, but had fallen into decay by the early nineteenth century. Magdalen Hall, neighbour of and closely allied to Magdalen College until 1602, migrated in 1813 to the site of the first Hertford College in Catte Street, later to be re-created in 1874 as the second Hertford College.

The older books belonging to the College today not unnaturally reflect these changes.[2] Halls, once numerous in Oxford, were neither wealthy nor the owners of large libraries, being very much the private kingdoms of their heads. Fortunately, in the later seventeenth century two principals were much concerned with establishing libraries for their halls. William Thornton, Principal of Hart Hall from 1688 to 1707, included a library room in a new gatehouse built with the help of several benefactors, such as Emmanuel Prichard, Janitor of the Bodleian Library and a member of the Hall. Henry Wilkinson, Principal of Magdalen Hall from 1648 to 1662, aimed to build up a good stock of books, so published a catalogue in 1661 designed both as an incentive to further study by his students and as a guide to old members who wished to fill gaps or replace books lost in the Civil War. This is the earliest published catalogue in a hall or college library. A study of the shelf-marks of the 646 entries has enabled a plan of the Library to be reconstructed; folio volumes were chained and there were nine manuscripts.[3]

[1] S. G. Hamilton, *Hertford College*, (College Histories, 1903); *VCH Oxon.* iii. 309–19.
[2] A history of the library appears in S. G. Hamilton, *op. cit.*, 157–60; and more fully by the same author, in *Hertford College Magazine*, no. 6, December 1912, 160–3. A summary is in *VCH Oxon.* iii, 315.
[3] F. Madan, *Oxford Books*, iii (1931), 145–6.

As in other Oxford libraries, the main source of accessions, apart from fees, has been the generosity of members and there were numerous gifts of one or two books each from individuals, a custom continued well into this century.[4] Henry Wilkinson began a *Catalogus Benefactorum Aulae Magdalenæ Bibliotheculæ* in 1656, the year the new library was opened, in which the compulsory fees, payable on admission or graduation, were entered as well as gifts of books or money. This register exists in two forms: a rough copy on vellum, whose latest entry is dated 1715, with a few memoranda at the end; and a fair copy on paper, made about the same time as the first, but used until 1746. The notes in the rough copy, made between 1657 and 1670, reveal a gradual growth; thus there were 332 books in 1657 when 386 chains were also acquired, while in 1661, 509 chains were counted at an inspection. Duplicates were exchanged and sold in 1657 and 1660. Both copies contain the library regulations made in 1656; apart from hours of opening and disciplinary rules, the various fees are particularised. Names of donors were not written inside their gifts as frequently as in other Oxford libraries, while there is a paucity of special bookplates or labels, so this Register is the main source of information on provenances for the earlier acquisitions.

What was in the first Hertford College library is obscure; today, Magdalen Hall books seem to outnumber those from other sources. The amalgamation of the two libraries naturally caused a certain amount of duplication and there was probably some discarding. Space must have been severely limited at all periods, so it is not surprising that more discarding took place in the years after the foundation of the second Hertford College in 1874, especially as several benefactions were received. Indeed, volumes with the distinctive 'Magdalene [*sic*] Hall Library' label at the bottom of the spine (added by J. D. Macbride, Principal 1813–1868) can now be seen on the shelves of several Oxford libraries. R. T. Gunther, the historian of Oxford science, has recorded that he bought books that once belonged to the Hall when he was an undergraduate in 1890.[5] In 1909–10 the old Chapel, consecrated for Hart Hall in 1716, was converted into a library and lastly, in 1963, its interior was reconstructed to provide two floors.

Manuscripts and Archives

Coxe[6] credits Magdalen Hall with six manuscripts; there are now over forty. A calendar of those not in Coxe is available in Duke Humfrey. There are three fifteenth-century manuscripts; a group of seventeenth-century English ones, mainly theological notes or sermons by Henry Wilkinson, or devotional works; about twenty volumes of late seventeenth-century notes on classical writers by P. Wesseling and others; and a few

---

[4] *Hertford College Magazine*, no. 3, May 1911, 68; no. 10, April 1921, 18.
[5] Gunther, 48. See also Anthony Wood, *Life and Times*. [Edited] by A. Clark. Vol. 2 (Oxford Historical Society, xxi, 1892), viii.
[6] Pt. II, section 10.

Oriental manuscripts, some presented by Robert Gandell (1818–1887), Professor of Arabic. More recently A. T. Lloyd gave nine letters of Charles James Fox.[7]

The manuscripts were deposited in 1890 in the Bodleian, where they may be consulted.

The archives are meagre, since Oxford halls were either part of a college or the personal property of the principal. A summary of what survives was given by S. G. Hamilton in the preface to his contribution to the College History series.[8] They include Magdalen Hall buttery books for 1661, 1663 and 1670–1862; similar books for Hertford began in 1879. A few Magdalen Hall records are among Magdalen College archives.

The archives are in the custody of the Bursar.

Printed Books

If any books were owned by either hall before the second half of the seventeenth century, they have not survived. Although there have been some subsequent dispersals, mentioned above, the older books now owned by the College form an interesting small collection, whose value lies in the fair number of seventeenth- and eighteenth-century works on economics and geography[9] but the bulk of the library consists of the usual theological and classical material common to so many colleges, though natural science is well represented here.

Henry Wilkinson's appeal for books for Magdalen Hall met with a good response; he himself gave nearly a hundred volumes, while many of his friends, such as Richard Baxter, and former members, made donations. These small gifts continued under Wilkinson's successors during the seventeenth and earlier part of the eighteenth century; a larger benefaction was of about two hundred volumes, mainly of medicine and surgery, from Samuel Thurnor in 1691. Most of the books acquired in this way were contemporary publications, though a few earlier ones were also given.

The most notable acquisition during the eighteenth century was the bequest to the first Hertford College in 1777 of the library of John Cale, of Barming, Kent, much of which is still on the shelves. J. D. Macbride, Principal of Magdalen Hall from 1813 to 1868, was also Professor of Arabic, a chair associated with the two halls from the seventeenth century, and he presented a number of books of Oriental interest. In 1855 Edward Phillips bequeathed both money and many modern works, greatly needed at the time.

Since the refoundation in 1874, there have been several further donations of modern works needed for teaching purposes, such as those from

7   F. M. H. M[arkham], 'Letters of Charles James Fox', *Hertford College Magazine*, no. 24, May 1935, 147–8.
8   See Note 1 above.
9   E. W. G[ilbert], 'Geography at Hertford', *Hertford College Magazine*, no. 46 (1958), 488–90.

the 12th Earl of Winchilsea (1851–1898) and George Sturton Ward (1828–1902); more recently bequests of historical books from C. R. M. F. Cruttwell (1887–1941), Principal 1930–39, and of mathematical works from L. Sadler in 1956 have been received.

A collection of forty-four broadsides and proclamations ranging from the reign of Charles I to Queen Anne, presented by F. D. S. Darwin in 1934,[10] contains several scarce items.

Catalogues of Printed Books

The earliest published Oxford college or hall library catalogue, *Catalogus Librorum in Bibliotheca Aulæ Magdalenæ*, printed by H. Hall in Oxford in 1661, has already been mentioned; it unfortunately omits dates and places of publication. Late seventeenth-century and late eighteenth-century catalogues have survived, with a third made about 1820. The shelf-list was inadvertently printed in 1888, entitled *Catalogue of Books in Hertford College Library*, and some copies distributed before the error was realized. A separate *Catalogue of Books . . . Printed in the Fifteenth and Sixteenth Centuries* appeared in 1910; in the Duke Humfrey copy, items not in the Bodleian are indicated. A typescript catalogue of books printed before 1800 was completed before 1947 and is available in Duke Humfrey, but the Oxford printed books at one time deposited in the Printer's Library are not included.

An interleaved Bodleian catalogue of 1843, annotated with Hertford holdings, is kept with the older books; while a card catalogue is now maintained.

---

[10]   Sir C. H. Firth, 'Notes on Mr. Darwin's Collection of Broadsides', *Hertford College Magazine*, no. 12, May 1934, 118–9.

JESUS COLLEGE
1571

Stock: 25,000   Incunabula: 44   *STC*: 789

Jesus College was founded by Queen Elizabeth on the petition of Hugh Price, Treasurer of St. David's Cathedral in Pembrokeshire; throughout its history there have been strong ties with Wales. The history of the College's library has already been described in detail, so it need only be summarised here.[1]

When a library was first started is unknown, but there seems to have been a small collection of books from its early days.[2] A separate building was erected while Sir Eubule Thelwall was Principal between 1621 and 1630, but this was pulled down in 1640 and the presses stored. A new library, paid for by the then Principal, Sir Leoline Jenkins, was completed in 1679, into which the old presses were fitted, with a gallery added resting on the cases along the east wall; and there the books remain today.

Chaining continued until at least 1765. A separate undergraduate library had been mooted as early as 1725 according to Hearne, but nothing came of the idea then though books for the 'Young Gentlemen' were bought in 1795. One was eventually established in 1865, called the Meyrick Library after a benefactor, Edward Meyrick (1636–1713), a former member of the College.

No College accounts survive earlier than 1631, but few books were bought in the seventeenth century and none at all between 1632 and 1663; as in other colleges, Jesus depended on benefactions and the piety of old members for the bulk of accessions. A Benefactors' Book was begun in the early eighteenth century which records a few donations from 1626. A register of books borrowed from 1774 to 1798 is kept with the archives.

Manuscripts

Coxe[3] lists 141 manuscripts, and only a few have since been acquired to bring the total to 149; all have been deposited in the Bodleian since 1886.[4]

---

[1]   The main basis for the present account is C. J. Fordyce and Sir T. M. Knox, 'The Library of Jesus College, Oxford. With an Appendix on the Books bequeathed thereto by Lord Herbert of Cherbury', *Oxford Bibliographical Society Proceedings and Papers*, v (1937), 49–115; this is also summarised in *VCH Oxon.* iii, 264.

[2]   Ker, 459, does not know of any there before 1600, but Fordyce and Knox, 59, mention the gift of an Aldine Bible of 1518 from Gabriel Goodman (d. 1601), possibly in 1589.

[3]   Pt. 2, section 7.

[4]   Sir E. Craster, *History of the Bodleian Library, 1845–1945* (1952), 189–90, where they are briefly described.

They form a very interesting collection; the medieval ones, mainly spoils from West Country religious houses, range from one tenth-century example to twenty-two of the twelfth, twenty-three of the thirteenth and eight of the fourteenth century. In the early 1620s, a list of forty-nine manuscripts was compiled, purporting to be those presented by Sir John Prise (*c.* 1502–1555) to the College; thirty-eight are still there. In fact, these should have gone to Hereford Cathedral and their actual provenance remains a mystery.[5]

Another group of unknown provenance consists of over twenty manuscripts, many of heraldic interest, formerly owned by Ralph Sheldon (1623–1684), and destined by him for the College of Arms;[6] Anthony Wood had some obscure connection with this diversion. The original draft of Lord Herbert of Cherbury's *History of the Reign of Henry VIII* (in the hand of Thomas Masters) belongs to this group and not to his bequest of printed books.

As might be expected from the Welsh connections of the College, there are some important manuscripts in that language, of which the most famous is the *Red Book of Hergest*, to translate the title, written in the fourteenth or fifteenth century and presented by Thomas Wilkins of Llanblethian, Glamorgan, in 1701. Others were given by Owen Owen in 1657 and by Griffin Lloyd in 1680.[7]

The modern manuscripts include some papers of two distinguished former members, John Richard Green (1837–1883), the historian, and T. E. Lawrence (1888–1935), 'of Arabia'.

Archives

Apart from the foundation charter, the bulk of the archives date from the early seventeenth century. Bursars' accounts go back to 1631 and buttery books to 1638; a transcription and detailed analysis of the latter was made by Andrew Clark in 1912–14 and given to the Bodleian.[8] Two benefactors' registers (in addition to the library one mentioned already) record gifts from 1571 to 1758. A register running from 1660 to 1752 gives particulars of elections and some resolutions of the governing body, but College minutes only survive from 1833. Deeds connected with land owned in Wales, the Midlands and Essex are also kept here, as well as the business papers of two principals, Francis Mansell (1579–1665) and Thomas Pardo (1688–1763). Junior Common Room minutes begin in 1905, but some of the five undergraduate societies represented are earlier, such as the Boat Club (1858) and the Debating Society (1892).

5 N. R. Ker, 'Sir John Prise', *The Library*, series V, x (1955), 6, 12–14, 17–20. Other books once owned by Prise were among those bequeathed by George Coningesby to Balliol in 1766.
6 I. G. Philip, 'Sheldon Manuscripts in Jesus College Library', *BLR*, I (1939), 119–123.
7 A full calendar will be found in 'Report on Manuscripts in the Welsh Language' (by J. Gwenogvryn Evans) in *Historical Manuscripts Commission* (1902), 1–90.
8 MSS. Top. Oxon. e. 128–44.

H. T. Riley wrote a short description for the Historical Manuscripts Commission.[9] Available in the archives are: a catalogue of records made in 1825, divided by counties with College material separate; a list made after reclassification in recent years by J. N. L. Baker associated with his quatercentenary history published in 1971; a third list, by Sir David Evans (1946), includes financial documents in greater detail as well as the formal records; and a brief list of battel books. The list by Sir David Evans is available in Duke Humfrey among the National Register of Archives reports. A handlist has also been compiled for the History of the University Project.

The archives are in the custody of the Bursar.

Printed Books

The library of printed books is strong, not only in classics and theology, but also in law, medicine and linguistics.

The earliest bequest, by Griffith Powell, Principal from 1613 to his death in 1620, was about a hundred volumes, the majority law books. During the next decade, money was given for the purchase of books by six people, probably London acquaintances of the next Principal, Sir Eubule Thelwall (1562–1630), such as Sir Thomas Myddleton (1550–1631) and Rowland Heylyn (1562–1631), both London businessmen of Welsh origins and patrons of Welsh culture and scholarship. Just under 200 volumes, mainly theology and fathers of the church, were acquired in this way. Another businessman, Lewis Roberts (1596–1640), of Anglesey origins and connected with the East India and Levant Companies, presented about fifty volumes not long before he died, chiefly classics and theology, but including a copy of his own *Merchantes Mappe of Commerce* (1638).

The Civil War and Commonwealth period saw a big increase in the size of the library. Edward, first Baron Herbert of Cherbury, was a good friend of Jesus, though he had been at University College in his youth, and apparently had expressed his intention verbally of leaving all his books to the College. After his death in 1648, however, it was found that only his 'Latin and Greek' books were bequeathed to Jesus 'for the inception of a library there', which caused some discontent among the fellows. Nevertheless, some English, French, Spanish and Italian books came among the some 900 volumes from this source.[10] The subject matter shows Lord Herbert's wide range of interests from mathematics, medicine and music to the physical sciences, history, theology and law; classical authors are poorly represented.

9   *2nd Report*, (1871), 130.
10   A catalogue, based on Michael Roberts' inventory of 1649, is printed in Fordyce and Knox, 75–115; it is arranged by size and subject in twenty alphabetical sequences with no author index; Herbert's notes of prices paid, usually written by him on title-pages, are included.

This bequest was probably engineered by Francis Mansell (1579–1665), Principal 1620–21 and 1630 to his ejectment by the Commonwealth visitors in 1649, when he gave his own scholar's library of about 600 volumes, chiefly classical and theological works, to the College. His Puritan successor, Michael Roberts (d. 1679), made a list of all the College's books in May 1649, on taking charge, showing there were now nearly 2,000 volumes. During the Commonwealth there were a number of smaller individual gifts, while Thomas Ellis (1625–1673), fellow 1649–67, presented a number of small-format sixteenth-century books.

After the Restoration, Mansell, who had returned to live in College in 1651, resumed the Principalship for two years and was then succeeded by Sir Leoline Jenkins (1625–1685), a distinguished lawyer and sometime a judge of the High Court of Admiralty, so that it is not surprising that his own extensive library, bequeathed to the College, had an emphasis on maritime law though containing works on many other scholarly subjects as well.

These various bequests naturally resulted in a certain amount of overlapping and at some time about the end of the seventeenth century, duplicate volumes were sold during the Principalship of Jonathan Edwards from 1688 to his death in 1712. Edwards, like so many of his predecessors and unlike most of his successors, bequeathed to the College his library which, besides the usual classics and theology, contained a good many tracts relating to the controversies of his own time.

Benefactions were not so numerous during the eighteenth century. Griffith Davies, a former fellow who died in 1724, left a large library of medical works dating from the sixteenth century to his own day, supplementing Lord Herbert's collection. Later in the century was received the general library of Henry Fisher (d. 1761), the University Registrar, and a member of Jesus, which is interesting for the contemporary printed items, bought as they came out, including some literature and science. Some Orientalia and classics, with a manuscript of the Koran, were received in the bequest of David Jones (d. 1750), a Magdalen Hall man.

Joseph Hoare, Principal from 1768 to his death in 1802, followed the example of the earlier holders of that office in leaving his large collection of books, but unfortunately the library was then passing through a period of neglect and they were not processed for many years.

No gifts of any consequence were received for over a century. The long interval was brought to an end in 1927 with the death of Charles Plummer of Corpus Christi College, who bequeathed his large collection of Celtic books; this is maintained as a live section, continually being added to and kept up-to-date.

An assemblage of books and manuscripts of J. R. Green (1837–1883), the historian, was presented by R. W. Stopford in 1947, and around it an *alumnus* collection is being built; it includes the original thesis of T. E. Lawrence on 'The influence of the Crusades on European military architecture'.

Catalogues

The inventory compiled in 1649 when Francis Mansell was ejected, has been mentioned above; it shows 430 items. Interleaved and annotated Bodleian printed catalogues of 1674, 1738 and 1843 survive, but are all imperfect and have been superseded by a card catalogue made in the 1950s. There are in addition separate card catalogues of works by Jesus *alumni* and of signatures noticed in the printed books. A shelf-list was compiled in 1801 and again *c.* 1850–56.

KEBLE COLLEGE
1870

Stock: 30,000   Current periodicals: 100   Incunabula: 89   *STC*: 237

Note: The Librarian stresses that anybody wishing to use this library must apply to him first for an appointment; any application requiring a written reply must be accompanied by a stamped, addressed envelope. This library is normally closed during University vacations and permission to consult the collections then is subject to the availability of library staff.

Founded as a memorial to John Keble, Keble College attracted many gifts of books and manuscripts during the first half-century of its existence. Though comparatively young as Oxford colleges go, its library is consequently of some importance and, in spite of the wealth of theological material in other colleges, the collections at Keble should not be overlooked by those concerned with liturgical problems, illuminated manuscripts, the history and literature of the Oxford Movement or of Port Royal.

Manuscripts

There are about sixty medieval manuscripts, mainly of a liturgical nature, and comprising the best collection of illuminated ones in an Oxford college. They were received in the bequests of Canon H. P. Liddon (1829–1890), Dr. J. E. Millard (1824–1894), and Canon C. E. Brooke (1847–1911). Liddon and Millard acquired theirs from various sources, while Brooke's formed the bulk, but not all, of the collection assembled by his elder brother, Sir Thomas Brooke (1830–1908), of Armitage Bridge House, Huddersfield.[1] There is also a number of Oriental and other manuscripts, mostly deriving from Liddon. The history of the collections will be discussed in the introduction to the forthcoming Catalogue, in the process of compilation by Mr. M. B. Parkes.

Besides these, there are two collections of great importance for the history of the Tractarian Movement. First are the correspondence and papers of John Keble (1792–1866), presented by his brother Thomas, in which most of his prominent contemporaries feature—Newman, Pusey, Froude, for instance. A typescript list with index of correspondents is available. The second collection comprises part of the correspondence and papers of H. P. Liddon, also important for Tractarian matters as well as for the history of St. Paul's Cathedral, Cuddesdon Theological College

---

[1]   *A Catalogue of the Manuscripts and Printed Books Collected by Sir Thomas Brooke and Preserved at Armitage Bridge House, near Huddersfield.* 2 vols. 1891.

and Oxford affairs; a typescript list, chronologically arranged, is available.[2] The notebooks of H. T. Morgan (1844–1910) concerned with his researches into the history of Port Royal are also worthy of mention.

Archives

There is a full series of archives, from the papers of the Keble Memorial Fund set up in 1866, onwards, such as Council minutes (recording details of College meetings), accounts, admissions, particulars of College servants and benefactions. Junior Common Room minutes begin in 1883, though records for some of the eight undergraduate societies preserved start earlier, such as the Boat Club in 1870, the Athletics Club in 1871, and the Debating Society in 1873. There is also a series of reports on the academic performances of individual undergraduates from 1873, rather similar to those at Oriel, which, however, go back to the 1830s.

A handlist has been prepared for use with the History of the University Project.

The archives are in the charge of the Librarian, except for the Admission Register, in the custody of the Senior Tutor.

Printed Books

Shortly after the granting of a charter in June 1870, the College received as the basis of its printed books the private library of John Keble, presented by his brother, Thomas. Essentially a working collection, it consists mainly of nineteenth-century theology and poetry, including many presentation copies, but Keble's wide range of interests in other subjects is also evident, and there are a few books from earlier centuries. A further gift of books from Keble's library was received in 1902.

A working library of English and German theology and history was bequeathed by Lord Richard Cavendish in 1874. Many individuals, including a large number of Anglican clergy, made small gifts over the years, such as E. B. Pusey in 1879 and Cardinal J. H. Newman in 1875; besides books, the latter gave the copyright and proceeds of his *Lyra Apostolica* for the purchase of medieval theology. The Oriental and Biblical scholar S. C. Malan (1812–1894) presented about 400 volumes of theological works in 1885, while a selection of scores and books on music was given in memory of R. M. Tamplin (d. 1887), a former College organist.

Besides his papers and early manuscripts, the College received with the bequest from H. P. Liddon in 1890, a large number of theological, liturgical and historical books.[3] Liddon had built up over the years an extensive library from many sources and there are numerous scarce and early

---

[2] More of Liddon's letters and papers are at St. Edmund Hall (of which he was Vice-Principal 1859–62) and Pusey House, in Oxford. His diaries from 1858 and further papers are at Liddon House, London, while his diaries for 1854–58, with about 100 letters and papers, are at Cuddesdon Theological College.

[3] Another large portion of Liddon's library was at Cuddesdon Theological College, but much has now been dispersed.

editions among his books. Liddon's large gift was shortly followed by two small but select ones; in 1893 W. Hatchett Jackson (1849–1924), then Praeceptor in Natural Science and later Sub-Warden 1915–21, presented an interesting group of early editions of and works on Aristotle formed by his father, William Jackson, sometime a fellow of Worcester College. In 1895, seventy-one volumes, including some incunabula, early service books and specimens of the work of fine printers such as Bodoni, were given by the widow of J. E. Millard (1824–1894), a former Master of Magdalen College School, Oxford. Two rather miscellaneous libraries, mainly theological, came from J. D. Chambers (1805–1893) and Richard Hooper (1821–1899), Vicar of Upton, Berks.; among these are several early printed books.

In 1903 H. T. Morgan gave his comprehensive collection of books by and about Port Royal, together with his notebooks; the donor, a Trinity man, considered that the older colleges had plenty of good special collections and that the newer foundations needed building up; a catalogue has been published.[4]

The next accession, possibly the most important so far received, was the bequest of Canon C. E. Brooke in 1911, containing about 300 printed books of liturgical interest besides the manuscripts already mentioned; many of the early printed books are of the greatest rarity and some are also in fine or interesting bindings.

During the last half-century, though there have been no large acquisitions of research interest, gifts of books for undergraduate use have continued to be given; for instance, many items from the library of C. H. Turner (1860–1930), a fellow of Magdalen College, and more recently from that of L. Rice-Oxley (1892–1960), were received by bequest.

Catalogues

The original manuscript author catalogue in ledgers is in the process of being transferred to cards. The *Catalogue* of the Port Royal collection has been mentioned above. There is also a separate sheaf catalogue of the Brooke bequest; most of the items are fully described in the published *Catalogue* of 1891, but it should be remembered that Sir Thomas Brooke continued to buy books, so not everything can be found in it.

A catalogue of the medieval manuscripts by Mr. M. B. Parkes, intended for publication, is nearly completed.

4 *Catalogue of the Port Royal Collection in Library of Keble College, Oxford.* 1905. See also H. T. Morgan, *Port Royal and other Studies* (1914), and *Lincoln Diocesan Magazine*, lxiv (1948), 113, 126–8.

LADY MARGARET HALL
1878

Stock: 42,000   Incunabulum: 1   *STC*: 42   Pre-1641 foreign books: 66
Wing: 59

Note: The Librarian stresses that this library is not accessible to the general public and bona fide *scholars must make arrangements with her in advance before they can be admitted.*

Lady Margaret Hall is the oldest of the Oxford women's colleges, having been founded about eight months earlier than Somerville. A collection of books was included from the beginning, based on gifts,[1] and a fund for a library building was started in 1904.[2] A library was placed in the new buildings completed in 1910[3] but it soon became too small. By 1930 there were more than 15,000 volumes[4] growing to over 18,000 in 1935[5] and double that figure by 1959 when a new building was planned,[6] opened in 1961. As in the early days of men's colleges, it was the custom for students going down to make a present of books.[7] Though the library is primarily a working collection, a certain number of interesting older books and a few manuscripts have been given over the years.

Manuscripts and Archives

A deed of 1559 relating to the grant by Christ Church of the site of Clattercole Priory, Oxfordshire, to Thomas Lee is the only early manuscript.[8] George Musgrave (d. 1932) bequeathed his papers, notes and correspondence dealing with his translation of Dante's *Inferno*,[9] with the original drawings by John D. Batten and E. H. New used for the illustrations. There are also the commonplace and notebooks of Mrs. Louisa Kathleen Haldane (1863–1961), widow of J. S. Haldane, as well as

[1] The first annual *Report* for 1879–80 states 'The nucleus of a Library has been formed by handsome gifts'. Later reports mention the more notable donations and from that for 1901–02, a more detailed section on the Library is included.
[2] Old Students' Association, [*Report*] 1904, 14.
[3] Gemma Bailey, *ed.*, *Lady Margaret Hall: a short History* (1923), 76–7.
[4] *Report 1930–1931*, 16.
[5] *Report 1935–1936*, 16.
[6] *Annual Report 1958–9*, 7.
[7] *e.g.*, *The Brown Book*, (1911), 33.
[8] Presented by Miss C. M. Borough.
[9] First published in 1893; a new edition mentioning this bequest was published in 1933.

56

manuscripts of her novels and short stories written under the pseudonym of 'Ludovic Keir', and of her reminiscences. Besides these there are some miscellaneous individual manuscripts or typescripts mainly concerned with the Hall or written by its members, such as Miss E. M. Jamison's account of Dame Elizabeth Wordsworth. Also kept in the Library are some archival documents such as various minute books about Hall business from 1879; minutes of Principal's meetings 1923–56; lists of benefactresses in 1901, and of contributors in 1954 to the Dorothy Everett Memorial Fund. There are also the minute books of the History Club, a society of Oxford women history dons, for 1899–1939.

A list of manuscripts in the Hall Library is kept in Duke Humfrey. The archives are in the custody of the Secretary.

Printed Books

The printed books at Lady Margaret Hall have mainly come through small, individual gifts and bequests, and consequently are of a rather miscellaneous character. John Ruskin, for instance, visited the Hall in 1884 and gave many of his own writings, followed up later with some botanical works.[10] Some early editions of Dante were bequeathed by Lucy Ethel Willcock in 1919 and supplemented by George Musgrave's comprehensive collection of Dante literature in 1932, when his manuscripts were received. Some interesting Spanish and Catalan books were left by Suzette M. Taylor in 1920. About 350 volumes, ranging from the sixteenth to the eighteenth century, of English, French and Italian books, together with some Nonesuch Press editions, were bequeathed in 1931 by Edward Hugh Norris Wilde, husband of a classics don.[11] Works on French history were left by Eleanor Lodge in 1936, while some early editions and works concerned with fifteenth-century French poetry have been given by Dr. Kathleen Chesney on several occasions.[12] Contemporary seventeenth-century English tracts have been presented in recent years by Dr. C. V. Wedgwood and Miss Mary Coate, and the English history section also greatly benefited from the generosity of Dr. Irene Churchill between 1938 and her death in 1960. Over one hundred volumes from the library of Harold J. Laski, the political economist, were received after his death in 1950.

In addition to these general subjects, there have been some interesting and more specialist gifts. Mrs. Maud Jennings in 1946 gave works on Buddhism formerly owned by her husband, J. G. Jennings (1866–1941);[13] Mrs. C. S. Orwin in 1955, presented an almost complete set of the writings of Charlotte M. Yonge;[14] and in 1966 a collection made by Professor

[10] Bailey, 46.
[11] *Report 1931–1932*, 16.
[12] *e.g.*, 300 volumes in 1962 (*Annual Report 1962*, 10).
[13] *Annual Report 1946–7*, 10.
[14] *Annual Report 1954–5*, 7.

Mary Barber of works concerned with William Blake was given by two of her friends.[15]

This library uses the Dewey classification scheme and there is a card catalogue.

[15]   *Annual Report 1966*, 10.

LINCOLN COLLEGE

1427

Stock: 25,000   Current periodicals: 50   Incunabula: 43   *STC*: 629

Lincoln College, founded by Richard Fleming, Bishop of Lincoln, to train graduates to combat the heretical doctrines of Wiclif, received its charter from Henry VI in October 1427. A library was included in the first buildings, completed in 1437, and books were acquired from the earliest days.[1] As in other older colleges, they were divided between those kept chained and those circulated among members—the *electiones librorum*. This loan system continued until at least 1596, although the number available for this purpose fell from ninety-two volumes in 1543 to eighteen in 1596; indeed, Lincoln is the only college to have definite evidence that *electiones* continued during the sixteenth century,[2] but it is unlikely that it was unique in this respect. Chains were purchased on several occasions, the last in 1663, and it is interesting that, when the library was reshelved in 1739, chaining continued so that it is probably the last library in Oxford with fittings designed in this way.[3]

The library was moved from its original home into the old Chapel before 1655;[4] von Uffenbach in 1710 thought it very restricted in space and 'very disorderly, as most of the *Bibliothecæ Collegiorum* are',[5] so its refitting in 1739 is hardly surprising, though using his gift of £500 angered the donor, Sir Nathaniel Lloyd.[6] The present building, incorporating the eighteenth-century shelving, was completed in 1906 with the help of a legacy from Thomas Fowler (1832–1904), President of Corpus and a former member of Lincoln.[7]

Accessions down the centuries have mainly come through gifts and bequests from members. There was no separate endowment for the purchase of books and there was no regular income until fellow-commoners were introduced in 1606, when it was laid down that their admission fees

[1] This account is a corrected version of one first published in *Lincoln College Record 1968–1969*, 6–12. A description of the library is in Andrew Clark, *Lincoln* (College Histories, 1898), 203–5; references to benefactions are made in each chronological section. This forms the basis of the briefer descriptions in Stephen A. Warner, *Lincoln College, Oxford* (1908), 62–5, and *VCH Oxon.* iii, 168.

[2] Ker, 465–6.

[3] Streeter, 250–5.

[4] Clark, 126.

[5] J. C. von Uffenbach, *Oxford in 1710*. Edited by W. H. and W. J. C. Quarrell (1928), 53–4.

[6] Clark, 176.

[7] Warner, 62; Streeter, 250.

could be spent on books or plate.[8] Only occasional book purchases, however, are recorded in the College accounts of the seventeenth and eighteenth centuries,[9] such as 44s. in 1618 to John Minsheu for his *Ductor in Linguas*, or 30s. for Loggan's *Oxonia Illustrata* in 1675, both still on the shelves.

The earliest record of a College librarian is in November 1687 when William Musson, a member, was elected *Bibliothecæ Præfectus*.[10] A Benefactors' Register was purchased in 1676,[11] but it has not survived. The earliest bookplate is dated 1703.[12]

During the nineteenth century, separate undergraduate libraries were established, first only for classics and philosophy, later extended to cover history and law.[13] These collections were amalgamated into a single Junior Library occupying the ground floor of the building finished in 1906, with the Senior Library for the older printed books kept on the modified eighteenth-century shelving above.

Manuscripts

Coxe[14] records one hundred and twenty Latin and thirty-seven Greek manuscripts; the former are mainly theological works from the fifteenth century, with some from the twelfth to the fourteenth centuries. The founder, Richard Fleming, gave twenty-five manuscripts about 1430, of which nine are still owned by the College, and he was followed by other benefactors, such as Thomas Gascoigne in 1432, eight of whose manuscripts survive,[15] including his *Liber de Veritatibus*, a commonplace book on Scriptural topics. By 1460 the College owned some sixty manuscripts,[16] but these were shortly afterwards augmented by the gift in 1465 of thirty-eight from Robert Fleming (d. 1483), Dean of Lincoln Cathedral. This was the most important donation since the foundation, and included several humanistic works such as Leonardo Bruni's *Isagogicon*, classical texts and Italian versions of Greek authors,[17] many still in the possession of the College. This made Lincoln College supreme in either university in the classical and humanistic fields, second only to the University Library, until the books of William Gray reached Balliol after his death in 1478.[18]

[8]   Clark, 61.
[9]   Andrew Clark's transcripts (Bodleian Library, MSS. Top. Oxon. e. 109–115), include a subject index.
[10] Bodleian Library, MS. Top. Oxon. e. 97, f. 29.
[11] Warner, 65.
[12] Ibid., 64–5.
[13] Clark, 205.
[14] Pt. I, Section 8.
[15] Clark, 18; *VCH Oxon.* iii, 168.
[16] Clark, 19.
[17] R. Weiss, 'The Earliest Catalogues of the Library of Lincoln College', *BQR*, viii, no. 94 (1937), 346. A. C. De La Mare, 'Vespasiano da Bisticci and the Florentine Manuscripts of Robert Flemmyng in Lincoln College', *Lincoln College Record 1962–63*, 7–16.
[18] Weiss, 343.

Robert Fleming's gift of an early Greek manuscript of the Acts and Epistles may represent one of the earliest Greek acquisitions by any college.

The early acquisitions were naturally of an anti-Wiclifite character, but an early fifteenth-century English translation of the Bible, attributed to John Wiclif, was received during the reign of James I. The Greek manuscripts were presented by Sir George Wheler (1650–1723), and had been collected during his Mediterranean travels with Spon in 1675–76, though the published account of their journey contains little about their provenance.[19]

Besides these early manuscripts there are some later ones not described by Coxe[20] which include some seventeenth-century theology from Richard Kilby, Rector 1590–1620; commonplace books and medical notes by John Smith (b. 1563), a Cambridge graduate; two volumes of English and Latin verse by Christopher Pitt (1699–1748), the translator of Virgil, presented by K. B. Foster in 1895; and the holograph scores of nos. 4 and 5 of Alban Berg's *Altenberg Lieder*, given by his widow in 1945.

The manuscript collections were deposited in the Bodleian in 1892, where they may be consulted, and the later accessions so far mentioned have been added to them.[21]

Some more recent acquisitions are kept in the College library, mainly comprising sermons by T. R. Halcomb (1834–1880), and the papers and diaries of two former fellows, W. Warde Fowler (1847–1921), the classical scholar and ornithologist, and Nevil Vincent Sidgwick (1873–1952), the chemist. There are also the manuscripts of Edward Thomas's book of essays on Oxford, and also an early draft of his poem, *Roads*, the latter presented by his widow in 1959,[22] and some ten letters of John Wesley, fellow from 1726 to 1751.[23]

Archives

Many of the older records of the College have been lost, possibly after the death of Paul Hood, Rector from 1621 to 1668,[24] since several classes, like buttery books and admission registers, only go back to 1670 and 1673 respectively. A memoranda book of College affairs, probably made in the seventeenth century, includes references back to the fifteenth, while accounts survive for 1455–6 and from 1476–7 onwards, with occasional gaps. H. T. Riley described only four items for the Historical Manuscripts Commission in 1871:[25] the memoranda book; admission register; register of deeds in the College Treasury; and copies of documents made by

19   Sir George Wheler, *A Journey into Greece in the Company of Dr. Spon of Lyons* (1682).
20   A hand-written list is kept in Duke Humfrey.
21   Sir E. Craster, *History of the Bodleian Library, 1845–1945* (1952), 190–2.
22   *Lincoln College Record 1959–60*, 7.
23   Ibid., *1968–69*, 18.
24   Clark, 141.
25   *2nd Report* (1871), 130–2.

William Vesey (d. 1755), a fellow. Land is, or has been, owned in Buck-inghamshire, Kent, Northamptonshire, Oxfordshire, Staffordshire, and Yorkshire.[26] A collection of the bills sent to one David Locock while an undergraduate between 1742 and 1744 throws light on student life.

Besides the copies of charters and chief deeds made by William Vesey and kept with the archives, extensive transcripts and extracts from the College records and accounts were made by Andrew Clark (1856–1922), the historian of the College, and are now in the Bodleian; the subject indices are particularly useful.[27]

A handlist was compiled in 1970 for the use of the History of the University Project.

The archives are in the custody of the College Archivist.

Printed Books

Though the library of printed books is not particularly large, it is strong in several subjects not so well represented elsewhere in Oxford, especially works connected with John Wesley (1703–1791) and Methodism, English plays of the late seventeenth and eighteenth centuries, scarce pamphlets and Hebrew books, both of the sixteenth and seventeenth centuries. There is also a small collection of works by or about former members, while older books on patristics and theology are present in force, as in other colleges of similar antiquity.

Lincoln has largely relied on gifts and bequests for accessions and there have been numerous small benefactions over the years. The first sub-stantial gift came from Edmund Audley, Bishop of Salisbury, in 1518; eighteen, mainly of patristic and theological interest, are still in the library, and some at least were formerly in the circulating collection.[28] There were no more large benefactions for over a century, though minor ones continued their flow; Richard Mansfield, for instance, in 1648 gave several volumes of English literature, including works by Ben Jonson and Beaumont and Fletcher; others, such as Daniel Hough (d. 1644), a fellow, left money for the purchase of books.

The move to a new room during the Commonwealth period, already mentioned, was perhaps not unconnected with the bequest in 1657 from Gilbert Wats, a fellow, of 'soe many books as cost me threescore pounds', to be selected by Thomas Barlow, then Bodley's Librarian.[29] This must have greatly enlarged the library, judging from the number of volumes with Wats's armorial stamp, or motto *Ruit Hora*, still on the shelves; he must have owned a good, general scholarly collection.

Another important bequest was received not long afterwards from Thomas Marshall (1621–1685), Rector from 1672 to his death; the first

26   *VCH Oxon.* iii, 163 n. 3.
27   Now Bodleian Library, MSS. Top. Oxon. e. 82–3, e. 97, e. 109–115.
28   Ker, 466.
29   *DNB*; Clark, 126.

choice was given to the Bodleian[30] but the College acquired over seventy bound volumes of tracts which include many rare items, together with some Hebrew books, mostly concerned with Biblical subjects, and other works.

Marshall's predecessor as Rector, Nathaniel Crewe (1633–1721), later Bishop of Durham, though a great benefactor to the College generally, did not give much to its library except, in 1674, a folio Bible and Book of Common Prayer, both magnificently bound for use originally in a Chapel Royal.[31]

No further large benefactions arrived until after the library had been refurnished in 1739. William Vesey (d. 1755), a fellow, bequeathed his books, a collection covering a wide range of subjects. A century later, in 1863, Edward William Grinfield (1785–1864), an *alumnus* who founded the Septuagint Lecturership in the University, presented his books on that subject, together with some miscellaneous books.

The latest important donation was a good collection of works by or about John Wesley and Wesleyan Methodism, made by the Rev. Albert F. Hall, sometime chairman of the Portsmouth Methodist District, and presented by his widow in 1951;[32] it includes many items formerly owned by the Rev. John Bell, of Leeds, in the early years of this century.

The library also houses a collection of seventeenth- and eighteenth-century English plays of uncertain provenance, some of which were possibly owned by Henry Fielding,[33] together with a small amount of English literature of the same period. Some good sets of eighteenth-century periodicals, like *The Tatler* and *The Spectator*, originally bought for the Common Room, survive. A small number of medical and scientific books from this century possibly indicate the influence of John Radcliffe (1650–1714), the famous physician. Besides the bound volumes of tracts bequeathed by Thomas Marshall, there are some similar ones filled with later seventeenth- and eighteenth-century pamphlets.

Catalogues

Two fifteenth-century lists, entered in the *Registrum Vetus*, have been fully described;[34] they comprise one of 1474 covering most of, but not all, the books in the library, and another of 1476 listing the contents of the circulating collection. In 1673 a 'catalogue book for the Library' was purchased and one Jones was paid £5 for making a catalogue,[35] but it has not survived. There is a copy of the 1843 Bodley catalogue annotated with the College holdings. The library was re-catalogued on cards in 1950; there are two separate alphabetical sequences, one for the Junior and one for the Senior Library.

30   W. D. Macray, *Annals of the Bodleian Library, Oxford*. 2nd edition (1890), 154; Clark, 156.
31   *Fine Bindings 1500–1700*, no. 221.
32   *Lincoln College Record 1951–2*, 10.
33   Appropriate holdings are recorded in Carl J. Stratman, *Bibliography of English Printed Tragedy, 1565–1900* (1960); Lincoln is the only Oxford college included. A catalogue of Henry Fielding's books, made in 1754, is in the College archives.
34   See Note 17 above.
35   Bodleian Library, MS. Top. Oxon. e. 112, f. 90v.

MAGDALEN COLLEGE
1458

Stock: 50,000   Incunabula: 139   *STC*: 577

Provision was made for a library at Magdalen College from its inception and about a hundred volumes were received before the founder, William of Waynflete, presented about eight hundred books in 1481, though hardly any of these now remain. As it is also one of the best documented college libraries, it has been described in print on more than one occasion.[1] The original statutes of 1479–82 laid down rules for a chained library, though books might be lent to fellows if not taken outside the College or Grammar School. Like the other pre-Reformation colleges, Magdalen was at first dependent upon donations, but after the changes imposed by Thomas Cromwell's Visitors in 1535, it was the earliest to pursue an active policy of book purchasing, spending over £73 between 1536 and 1550.[2] Moreover, a larger proportion than in other colleges is still present. Magdalen also pioneered in another respect at this period since, in 1550, it was the first to pay a fellow especially to look after the library.

Although the library was quiescent and illustrative of Gibbon's famous dictum in the eighteenth century, during the seventeenth and nineteenth centuries several important benefactions were received of both books and manuscripts, while as the neighbour of the Botanic Garden the holdings reflect an interest in botany, besides the usual relics of older curricula.

There is much material available for the history of the library, with records of book purchases in the College accounts from 1500. A Benefactors' Book has been kept since about 1680, though earlier gifts were entered at the beginning.[3] A set of library book bills starts in 1671, separate accounts in 1865, while regular Reports begin in 1863. Chains were not removed until 1799,[4] Magdalen being the last college to retain them.

Manuscripts

Coxe[5] describes 247 manuscripts, while there is a typescript calendar of additions, of which a copy is kept in Duke Humfrey. Some more, omitted

---

[1] W. A. B. Coolidge, 'The Library of Magdalen College, Oxford', *Notes and Queries*, series 6, vii (1883), 361–64, 421–23, 441–43. This is chiefly concerned with descriptions of outstanding manuscripts and incunabula. [G. R. Driver *et al.*], 'Magdalen College Library', *Oxford Bibliographical Society Proceedings and Papers*, ii (1929), 144–49.
[2] Ker, 480–1.
[3] *Fine Bindings 1500–1700*, no. 224.
[4] Streeter, xiv.
[5] Pt. 2, section 2.

by Coxe or missing when he was making his list, were described by W. D. Macray.[6] There are sixteen Greek besides Italian and English manuscripts. Among the Latin ones may be mentioned William of Malmesbury's autograph *Gesta Pontificum*; several liturgical items, such as an illuminated Gospels (the companion to the Lectionary at Christ Church) made for Cardinal Wolsey, once a fellow, about 1528; an illuminated fifteenth-century copy of Gower's *Confessio Amantis*; chronicles, like a twelfth-century *Historia Ecclesiastica* by Bede; and several patristic manuscripts. There is also one in Chinese and a fourteenth-century Trivet's *Annales* in French. Several, of theological interest, came with Sir Arthur Throckmorton's gift in 1626. Other former owners, besides the Founder, include John Dygoun (*fl.* 1497–1512), Sir John Fastolf (*c.* 1378–1459) and Henry, Earl of Rutland (d. 1563).

Among the papers of later date should be mentioned those of George Horne, Bishop of Norwich (1730–1792), President 1768–90, and some from his long-lived successor, Martin Routh (1755–1854),[7] whose printed books are now in the University of Durham. Another interesting group is that bequeathed by J. R. Bloxam (1807–1891), the high churchman and historian of the College, whose affairs naturally loom large though there is also interesting correspondence with contemporaries, such as Cardinal Wiseman and Ambrose Lisle Phillips.[8]

Archives

Magdalen College owns, or has had connections with, property in several parts of Britain, so its muniments are extensive. W. D. Macray, who spent many years arranging and indexing them, estimated there were about 13,000 deeds dating from the twelfth to the sixteenth centuries, but mainly between the thirteenth and the fifteenth. He described some of the charters for the Historical Manuscripts Commission[9] and also summarised the chief collections by place.[10] In addition, he published in 1882 his *Notes from the Muniments*, which include useful indices of persons, places, seals and unusual words found in these documents. Macray, furthermore, edited for publication documents concerned with Selborne Priory,[11] while those relating to Basingstoke have also been printed.[12] The cartulary of the Hospital of St. John the Baptist, together with many associated Oxford deeds, has been published,[13] and a supplementary index by N. Denholm-Young, available in the College, has since been

---

6   *A Register of the Members of St. Mary Magdalen College, Oxford.* New series, ii (1897), Appendix II.

7   Macray, *Register* ii, Appendix II.

8   Macray, *Register* vi, 114–6.

9   *4th Report* (1876), 458–65.

10   *8th Report* (1881), 262–9.

11   *Calendar of Charters and Documents relating to Selborne and its Priory Preserved in the Muniment Room, Magdalen College.* 2 vols. (Hampshire Record Society 1891, 1894.)

12   F. J. Baigent and J. E. Millard, *A History of Basingstoke* (1889), pp. 651–9.

13   Edited by H. E. Salter. 3 vols. (Oxford Historical Society, lxvi, lxvii–lxviii, 1914–17.)

compiled. The College accounts, which begin in 1481, were used by J. E. Thorold Rogers for his *History of Agriculture and Prices in England*.[14] Records of battels survive for 1477–1509 and from 1653 onwards. Admission registers begin about the same time, in 1480, but details of College government only date from 1547. Senior Common Room wine accounts exist from 1763. There are also numerous papers concerned with the struggle with James II in 1687. Records of music performed in the College Chapel survive only from 1850.[15]

A manuscript calendar and slip-index to the whole collection, made by Macray, is available for consultation; there is a continuation on cards. A handlist has been compiled for the use of the History of the University Project.

The archives are in the custody of the College Archivist.

Printed Books

Magdalen was the first college, as has already been mentioned, to purchase books after 1535 using income for the purpose, so the library is strong in the large folio volumes of the Fathers, theology and classics printed on the Continent, especially Basle, in the sixteenth century. Other subjects were not entirely neglected; a copy of Vesalius, *De Humani Corporis Fabrica*, printed in Basle in 1543, was bought as early as 1545. In the latter part of the century much Protestant theology was purchased,[16] including the notable acquisition in 1572 of the library of John Jewel, Bishop of Salisbury, for the then large price of £120; no other Oxford college spent such a considerable sum on books at one time during this century.

During the seventeenth century the College received several notable benefactions, the bulk of which are still on the shelves. The earliest was the bequest in 1608 by Nicholas Gibbard, an Oxford physician, of his 'rewe of books';[17] thirty of the thirty-three volumes now remain. A much larger gift came in 1626 from Sir Arthur Throckmorton, son of Sir Nicholas, the Elizabethan diplomatist, which includes many Continental printed books on a wide variety of topics, acquired both by the father and the son, but Sir Arthur's widow kept the English books she liked best. Dr. A. L. Rowse has analysed their tastes; Sir Nicholas was interested in politics and religion, Sir Arthur in literature, voyages, the military arts, natural sciences and things Italian generally.[18]

Possibly the most important gift in this century came from John

[14]   7 vols. 1866–1902; see Preface to vol. 6, p. v.
[15]   B. Rose, 'Magdalen College Chapel', *Oxford Magazine*, 13 Nov. 1970, 74.
[16]   R. T. Gunther, 'The Circulating Library of a Brotherhood of Reformers of the Sixteenth Century at Magdalen College, Oxford', *Notes and Queries*, series 13, i (1923), 483–4.
[17]   R. T. Gunther, 'The Row of Books of Nicholas Gibbard of Oxford', *Annals of Medical History*, iii (1921), 324–6. See also his *Daubeny Laboratory Register 1916–23*, iii (1924), 382–8.
[18]   *Ralegh and the Throckmortons* (1962), 336–9.

COLLEGE AND HALL LIBRARIES

Warner (1581–1666), Bishop of Rochester, a former demy and fellow 1604–10, who gave £300 about 1642 and £1,000 after 1660 for the purchase of books; a large number now bear an inscription recording that they were bought from these funds. In 1664 a Hampshire physician, John Goodyer (1592–1664), bequeathed 239 printed books and some manuscripts, mostly concerned with botany.[19] This bequest is interesting for the historian of the book-trade since Goodyer noted on the fly-leaves the prices paid for the books and for binding, with dates.

Another collection containing many Italian and French printed books on several subjects came in 1699 after the death of John Fitzwilliam, a non-juror and fellow 1661–70, who also left some to the Bodleian.

Little happened in the eighteenth century. At some time in its early years Edward Maynard (1674–1740), an antiquary and former demy, presented a large number of tracts relating to the theological controversies of the late seventeenth century, similar to those found in most of the older colleges. At the end of the century, in 1799, chains were removed and library rules printed, while a regular income for acquisitions was obtained from the sums paid by gentlemen commoners *loco convivii*, together with a legacy from William Clement, a former demy.[20]

J. R. Bloxam, already mentioned, was Librarian for many years while he was a fellow (1836–63) and he began a collection of works written or edited by *alumni* which is still actively maintained; indeed it is one of the best of this kind to be found in Oxford today. All periods are well represented, from Colet and Lily in the sixteenth century, through William Camden, James Mabbe, Addison, Routh, Charles Reade, J. A. Symonds and the Duke of Windsor to the present day, to name only a few. There are also several books from Edward Gibbon's library.

At some time in the nineteenth century was presented a curious collection of early printed and later Swedish liturgical and homilectical works. Coolidge[21] thought William Palmer (1811–1879) the Tractarian, was the donor, but Macray, who gave a list of the earliest specimens,[22] could find no evidence for this assertion. A fair collection of tracts on nineteenth-century religious controversies, comparable to those in several other colleges, was given by John Rigaud (1821–1888).

C. G. B. Daubeny (1795–1867), Professor of Chemistry from 1822 until 1855 and founder of the laboratory that bears his name, left his books to the College; this important bequest reflected interests in botany, geology (especially volcanoes) and rural economy besides pure science.[23] Another

---

[19] R. T. Gunther, 'Goodyer's Library', in his *Early British Botanists* (1922), 197–232. Besides a catalogue, the items are also listed by date of acquisition with details of former owners.

[20] Macray, *Register*, v, 33.

[21] Op. cit., 361.

[22] *Register*, ii, 221–3.

[23] R. T. Gunther, *The Daubeny Laboratory Register 1916–1923*, iii (1924), 382–8, where the scientific collections in the College and their donors are discussed. See also his *Oxford and the History of Science*, 44.

scientific collection was given by the widow of Edward Chapman (fellow 1882–1906), while the library of mathematical books formed by H. T. Gerrans was presented in 1924 for the use of the Waynflete Professor of Pure Mathematics and accredited students from outside the College; much of this collection was transferred to the Mathematics Institute in 1966.

Two smaller benefactions from former members also deserve notice: a fair collection of nineteenth-century colour-plate books, mainly Acker-mans, given by Reginald Dalton Pontifex (1857–1930), and a few original editions of Elizabethan and Jacobean plays from the writer E. H. W. Meyerstein (1889–1952), an *alumnus*.

Music used in the College chapel only goes back to the late nineteenth century, and is kept in the Song School; the early part-books inventoried by Macray do not appear to have survived.[24]

A small number of books in Oriental languages has been deposited in the Oriental Institute library.

Magdalen is fortunate in that so much remains of the acquisitions and gifts of four centuries,[25] though some duplicates have been discarded, including a batch sent to the University of King's College, Windsor, Nova Scotia (now in Halifax) in 1803.

Catalogues

The earliest manuscript catalogue was made in 1550, and the next in 1630 at the cost of 1s. 6d.; by 1656, when another was compiled, the price had risen to £2. Three two-volume library catalogues, probably made in the early eighteenth century, have survived. In the nineteenth century, a catalogue made by E. M. Macfarlane was printed in three volumes between 1860 and 1862 and contained an interesting appendix listing the *alumnus* collection.[26] Between 1926 and 1929 the whole library was catalogued on cards, which form the basis of the existing tool; at the same time opportunity was taken to publish a list of foreign and British early printed books not in the Bodleian, together with short references to *STC* items in both libraries.[27] A separate, typescript catalogue of the pre-1641 books is kept in the Old Library; the card catalogue, however, covers all holdings.

---

[24] *Register*, ii, 198–9; B. Rose, *loc. cit.*
[25] Notes on many of the earlier books 'bearing memoranda of ownership, donorship, etc.' are included in Macray, *Register*, ii, Appendix II.
[26] *Catalogus Librorum Impressorum Bibliothecae* . . . . 3 vol. (1860–62).
[27] 'Magdalen College Library [with a] List of Books printed before 1641 . . . not in the Bodleian Library', *Oxford Bibliographical Society Proceedings and Papers*, ii (1929), 151–200; unfortunately there are some omissions in this list.

MAGDALEN COLLEGE SCHOOL

STC: 22

Magdalen College School, founded at the same time as the College, has a small collection of works published by masters and *alumni*, including such persons as John Stanbridge (d. 1474); John Colet (d. 1519), Dean of St. Paul's; William Camden (1551–1623); and Daniel Featley (1582–1645), the controversialist, as well as later members.

MANCHESTER COLLEGE
1757

Stock: 75,000   Incunabula: 2   *STC*: 484

Manchester College is a descendant of the nonconformist academies founded to provide higher education for those denied entry to the universities through their inability to subscribe to the tenets of the Church of England. Though open to members of every religious denomination, it has had close relations with Unitarians throughout its history. It traces its ancestry back to the Warrington Academy, founded in 1757, then through the re-foundation at Manchester in 1786 as Manchester Academy which moved successively to York in 1802 as Manchester College, back to Manchester in 1840 as Manchester New College, to London in 1853 and finally to Oxford in 1889. 'New' was dropped from the title in 1893. The present building was completed in 1893 and added to in 1915.[1] Since 1840 the College has been a recognised part of the University of London and was granted the status of 'an institution for higher study' by the University of Oxford in 1966.[2]

There is an extensive library, rightly described as 'one of the two best theological libraries in Oxford'[3] and also as 'one of the College's greatest assets',[4] containing much material not found elsewhere in Oxford ranging over several fields of knowledge. Its foundations are based on the remains of the libraries of several dissenting academies: those of Warrington (though eighteenth-century students there also used Chetham's Library in Manchester[5]), Exeter and Taunton.[6] The libraries of some congregations are also here: Renshaw Street (later in Ullet Road), Liverpool, and Hinckley, Leicestershire.

In the early nineteenth century, books were borrowed by particular

---

[1] V. D. Davis, *A History of Manchester College from its Foundation in Manchester to its Establishment in Oxford* (1932), *passim*. See also L. A. Garrard, 'Manchester College', *Oxford Magazine*, n.s. ii (1962), 338–9. The library is frequently mentioned in the published annual *Reports*; the College's set begins in 1817, the Bodleian's not until 1861.

[2] *Report of Manchester College 1966*, 4; under Statute Tit. VIII, sect. XIII, cl. 3.

[3] Garrard, 338; its unnamed co-equal must be Pusey House library.

[4] *Report 1960*, 10.

[5] Davis, 27, 65.

[6] Some Warrington, Exeter and Taunton Academy books are also in the Unitarian College Library, Manchester; see H. McLachlan, *The Unitarian College Library* (1939), 103–6.

classes, not by individuals,[7] and a salary for a librarian appears in the earliest published accounts for 1816–17.[8] A certain amount was spent each year on purchasing new books and as early as 1821–22 over £100 was expended in Germany.[9] At the same time there were frequent gifts and bequests of books and manuscripts; the principal items were listed in the annual *Reports* of the College published between 1843 and 1907, since when a smaller selection only has appeared. During the librarianship of Lucy Toulmin Smith (d. 1911), lists of books bought between 1896 and 1906 were also included. In 1859 it was felt that the library was not keeping up to date with the 'scholarship of the age', so the grant for purchases was increased and an appeal made for gifts.[10] During the London period, between 1883 and 1889, a library fee of 10s. 6d. a year was charged, since many readers were not members of the College, a practice revived in 1967.

After the removal to Oxford, the munificence of Sir Henry Tate, bart., enabled a fine room to be built to house the library, named after him; it then contained about 17,000 volumes. It serves as the main reading room today, though inevitable growth has caused much of the less-used material to be stored elsewhere.

Duplicates were sold in 1874, 1887, 1947 and 1949; others were given to Dr. Williams's Library, London, to the Unitarian College library, Manchester, in 1901 and 1949, and to Carmarthen College in 1947.

The wealth of research material in Manchester College library is too little known; it is well worth exploring not only for theology or the history of dissent in Britain, which are not unexpectedly strong, but also for early printed books, Oriental studies, anthropology, literature, medicine[11] and comparative religion, while the succession of students from the Unitarian churches of Transylvania from the London period[12] until 1939 has ensured the presence of many Hungarian works not found elsewhere in Oxford. One can find here anything from the Complutensian Polyglot Bible to first editions of Newton's *Principia* and of the *Lyrical ballads*.

Manuscripts

The manuscripts at Manchester College form one of the most important collections in existence for the history of British dissent, with the emphasis on Unitarianism. They consist mainly of the papers and correspondence of prominent nonconformists, many of them teachers at the College, together with notes taken at lectures or sermons, and only a selection can be named here.

---

[7]   H. McLachlan, *English Education under the Test Acts; being the History of the Nonconformist Academies 1662–1820*. (Publications of the University of Manchester, no. ccxiii; Historical series no. lix, 1931), 269.
[8]   *Report 1817*, 6.
[9]   *Report 1822*, 9.
[10] *Report 1859*, 6.
[11] R. Guest-Gormall, 'The Warrington Dispensary Library', *Medical History*, xi (1967), 288–9.
[12] Davis, 101–2, 154.

Chronologically, the oldest manuscripts are probably a volume of sermons by Philip Henry (1631–1696), presented by Miss Harrold,[13] and shorthand notes of sermons by Thomas Dickenson (1694–1696), presented by S. A. Steinthal in 1897.[14] The eighteenth and nineteenth centuries are well represented by notebooks and letters of Joseph Priestley (1733–1804),[15] Theophilus Lindsey (1723–1808), John Seddon (1725–1770), John Taylor of Norwich (1750–1826)[16] and Richard Wright (1764–1836). There is a separate handlist to the letters, sermons and journals of Joseph Blanco White (1775–1841), covering 1822–39 intermittently. The collections of William Shepherd of Liverpool (1768–1847) comprise twenty-six guard-books of letters, sermons and papers,[17] including many from Lord Brougham and some of American interest.

Of former principals, there are letters of Charles Wellbeloved (1769–1858) and many of James Martineau (1805–1900), together with the latter's family papers and correspondence, including some of his sister Harriet.[18] J. E. Carpenter, Principal 1906–15, presented the correspondence and papers of his grandfather, Lant Carpenter (1780–1840).[19] Letters and papers formerly belonging to Sir John Bowring (1792–1872) were given by Thomas Bowring in 1910.[20] Among these various collections can be found letters of eminent contemporaries, such as Charles Dickens and Cardinal Newman.

Notes made at College lectures up to 1820 have been previously described in print;[21] some later ones include Samuel Bache's notes of John Kenrick's lectures 1827–28, 1830–31,[22] and the lecture notes made by Charles Wicksteed (d. 1885).[23] There are also nineteen manuscript notebooks of N. T. Heineken presented in 1847.[24]

Besides these original documents, there is a notable collection on the history of nonconformity in England made by Robert Brook Aspland (d. 1869), consisting of twenty-four volumes of notes on Puritanism and dissent in Cheshire, Derbyshire, Lancashire, Leicestershire, Sussex, Worcestershire and Yorkshire, as well as Hackney, Manchester and Nottingham, with twenty-six further volumes of historical notes. There is also a set of historical notebooks left by John Collins Odgers (d. 1928).

13   *Report 1926*, 8.
14   *Report 1898*, 27.
15   *Cf. Report 1901*, 36.
16   *Report 1903*, 27–9.
17   F. Nicholson and E. Axon, 'The Manuscripts of William Shepherd at Manchester College, Oxford', *Transactions of the Unitarian Historical Society*, ii (1922), 119–30. Items of American interest are listed in B. R. Crick and M. Alman, *A Guide to Manuscripts relating to America in Great Britain* (1961), 399–400.
18   *Report 1906*, 28; *1961*, 10.
19   *Report 1900*, 38; *1906*, 27.
20   *Report 1910*, 32.
21   McLachlan, *English Education*, Appendix I.
22   *Report 1912*, 32.
23   *Report 1946*, 15.
24   *Report 1847*, 7.

There are some topographical notes on Sidmouth, the Isles of Aran and Barnton (Lancs.) made by Arthur William Fox (1863–1945), together with some of his verse.

There is a sheaf catalogue of the manuscripts, but it does not yet cover all the collections.

Archives

The archives go back to the College's ancestors, the Warrington Academy, whose Trustees' minute books for 1757–86 are here,[25], together with the minutes of the Committee of the Manchester Academy 1786–1820, as well as cash accounts and bills.[26] Apart from a few items like a cash book for exhibitions covering 1799–1803, and lists of subscribers from 1839, the bulk of the archives, such as Committee minutes, date from 1885; a brief list has been prepared for the National Register of Archives and can be consulted in Duke Humfrey.[27] The minute book of the Taylerian Society 1874–90, a College society, is kept with the archives.

For permission to consult, application must be made to the Principal.

Printed Books

The remarkable collection of printed books at Manchester College reflects its history. As has already been mentioned, its foundations rest on the remains of dissenting academy libraries, beginning with Warrington Academy, once described as 'probably the most extensive academy library in England'.[28] A library was started by the tutors soon after its beginning in 1757; later it was enriched with the books of Benjamin Grosvenor (1676–1758), given by his son Richard, and a wide-ranging collection formed by Samuel Stubbs (1715–1753), first placed on loan by his brother and in 1786 presented outright by his widow;[29] the surviving volumes can be identified from Stubbs' signature. The books so far acquired, about 3,000 in all, moved with the College to York in 1803 and were augmented in 1813 with the second and third Exeter Academy library, which included the books of William Harris (1675–1740).[30] Among the Exeter books were valuable ones removed from the Taunton Academy in 1759, which had been housed in several academies; a Suetonius printed by Aldus in 1516, once owned by Jean Grolier, was probably one of these.[31] While at York, the library was increased by addi-

---

25 Used in H. A. Bright, 'A Historical Sketch of Warrington Academy', *Transactions of the Historic Society of Lancashire and Cheshire*, xi (1858), 1–30; and in J. F. Fulton, 'The Warrington Academy (1757–1786) and its Influence upon Medicine and Science', *Bulletin of the Institute of the History of Medicine*, i, no. 2 (1933), 50–80.

26 McLachlan, *English Education*, 312.

27 R. 7. 62/3, ff. 194a–d; there is also a brief summary in the Preface to Davis.

28 Fulton, 64.

29 McLachlan, *English Education*, 228, 261.

30 Ibid., 234–5, 269.

31 British Museum, *Bookbindings from the Library of Jean Grolier; a Loan Exhibition* (1965), no. 137.

tions from the books of Theophilus Lindsey (1723–1808), Joseph Bretland (d. 1819) and Samuel Shore (d. 1828) among others.[32]

It is not surprising, therefore, in view of the provenances so far mentioned, that the library is particularly strong in eighteenth-century English works concerned with dissent or written by dissenters. But Europe was not ignored; German and French works were steadily acquired so that, for instance, the Encyclopédistes are well represented by early editions of Rousseau and Diderot, and the Kehl edition of Voltaire, 1785–89, edited by Beaumarchais.[33] More eighteenth-century books, mainly theological, came with the bequest of Charles Wellbeloved (1769–1858), Principal 1803–40.

A significant development affecting the library happened during the London period of the 1860s, when close contacts were made with the old-established Unitarian Church of Transylvania in Hungary, which continued until 1939.[34] A Hungarian student joined the College in 1860[35] and for the next seventy years there came to the library[36] a steady flow of books by Transylvanian Unitarians and theological works in Hungarian of a sort not to be found elsewhere in Oxford.

During the London period there were also some interesting donations. A large number of theological works were bequeathed by John James Tayler (1797–1869), sometime Principal. William James Lamport (d. 1873) of Liverpool left some sixteenth-century theology and contemporary books on Eastern religions;[37] a miscellaneous group of philological and linguistic books was bequeathed by John Colston in 1878;[38] and a large collection of nineteenth-century theology, containing many works in French and German with a few early printed items, from the library of J. R. Beard (d. 1876) was presented by his widow.[39] The next important donation, after the move to Oxford, also came from the same family when, in 1893, over a thousand volumes concerned with Luther, the Reformation in Germany and the history of Port-Royal, formerly owned by Charles Beard (1827–1888), were received.[40]

In 1896 two smaller groups, both containing seventeenth- and eighteenth-century printed books as well as nineteenth-century dissenting history, were given by G. Rayner Wood, and by Francis Taylor selected

[32] McLachlan, *English Education*, 269.
[33] H. L. Short, 'A Link with the Encyclopédistes', *Oxford Magazine*, n.s. i (2 March 1961), 266.
[34] Davis, 102, 154; Garrard, 339.
[35] *Roll of Students Entered at the Manchester Academy 1786 . . . [to] 1867* (1868), *sub anno* 1860.
[36] Besides those listed in the annual *Reports* from 1862 onwards, some special gifts are mentioned in J. J. Tayler, 'Narrative of a Visit to the Unitarian Churches of Transylvania', *Theological Review* vi (1869), 12–3, 15.
[37] *Report 1875*, 6–7, 11–2.
[38] *Report 1879*, 12.
[39] *Report 1878*, 10–12.
[40] *Report 1890*, 9; *1893*, 16.

from the library of Sir Jerom Murch of Bath.[41] After the death in 1900 of James Martineau, Principal 1869–85, his family gave a number of philosophical and theological works, mostly contemporary, while the widow of his son, Russell Martineau (1831–1898), presented some of his books, chiefly relative to Oriental philology and linguistics.[42] With the gift of Mrs. Alfred Currey in 1902 of the papers of her forebear, John Taylor of Norwich (1750–1826), were some of his printed books, mostly theology published in his lifetime.[43]

The benefaction of James and Isabella Arlosh, who both died in 1904,[44] included their books; they were the last survivors of the ancient family of Losh (later Arlosh), of Woodside, Carlisle; thereby Manchester College acquired a good, small, general library of the English country-house kind, similar to what was sometimes left to the older Oxford colleges.

In 1921 J. E. Carpenter, Principal 1906–15, gave his large collection of books on comparative religion, subsequently increased by another 1,500 volumes after his death in 1927.[45] This important group is kept as a separate entity; works on every kind of western or eastern religion can be found in it. Dr. Carpenter also owned some rare early printed books and had been a particularly generous benefactor of the library over a long period of years.

Apart from the main donations and bequests so far mentioned, there has been a steady flow of smaller gifts from members of the teaching staff, former students and friends, as the lists printed in the *Reports* reveal; F. W. Newman, brother of the Cardinal (1805–1897), who taught in the College; James Edwin Odgers (1843–1925) and J. M. Connell (*fl.* 1880–1946) are, perhaps, worthy of commemoration in this connection.

Another feature of the library is the great collection of tracts, mostly bound up in chronological order and numbering about 11,000; it is particularly strong for the seventeenth, eighteenth and nineteenth centuries, and naturally has some rare pamphlets of Unitarian interest.[46]

Catalogues

There is a good series of catalogues that illustrate the growth of the library, while, as in so many Oxford libraries, the books have been re-arranged, re-classified and re-catalogued at fairly frequent intervals.

A catalogue of the books in Warrington Academy was printed locally in 1775;[47] and there is an undated manuscript one now in the College

[41]   *Report 1897*, 12–3, 22, 25–6.
[42]   *Report 1901*, 21–32.
[43]   *Report 1903*, 27–9.
[44]   Davis, 177–8.
[45]   *Report 1922*, 7; *1928*, 10.
[46]   H. McLachlan, 'Seventeenth Century Unitarian Tracts', *Transactions of the Uniterian Historical Society*, ii, pt. 4 (1922), 141–2; reprinted in his *The Story of a Nonconformist Library*. (Publications of the University of Manchester, no. clvi; Historical series no. xli, 1923), 57–8.
[47]   McLachlan, *English Education*, 225; copy in College, but not in the Bodleian.

library, with a later manuscript version containing additions and alterations. A catalogue of the Exeter Academy books, dated 1800, contains
references to 1,881 items.[48] There are also two other manuscript catalogues
without any institutional name; one possibly dates from the first Manchester period, about 1800, and the other, with between five and six
thousand references, to the subsequent York period.

A proposal to print an 'arranged catalogue' was put forward in 1820,[49]
but it never apparently reached publication. A new shelf-catalogue was
made after the move to London in 1853.[50] Money was spent on 're-
arranging' books regularly between 1873 and 1883, and paper for the
catalogue was bought in 1884 and 1885. A change was made to cards in
1886 and author and shelf-catalogues were completed in 1888,[51] shortly
before the move to Oxford. The library was again re-arranged and the
pamphlets were catalogued and classified at the same time.[52]

This arrangement apparently was considered satisfactory until problems
of space in the early 1930s caused much material to be moved to store-
rooms, when the books were re-arranged and re-catalogued, this time in
sheaf form, a process that occupied the years 1934 to 1939.[53] During the
Second World War, the College was commandeered and the older books
evacuated to the Bodleian; on their return in 1946, they were shelved in
chronological sequence by size.[54] At the same time a subject catalogue was
started[55] and the modern books re-catalogued and re-classified.[56]

The current catalogues comprise:

(i) An index of manuscripts, begun in 1950[57] and as yet incomplete;
it is subdivided into: Miscellaneous manuscripts; Note books;
Lectures; Letters; Sermons; Varia.

(ii) Author catalogue.

(iii) Subject catalogue of books in the main reading room.

(iv) Two catalogues of pamphlets:

(a) entries, written mostly in the 1880s, pasted into four guard
books, the first two arranged A–Z by known authors, the other
two with anonymous works arranged by subject.

(b) A sheaf catalogue, begun in the 1930s, of the first two volumes of
(a), but omitting shelfmarks.

(v) Catalogue of the Carpenter Library of Comparative Religion, in
two parts: author and classified by subject.

The first three and the fifth are sheaf catalogues.

48   Ibid., 235.
49   Report 1820, 2.
50   Report 1854, 6.
51   Report 1887, 23; 1889, 9.
52   Report 1888, 10.
53   Report 1932, 6; 1934, 10; 1936, 10; 1937, 10; 1939, 16.
54   Report 1947, 10; 1948, 12.
55   Report 1947, 10.
56   Report 1949, 13.
57   Report 1950, 13.

MANSFIELD COLLEGE

1955

Stock: 30,000   Current periodicals: 46   *STC*: 88

Mansfield College had its origins in Spring Hill College, founded in Birmingham in 1838 for the training of Congregational ministers, mainly through the benefactions of George Storer Mansfield and his sisters. After the religious tests connected with the universities were abolished in 1871, the transfer of the College to Oxford began to be considered, eventually taking place in 1886, when it was re-named after its founders. The present College buildings were opened in 1889.[1] As it was the first nonconformist college to come to Oxford, its objects were broadened to train ministers of any dissenting denomination and to provide a Free Church faculty in theology.[2] In 1955 its status was further changed when it was recognised by the University as a permanent private hall and began to receive undergraduates reading for any school.[3]

Books were purchased regularly from its earliest days,[4] while there have been many small gifts throughout its existence;[5] in 1872 the Committee appealed for more people to 'emulate the generous examples recorded'.[6] After the move to Oxford, the local custom of students presenting a volume on leaving was adopted.[7] There was also a separate Junior Common Room library between about 1900 and 1930.

As in other Oxford colleges, space has been and is a problem as the library has grown. Duplicates and surplus books were discarded between 1917 and 1920 after a thorough overhaul to alleviate the overcrowding.[8] By 1961 it was estimated that the library contained some 20,000 volumes, while opening the College to students reading a wider range of subjects meant that their book needs had to be catered for.[9] This again caused

---

[1]   R. W. Dale, 'The History of Spring Hill College', in *Mansfield College, Oxford: its Origin and Opening* (1890), 3–28.

[2]   W. B. Selbie, 'Mansfield College', ibid., 29–45.

[3]   John Marsh, 'Mansfield College, 1886–1959', *American Oxonian*, xlvi (1959), 114–7; and his 'Mansfield 1962', *Oxford Magazine*, n.s. ii (1962), 370–1. See also W. T. Pennar Davies, *Mansfield College: its History, Aims and Achievements* (1947).

[4]   A sum for books bought regularly appears in the accounts published in the annual *Report of the General Committee of Spring-Hill College* from 1838.

[5]   Entered from 1838 to 1889 in the Donors' Register; a selection is listed in the annual *Reports* between 1863–4 and 1922–3.

[6]   *Report 1871–2*, 16–18.

[7]   *Report 1887–8*, 9.

[8]   *Report 1916–17*, 7; *1917–8*, 6–7; *1918–9*, 7; *1919–20*, 7; some were presented to the London Missionary Society and the Radcliffe Library, Oxford.

[9]   *Mansfield College Magazine*, xi, no. 159, 88–9.

problems of accommodation, so in recent years a considerable quantity of the older or less-used material has been dispersed; some of the rarer items have gone to the Bodleian.

Although the library at Mansfield is primarily a working collection and in spite of the scattering of so much of its stock, it contains valuable material for the study of Congregational history, especially accounts of local churches and early ordination sermons, as well as a small number of books from the sixteenth to the nineteenth centuries.

Manuscripts and Archives

There is a small group of manuscripts kept in the College library; a summary list is available in Duke Humfrey. The oldest is an imperfect early fifteenth-century Italian missal, presented by J. L. Cherry in 1909. Apart from some papers concerned with nonconformist history written by Arthur Gwynne Matthews (1881–1962), the editor of Calamy and Walker, the remainder relate to the College. There is an interesting series of replies, all dated 1841, from other dissenting colleges, to questions sent by the students of Spring Hill on the anti-slavery movement, the value of prayer meetings and similar topics. College societies are well represented, beginning with the Debating Society correspondence in 1839, then there are the minutes of the Discussion Society 1847–66 and its probable successor, the Debating Society 1866–77. More modern are the papers of the Fairbairn Society 1908–61; the Oxford University Congregational Society Committee minutes 1950–57, and the papers of the Free Church Fellowship 1950–65. There is also a College magazine in manuscript of 1873.[10] The Donors' Register (called *Contributions to the Library and Museum of Spring Hill College*) covering 1838 to 1889 has survived. Other library documents are a borrowers' register 1916 to 1926 and an interesting diary kept by the librarian between September 1945 and June 1946, when the College regained possession of its buildings after the War.

The records of the Frilford and Longworth Home Mission for the years 1854 to 1939 have been deposited.

The archives proper are in the custody of the Bursar. They comprise mainly the minutes of the governing body, cash books and minutes of the Board of Education (*i.e.* the committee responsible for the syllabus) dating from the foundation of Spring Hill College in 1838.

Printed Books

The printed books now in the College library are the residue of purchases and benefactions of all sizes from the foundation in 1838. Though much has been discarded, remnants of the gifts of the donors named below can still be found on the shelves. The books received in the early years were nearly all theological or classical, such as seventy volumes given by Mrs. W. Beaumont and a bequest from John Riland in 1863.[11] One of the

---

10 The printed *Mansfield College Magazine* began in March 1895.
11 *Report 1863–4*, 10–1.

most important gifts the library ever received came in 1872 when Thomas Smith James (1809–1874), son of John Angell James (1785–1859), presented over 600 volumes concerned with the classics, classical antiquity, history, literature, philosophy and theology; included were many formerly owned by the father, one of the moving spirits behind the College from its start. Nearly 400 volumes on theology and philology were presented by W. Froggatt in 1881.[12] Of a different character was a gift of books on temperance from Mrs. Thomas Avery in 1882.[13]

After the move to Oxford, the first important benefaction was a collection presented by J. B. Paton (1830–1911) in 1902 of books dealing with early church history, augmented by a further gift in 1907,[14] mainly of nineteenth-century theological works. A condition of acceptance was that the collection should be generally available 'for the use of ministers and students especially of the free churches in England and Wales' and to that end a catalogue was published in 1909[15] and it was separately housed. The survivors are now incorporated into the rest of the library. With the Paton Library were kept originally the books presented by A. M. Fairbairn, Principal 1885–1909, upon his retirement, but they were distributed after his death in 1912.[16] Besides the early missal already mentioned, J. L. Cherry gave in 1909 a small collection of seventeenth-century Bibles and theological works, mostly still in the library. An interesting group of about 200 volumes of palaeographical works was presented by Canon W. Sanday of Christ Church in 1919, and further books were received after his death in the following year.[17] The latest substantial bequest came after the death of J. Vernon Bartlett, a member of the teaching staff, in 1940.

Besides these larger collections, there has been a continuous stream of donations of individual books from teachers, students, old members and friends, as well as the regular purchases. The weeding-out mentioned must be borne in mind; books known to have been presented, or to have belonged to a particular benefactor, are not necessarily still present.

Catalogues

The earliest catalogue surviving is a ledger used from c. 1860 to the early years of this century, judging from imprint dates. The published Paton Library catalogue of 1909 has been mentioned above. A separate catalogue of the Junior Common Room library was made in 1914.

There is now an author and title catalogue on cards, while in each bay is a wall slip index of books kept in that position.

12   *Report 1876–7*, 11; *1881–2*, 9.
13   *Report 1882–3*, 9–10.
14   *Report 1902–3*, 8; *1907–8*, 9.
15   *Mansfield College, Oxford. Paton Library for the Use of Ministers and Students especially of the Free Churches in England and Wales. Catalogue and Rules* (1909).
16   *Report 1909–10*, 8–9; *1912–3*, 9.
17   *Report 1919–20*, 7; *1920–21*, 9.

MERTON COLLEGE
1264

Stock: 40,000　Incunabula: 112　*STC*: 527

Note: The Librarian stresses that this library is not accessible to the general public. Special collections can be consulted by bona fide *scholars under arrangements to be negotiated with the Librarian in advance. The library is shut during August.*

The library of Merton, as one of the oldest surviving buildings designed for housing books in England, is probably visited by more people than any other Oxford library except the Bodleian. Lying on the south and west sides of Mob Quad, very likely the oldest quadrangle in Oxford, the present library, built between 1373 and 1378, was largely paid for by William Rede, Bishop of Chichester (d. 1385), to supersede an earlier building where books had been kept as early as the 1330s. Not unnaturally the history of this library has received much attention from both the architectural and bibliographical angles, especially for its earlier periods.[1]

Besides a library of chained books, like the other older colleges, Merton also owned a large collection of manuscripts kept for circulation among the fellows and probably stored in chests when not in use; a detailed study has been published.[2] Dispersed in the first half of the sixteenth century, only one or two are known still to exist, none at Merton now.[3]

As elsewhere, the library was built up by benefactions from fellows and others from the earliest times, but there was no Benefactors' Register until the late seventeenth century, when one was drawn up according to Griffin Higgs' instructions. Indeed, the duty of fellows to bequeath their books was insisted upon in 1276.[4] By the time the Marian Commissioners visited the University in 1556, there were a little over 300 medieval manuscripts and a little under 200 printed books,[5] covering all subjects except law. Until the 1540s, there was a strange absence of printed books, possibly

---

[1]　George C. Brodrick, *Memorials of Merton College* [to 1750] (Oxford Historical Society, iv, 1886), *passim;* B. W. Henderson, *Merton College* (College Histories, 1899), pp. 224–42; P. S. Allen, 'Early Documents Connected with the Library of Merton College, Oxford', *The Library*, series IV, iv (1924), 249–76; H. W. Garrod, 'The Library Regulations of a Medieval College', *The Library*, series IV, viii (1927), 312–35; Streeter, 130–49. J. R. L. Highfield, 'Aspects of College Libraries, no. 1: Merton', *Oxford Magazine*, 1 Nov. 1968, 35–38, gives a general account up to 1613; a summary can be found in *VCH Oxon.* iii, 101–6.

[2]　Sir F. M. Powicke, *The Medieval Books of Merton College* (1931).

[3]　Ker, 464.

[4]　Henderson, 239; Powicke, 3.

[5]　Ker, 483; *Oxford College Libraries in 1556*, 6, 15.

caused by stagnation after Richard Fitzjames resigned the Wardenship in 1507. Some printed books seem to have been purchased in that decade using money obtained by selling College plate; Merton was unusual in this respect, and only a minority of the colleges spent money regularly on books before the seventeenth century.[6]

Sir Henry Savile, Warden 1585–1622, made many reforms within the College, including the library, which was refurnished[7] and a policy of book purchase adopted, especially by fellows when travelling abroad.[8] During the seventeenth century generally, there was a number of important gifts and bequests; that of Griffin Higgs in 1659 included an endowment for a College Library Keeper, an office previously performed by the Sub-Warden.[9] The eighteenth century was one of neglect; indeed one librarian was wont to hurl works on logic out into the Quad to spite the Warden.[10] Chaining continued until 1792 when all were removed except for two volumes.[11] Undergraduates were first admitted to the library in 1827 for one hour each week, extended to three hours daily by 1899[12] and now, of course, the whole day.

During the brief life of the scheme, about 1870, for colleges to specialise in different subjects, Merton pledged itself to spend £20 a year on modern history. In more recent years, the library has spread to the ground floor under the original rooms and into other parts of the College.

To some people, it is perhaps a disappointment to see modern books on the old shelves, as in neighbouring Corpus, but it means that libraries like these continue to live and be used, avoiding the ossification of a preserved ancient monument.

Manuscripts

The position of the rich collection of medieval manuscripts at Merton relative to other colleges has been succinctly put by a former distinguished College librarian: 'In wealth of medieval manuscripts Merton yields to Corpus, rivals Balliol and—leaving out of account the Queen's manuscripts . . . surpasses all other colleges'.[13] The disappearance of those from the circulating collection has already been mentioned. Coxe[14] records 337 western and nineteen Oriental manuscripts, but the majority of these did not reach the College till the fifteenth century; over eighty were received between 1264 and 1374, while the remainder came at later

6   Ker, 468, 483–4; *Oxford College Libraries in 1556*, 17.
7   Ker, 506–8; Streeter, 139–42.
8   Henderson, 240; Ker, 508–9.
9   P. S. Morrish, 'Dr. Griffin Higgs, 1589–1659', *Oxoniensia*, xxxi (1968), 129; Henderson, 241.
10 Henderson, 241.
11 Illustrated in Streeter, 143.
12 Henderson, 240.
13 H. W. G[arrod], 'Curiosities of Oxford Libraries, II: Merton', *Oxford Magazine* xliv (1925), 30.
14 Pt. 1, section 3.

periods.[15] Forty-three were part of the bequest of William Rede, Bishop of Chichester, who owned no fewer than 370 books when he died in 1385, which must have been one of the largest libraries in private hands at the time.

The duty of members of the College to give or bequeath books is reflected in the large number of donors who gave only one or two items. Among those who gave more were Richard Fitzjames, Warden 1482–1507; Henry Sever, Warden 1455–71; John Raynham, Bursar in 1330; Thomas Bloxham, fellow and a notable physician who died in 1473; and Robert Huntington, a fellow, sometime chaplain in Aleppo and later Bishop of Raphoe, who died in 1701.

Later manuscripts are perhaps not so numerous as in some other colleges, and are mainly connected with former members. For example, there are two of the four surviving diaries of Griffin Higgs;[16] some of the papers of James Harris, first Earl of Malmesbury (1746–1820),[17] and those of G. C. Brodrick (1831–1903), which relate largely to academic matters during his Wardenship or to his volume in the Methuen Political History of England; some letters of George Saintsbury (1845–1933), the critic; the autobiography of Edward Nares (1762–1841), the theological and historical writer, and the papers of F. H. Bradley (1846–1924), the philosopher. Manuscripts from the twentieth century include a few letters (chiefly of a formal kind) of T. S. Eliot (1888–1965), who spent a short period in 1914 at Merton, and the original musical score of *The Music Makers* of Zoltán Kodály (1881–1967), together with four letters of his written in 1964.

A typescript shelflist provides a guide to both those manuscripts calendared by Coxe and those not included by him, the latter being given full descriptions; a copy is available in Duke Humfrey.

Recently the College has established a Beerbohm Room, in the custody of the Librarian, to commemorate Sir Max Beerbohm (1872–1956), an *alumnus*, and has filled it with his manuscripts, drawings and first editions.[18]

Archives

The muniments of Merton are very extensive. They were fully calendared by W. H. Stevenson in the 1890s; a supplementary catalogue was compiled in 1920 and both, reproduced by the National Register of Archives, are available in Duke Humfrey. Indexes of deeds, surveys and maps made in the early nineteenth century, with modern handlists, are available in the College. A summary handlist was made in 1970 for the History of the University Project.

15 Powicke, 6.
16 For 1637 and 1638; those for 1630 and 1632 are in the Bodleian.
17 A list of his correspondence, prepared for the National Register of Archives, is kept in Duke Humfrey.
18 J. R. L. Highfield, 'Max at Merton', *The Postmaster* 1, iv (1960), 12–6; and his 'Opening of the Beerbohm Room' in the same journal, 2, v (1961), 9–11.

H. T. Riley gave a brief outline with a rather haphazard selection of transcripts for the Historical Manuscripts Commission in 1877.[19] The pre-1300 documents relating to Oxford have been published[20] as well as a selection of thirty-eight manuscripts from the middle of the twelfth century to 1519.[21]

College records go back to its foundation in 1284, including bursars' rolls and inventories of goods; wardens' accounts begin in 1299, but there are gaps. Registers recording College government begin in 1482, while battels and buttery books date from 1521 and 1561 respectively. Admission records, except for fellows, do not start until 1758. The register and accounts of the College musical club for 1712 to 1719 have survived here—one of the earliest extant records of a student society. There are also Anthony Wood's collections for a history of Merton, made in the seventeenth century, and Stephen Edwards' similar collection from the late nineteenth century. In addition, there are various biographical accounts of fellows.

Land is or has been owned in several parts of Britain and relevant deeds cover no less than eighteen counties, mainly in the South-East, the Midlands and Yorkshire. A calendar of those relating to Basingstoke has been published,[22] as well as pre-1300 rolls of Surrey manors.[23] More recently a calendar of records concerning the City and University of Oxford has been prepared and a copy placed in Duke Humfrey. A map of Cuxham, made in 1767, was published by J. L. G. Mowat.[24]

The archives are in the custody of the College Archivist.

Printed Books

There seem to have been few printed books in Merton until the later 1540s, when a large number of standard works in all subjects taught, except law, was purchased;[25] the 1556 catalogue shows just under 200 printed items. The custom of members presenting or bequeathing their books to the College, already mentioned in connection with manuscripts, continued into the printed book era; gifts of single items are not uncommon. A larger bequest of fifty-four volumes was received in 1583 from William Marshall, a fellow with Roman Catholic sympathies and an interest in the medieval schoolmen;[26] his will stipulated that the books were to be chained.

[19] 6th Report (1877), 545–49.
[20] J. R. L. Highfield, The Early Rolls of Merton College, Oxford (Oxford Historical Society, new series xviii, 1964).
[21] P. S. Allen and H. W. Garrod, Merton Muniments (Oxford Historical Society, lxxxvi, 1928). Some were also printed in [J. Kilner], The Account of Pythagoras's School in Cambridge, [1797].
[22] F. J. Baigent and J. E. Millard, A History of Basingstoke (1889), 593–650.
[23] H. M. Briggs and Sir C. H. Jenkinson, Surrey Manorial Accounts: a Catalogue and Index of the Earliest Surviving Rolls (Surrey Record Society, xv, no. 37, 1935).
[24] Sixteen Old Maps of Properties in Oxfordshire in the Possession of Some of the Colleges (1888), Map 8.
[25] Ker, 482–84.
[26] Ibid., 505–6.

Sir Henry Savile's reforms included a policy of purchasing books, unusual at that time. The Warden's brother, Thomas, for instance, bought books in Italy in 1589 and at the Frankfurt book fair in 1591.[27] A wide range of subjects was covered. Bequests continued; that of James Leeche (d. 1589), a former fellow, of 200 volumes of which 100 survive today, included many law books, filling the gap noticeable in the 1556 catalogue. The purchase of the law books of John Betts, of Trinity Hall, Cambridge, in 1599 also helped; about 150 are in the library now.

From 1549, two medical lecturerships had been held at Merton, endowed from the bequest of Thomas Linacre (d. 1524), the physician, so it is not surprising that the College library is strong in early books in this field, some purchased and some bequeathed. The first Linacre Lecturer, Robert Barnes (d. 1604), a fellow for over sixty years, gave a large number in 1594, and Roger Gifford (d. 1597), another former fellow and Linacre Lecturer, bequeathed more. This good collection of mainly sixteenth-century medical works was further increased in 1635 by the gift of the library of Theodore Gulston (d. 1632) from his widow, Helen; he had been an eminent physician as well as a former fellow.

Merton library received one of its largest bequests during the Commonwealth, when Griffin Higgs (1589–1659) left 'all my printed books and all my mappes' besides money for buying divinity books and endowing a salary for a library keeper.[28] Higgs, sometime a fellow, had been a chaplain to Elizabeth of Bohemia during her exile in the Netherlands, and later became Dean of Lichfield; while abroad between 1627 and 1638 he purchased many books at auction sales, as can be seen from his marked and annotated catalogues still in the library; the majority of this series are unique copies.[29]

A group of Oriental books was sent to the College in 1673 by Robert Huntington (d. 1701), a fellow, while serving as a chaplain to the English Factory in Aleppo; his own collection of Oriental manuscripts was bought by the Bodleian in 1693. The Wardenship of Sir Thomas Clayton from 1661 to his death in 1693 is notorious for internal disputes and strife, and the Warden himself, sometime Regius Professor of Medicine, was an unattractive character, but he had an extensive library covering many subjects other than medicine which was bequeathed to the College and is still there.

The eighteenth century was a period of neglect in Merton, but two bequests of some importance were received. The first was from Thomas Herne (d. 1722), a former fellow and tutor to the family of the Dukes of Bedford; among his books are about thirty volumes of tracts concerned

27 Ibid., 508–9; Brodrick, 62; Henderson, 240.
28 Morrish, loc. cit.
29 For a list of these catalogues, see P. S. Morrish, 'A Collection of Seventeenth-Century Book Sale Catalogues', *Quaerendo*, 1 (1971), 35–45; see also G. Pollard and A. Ehrman, *The Distribution of Books by Catalogue . . . to A.D. 1800* (Roxburghe Club, 1965), 222–25.

with the Bangorian and other controversies of the time, in which Herne himself had taken part. The second was towards the end of the century when Henry Kent, a commoner in 1737, left over 800 volumes of a miscellaneous nature.

Though small gifts from individuals continued, there have been no further outstanding benefactions of printed books to the library. With F. H. Bradley's papers, already mentioned, came in 1924 his printed books, to form the basis of the Bradley Memorial Library of Philosophy, open to all graduates reading philosophy on application to the Librarian. The latest collection of note is the group of early editions of Sir Max Beerbohm, kept with his manuscripts and drawings in the Beerbohm room.

The Librarian also now has charge of the Warden's Library, formerly kept for the use of the head of the house in his lodgings; it is a small collection of printed books, the majority of eighteenth-century date, strong in classical authors and British topography.

Compared with the other ancient college libraries, the contents of Merton are very similar; it is rather weak in the *editiones principes* of the classics, a gap that Sir Basil Blackwell, an *alumnus*, is currently trying to fill. Its strength lies in the large number of books printed in the sixteenth century that have survived, while, from the subject point of view, medicine and law are especially well covered. Then there is a fair number of tracts from the sixteenth to the eighteenth century, many rare.

Catalogues

The catalogue prepared for the Marian Commissioners in 1556 shows just under 200 printed books.[30] Interleaved copies of the Bodleian printed books catalogues of 1674 and 1843 have survived, annotated with Merton holdings. A catalogue of the printed books was published in 1879, to which supplements were issued in 1883, 1890 and 1899.[31] This has now been superseded by a card catalogue. There is a separate typescript catalogue of the Warden's Library, but the books are also entered in the main card catalogue.

[30] Ker, 483; *Oxford College Libraries in 1556*, 15.
[31] Used as a source for locations by Donald Wing in his *Short-title Catalogue ... 1641–1700*.

NEW COLLEGE
1379

Stock: 70,000   Current periodicals: 120   Incunabula: 332   *STC*: 556

The provision of books was envisaged by the Founder, William of Wyke-ham, in his detailed statutes and, as in other existing colleges, the volumes acquired were to be divided between a loan collection and a chained library. Similarly, there was to be an annual distribution of loans, though here jurists could keep their texts for the whole period required to study for a degree.[1] The Founder presented over 240 volumes himself,[2] followed by a succession of other benefactors so that in 1480 a new room for law books was added, and a century later the attics over the main library were converted to house those chained.[3] From the early sixteenth century onwards (except for the decade 1509–19), books were purchased from corporate revenue; by 1558 there was an impressive collection of canon and civil law books, acquired by purchase and gift.[4]

A library Benefactors' Book was compiled in 1617, recording previous donations,[5] slightly later than All Souls and Christ Church but earlier than the majority of Oxford colleges; it was used till 1909. A general Benefactors' Book opens with the bequest of Michael Woodward, Warden from 1658 to his death in 1675, and has entries up to 1757. The library obtained its first specific income from the Woodward bequest: an annuity of £5. Begin-ning in 1735, degree money was used for the purchase of books, averaging about £100 a year. Two further bequests of £5 annuities were received from Osborne Wright, elected fellow in 1775, and Samuel Gauntlett, Warden 1794–1822.[6] Separate library accounts for 1820–82 are in the College archives, together with a list of books added 1861–75. Interesting survivals from the seventeenth and eighteenth centuries are a register of books borrowed between 1665 and 1698, with a similar record for 1841–89 at

---

[1]   *VCH Oxon.* iii, 158; A. H. Smith, *New College, Oxford, and its Buildings* (1952), 51; Streeter, 7. There is no good published account of this library; Hastings Rashdall and Sir Robert S. Rait, *New College* (College Histories, 1901), 250–1, give a rather deprecatory outline listing the major benefactions. *VCH Oxon.* iii omits the separate section on the library given for other colleges. Since 1903, the annual *New College* [*Report*] (*New College Record* from 1929) has listed donations, with occasional other details about the library.

[2]   A. F. Leach, *Wykeham's Books at New College* (Oxford Historical Society, xxxii; Col-lectanea series 3, 1896), 215.

[3]   *VCH Oxon.* iii, 148; Ker, 490.

[4]   Ker, 468, 482, 504.

[5]   Leach, 217, note 1.

[6]   Rashdall and Rait, 250–1.

the other end of the notebook, and another from 1753 to 1766, with the library rules of April 1753 at the front. Another register in the archives lists books borrowed between 1840 and 1845.

The two parts of the library, the Upper and the Lower, were included in the structural alterations made by Wyatt between 1778 and 1780; a new library building was completed in 1939,[7] but though restored and mainly used for other purposes, the two original rooms continue to house most of the printed books acquired before 1800.[8] This library has grown so steadily over the centuries, partly by purchases and rather more by benefactions and bequests from members, (though duplicates were sold or exchanged in 1691, 1739, 1809 and 1849,) that books now have to be stored in different parts of the College.

Manuscripts

Coxe[9] lists 344 manuscripts and they now number 361. A large proportion of those owned in medieval times have since been dispersed, many during the religious troubles of the sixteenth century;[10] only twenty-seven of the 240 given by the Founder, for instance, have survived.[11] Other donors well represented in the existing collection include William Rede, (d. 1385), Bishop of Chichester; Richard Andrew (d. 1477), Dean of York; Thomas de Bekington (d. 1465), Bishop of Bath and Wells; William Warham (d. 1532), Archbishop of Canterbury; Thomas Cranley (d. 1417), Bishop of Durham.[12] The purpose of the College as a place of education for priests and lawyers is reflected in the subject matter of these manuscripts, though one written by Thomas Chaundler, Warden 1454–75, and given by Bishop Bekington, contains an interesting illustrated account of the early days of the College.[13]

Several Greek manuscripts came in the bequest of Reginald Pole (d. 1558), Cardinal Archbishop of Canterbury, some of which had formerly belonged to Christopher Longolius.[14] Since Coxe's catalogue was published there have been further acquisitions[15] including two medieval Gospels; a thirteenth-century St. Albans Psalter; Daniel Vivian's 'Report of a Voyage begun at Bristow, 1636'; four volumes of chronological and theological papers of Sir Isaac Newton (1642–1727), presented by the

7   *New College Record 1939–1940*, 11; *1940–1941*, 5–6.
8   Smith, 159–67; *New College Record 1955–1956*, 5–8.
9   Pt. 1, Section 7.
10 Rashdall and Rait, 113.
11 Leach, 217; Coxe, however, attributes thirty-six to Wykeham.
12 The principal donors are named in Rashdall and Rait, 250–1, and Leach 230–1, besides, of course, Coxe.
13 Reproduced in *The Chaundler Manuscripts*. Edited by M. R. James (Roxburghe Club, 1916).
14 E. Lobel, 'Cardinal Pole's Manuscripts', *Proceedings of the British Academy*, xvii (1931), 97–101; Ker, 488–9.
15 Brief hand-written descriptions have been added to the Duke Humfrey copy of Coxe.

Ekins family;[16] and the papers of Alfred Milner, first Viscount Milner (1854–1925), important for the history of South Africa, presented by his widow in 1933.[17] Those who wish to obtain the Librarian's permission to consult the Milner papers in the Bodleian should write to him at least one week before they expect to begin work; permission cannot be obtained by visiting the College library.

The manuscripts so far mentioned are deposited in the Bodleian,[18] where they may be consulted, but there are some other groups worthy of mention. The Chapel Choir music books, containing scores in use from the seventeenth to the mid-nineteenth centuries, mainly in the hands of eighteenth-century copyists, were given to the Bodleian in 1919.[19] Some modern manuscripts are kept in the College library or its archives; the most significant is probably a collection of about 500 letters of Sydney Smith (1771–1845), presented by the second Viscount Knutsford and augmented by their editor, Nowell C. Smith.[20] The papers of J. E. Sewell (1810–1903) and W. A. Spooner (1844–1930), successive wardens between 1860 and 1930, include personal correspondence, notes and diaries. In addition, there is a commonplace book made by William Smith (1711–1787), Dean of Chester; two series of *Carmina Wykehamica* covering most of the eighteenth and the first quarter of the nineteenth century; papers of Alfred Robinson (1841–1895), sometime senior Bursar; the correspondence of Hastings Rashdall (1858–1924) from 1866 to 1923; diaries and reminiscences of P. E. Matheson (1859–1946), the classical scholar; travel diaries of H. W. B. Joseph (1867–1943), the philosopher; some musical notebooks of Sir Hugh Allen (1869–1946), and some scores by R. O. Morris (1886–1948); and notes on naval history made by the late Captain A. C. Dewar, R.N.

Archives

The historians of New College have written that there is an 'enormous mass of documents relating to its property and finances, but very little besides', and 'an immense collection of court rolls of manors, farmers' accounts, and the like'.[21] Actually, there is a substantial number of documents concerned with College administration and business from the original Statutes onwards, such as bursars' rolls from the foundation.

16 Described in Sir David Brewster, *Memoirs of Sir Isaac Newton*, ii (1855), 341–2; and in F. Shirras, 'Sir Isaac Newton and the Currency', *Economic Journal*, lv (1945), 218–20.

17 *New College Record 1933–1934*, 6; a separate calendar is available in Duke Humfrey.

18 First in 1907; see *New College [Report] 1906–1907*, 2; and Sir E. Craster, *History of the Bodleian Library, 1845–1945* (1952), 192–3.

19 Now Bodleian Library MSS. Mus. c. 46–51; d. 149–69; e. 22–5; f. 32.

20 These letters were published in *The Letters of Sydney Smith*. Edited by Nowell C. Smith. 2 vols. 1953.

21 Rashdall and Rait, ix–xi, where a summary can be found; *VCH Oxon*. iii, 144 also gives a good summary of existing muniments. For the catalogue now in preparation, see end of this section.

The accounts were used by J. E. Thorold Rogers for his *History of Agriculture and Prices in England*, published 1866–1902. Hall books, containing names of all members receiving commons, and often their guests, date from 1386, though there are gaps; incidentally, the College cook's recipe book of 1719 has survived. The early records of the acts of the Warden and senior fellows are lost, and only date from the eighteenth century. The 'Register of Protocols', or admission registers in which each scholar's and fellow's entry has to be attested by a notary public, begin in 1450. Buttery books only go back to 1660, as does the large Chapel archive mainly concerned with structural repairs and restoration.

H. T. Riley described very few of these for the Historical Manuscripts Commission, but did give long extracts from selected documents on College affairs in the fourteenth and fifteenth centuries;[22] he also published an inventory of goods handed over by Thomas Cranley when resigning the wardenship in 1396.[23]

Compared with those in other colleges, the surviving records of New College clubs and societies are meagre; there are those of the Glee Club 1839–67, the Essay Society 1868–1947 and the Boat Club 1908–13.

The large collection of estate records chiefly refer to property in Berkshire, Buckinghamshire, Kent, Norfolk, Oxfordshire and Wiltshire; a catalogue of manorial documents has been published,[24] as well as a list of those relating to Hornchurch Priory in Essex.[25] Evidences and copies of leases have been transcribed in a long series of registers dating from 1480 to the present day. An interesting group of documents is concerned with the progresses (or tours of inspection) of the Warden around College properties; the earliest survivor was made in 1542 and they continue into this century. Many of the notes made by Michael Woodward, Warden 1658–75, have been published.[26]

There are naturally many records connected with the sister foundation at Winchester, mainly concerned with elections at both colleges. New College also had close ties with the grammar schools at Thame and Bedford, reflected in the archives.

A detailed catalogue, which it is hoped to publish, has recently been completed by Mr. Francis W. Steer; it will then be the first full description of an Oxford college's archives to appear in print since Trice Martin's catalogue of those at All Souls nearly a century ago.[27]

22   *2nd Report* (1871), 132–6.
23   'Inventory of Goods Belonging to a Warden of New College, A.D. 1396', *Archaeological Journal*, xxxviii (1871), 232–4.
24   T. F. Hobson, *A Catalogue of the Manorial Documents Preserved in the Muniment Room of New College, Oxford* (Manorial Society's Publications, no. 16, 1929.)
25   *Hornchurch Priory; a Kalendar of Documents in the Possession of the Warden and Fellows of New College, Oxford*. With an Introduction by H. F. Westlake (1923).
26   *Records of Buckinghamshire*, xiii (1934–40), 77–137; *Oxfordshire Record Society*, xxvii (1949); *Norfolk Record Society*, xxii (1951), 85–115; *Wiltshire Archaeological and Natural History Society, Records Branch*, xii (1957).
27   I am greatly indebted to Mr. Steer for permission to consult his catalogue in manuscript.

The archives are in the custody of the College Librarian, who is also Keeper of the Archives.

Printed Books

The printed books purchased or given over the centuries have survived in large numbers, so that New College now owns an extensive library reflecting former courses of study, particularly law, theology and the classics, but also strong in early medical and scientific works[28] besides some interesting donations of specialist collections such as those of Lord Donoughmore and R. W. Seton-Watson mentioned below.

Books were acquired in the very early days of printing; Richard Mason, for example, gave a copy of Albertus de Eyb, *Margarita Poetica* in 1479.[29] The impressive collection of early law books was built up partly by purchase and partly by gift; patristic and classical works were similarly acquired.

The bequest of Cardinal Pole, received in 1558, included printed books as well as manuscripts; twenty-five have survived, identified by the inscription *RPC* or from the signature of Christopher Longolius.[30] The practice of recording donors' names on title-pages or elsewhere in the books given was not observed so frequently in New College as in some other Oxford libraries. An interesting bequest of about thirty volumes from Thomas Martin, a fellow who died in 1584, included French translations of Thucydides, Livy and Aristotle,[31] while that from Walter Bailey (d. 1592), sometime Regius Professor of Medicine, contained some medical and scientific books among other subjects. About four hundred medical books came in 1617 from Thomas Hopper (d. 1624), while in that same year Arthur Lake, Warden 1613–16, presented about five hundred volumes, mainly of contemporary theology. Lake's successor in office, Robert Pinke, Warden 1617–47, bequeathed a general scholarly library of about 170 items. Another Warden, Michael Woodward, left in 1675 a general collection of about six hundred volumes in addition to the annuity mentioned above.

The only notable benefaction received during the eighteenth century was that given in 1776 by Martin Wall (1747–1824), Litchfield Professor of Clinical Medicine, a man interested in the history of his subject and acquainted with Dr. Samuel Johnson; it included numerous tracts and works by Robert Boyle (1627–1691). Dating from this century, but not handed over to the library until much later and as yet not fully sorted, is a collection of music formerly used in the College Chapel.

---

[28] Gunther, 43.

[29] Rashdall and Rait, 259, wrongly state that it was printed at Rome in 1475; it was actually printed at Strasbourg by Georg Husner about 1479 (F. R. Goff, *Incunabula in American Libraries* (1964), no. E-173).

[30] Listed by A. B. Emden in *Opus Epistolarum Des. Erasmi denuo recognitum per P. S. Allen*, xi (1947), Appendix xxviii, 379–83.

[31] Ker, 504–5.

The nineteenth century was similarly devoid of important accessions beyond the usual flow from purchases and small gifts; several, however, have been received in this century. A bequest from Edward Charles Wickham (1834–1910), sometime fellow and later Dean of Lincoln Cathedral, was strong in classical literature; in the same year another bequest of twenty bound volumes of pamphlets came from Hereford B. George (1838–1910), the historian.[32] After the death of Hastings Rashdall in 1924, his widow presented his collection of works on the history of universities, complementing that given by Robert Laing (later Cuthbert Shields) to Queen's College.[33] Of genealogical interest are the printed papers collected by the sixth Earl of Donoughmore while Chairman of the Committee of Privileges of the House of Lords between 1911 and 1931, many with his marginal notes;[34] these overlap to a large extent the similar group given to All Souls by Granville Proby in 1932. Philosophical books were bequeathed by H. W. B. Joseph in 1943,[35] while works on British history of the nineteenth and twentieth centuries were left to his old College by the fifth Earl of Onslow in 1945.[36] A more unusual collection, on the history of Eastern Europe, especially Czechoslovakia in this century, including a large number of pamphlets, was presented by R. W. Seton-Watson in 1949.

In more recent years, the generosity of members of the College to its library has continued, but the benefactions have been working collections on specific subjects rather than of more recherché material, though that of Sir John Myres (1869–1954), Wykeham Professor of Ancient History, contained some older volumes, and Captain A. C. Dewar's gift included specialist naval books.

Catalogues

A remarkably complete series of catalogues, both of manuscripts and of printed books, has survived in the library of New College. Besides those of the former class already mentioned, there are two others, compiled in 1624 and 1729 respectively. The earliest extant catalogue of printed books is one of the medical section made about 1623, with about 500 entries. A shelf-list, styled *supra catalogus librorum*, was written about 1640. Two volumes, compiled about 1655, describe in one the theological, and in the other the classical, historical and scientific books; the latter contains additions to about 1690.[37] A new catalogue was then made, which has

---

[32] *New College [Report] 1910–1911*, 5.
[33] Mrs. Rashdall presented in addition some 300 volumes of general history and theology in 1941 (*New College Record 1941–1942*, 6).
[34] Ibid., *1932–1933*, 5.
[35] Ibid., *1943–1944*, 6.
[36] Ibid., *1944–1945*, 4.
[37] This may be the catalogue dated *c.* 1625 in Sears Jayne, *Library Catalogues of the English Renaissance* (1956), 72, but the earlier entries include some books published in the 1650s.

survived with a companion volume covering additions up to the mid-eighteenth century. Catalogues of the books in the Lower Library, and of the Side Room, were compiled in 1822, and one of those in the Oaken Cabinet about 1832. Another catalogue is dated 1842; these last four were presumably used till about 1884 when both the Upper and the Lower Library were re-classified and the preparation of a new catalogue was announced.[38] This probably referred to the annotation of a copy of the Bodleian General Catalogue of 1843 now in the library, superseded by a card catalogue in recent years.

[38]   Printed notice, signed by W. L. Courtney, Sept. 1884, in Bodleian Library scrapbook, G. A. Oxon. c. 282, f. 15.

NUFFIELD COLLEGE

1937

Stock: 105,000   Current periodicals: 600   *STC*: 4

Nuffield College, endowed in 1937 by W. R. Morris, 1st Viscount Nuffield, received a charter in 1958 and full collegiate status in 1963. It concentrates on postgraduate research into social, economic and political problems, while acting as a meeting-place for academics and men of affairs, according to the Founder's wishes. An excellent library has been built up to serve these special interests, including manuscript and scarce printed material of considerable value. Since 1963, a duplicated list of accessions has been regularly issued, available in Duke Humfrey.

Manuscripts

There are several groups of papers concerned with modern British political and economic history. The oldest are those of William Cobbett (1762–1835), most of which were originally acquired by G. D. H. Cole (1889–1959), whose library was bought by the College; a calendar prepared by M. L. Pearl is available in Duke Humfrey. The Political Records Project, designed to locate and preserve the archives of twentieth-century politicians, was based on Nuffield College before moving to the London School of Economics in 1970,[1] though the section on Cabinet Ministers' Records is remaining until September 1972. The papers of the following have been deposited or given: Sir Stafford Cripps (1889–1952); Alfred Emmott, 1st Baron Emmott (1858–1926); F. A. Lindemann, 1st Viscount Cherwell (1886–1957); J. E. B. Seely, 1st Baron Mottistone (1868–1947); and J. A. Pease, 1st Baron Gainford (1860–1943). Lists have been prepared for the papers of Lords Emmott and Mottistone. The minute books and numerous other records of the Fabian Society are here, and also a few papers of Herbert Morrison (d. 1965),[2] and some relating to the Donovan Commission on Trade Unions and Employers' Associations, 1965–68, and C. W. Guillebaud's Committee on Railway Pay, 1959.

Two groups are appropriately concerned with Lord Nuffield's philanthropic activities: the records of the Nuffield Trust for the Special Areas[3] and of the Nuffield College Social Reconstruction Survey, organised and directed between 1940 and 1944 by G. D. H. Cole, Sub-Warden.[4] It

---

[1] *Library Association Record*, lxxii (Nov. 1970), *Liaison*, 69.
[2] Application to consult must first be made to the Warden of the College.
[3] P. W. S. Andrews and E. Brunner, *The Life of Lord Nuffield* (1955), 279–81.
[4] Ibid., 314–5.

is also appropriate that the College has the custody of the papers of two former heads: Sir Henry Clay (1883–1954), Warden 1944–49, and his successor, Alexander Loveday (1881–1962), Warden 1950–54.

Printed Books

A certain amount of scarce printed material can be found in Nuffield College library; older tracts and nineteenth-century works useful for economic history are among the books acquired after the death of G. D. H. Cole. A good collection of the publications of William Cobbett[5] to which additions are occasionally made, supplements his manuscripts, while Robert Owen (1771–1858) is strongly represented. More recent printed material can be found in the good sets of post-1945 British election addresses, gathered during the preparation of the surveys edited by Dr. David Butler.

As a whole, this library is well stocked with the secondary works necessary for research into economics, current politics and sociology, especially though not exclusively British. The range of periodicals taken is very wide and includes trade union journals not found elsewhere in Oxford.

Catalogues

There is an author catalogue on cards and a dictionary subject one on slips; Library of Congress classification is used. There is a separate card index to British government publications.

[5]   Holdings are marked with an asterisk in M. L. Pearl, *William Cobbett; a Bibliographical Account of his Life and Times* (1953).

ORIEL COLLEGE

1326

Stock: 37,000   Current periodicals: 82   Incunabula: 34   *STC*: 509

Oriel, originally styled 'The House of Blessed Mary the Virgin in Oxford', was projected by Adam de Brome in 1324, receiving its foundation deed and statutes in 1326 from Edward II, who can be regarded as the co-founder. It was very soon nicknamed Oriel after a large tenement, La Oriole, acquired in 1329. Until the reign of Elizabeth I, it was primarily a graduate college, rather like All Souls today.

Books were acquired from the earliest period,[1] though Adam de Brome was unsuccessful in retaining Thomas de Cobham's books. Those known to have been in the College in 1375 have been dispersed, but over seventy of the surviving manuscripts come from the medieval library. Oriel resembles other colleges in that its books were divided into those for lending and those kept chained; the latter practice continued until 1755. As elsewhere, also, reliance for acquisitions was placed on benefactions rather than purchases, though plate to the value of £30 was sold to provide money for theological works in the 1540s.[2] Later, fees from *commensales* were used for library purposes.[3] Regulations for the library were made in 1589 and revised in 1625 and 1655.[4] A Benefactors' Book, started about 1680, recording earlier gifts including books and fees for the library, was used until the early nineteenth century. A good series of library accounts begins in 1662, when admission fees began to be spent on books; registers of borrowings begin in 1766, but there are gaps up to 1837.

Oriel's most famous period was between 1780 and 1850, culminating with its association with Keble, Newman and the Tractarian Movement. At its outset, the library was doubled in size by a bequest from Edward, fifth Baron Leigh (1742–1786), which necessitated a new library building, now known as the Senior Library.[5] The neighbouring St. Mary Hall (with its small collection of books) was absorbed by Oriel in 1902; its former chapel was later converted into a reading room for undergraduates, previously only served by a cupboard of books kept in a lecture room. In

---

[1] The best general account is in *VCH Oxon.* iii, 124; that by D. W. Rannie in *Oriel College* (College Histories, 1900), 230–33, is unnecessarily deprecatory.

[2] *Oxford College Libraries in the Sixteenth Century*, 6–7; Ker, 481.

[3] C. L. Shadwell, *Registrum Orielense*, i (1893), xi, xiv.

[4] *The Dean's Register of Oriel, 1446–1661.* Edited by G. C. Richards and H. E. Salter (Oxford Historical Society, lxxxiv, 1926), 260, 334, 344–5.

[5] Antony Dale, 'The History of the Library', *Oriel Record*, 1951, 13–8.

1921 they were admitted as readers into the Senior Library and in the following year the two rooms were connected by a bridge.[6] A fire, spreading from the roof of the Senior Library on 7th March 1949, resulted in the total loss of just over 300 printed books and the few manuscripts on exhibition, but over 3,000 were able to be repaired; many had only their spines affected while a large number escaped serious damage altogether.[7]

Oriel library has several distinctive features. The medieval manuscripts reflect its antiquity while in some ways it still bears the impression of a country house library from the Leigh bequest. The early medical books are important, as are the connections with the Tractarians; no-one interested in any former member can afford to neglect its *Orielensia*—one of the best *alumnus* collections in Oxford.

Manuscripts

Coxe[8] lists eighty-two manuscripts, the majority medieval ones concerned with theology, philosophy, patristics, law and liturgy, mostly single gifts from individuals. Some of the more notable are listed by D. W. Rannie, with their donors,[9] ranging from an Aurelius Prudentius of possibly the tenth century, presented by John Taylor, Provost 1479–93, to a Piers Plowman, and John Capgrave's *Postilla* of the fifteenth. A Greek manuscript of the Gospels, a fifteenth-century copy of the second version of Wiclif's Gospels and a document written by William Prynne (1600–1669) have been acquired since Coxe's day. These manuscripts are deposited in the Bodleian Library where they may be consulted.

An extensive collection of nineteenth-century letters and papers, well arranged and indexed, is housed in the College library, connected with the leading figures of the Tractarian movement, such as John Keble (1792–1886)[10] and J. H. Newman (1801–1890), of whose letters there are more than a hundred.[11] Edward Hawkins (1789–1882), Provost from 1828 to his death, is represented by some three hundred letters written to him, including over sixty from W. E. Gladstone.[12] More Hawkins papers are kept with the College archives. There is also much to do with the Hampden controversy. The College maintains a policy of acquiring letters and documents concerned with this period of its history.

[6]   Ibid., iv, no. 4 (1922), 104–5.
[7]   R. R[obinson] and C. S[eton]-W[atson], 'The Fire in the Senior Library', ibid., 1949, 8–9; R. R[obinson], 'The Restoration of the Senior Library', ibid., 1951, 10–2; the final figures are given in the 1957 number, p. 7.
[8]   Pt. 1, section 5; see also the comments by C. C. J. Webb in *The Dean's Register*, x–xi, and a calendar of MSS. 1–83 on pp. 386–97, which adds full transcriptions of entries with donors' names, not given by Coxe.
[9]   Pp. 231–2.
[10]   W. D. R[oss], 'Some Unpublished Letters of Keble', *Oriel Record*, vii, no. 2 (1938), 251–5.
[11]   Ibid., viii, no. 1 (1939), 5–10; and 1954, 16–9.
[12]   J. W. G[ough], 'Provost Hawkin's Correspondence', ibid., 1958, 15–20.

Archives

The early archives of Oriel have been well studied in the past and a high proportion published. C. L. Shadwell (1840–1918), Provost 1905 to 1915, printed a few copies only of his *Catalogue of Muniments* in ten fascicles between 1893 and 1905,[13] dealing with those of the fourteenth and fifteenth centuries. He also provided a good summary in his *Registrum Orielense*.[14] With a few gaps in the earlier years, accounts survive from 1409, caution books from 1617–30 and buttery books from 1642. The records of the fourteenth and fifteenth centuries have also been published in fuller form.[15] An interesting document, the Dean's Register, which is concerned with many aspects of the College, has also been published for the years 1446 to 1661;[16] the original continues until 1889 when supplanted by College minutes. From later centuries, only an eighteenth-century memorandum book with Provost Carter's drafts of letters and notes on College affairs, *c.* 1710–27, with a record of leases granted in the 1760s added,[17] together with two Oxfordshire estate maps[18] have been published, though the Common Room accounts of the mid-eighteenth century have been described.[19] The College accounts were used by J. E. Thorold Rogers for his *History of Agriculture and Prices in England* (1866–1902).

A collection of later date is the accumulation of correspondence sent to L. R. Phelps (1853–1936), Provost from 1914 until his death; nothing, however trivial, was ever thrown away, but these papers have unfortunately been weeded since coming to the College. The only letters actually written by Phelps are those to H. R. Boyce, a schoolfellow, returned after the latter's death. The interest lies mainly in the large number from former members of Oriel and of St. Mary Hall, of which Phelps was sometime Vice-Principal.[20]

Records of undergraduate clubs chiefly date from the later nineteenth century, such as the Debating Society from 1888, but the Boat Club's diary begins in 1842.

---

13 Well summarised in *VCH Oxon.* iii, 119 note 1. The account by H. T. Riley in Historical Manuscripts Commission, *2nd Report* (1871), 136–7, is unsatisfactory, mostly devoted to extracts from early accounts and a description of the Dean's Register.

14 i (1893), xii–xv.

15 C. L. Shadwell and H. E. Salter, *Oriel College Records* (Oxford Historical Society, lxxxv, 1926), follow the order in Shadwell's *Catalogue*, but exclude his fascicles 6–8 describing deeds for properties outside Oxford; all those not in common form are printed in full.

16 See note 4 above.

17 F. J. V[arley], 'The Provost of Oriel's Memorandum Book', *Oriel Record*, viii, no. 4 (1941), 155–7.

18 Those of Dean and Chalford, both 1743, in J. L. G. Mowat, *Sixteen Old Maps of Properties in Oxfordshire in the Possession of Some of the Colleges* (1888).

19 *Oriel Record*, iv, no. 10 (1925), 304–8; C. S. Emden, *Oriel Papers* (1948), 193–5.

20 P. C. Lyon, *Report on the Correspondence of the Rev. L. R. Phelps, D.C.L., from 1877–1936* (1939).

Most of the surviving archives of St. Mary Hall are now kept in Oriel[21] though a few are in the University Archives.[22] Besides a late seventeenth-century Benefactors' Register, used till 1804, and building accounts and memoranda of the eighteenth and nineteenth centuries, there are battels books from 1728 and a few earlier documents. There is also material about this Hall among the Phelps papers.

The archives are in the custody of the Treasurer, who maintains a card index arranged by place for estate affairs, and another card index for general matters. A handlist has been prepared for the History of the University Project.

Printed Books

There does not seem to have been any substantial donation of printed books before the end of the sixteenth century when Thomas Cogan, fellow 1563–74, a physician who later became Master of Manchester Grammar School, presented in 1595 some medical works including a set of Galen (1562).[23] Not long afterwards in 1600, a larger collection of medical and botanical books was bequeathed by John Jackman, a former fellow expelled for marriage in 1586, who practised as a physician in Oxford.[24] Further medical and scientific works were presented by John Sanders, Provost 1644–53, while Samuel Desmaitres, a fellow who died from smallpox in 1686, left all his books, many on medical topics.[25]

Most of these donations are still on the shelves, but part of another, on different subjects, does not seem to have reached the College; William Prynne (1600–1669), an *alumnus*, left 'Ocham upon the Sentences [Lyons, 1494], Saint Brigets Revelations [Nuremberg, 1521], Laurentius Surius his councils in four tomes, and one of each sort of my own printed books which they yet want',[26] but only the first two named are now in Oriel; most of the large number of Prynne's printed outpourings have been acquired more recently.

George Royse, Provost 1691–1708, bequeathed a general collection which included one of the two books now in the College once owned and annotated by John Donne.[27] With the legacy of Edward, fifth Baron Leigh (1742–1786), an *alumnus* who later became Chancellor of the University, of all his books at Stoneleigh Abbey, Warwickshire, the College

---

21   For a summary, see *VCH Oxon.* iii, 121, note 17.

22   Nominations of principals and some inventories of plate.

23   R. T. Gunther, 'Thomas Cogan', *Oriel Record*, vii, no. 1 (1935), 15–8; *Dean's Register*, 208. The Vesalius is no longer in the library.

24   Gunther, 'T. Cogan', 16.

25   Gunther, 42.

26   C. L. Shadwell, *Registrum Orielense*, i, 152; Sir G. N. Clark, 'William Prynne, Commoner 1616–1621', *Oriel Record*, iv, no. 1 (1921), 12.

27   Antonius Clarus Sylvius, *Commentarius ad Leges Romani Iuris Antiqui* (Paris, 1603). The other is M. Parker, *De Antiquitate Britannicæ Ecclesiæ* (Hanoviae, 1605), described by Percy Simpson in *Oriel Record*, vi, no. 14 (1935), 427–9, and Sir Geoffrey Keynes, *A Bibliography of John Donne*, 3rd ed. (1958), 218, no. L 137.

library was immediately doubled in size, entailing a new building. This collection gives Oriel's books a country house flavour; it comprises all the works one would expect a cultured nobleman to own, from the classics, history, law and the natural sciences to music and the fine arts; there are many large, illustrated volumes, while the influence of the Grand Tour is apparent;[28] some of the older books formerly belonged to the antiquary A. C. Ducarel (1713–1785).

There were no more substantial donations for nearly a century, though smaller gifts continued; Cardinal Newman, for instance, in 1876, presented his copy of the *Remains of William Ralph Churton*, privately printed in 1830, 'To Oriel College Library, with grateful recollection of its stores'. About a thousand volumes were given by the widow of Edward Hawkins (1789–1882), Provost 1828–1882, mostly theological but including many presentation copies from authors and a good number of pamphlets on the Oxford controversies that raged between 1830 and 1878.[29] Another large donation of historical and theological books, strong in German publications and works on Dante, came from the widow of R. W. Church (1815–1890), sometime Dean of St. Paul's.

Like other halls, St. Mary Hall apparently did not own a large library, so when it was taken over by Oriel in 1902 only a small number of books was transferred, of no outstanding interest. An important collection of more than a thousand volumes relating to comparative philology and comparative mythology was bequeathed by D. B. Monro (1836–1905), Provost from 1882 to his death; some have since been transferred to the Taylor Institution, Ashmolean Museum and other Oxford libraries. Provost Monro had put into effect the suggestion made by James Bryce, then a fellow, about 1870, that college libraries should specialise in different subjects,[30] Oriel adopting the Provost's own specialities of comparative philology and mythology; a *Catalogue* with three *Addenda* was published between 1880 and 1903, showing that close on a thousand works were acquired,[31] but the scheme failed here as elsewhere.

Since the Monro bequest, there has been a series of interesting though less substantial benefactions. In the early 1920s about two hundred volumes were selected from the parochial library at Tortworth, Gloucestershire, founded on the collection of Henry Brooks (or Brooke, d. 1757), an Oriel man, afterwards augmented with those of John Bosworth; an Aldine *De Re Rustica* (1514) and a Coverdale Bible of 1540 were among those taken for the College.[32] Between 1929 and 1939 F. J. Varley, an

28 Rannie, 162; F. J. Varley, 'Lord Leigh's Benefactions', *Oriel Record*, viii, no. 6 (1943), 216–7.
29 Rannie, 232.
30 J. Cook Wilson, *David Binning Monro; a Short Memoir* (1907), 11–2, 14.
31 *A Catalogue of the Books in the Library of Oriel College Connected with the Studies of Comparative Philology and Comparative Mythology* (1880); and *Addenda*, 1885, 1891, 1903.
32 W. D. R[oss], 'The Tortworth Library', *Oriel Record*, iv, no. 4 (1922), 105–6; Central Council for the Care of Churches, *The Parochial Libraries of the Church of England* (1959), 102, 112.

alumnus, presented a number of views of Oriel[33] and a collection of books on Oxford and Oxfordshire,[34] besides providing financial help for repairing the older volumes.[35] Other gifts include a quantity of books to do with seventeenth-century English history from the library of Sir C. H. Firth (1857–1936), presented by his widow[36]; a small collection of works on India once owned by Sir Hopetoun Stokes (1873–1951), a former Indian civil servant; some eighteenth-century English literature from H. M. Margoliouth (1887–1959); and a small group, mainly of eighteenth-century classical and English literature, from Edward R. Marshall, received in 1968. Sir David Ross (1877–1971), Provost 1929–1947, bequeathed an interesting collection of early editions of and later works about Aristotle.

Oriel has an extensive collection of works written by or relating to, *alumni*, comparable only to that at Magdalen for size and completeness; St. Mary Hall men are included. Besides the main figures of the Tractarian Movement, many other famous and less famous men are represented, such as Sir Walter Raleigh (d. 1618), Gilbert White of Selborne, Matthew Arnold, A. H. Clough, T. E. Brown, Cecil Rhodes and Richard Hughes, to take a few names not already mentioned.

Catalogues

Of the earliest catalogue of 1375, only one leaf was saved from the 1949 fire, but it had fortunately been published in full;[37] it described about a hundred volumes of theology and the schoolmen then necessary for a university course.

Catalogues made in 1625 and 1684, bound together now, have survived, together with an undated seventeenth-century one, possibly earlier. There are also two undated catalogues compiled during the eighteenth century and another made about 1800. It was presumably this last one which was made by Henry Beeke, a fellow, with the assistance of Joseph Pickford, about which Thomas Mozley gave an amusing account.[38] A late seventeenth- or early eighteenth-century catalogue of St. Mary Hall's library, in a volume listing fees paid by members, has survived among its archives. There are also various special catalogues; *Incunabula in Oriel College Library*, with thirty-four entries, printed about 1919; an unfinished manuscript *Catalogue of Books in Oriel College Library which are not Inserted in the Bodleian Interleaved Catalogue*, covering A–E only and made about 1800, is kept in Duke Humfrey; the special catalogue of comparative philology and

---

[33]   *Oriel Record*, v, no. 4 (1929), 88–91; vii, no. 2 (1936), 60–1; vii, no. 8 (1939), 352–3.
[34]   Ibid., vii, no. 2 (1936), 58; vii, no. 3 (1936), 106.
[35]   Ibid., vii, no. 6 (1938), 244.
[36]   Ibid., vii, no. 5 (1937), 223.
[37]   *A Catalogue of the Library of Oriel College in . . . 1375 A.D.* Edited by C. L. Shadwell (Oxford Historical Society, v, 1885), pp. 57–70.
[38]   *Reminiscences chiefly of Oriel College and the Oxford Movement.* 2nd ed., i (1882), 67–70; but *cf. VCH Oxon.* iii, 124, note 9.

comparative mythology has been mentioned above; and the typescript catalogue of books printed before 1641, prepared for the Inter-Collegiate Catalogue scheme.

There are three current card catalogues: for the Senior Library (including the manuscripts, excluding the nineteenth-century letters kept there); for the *Orielensia* collection; and for the Junior Library.

PEMBROKE COLLEGE

1624

Stock: 27,000   Current periodicals: 70   Incunabula: 103   *STC*: 197

Note: The Librarian stresses that this library is not open to the general public and bona fide *scholars must make arrangements with him in advance before they can be admitted.*

At its foundation, mainly from the endowments of Thomas Tesdale (1547–1610) and Richard Wightwick (1547–1630), Pembroke College absorbed the then flourishing Broadgates Hall, a medieval institution and, although the latter's library had been lost during the reign of Edward VI,[1] its refectory was converted into the present library about 1850, after the modern hall had been built. Books were acquired from the beginning and records of chaining in 1654 survive.[2] A Benefactors' Register was begun about 1692, recording earlier gifts.[3] During the eighteenth century commoners and scholars paid a fee on admission for the use of the library, but this was discontinued before 1875.[4] Registers of books borrowed begin very early, in 1733.[5] A separate undergraduates' library was established in 1883.

Although only a comparatively small library, there is a certain number of early books and manuscripts, and its holdings should not be neglected, particularly by anyone interested in Aristotelian studies or former members of the College.

Manuscripts

There are about twenty medieval manuscripts (not described by Coxe), including six of medical interest, and a miscellaneous group of later ones. Among the latter are: a manuscript of the *Religio Medici* of Sir Thomas Browne (1605–1682), an *alumnus* of Broadgates Hall; fourteen packets of private devotions and other documents by or connected with Samuel Johnson (1709–1784), the lexicographer, whose brief career at the College

---

[1]   Douglas Macleane, *A History of Pembroke College, anciently Broadgates Hall* (Oxford Historical Society, xxxiii, 1897), 18. A summary of the library's history is in *VCH Oxon.* iii, 297.

[2]   Macleane, 279.

[3]   *Fine Bindings 1500–1700*, no. 231.

[4]   Macleane, 499.

[5]   J. L. Lundin, 'T. L. Beddoes at Pembroke College', *Studia Neophilologica*, xli (1969), 346–58, describes his borrowings when an undergraduate, based on the register used 1812–64.

is well known.[6] An unusual item is the log-book of T. Atkinson, Master of *The Victory*, for 1804–5, which unluckily ends just before the Battle of Trafalgar. A volume of the correspondence of G. W. Hall, Master 1809–43, is interesting for the letters of Keble and other Tractarians, while the letters of Sir Peter Le Page Renouf (1822–1897) include some from J. H. Newman, Lord Acton and other eminent contemporaries. Among more recent accessions are some Syriac manuscripts given by Athelstan Riley (1858–1945) and the theological notebooks of St. George Stock.

Archives

The archives of Broadgates Hall have disappeared. Those of Pembroke were briefly described by H. T. Riley for the Historical Manuscripts Commission.[7] A typescript calendar is available, but a fuller one is being prepared. A handlist has been made for the History of the University Project.

Statutes go back to the foundation in the seventeenth century, but the acts of the governing body only start in 1712. General accounts date from 1651, battels from 1664 and bursarial accounts from 1672. Admission records begin in 1678. The Junior Common Room Book for 1794–1808 has survived, but the main series does not start until 1844. There are also some records connected with various undergraduate societies, such as the minutes of the Boat Club from 1842 and of the Debating Society from 1852, among others.

Printed Books

There is a small collection of sixteenth- and seventeenth-century printed books, chiefly of theological interest, deriving mainly from the bequest of John Hall (1633–1710), Bishop of Bristol, Master of the College 1664–1710, and an *alumnus*. A bequest from Francis Wightwick (d. 1776) of Wombridge, Berks., a descendant of the Founder, included his books which reflect the interests of a family of country squires.

Probably the most important feature of the library is the H. W. Chandler collection of books connected with Artistotle, particularly his Nicomachean Ethics. Chandler (1828–1889), a bachelor and sometime Waynflete Professor of Moral and Metaphysical Philosophy, actually bequeathed his books to Mrs. Evans, wife of the then Master, but she promptly passed them over to the College, stipulating that they should be kept together and that an annual report on their condition should be sent to the Curators of the Bodleian Library. This collection is remarkable for its early editions of Aristotle and other philosophers, many in interesting bindings. A *Catalogue* has been published;[8] a revised, corrected shelf-list is kept at the College.

---

[6] Brief descriptions of each, with details of where published, are given in J. D. Fleeman, *A Preliminary Handlist of Documents and Manuscripts of Samuel Johnson* (Oxford Bibliographical Society, Occasional Publication no. 2, 1967).

[7] *6th Report* (1877), 549–51.

[8] *Catalogue of the Aristotelian and Philosophical Portions of the Library of the late Henry William Chandler* (1891).

Of later gifts, the most substantial was that of George Birkbeck Hill (1835–1903), the editor of Boswell and Johnson, so not surprisingly it is strong in eighteenth-century English literature including his own annotated Boswell.

Although there is no specific *alumnus* collection, it should be remembered that several men of literary interest were educated here and may have used the books still on the shelves, even if this library is not strong in their own writings. During the first half of the eighteenth century, William Shenstone, Richard Jago, Richard Graves and William Hawkins were all at Pembroke, while during the nineteenth century R. S. Hawker, T. L. Beddoes[9] and R. W. Dixon were at this College, to name some.

Catalogues

Three catalogues from the eighteenth century have survived—1718, 1727 and 1739. The current author catalogue, on cards, is divided into two sections: for books printed before 1641, and for later works. This does not include the Chandler Collection, for which the printed *Catalogue* is used as a guide.

[9] See note 5.

THE QUEEN'S COLLEGE
1340

Stock: 130,000   Incunabula: 286   *STC*: 2,552

Although founded in the fourteenth century, practically nothing of the older collections of manuscripts or printed books remains, yet the library of Queen's College is one of the largest and most magnificent in Oxford. The early benefactors and what they gave have been fully described,[1] but these had almost completely vanished by 1555 from the depredations of John Bale and others.[2] There is no evidence of any action being taken after Cromwell's Visitors in 1535, though a little was spent on patristics between 1560 and 1580.[3] The bequest of Thomas Barlow in 1691, followed shortly afterwards by that of Sir Joseph Williamson, with a consequent new building, marked a transformation and the real beginning of the present library; later bequests and gifts have maintained its status. Lists of books bought in the late seventeenth century, and library accounts for 1692 and then from 1716 onwards have survived.

This library was one of those described on Dibdin lines in *Notes and Queries*[4] in the last century, while its annals can be found scattered through Magrath's history. Unfortunately for the present purpose, the more recent account by R. H. Hodgkin[5] hardly touches the subject and does not bring the record up to date.

A Benefactors' Register was started in 1621 which listed previous principal gifts to that year and it was maintained in detail until 1692. The original Register no longer exists, though there is a revised transcript made in 1659, accounting for about 1,200 volumes, mostly seventeenth-century works with a few earlier ones, while at the end there is a list of manuscripts made in 1689, probably by Hugh Hartley, a fellow. An alphabetical list, based on this document, was made by Edward Rowe Mores between 1752 and 1761 which was later acquired with other notes on the College archives by Richard Gough, the antiquary; these are now in the Bodleian and were published by Magrath.[6] There is also a *Liber*

[1]   J. R. Magrath, *The Queen's College*, i (1921), 76–80, 126–29, 161–62, 359–60.
[2]   Ker, 490–1.
[3]   Ker, 484; *Oxford College Libraries in 1556*, 7.
[4]   By R. L. Clarke, series 6, iv (1881), 441–3, 461–3. Reprinted with corrections in Magrath, ii, 257–66.
[5]   *Six Centuries of an Oxford College* (1949).
[6]   ii, 266–78. See also E. R. Mores, *A Dissertation upon English Typographical Founders and Founderies*. Edited by H. Carter and C. Ricks (Oxford Bibliographical Society publications, new series ix, 1961), xv–xvi.

Albus Benefactorum, magnificently bound by Samuel Mearne, presented by Sir Joseph Williamson about 1680; all types of gifts received between 1560 and 1815 are entered in it.[7]

Quarto and folio volumes were chained until 1780 when accounts show that 5,717 books were unfettered; the smaller works were kept in the locked cupboards still present at the tops of the book-cases.

Until 1938, the use of the library was confined to fellows, or M.A.'s introduced by them on payment of a fee, while undergraduates had a separate reading room till that date. From at least 1625 until about 1840 the taberdars (*i.e.* senior scholars) also had a separate library, chiefly maintained by gifts from among their own number; a few of their books have survived, together with a late seventeenth-century catalogue.

Manuscripts

Coxe[8] describes 390 manuscripts, while there is a brief typescript list of additions, of which there is a copy in Duke Humfrey. None of those belonging to the College in medieval times has survived, though there are a few patristic manuscripts acquired during the sixteenth century. Of the rest, there are two main sources: firstly, 106 that came with the bequest of Thomas Barlow (1607–1691), Provost 1657–58, Bodley's Librarian 1652–60, Lady Margaret Professor of Divinity 1660–75, and finally Bishop of Lincoln; these are mainly of a theological nature, including his own notes, writings, and working texts. Secondly, and more numerous, the bequest of Sir Joseph Williamson (1633–1701), an *alumnus* and the College's greatest benefactor who had held several political and diplomatic posts. From 1666 to his death, a stream of gifts of money, plate and books poured from him; among the manuscripts are many documents and letters of political interest, but perhaps the most important is a large group of heraldic and genealogical manuscripts that include drafts of visitations by several members of the College of Heralds, such as Richard St. George (d. 1635), Joseph Glover (*fl.* 1580) and Ralph Brooke (1553–1625), and many formerly owned by Sir Thomas Shirley (*c.* 1590–1654).

The remainder, not mentioned by Coxe, are principally of College interest and from individual donors. There are verses by Theophilus Metcalfe which came with his gift of printed books in 1757, and poems by Edmund Blunden, an *alumnus*. J. R. Magrath (1839–1930), Provost 1878–1930, bequeathed his collections of genealogical and topographical manuscripts, including several formerly owned by Joseph Smith (1670–1756), Provost 1730–56, among which are some concerning Smiths' family and lands chiefly of Oxfordshire interest.[9] There are also some memoranda and notes connected with his theological studies by B. H. Streeter (1874–1937), Provost 1933–37.

---

[7]   Magrath, i, 283; *Fine Bindings 1500–1700*, no. 219.
[8]   Pt. 1, section 6.
[9]   Briefly described in Magrath, i, 299.

The records of only two under graduatesocieties have survived, unlike other colleges; those of the Halcyon Club 1869–96, and of the Addison Society 1876–1931. On the other hand, there is a large group of student notebooks from the seventeenth to the nineteenth centuries, such as notes taken by Jeremy Bentham at Sir William Blackstone's lectures about 1760, or theological questions discussed by the tabardars in 1823.

Archives

The archives were first classified and calendared by Edward Rowe Mores (1731–1778) about 1750; those relating to the history of the College were summarised by Magrath.[10] Some others, concerned with property in several parts of the country, especially Sherbourne Priory and the Hospital of St. Julian, Southampton, were described by H. T. Riley for the Historical Manuscripts Commission.[11] A full typescript calendar by N. Denholm-Young and E. M. Snodgrass was completed in 1931, and a copy, made available by the National Register of Archives, is kept in Duke Humfrey. A handlist has also been prepared for the History of the University Project, providing summary descriptions of archive material in both the typescript calendar and the list of additions to the manuscript collections

Records concerning the foundation in the fourteenth century have survived, while annual accounts begin in 1348. Registers with details of College government date from 1565; before the opening of the eighteenth century, only the battels book for 1570 has survived. Admission registers date from 1635. There are also documents concerned with the College's relationships with various schools from the sixteenth century onwards, notably in the north-west of England such as those at St. Bees, Kendal and Appleby, but also with some nearer Oxford such as Abingdon, Bampton and Northleach.

The medieval deeds, except for those in volumes 3 and 4 of the Denholm-Young and Snodgrass calendar, have been deposited in the Bodleian; the remainder are in the custody of the Bursar.

Printed Books

It has already been indicated that the printed books collection of Queen's College is one of the richest and largest to be found among the Oxford colleges, in spite of the disappearance of the earlier library. Depredations continued into this century, for in 1938 a large number of books, chiefly on bibliography and travel, were sold to make room for current accessions, and many gifts were depleted. As in most other colleges, there are many gifts from individuals, while in the seventeenth century one of the main sources of library income was money received from members on the foundation as they graduated; at Queen's, this money was derived from

[10] i, 278–83.
[11] 2nd Report (1871), 137–42; 4th Report (1874), 451–8.

fees received in lieu of celebrations in hall and averaged about £80 a year.

Of the books still present, the earliest notable group is a theological collection of about eighty volumes bequeathed by Edmund Grindal (1519?–1583), Archbishop of Canterbury. In the sixteenth century came, too, smaller gifts from men like John Rainolds (1549–1607), who taught in the College while in exile from Corpus between 1586 and 1598, and Henry Cuffe (1563–1601), Professor of Greek 1590–96, who was later hanged for complicity in the Essex rebellion. During the seventeenth century numerous smaller groups were presented; most of the provosts and many fellows gave books, such as Gerard Langbaine, Provost 1646–58, and his predecessor Christopher Potter, Bernard Robinson, Thomas Lough and Antony Tonstall. Some law books came from Robert Nicholas (1595–1667), an *alumnus*, later a Baron of the Exchequer, while an anonymous Cumberland man presented forty-six folios in 1638.

Until the end of the seventeenth century, this College library resembled others of like size, but with the bequest of Thomas Barlow in 1691 a transformation began. After a few personal gifts and the 'choyce of such ... wch shall appear to be wanting' to the Bodleian, the remainder of his extensive library, strong in theology, church history, ecclesiastical law and sixteenth- and seventeenth-century tracts, was left to Queen's. It was the most important accession so far and was the cause of the rebuilding of the library between 1691 and 1695, under the aegis of Timothy Halton, then Provost and himself a benefactor.

Within a decade, many more books arrived in the bequest of Sir Joseph Williamson; law, history and science were well represented and particularly noteworthy are the sets of English, French and Dutch proclamations, the *London Gazette* 1665–1688, and a Shakespeare third folio. This bequest, together with that of Barlow, immediately made Queen's one of the most important college libraries in Oxford.

During the eighteenth century two smaller bequests were made: those of Sir John Floyer and Theophilus Metcalfe. Floyer (1649–1734), an *alumnus*, was a prosperous physician in Lichfield, whose library was mainly medical and scientific, while Metcalfe (1690–1757) was another medical practitioner, but of Hart Hall with no obvious connections with Queen's. The latter had a valuable collection (the original catalogue survives among the College manuscripts), strong in alchemical and chemical works besides medical ones. R. T. Gunther[12] thought that these two gifts 'together include most of the sixteenth- and seventeenth-century writers (on medicine) and are deserving of special study'.

Apart from donations of individual items, there were no further large benefactions until 1841, when the College received £30,000 under the will of Robert Mason, an *alumnus*, on condition the whole sum was spent on books within three years. Mason left £40,000 to the Bodleian in like

12   Gunther, 42.

fashion. Some of this money was used to convert the ground floor of the library, then an ambulatory, into a room suitable for books, and most was spent by 1847. Henry Hayton Wood, a geologist and antiquary, then the fellow in charge of the library, supervised the purchasing very wisely, using Brunet's *Manuel* and the advice of Oxford experts, including Bulkeley Bandinel, then Bodley's Librarian. The books acquired defy summary; among them were the first, second, and fourth Shakespeare folios (Williamson had had a third), many incunabula, an almost complete set of Aldines and many colour-plate works on natural history; there is little doubt that after 1847 Queen's had become one of the best, if not the foremost, college library in Great Britain.

Since this magnificent bequest, there have been several more of a less spectacular nature. In 1908 was received a considerable number of books connected with the history of universities from Robert Laing (later Cuthbert Shields), a fellow of Corpus who had looked after Queen's library in the 1870s. In the next year W. R. Morfill (1834–1909), Professor of Russian, left about 8,000 volumes on the Slavonic languages, literature and history, while in 1916 Edward Moore (1864–1916), Principal of St. Edmund Hall, bequeathed about 1,000 items connected with Dante; both these collections are now deposited on permanent loan in the Taylor Institution.[13] Shortly afterwards, William Sanday (1843–1920), Lady Margaret Professor of Divinity and Canon of Christ Church 1895–1919, left a large collection of theological pamphlets and offprints, mainly contemporary in date with him.

A. H. Sayce (1845–1933), Professor of Assyriology and an *alumnus*, bequeathed some incunabula and a good group of books on Chinese and Japanese art. More recently, between 1963 and 1965, Allan Nevins, of Columbia University, who was Harmsworth Professor of American History in Oxford 1940–41, and again in 1964–65, presented a good working library for the use of holders of that chair, which carries a fellowship at Queen's. In 1966, a gift of books on American history and politics, made in memory of George W. Oakes, an *alumnus*, supplemented this collection, which is housed separately from the College library though it is open to accredited students on application. There is a similarly organised Egyptological library in memory of Thomas Eric Peet (1882–1934), an *alumnus*.

Catalogues

The earliest catalogue is a list of twenty-five manuscripts made in 1378, which was published by Magrath with annotations.[14] The next dates from 1663.[15] Another was compiled about 1822, but the Mason bequest made a new one imperative. With reclassification, the work took Edward

[13] Magrath, ii, 279–80.
[14] Ibid., i, 126–29.
[15] Used by Clarke, 441, but untraced by Magrath, ii, 258, although it is still among the library records.

Edwards, the pioneer of public libraries, six years to complete between 1870 and 1876.[16] This catalogue, written by Edwards himself on slips pasted into guard books, is an interesting historical document. In 1938, after the thinning-out already mentioned, a fresh catalogue on cards was begun, but does not quite cover all the contents of the library yet.[17]

[16]   W. A. Munford, *Edward Edwards, 1812–1886* (1963), 178–84.
[17]   I am much indebted to Mr. R. Drummond-Hay, formerly Assistant-Librarian, for letting me use his unpublished account of this library.

REGENT'S PARK COLLEGE
1957

Stock: 10,000   Current periodicals: 13   *STC*: 385

Note: The Librarian stresses that this library is not open to the general public and bona fide *scholars must make arrangements with him in advance before they can be admitted.*

Regent's Park College was actually first established at Stepney in 1810 to train ministers for the Baptist churches; from 1856 it was in Regent's Park, London—hence its name today. A house was acquired in Oxford in 1927, with teaching continuing in London as well until 1940, when it was completely transferred to Oxford. Since 1957 it has been recognized by the University as a permanent private hall.[1]

As might be expected, its library is particularly important and valuable for any study connected with the history or thought of Baptists in this country, or any author who was of that denomination. The library is in two sections: the Main Library, chiefly a working collection associated with teaching; and the Angus Library, based on the material gathered by Joseph Angus, Principal 1849–93.

Manuscripts and Archives

The Angus Library includes papers from several prominent Baptists, of whom William Newman (1773–1835) and Daniel Turner (1710–1798) are examples, and also records of some early congregations, like the minute book from 1680 to 1740 of the Glasshouse Yard General Baptist Church in London. There are also the papers of Joseph Angus himself and those of W. T. Whitley (1861–1947), the Baptist bibliographer and historian. Manuscripts are not kept separately from printed books.

The archives of the College are in the custody of the Librarian.

Printed books

The richness of the Angus Library lies in the extensive collections of the writings of Baptist authors made by Joseph Angus (1816–1902), whose antiquarian tastes saved much that was not preserved elsewhere.[2] Since

---

[1] R. E. Cooper, *From Stepney to St. Giles'; the Story of Regent's Park College, 1810–1960* (1960), 35, 60, 87, 92, 105.

[2] W. T. Whitley, *A Baptist Bibliography; being a Register of the Chief Materials for Baptist History, whether in Manuscript or in Print, Preserved in Great Britain, Ireland and the Colonies.* Vol. 1: 1526–1776 (1916), v.

his death, it has been added to considerably and kept up to date to some extent. It is particularly good for seventeenth-century authors, locally-printed Baptist writings and histories of all periods. There is also a fair series of editions of John Bunyan. Other principals besides Angus have bequeathed books to the College: a predecessor, William Newman, Principal 1811–26, did so and more recently G. P. Gould (1848–1921), Principal 1896–1920, as well as the Old Testament scholar, H. Wheeler Robinson (1872–1945), Principal 1920–42.

Besides the wealth of material for Baptist history, there is a small collection of grammars, dictionaries and Bible translations in various Indian languages prepared by William Carey (1761–1834), the first Baptist missionary to India, founder of the Serampore Mission and Press, who became a distinguished orientalist.

There are also three Baptist church libraries on deposit at the College, of interest since so few nonconformist libraries are still in existence compared with the numerous Anglican survivors. That belonging to the church at Bourton-on-the-Water, Glos., consists basically of the books of Benjamin Beddome (1717–1798), the hymn-writer, minister there from 1740 to 1789, with a very few additions; it contains mainly theological works of the seventeenth and eighteenth centuries, some very scarce.

The library deposited by the Baptist Church at Abingdon, Berks., has many eighteenth-century theological and patristic works collected by Harding Tomkins, of Hackney, and his son-in-law, William Tomkins, of Abingdon, presented in 1755,[3] supplemented by a legacy of 1778. In the late nineteenth century, a small number of books were added from the bequest of the Rev. E. S. Pryce.

The third deposited library, that from the Broughton Baptist Church, Hants., has a surprisingly wide range of subject matter. It would seem that this collection of mainly eighteenth-century books was intended more as a general library for all members rather than a specialist library for the minister's use, as at Bourton and Abingdon. Although not very numerous, the Broughton books include literary and historical works as well as devotional and theological ones; many volumes are in the original boards in which they were first published.

Catalogues

A catalogue of the Angus Library was published in 1908;[4] it leaves something to be desired by modern standards. An interleaved, annotated and corrected copy is kept in the library.

The Angus Library is one of the main locations cited in W. T. Whitley's *Baptist Bibliography*[5] covering works published between 1526 and 1837;

[3]   E. A. Payne, *The Baptists of Berkshire through three Centuries* (1951), 76.
[4]   *Catalogue of the Books, Pamphlets & Manuscripts in the Angus Library at Regent's Park College, London.*
[5]   See note 2 above; a second volume, covering 1777–1837, was published in 1922.

it is also occasionally quoted as a location in the complementary bibliography by Edward C. Starr,[6] bringing the period covered to the present day. G. P. Gould's books, bequeathed in 1921, are also mentioned by Whitley.

There is a card catalogue of the Main Library.

[6] Edward C. Starr, *A Baptist Bibliography; being a Register of Printed Material by and about Baptists, including Works Written against the Baptists.* Pt. A–, 1947– *in progress.*

ST. ANNE'S COLLEGE
1879

Stock: 47,000   Current periodicals: 88   Incunabula: 6   *STC*: 26
Pre-1641 foreign books: 26   Wing: 28

St. Anne's College began in 1879 as the Society of Oxford Home-Students, changing its name to St. Anne's Society in 1942 and achieving collegiate status in 1952. At first there were no books, but from 1895 members made use of the Nettleship Library, founded as a result of a gift from the widow of Henry Nettleship (1829–1893), sometime Corpus Professor of Latin, of part of his collection of classical books, and designed as an inter-collegiate women's library. Money from a memorial fund was also made available. By 1920 it had grown to about 8,000 volumes. Two years later, in 1922, Mrs. Emily Geldart gave the law books of her late husband, William Martin Geldart (1870–1922), Vinerian Professor of Law 1909–22, for a women's law library, together with a capital sum. In the years between the two wars, though, the other four women's colleges were building up collections of their own so that their students had less need for an inter-collegiate library, while the Home-Students were also acquiring books. Consequently in 1934, the other four colleges relinquished their claims to the Nettleship Library, by now containing about 14,000 volumes, and a library building was planned for the Home-Students which was opened in 1938.[1] The Geldart Library, devoted to law and open to all Oxford women law students, is now housed in the same building at St. Anne's.

Manuscripts and Archives

The College has a collection of letters concerned with the struggle to obtain the right for women to take degrees, and also the manuscript of Annie M. A. H. Rogers, *Degrees by Degrees: the Admission of Oxford Women Students*, published in 1938. These, with the College archives, are in the custody of the Librarian.

Printed Books

Although primarily a working collection, a certain number of older and rarer books has accumulated over the years, mainly from gifts and bequests. Grace Hadow (Principal 1929–40) wisely wrote in 1930: 'I will *not* refuse fifteenth-century printing. It is very good for our young to

---

[1] Full details of the history of this library can be found in Ruth F. Butler, *A History of St. Anne's Society, II: 1921–1946* (1949), 53–64.

feel they own such things', when offered by the mother of an old student a few incunabula and sixteenth-century books from the library of Henry Rose Barrow.[2] An interesting collection of books concerned with European literature, including Russian, came with the bequest of Marianne Cécile Gabrielle Hugon (1881–1952), a former tutor in modern languages whose father had taught at the Lycée in Moscow.[3] St. Anne's was given the first chance to make a selection of books under the terms of the will of Claude Jenkins (1877–1959), Canon of Christ Church and sometime Professor of Ecclesiastical History—a totally unexpected bequest; about 7,000 volumes were taken, including some early printed items and much theology and church history.[4] A number of books on art and archaeology, some rare, came in the bequest of the Hon. Eleanor Plumer (1885–1967), Principal 1940–53.[5]

There is a name catalogue on cards; an amended version of the Dewey classification scheme is used.

[2] Ibid., 55 note 2.
[3] *The Ship*, no. 42 (1952), 25–8.
[4] Ibid., no. 49 (1959), 14–15; other beneficiaries were, in order of choice, St. Peter's College, Pusey House, St. Catherine's College, the Union Society and Oxford City Libraries.
[5] Ibid., no. 57 (1967), 7.

ST. ANTONY'S COLLEGE

1948

Stock: 60,000   Current periodicals: 300

St. Antony's, a graduate college, was founded in 1948 through the muni-ficence of M. Antonin Besse, attaining full collegiate status in 1963. The College is chiefly concerned with the study of, and research into, the history, political and international, from the early nineteenth century to the present day, of Western Europe, Russia and Eastern Europe, the Far and Middle East, and Latin America. The library, planned to serve these subjects, is divided into several parts on a regional basis. The main acquisitions and developments have been described regularly since 1963 in the *College Record*. It is open to those who are not members of the College on suitable introduction.

The *Main Library* contains the general collection and works on Eastern Europe. It includes some important groups, such as printed books on modern German history presented by Sir John Wheeler-Bennett, fellow 1950–57; papers connected with the occupation of the Rhineland in the 1920s from Major-General Sir Neill Malcolm (1869–1953);[1] microfilms of the archives of the German Auswärtiges Amt between 1870 and 1920; and photocopies of the Italian diplomatic files from the Foreign Office, *c.* 1920–1940, on loan from that Office. The Library of Congress classi-fication scheme has been adopted and there is a card catalogue.

The *Far East Centre* mainly has works in modern European languages, but it also houses a good collection of Chinese newspapers up to about 1966. There is a separate card catalogue.

The *Latin American Centre* concentrates on the basic twentieth-century history, politics and sociology of its area, leaving earlier periods to be cared for by the Bodleian, and literature by the Taylor Institution. The Centre houses the union card catalogue for Latin Americana in Oxford.

The *Middle East Centre* is concerned with the history, politics, economics and literature of all Arabic-speaking countries, including North Africa, Israel, Iran and Turkey, and to a lesser extent, Afghanistan and Ethiopia. A union catalogue exists for books on the modern Middle East in Nuffield College; the Department of Eastern Art, Ashmolean Museum; Rhodes House Library; the School of Geography; the Institutes for Research in Agricultural Economics, of Commonwealth Studies, of Economics and Statistics, of Social Anthropology; and the Oriental Institute; copies are deposited in the Oriental Reading Room of the Bodleian and the Oriental

[1] *College Record 1964–65*, 7.

Institute as well as this Centre. Since January 1964 combined accessions lists of works in European languages, including Turkish, received in the libraries mentioned, and all Arabic works received by the Middle East Centre, have been issued at intervals. *Arabic Periodicals in Oxford: a Union List*. Compiled by D. Hopwood [1968], lists Arabic periodicals held in the Centre, the Oriental Institute and the Bodleian.

Besides printed material the Centre also collects manuscripts in the form of private papers on British policy and administration in the Middle East, mainly dealing with the twentieth century; a small number of earlier periods is also held.[2] The papers of the Ionian Bank Ltd., active in the Middle East, have been deposited in the College library. Details are sent to the National Register of Archives which has, since 1966, put them into its *Lists of Accessions to Repositories*. The Centre also maintains an index of similar papers in other British and foreign institutions.

The *Russian Centre*,[3] besides works on the recent history of Russia and Eastern Europe, also has a loan collection of Slavonic books containing some early printed works as well as a quantity of Soviet literature of the twentieth century. A representative collection of Soviet and East European newspapers and journals is maintained. There is a separate catalogue.

The *Iberian Centre* has a small working library relevant to its speciality.

[2]   Described, together with the Colonial Records Project at Rhodes House, in 'Private Thoughts on Public Policy', *Times Literary Supplement*, 19 May 1966, 464.

[3]   Anne Abley, 'The Russian and East European Library at St. Antony's College, Oxford', *Solanus*, iv (1969), 4; J. S. G. Simmons, 'Slavica Tayloriana Oxoniensia', *Cahiers du Monde Russe et Soviétique*, x (1969), 537–8.

ST. EDMUND HALL
Thirteenth century

Stock: 20,000   Current periodicals: 40   Incunabula: 3   *STC*: 150

St. Edmund Hall is the last and oldest of those halls which originally provided accommodation for scholars in Oxford. It has had a continuous existence as an academic hall since the thirteenth century, with Oseney Abbey as its landlord until the Reformation and, thereafter, Queen's College, which acquired the site and buildings in 1553, and continued to nominate principals until 1937. It was then controlled by the Principal and Trustees for twenty years before receiving a Royal Charter in 1957 as an independent college, though keeping its original name.

Apart from a fourteenth-century manuscript rather doubtfully inscribed in the seventeenth as a gift to the Hall from Henry VIII in 1422 (!),[1] there is no record of a library during the first four hundred years of its existence;[2] lack of means and liability to close at any time, as in other halls, prevented the acquisition of books. Thomas Tullie, Principal 1658–76, began forming a collection in the reign of Charles II about the time that Henry Wilkinson was doing the same at Magdalen Hall. Tullie adopted the custom, frequently found in Oxford, of upper commoners making a leaving present of plate or books. The foundation stone of the present chapel and library building was laid in 1680 and finished some ten years later. It is believed to be the first college library to have book cases backed against the wall and the last one to be built on the wall-system with the idea of chaining the books;[3] the chains were removed about 1760. Only the gallery now remains of the original furnishings.[4]

In 1685, during the principalship of Stephen Penton, a finely bound Benefactors' Register was purchased, similar to those at Magdalen and University Colleges;[5] gifts received from 1659 were copied into it and it

---

[1]   Now Bodleian Library, MS. Rawl. C.900; N. R. Ker, *Medieval Libraries of Great Britain*, 2nd ed. (1964), 149.

[2]   The architectural history of the library, how the collection was built up and details of benefactors are fully described in A. B. Emden, *An Account of the Chapel and Library Building, St. Edmund Hall, Oxford* (1932), from which much of this account is taken. Dr. Emden omits to mention his own numerous gifts, but a large number presented on his retirement are listed in *St. Edmund Hall Magazine*, vi, no. 1 (1951), 11–13. The present account owes much, also, to his personal interest.

[3]   Streeter, 74. The cases erected in 1739 at Lincoln College stood out from the walls.

[4]   Emden, 23–4.

[5]   Emden, 13–4 and Plates II–III; *Fine Bindings, 1500–1700*, p. 128, no. 226; also illustrated in *St. Edmund Hall Magazine*, i, no. 4 (1922/3), 15–6, and *The Library*, 5th series, x (1955), Plate IV.

was used until 1861; a new book was acquired in 1926.[6] Occasional book purchases are entered in the Principal's Leiger Book from 1685 onwards. Penton's successor, John Mill, Principal 1685–1707, instituted the practice of collecting 20s. library money from every member of the Hall; it is interesting to note that the library was open to all, whether graduates or not, unlike most college libraries at this time. A borrowers' register covering 1666 to 1674[7] and others for 1838–1848[8] and 1867–1881 have survived. For over fifty years after Mill's death in 1707 few books were either presented or bought, but after George Dixon became Principal in 1760, the chains were removed and several bequests received. Growth continued slowly during the nineteenth century. In 1920 modern books were moved to a ground floor room at the bottom of No. 1 Staircase, the following year to the second floor of the Principal's lodgings, and in 1927 to quarters in the new building (now the Emden Room). This was used until 1958, when another move took the books to the Besse Building. Finally, the conversion of the disused church of St. Peter-in-the-East into a modern library was completed in 1970. The original library building continues to house the older volumes and the earlier *alumnus* collection, a section considerably augmented by Dr. A. B. Emden both during his Principalship 1929–51 and since.

Major acquisitions and works by former members have been mentioned since 1923 in the annual *St. Edmund Hall Magazine*, which also frequently contains articles on manuscripts and books, notably between 1926 and 1951.

Manuscripts

St. Edmund Hall owns a small and rather miscellaneous group of manuscripts, now all deposited in the Bodleian[9] where a typescript catalogue is kept in Duke Humfrey. Apart from one or two pre-Reformation ones, such as a *Tractatus de sacramentis* of *c.* 1435, the majority are connected with former members. There are the papers of Henry Dodwell (1641–1711), sometime Camden Professor of History,[10] a non-juror like his friend Thomas Hearne (1678–1735), who lived in the Hall for so long; one of his letters is here.[11] Of antiquarian interest are five volumes of collections towards the history of the Diocese of Carlisle made by Hugh Todd (1658–1758), presented by Dr. A. B. Emden. There are various theological notebooks of John Mill, Principal 1685–1707, and some documents concerned with Thomas Tullie, Principal 1658–76.[12]

---

6   Emden, 38–9.
7   Ibid., 7–8. Among the papers of White Kennet, Vice-Principal 1691–95, and now British Museum, Lansdowne MS. 697.
8   Emden, 30.
9   *St. Edmund Hall Magazine*, vii, no. 5 (1960), 7; the manuscripts mentioned there as remaining at the Hall have since been deposited in the Bodleian.
10 Ibid., iii, no. 3 (1933), 23–9; no. 4 (1934), 87–93.
11 Ibid., vii, no. 5 (1960), 8.
12 Ibid., v, no. 3/8 (1948), 40.

Of eighteenth-century interest are some papers of Peter Newcome, Vicar of Hackney 1703–38, a private account book kept between 1746 and 1758 by Thomas Secker (1693–1768) while Bishop of Oxford,[13] and some papers collected by John Higson, then Vice-Principal, concerned with the expulsion of six students for Methodist practices in 1768.[14]

The vice-principals of the nineteenth century are well represented; there are twenty-three volumes of diaries for 1803–1805 and 1820–1855 of John Hill (1787–1855), Vice-Principal 1812–51;[15] and letters and papers of H. P. Liddon (1829–1862).[16] Some notebooks and papers of Edward Moore (1835–1916), Principal 1864–1913, well known as a Dante scholar, are also here. The remarkable collection of transcripts of documents relating to the Hall or its members made by Dr. A. B. Emden was presented by him to the Hall on his retirement in 1951[17] and not transferred to the Bodleian with the other manuscripts.

Archives

Since halls were either under the control of a college or the private concern of the Principal, it is not surprising that the records at St. Edmund Hall are meagre, in spite of its antiquity. The buttery books go back to 1694 (with gaps) and there are good runs of minutes and accounts of various Hall societies, the most elusive type of record: both the Boat Club and the Debating Society go back to 1869, for instance.[18] The Books of Benefactors have already been mentioned; there is also the Principal's Leiger Book begun in 1684 and covering all aspects of Hall affairs, including finance. Dr. Emden's transcripts contain much archival material now in the Bodleian, British Museum and Queen's College.

A handlist has been prepared for the History of the University Project. The archives are in the custody of the Librarian.

Printed Books

Apart from some short runs of late eighteenth-century newspapers, such as *The London Chronicle*, 1757–1759, and *Lloyd's Evening Post* for 1765–1780, the interest of the printed books lies in the large number of works by past and present members of the Hall. Since Thomas Tullie began to form a library in the late seventeenth century, there have been many donations, mostly single items or small in number, from *alumni*; Robert Burton, for example, gave a few medical books in 1697 which remain, and Nathaniel Ellison presented some theological works while he was Archdeacon of

---

13   Ibid., i, no. 2 (1921), 13.

14   Ibid., i, no. 6 (1925), 11.

15   Ibid., i, no. 6 (1925), 11–12; these diaries were used in J. S. Reynolds, *The Evangelicals at Oxford, 1735–1871* (1953).

16   *St. Edmund Hall Magazine*, i, no. 6 (1925), 11; ii, no. 1 (1926), 6; ii, no. 5 (1930), 14–5.

17   Ibid., vi, no. 1 (1951), 12–3.

18   Ibid., ii, no. 2 (1927), 18; iii, no. 4 (1934), 32–3; iii, no. 5 (1935), 13, for lists of club records.

Stafford in the 1680s. The initial impetus under Tullie, Penton, and Mill from the reigns of Charles II to Anne, however, died down and the collection did not grow until George Dixon became Principal in 1760, notably with gifts from John Berriman (1691–1768) in 1762 and 1765, and Dixon himself in 1780 and 1781. Isaac Crouch, Vice-Principal 1783–1807, was also a conspicuous donor. A bequest of £100 to be expended on books came from Lawrence Heapy in 1842, and a good general library was given by Tullie Cornthwaite in 1847, which included books once owned by the former Principal, Thomas Tullie, and other members of his family. Some theology was bequeathed by William Borrows (1781–1852).

The acquisitions so far mentioned must have built up a small, general library with an emphasis on theology, suitable for the training and reading of the Anglican clergymen so many members became. During the present century, especially since 1919 when Dr. A. B. Emden began his connection with the Hall, this emphasis has shifted, so far as the older books are concerned, to works written by, about, or formerly owned by Aularians, or having something to do with the Hall's history. Dr. Emden has presented a great many of these volumes himself. Thus there are good sequences of the editions of John Oldham (1653–1683)[19] and Sir Richard Blackmore (d. 1729), both poets; several books once owned by Thomas Hearne, and most of the texts he edited; a set of Edward Chamberlayne's *Angliae notitia*; a copy of Lewis Carroll's *Game of Logic* (1887) inscribed by the author to H. P. Liddon, and so forth. This *alumnus* collection is divided into pre- and post-1900 sections, and in the notable absence of statistics from Oxford libraries, it is interesting that in the latter part in 1968 there were 683 items by 201 authors.[20] In these two sections lies the strength of the library for research. Works by Hall men have been listed since 1923 in the annual *St. Edmund Hall Magazine*

Naturally, there have been gifts and bequests of other types of books as well, suitable for current work in the new library. For example, two hundred volumes came from the library of William Sanday (1843–1920) in 1921;[21] law reports from J. Serrell Watts in 1957;[22] education books from W. L. Freeman in 1942.[23] H. H. Williams (1872–1961), Principal 1913–20 and later Bishop of Carlisle, also gave a good collection to the Hall.[24]

A set of the 'Littlemore' *Lives of the English Saints* formerly owned by the publisher, James Toovey, was presented in 1933;[25] letters received from contributors (including fifteen from J. H. Newman) had been

[19]   Ibid., iii, no. 3 (1933), 20.
[20]   Ibid., ix, no. 3 (1968), 12.
[21]   Ibid., i, no. 2 (1921), 8.
[22]   Ibid., vii, no. 2 (1957), 4–5.
[23]   Ibid., v, no. 2 (1942), 32.
[24]   Ibid., vi, no. 3 (1951), 10–11; viii, no. 2 (1962), 5.
[25]   Ibid., iii, no. 3 (1933), 19–20, 78–88; the letters have now been extracted and kept separately.

inserted in it, solving some of the problems of authorship of certain *Lives*. Finally, it is appropriate that seven volumes formerly in the parish library of St. Peter-in-the-East, recently converted into the Hall library, should have been added to the Hall's books.[26]

Catalogues

The earliest surviving catalogue was made about 1699 by Thomas Hearne; there are both rough and fair copies[27] showing 421 volumes. By 1774–76, when the next one was compiled, the number had risen to nearly a thousand and growth was such that another was required in 1792, which lasted until 1843,[28] when a two-volume catalogue was acquired. Re-cataloguing was begun again in 1925[29] and in 1939 it was recatalogued on to cards.[30] Separate sequences are maintained for the two parts of the library.

26   Ibid., ix, no. 4 (1969), 12–3.
27   Now Bodleian Library MSS. Rawl. D.39, ff. 13–47 and C.851, ff. 87–115; see also Emden, 29.
28   Emden, 30.
29   *St. Edmund Hall Magazine*, i, no. 6 (1925), 13.
30   Ibid., iv, no. 4 (1939), 13–4.

ST. HILDA'S COLLEGE
1893

Stock: 23,000   Incunabula: 2   *STC*: 22   Pre-1641 foreign books: 31
Wing: 32

St. Hilda's was founded in 1893 by Dorothea Beale, Principal of Cheltenham Ladies College, and the two establishments have maintained their
association.[1] The library is principally a working collection, and although
there have been numerous and generous donors from the beginning, no
really outstanding gifts or bequests have been received. There are, however, a few manuscripts and early printed books of some interest. Important
accessions can be found listed in the *Annual Report* from 1900 to the
present day.

Manuscripts and Archives

A fifteenth-century Dutch Book of Hours was bequeathed by Helen Smith
in 1944. Mrs. E. Dorothy Chambers presented in 1952 several manuscripts written by her mother, Emily Sophia Capper, and Graily Hewitt
in the early part of this century, besides an interesting scrap-book compiled by the Capper family, 1887–1943. There are also several other
miscellaneous manuscripts; a list is available in Duke Humfrey.

The archives, in the custody of the Librarian, include correspondence
and papers concerning the early struggles for women's education in
Oxford in addition to strictly college affairs. There is a card index.

Printed Books

There is a small collection of older printed books of a miscellaneous
nature, built up mainly by gifts from members of the College. Miss Beale
bequeathed several sixteenth- and seventeenth-century Italian books,
while there are also a few French and English books of the sixteenth,
seventeenth and eighteenth centuries. The library is classified by a modified Dewey system and there is a card catalogue.

[1]   K. Major, 'St. Hilda's College', *American Oxonian*, xlviii (1961), 57–60.

ST. HUGH'S COLLEGE
1886

Stock: 36,000   Incunabula: 25   *STC*: 40   Pre-1641 foreign books: 65
Wing: 47

Dame Elizabeth Wordsworth, the first Principal of Lady Margaret Hall, founded St. Hugh's Hall in 1886 as a women's place of residence with lower fees than elsewhere. At first in Norham Gardens where the library was merely a few text-books, the College moved to its present site in 1916. A library building was opened in 1936 which now houses more than 30,000 volumes.[1] Although St. Hugh's had no endowments before 1930 and is still one of the poorer colleges, the generosity of friends and former members has built up a library well worth consulting, especially by those interested in ornithology or French history and literature of the eighteenth century.

Printed Books

The library was primarily a working collection for the first forty years of the College's history, but important gifts or bequests began to arrive in the 1920s. In 1926, Mary, a daughter of R. W. Church, sometime Dean of St. Paul's, bequeathed her books, including several fifteenth- and sixteenth-century Italian works; most of these had formerly been owned by her father, much of whose library is now at Oriel.

In 1937, three most interesting collections were received. Eleanor Frances Jourdain, Principal from 1915 to her death in 1924, had left her extensive library to the College, but it was not received until this year; it is especially strong in eighteenth-century French literature and history. Canon B. H. Streeter (1874–1937), Provost of Queen's, also bequeathed a small group of early English printed books, mainly Bibles and theology. The third is one of the most unusual collections to be found in an Oxford college; it consists of the library of Mary du Caurroy (1865–1937), wife of the eleventh Duke of Bedford, and reflects the donor's great interest in ornithology, natural history and the fine arts. There are numerous large paper copies, limited or specially illustrated editions, mostly published during the Duchess's lifetime, and all in fine bindings of the period.

In more recent years, Cecilia Ady, the historian, a former fellow, bequeathed in 1958 about 400 volumes chiefly concerned with Italian history and art, including some early printed books. As a parting gift on her

---

[1]   *VCH Oxon.* iii, 347–8; E. E. S. Procter, 'St. Hugh's College', *American Oxonian*, xlviii (1961), 10–15.

retirement as Principal, Miss E. E. S. Procter gave, in 1962, a collection of historical works that includes several interesting seventeenth-century items.

As in other college libraries, there have been many benefactresses amongst whom *alumnae* figure prominently, and though it is invidious to single out individuals, mention should perhaps be made of two who have made gifts of books of archaeological and historical interest: Mrs. Winifred Haverfield, wife of F. J. Haverfield, the Romano-British historian, and Dr. Joan Evans, an honorary fellow.

Catalogues

There is a name catalogue on cards, covering all the library; there is also a separate card catalogue of the Jourdain bequest. The whole library is classified by Dewey, irrespective of provenance.

ST. JOHN'S COLLEGE
1555

Stock: 60,000   Current periodicals: 80   Incunabula: 145   *STC*: 1,077

Sir Thomas White, a prosperous merchant taylor, received Letters Patent for the foundation of St. John's College in May 1555, shortly after Sir Thomas Pope had obtained the old Durham College for Trinity. The site of St. Bernard's College in St. Giles was acquired from Christ Church, and the College began in earnest in 1557.

There was no library room as such at first,[1] but about one hundred and fifty volumes were presented by the Founder and his friends in 1555.[2] The first statutes laid down, as did those of older colleges such as Merton and All Souls,[3] that there should be an annual distribution of books among members, with borrowings recorded.[4] A special room for a library was in existence by 1583, however, when books were chained,[5] a practice that continued until at least 1744.[6] Between 1596 and 1598 a new library was built, on a·scale then surpassing all other Oxford colleges, and probably the first (after Merton's adaptation) to substitute stalls for lectern desks, closely followed by All Souls and Queen's:[7] this building and its fittings are still in everyday use.

The period between the completion of this room and the Civil War was one of great expansion in the College generally, especially during and after Laud's Presidency from 1615 to 1621. Benefactions to the library also came in such quantity that more space was needed. A north wing was added, as part of Laud's plan for enlarging the College, destined for manuscripts, small-sized books and mathematical instruments; the original cases were replaced in 1837.[8] Undergraduates, previously only allowed to

---

[1]   A general account of the history of the library can be found in W. H. Hutton, *S. John Baptist College* (College Histories, 1898), 235–44. Details of the building erected 1596–97 are in W. H. Stevenson and H. E. Salter, *The Early History of St. John's College, Oxford* (Oxford Historical Society, new series, i, 1939,) 293–302; its fittings are described in Streeter, 183–98. W. C. Costin, *The History of St. John's College, Oxford, 1598–1860* (Oxford Historical Society, new series xii, 1958), gives particulars of the library in each chronological section. *VCH Oxon.* iii, 258 only gives a brief outline based on the first two works listed above.

[2]   Ker, 486–7.

[3]   See pp. 3, 80.

[4]   Ker, 465.

[5]   Ker, 486, 511; Stevenson and Salter, 754.

[6]   Stevenson and Salter, 302.

[7]   Ker, 510–1; Streeter, 127–8.

[8]   Costin, 76–8.

borrow, were admitted as readers in 1933. Although there have been many large bequests and gifts since, these two rooms still form the main core of the library, though various store- and reading-rooms have been taken into use.

The office of *Custos Bibliothecae* was instituted in 1603.[9] A library Register was begun about 1620; its last entry is dated 1756.

Though a few printed books were purchased between 1618 and 1638,[10] the main source of accessions seems to have been gifts and bequests. A most interesting letter, written in February 1598, from Sir Thomas Tresham (d. 1605), a Recusant and benefactor, to John Case, has survived[11] in which suitable books for the College library are listed (mainly patristics), and how they should be bound in leather with cloth coverings. As elsewhere, the custom of each commoner presenting a book was observed; seventeen did so between 1620 and 1635.[12] From 1658, a more regular source was laid down by decree: £10 was allocated from rents for book purchases, while each gentleman commoner had to give £1 and each commoner 10s., or books of equivalent value, on admission. Later, in 1718, it was ordered that £1 should be given to the library by everyone who took a degree or moved to a higher table in hall. The College records contain numerous entries for library expenditure—book purchases, chains, binding and cataloguing during the later seventeenth and eighteenth centuries,[13] besides references to library discipline and management, including a detailed set of regulations in 1736.[14] During the nineteenth century the library fell into some confusion, not helped by a committee replacing the librarian from 1871 to 1883, but it was later re-organized by A. T. S. Goodricke.[15]

In 1612, about 250 volumes of duplicates were sold and the donors' names inscribed in them transferred to the books bought with the proceeds; there was further discarding in at least 1636 and 1731.[16]

Manuscripts

Coxe[17] lists 212 manuscripts, to which a typescript supplement has been compiled describing nos. 213 to 310; a copy is available in Duke Humfrey. A list of the majority of donors can be found in Coxe and Hutton;[18] the principal ones during the sixteenth and seventeenth centuries were the Founder and his brother, John White; Richard Butler (d. 1612), Archdeacon of Northampton; Sir William Paddy (of whom more later);

9   Ibid., 72.
10 Ibid., 74.
11 Ker, 512–5.
12 Costin, 74.
13 Ibid., 72–8, 115–6, 148–9, 177–81, 191, 238.
14 Ibid., 191.
15 Hutton, 236–7.
16 Costin, 73, 76, 191.
17 Pt. II, section 6.
18 238–9.

John Stonor of North Stoke, Oxon.; and Archbishop Laud, while there were also numerous small gifts from individuals. These earlier accessions include a certain number of medieval manuscripts from despoiled monasteries and contemporary theological writings.

During the eighteenth century there were further donations and bequests. Among the more interesting are a gift from Richard Rawlinson (1690–1755), the antiquary and non-juror, of a number of transcripts of older documents[19] and the bequest from John Pointer (1668–1754), a fellow of Merton, of his collection of curiosities which included some scrap-books with a few older manuscripts.

Nineteenth- and twentieth-century manuscripts are not so numerous; there are seven letters of Jane Austen; a commonplace book and some correspondence of W. H. Hutton (1860–1930), the historian, sometime fellow; and various papers of W. H. Stevenson (1858–1924), connected with his archival and historical studies. There are also a few letters of A. E. Housman (1859–1936), the poet and classical scholar, an *alumnus*.

Archives

H. T. Riley, in his account of the College archives prepared for the Historical Manuscripts Commission,[20] misleadingly wrote that they were 'comparatively few in number', dealing briefly with the numerous account books, registers and deeds, and devoting most of his space to quotations from Richard Rawlinson's transcripts, mentioned above. A complete catalogue is in preparation, but a summary guide, dated 1961, has been made available through the National Register of Archives and a copy is in Duke Humfrey. A handlist has also been made for the History of the University Project.

Though documents relating to St. Bernard's College are few, there are a great many deeds concerned with the site of the present College and its extensive Oxford properties. Sixteenth-century records of St. John's College have been published in W. H. Stevenson and H. E. Salter's *Early History* already referred to.[21] The Chapel register covering the years 1695 to 1752 has also been printed,[22] as well as a list of sixty-six manuscript maps of places in ten counties, the majority in Berkshire and Oxfordshire.[23] Some College accounts were used by J. E. Thorold Rogers in his *History of Agriculture and Prices in England*.[24] Annual accounts survive from 1568, though bursarial records begin in 1562 and battels only in 1598. There are some minute books of undergraduate societies including an Essay Club from 1880. William Derham, President 1748 to 1757, filled

---

[19] The bulk of his collections is in the Bodleian.
[20] *4th Report* (1874), 465–8.
[21] See Note 1 above.
[22] Edited by E. Jermyn from Rawlinson's transcript now in the Bodleian, in *Miscellanea Genealogica et Heraldica*, series V, vol. 1 (1914), 121–8.
[23] H. M. Colvin, 'Manuscript maps belonging to St. John's College', *Oxoniensia*, xv (1950), 92–103.
[24] 7 vols., 1866–1907; see Preface to vol. 6, p. v.

several notebooks with information about various properties gathered from the documents and these can be used as a guide to the archives at that date.

The archives are in the custody of the Librarian.

Printed Books

St. John's was given printed books by the Founder, Sir Thomas White, and his friends although there was no special room set apart as a library for over twenty years. Among the principal early donors were Henry Cole (c. 1500–1580), sometime Dean of St. Paul's; Gabriel Dunne (d. 1558), a Canon of St. Paul's; and Thomas Paynell (d. 1564), a former Canon of Merton, all of whom had Roman Catholic sympathies, so giving St. John's books a different character from that of its near contemporary, Trinity, with its Protestant tendencies. The gifts of Cole and Dunne were strong in books in Greek—the Fathers, Homer, Plato, Aristotle and so on—rivalling Corpus;[25] few editions of the Fathers in Latin were received before the end of the sixteenth century. About a third of these two benefactions were discarded as duplicates in 1612 and later. Paynell[26] bequeathed about 150 volumes, of which over a hundred survive in the College today, at once doubling the number of printed books; the subjects covered were law, theology, philosophy, grammar and a little patristics, resembling to some extent David Pole's books at All Souls. Paynell had the habit of writing on many of his books *Sum Thomæ Paynelli et amicorum*, like Grolier and other sixteenth-century book collectors.

After the completion of the new library in 1598, there was an important series of donations. In 1600 were received 256 volumes from a bequest by Henry Price, a former fellow and librarian, containing much Latin literature, duplicating texts already held, so much of it was sold in 1612.[27] About the same time, about eighty volumes of Protestant theology, hitherto conspicuously lacking, were given by Robert Lee and Laurence Holiday, two London mercers, while Sir Thomas Ducket presented about thirty volumes of historical works, some of which were in Italian. In 1600 also Sir William Paddy (1554–1635), an *alumnus* who became physician to James I, began giving books to the library, a practice that continued for the rest of his life; in all about 800 books came in this way,[28] mainly of medical interest though other subjects, such as geography, are well represented. Paddy can be regarded as St. John's most substantial benefactor in the early seventeenth century.

Sir Thomas Tresham not only suggested suitable books for acquisition, as already mentioned, but also gave, between 1598 and his death in 1605,

---

[25] Ker, 503.
[26] Stevenson and Salter, 133–6; Ker, 503.
[27] Costin, 71–2; Ker, 512.
[28] Costin, 38; Ker, 512; Carleton B. Chapman, 'A Comment on the Holdings in Old Medicine and Science in St. John's College Library', *American Oxonian*, li (1964), 116–7.

copies of all those he named,[29] chiefly the Fathers of the Church, totalling nearly 200, most of which are still on the shelves.

Laud, who did so much for St. John's as a whole, presented more to the Bodleian than to his College in the way of books and manuscripts, but various gifts are recorded from him between 1620 and 1642, such as his interleaved and annotated Vulgate, printed at Lyons in 1566, and some Orientalia.[30] At this period the bulk of accessions were donations from individuals of books connected with the curriculum; during the Commonwealth period, purchases became the main source.

After the Restoration, the flow of benefactions from *alumni* was resumed. Griffin Higgs (1589–1659), who left his own books to Merton, bequeathed £50 for buying books to St. John's, mostly spent on theology and Orientalia. The library of Archbishop Juxon (1582–1663), Laud's successor as President 1621–33, and later Archbishop of Canterbury, was given to the College by his son, Sir William Juxon, in 1664, entailing the purchase of half a gross of chains;[31] it was mostly theological and historical in character. There were also monetary gifts to the library as well as smaller collections presented during this period: Charles Perrott (d. 1686), for instance, a lawyer, left a very general collection. During the first half of the eighteenth century, several substantial and important bequests were received. The earliest was that of William Brewster (1665–1715), a prosperous physician in Hereford, who divided his books between All Saints' Church, Hereford, the Bodleian, and his old College; about 200 volumes came to St. John's, over half of them medical and the others concerned with antiquities, travel and the classics.[32] The second bequest, larger and more important, came from another physician, Nathaniel Crynes (1688–1745), sometime fellow and Superior Bedel of Arts, who later practised in his native Coventry. Although the first choice was given to the Bodleian (some of whose duplicates he had bought in 1727[33]), over one and a half thousand volumes were taken by the College, and his books can now be found in all parts of the library. Crynes, like Paddy, belongs to that long line of book-collectors who are also medical men; his books cover many subjects and include many rare works; the eleven Caxtons form perhaps the most outstanding group, but it is invidious to single out particular items in such a collection. Shortly after the Crynes legacy, another general library was bequeathed by Daniel Lombard (1678–1748); it is strong in Italian literature of the seventeenth and eighteenth centuries.

The latter half of the eighteenth century did not see such munificence. The *Museum Pointerianum* which came in 1754, already mentioned for its

[29] Ker, 512–5.
[30] Hutton, 241–2, lists twelve donations, but there are many more.
[31] Costin, 128–9, where it is stated the books were part of the Archbishop's bequest, but the inscriptions state they were presented by Sir William Juxon in 1664.
[32] F. C. Morgan, 'Dr. William Brewster of Hereford (1665–1715)', *Medical History*, viii (1964), 141–2.
[33] W. D. Macray, *Annals of the Bodleian Library, Oxford*. 2nd ed. (1890), 220–1.

manuscripts, has some rare printed scraps. Books were also received from Richard Rawlinson, though the bulk of his great collections went to the Bodleian. Later in the century, a general private library, strong in early medical works and the literature of spas, was given by John Merrick (1705–1764?), a former fellow who practised as a physician in Reading.

Since this period, although there have been numerous small benefactions, the number of large ones has diminished. R. M. Casberd (1772–1842), sometime a Member of Parliament and a judge, gave some law books in 1832. An excellent collection of Spanish literature, made by H. Butler Clarke (1864–1904), sometime Fereday fellow, was acquired through the generosity of W. H. Hutton in 1905; it is now deposited in the Taylor Institution.[34] W. H. Hutton (1860–1930) also bequeathed his own general library to the College; it is particularly useful for works on seventeenth-century history and for English literature from that century to the nineteenth. A recent bequest of Lacy C. Collison-Morley (1875–1958) of Italian literature and books written in English on Italy, strengthens that of Daniel Lombard two hundred years earlier.

There is no *alumnus* collection such as is found in some other college libraries, though St. John's is rich in the writings of former members; the library also has a group of classical books formerly owned by A. E. Housman (1859–1936), some with marginalia.

The strength of St. John's lies particularly in its holdings of books on medicine and the natural sciences,[35] but it is also important for its Orientalia and liturgical works, while its collections of scarce tracts should not be neglected. Fortunately there does not seem to have been much discarding down the centuries, so that the theological, classical and legal books necessary for past curricula are mostly still present.

Catalogues

A catalogue was started in 1604.[36] One made of the Inner Library about 1670 has survived, but this was supplanted by the 1674 Bodleian printed catalogue interleaved and annotated with St. John's holdings. William Derham, President 1748 to 1757, made notes on the College's printed books in the same way as he dealt with its muniments, and these form virtually shelf-lists for the period. During the Presidency of William Holmes, 1728 to 1748, a new catalogue was begun,[37] completed in two volumes in 1761 and used until 1844,[38] when H. O. Coxe, later Bodley's Librarian, was engaged to make a new one. This was done by interleaving and annotating the 1843 Bodleian catalogue, to which new accessions were added until 1925 when the entries were transcribed on to typewritten cards. The current author catalogue is divided into two parts: books printed before 1643, and books printed 1643 to date.

34   St. John's College, [*Annual Report*] *1905–06*, 3.
35   Gunther, 46–7; Chapman, 115–23.
36   Costin, 72.
37   Ibid., 191.
38   A. T. S. Goodricke, *Report on the College Library*, [c. 1889], 1; Costin, 238.

SOMERVILLE COLLEGE
1879

Stock: 70,000   STC: 32

Started as a hall for women students in 1879, Somerville was first styled a college in 1894, receiving official recognition by the University in 1910 and reaching full status in 1960 with the other women's colleges. Gifts of books were made from the beginning and since 1895 a section of each *Annual Report* has been devoted to library matters.[1] From these it will be seen that a fluid policy has always been adopted, with no hesitation about the sale of duplicates or books unwanted at the time. The object has obviously been to try to satisfy the immediate needs of senior and junior members as far as possible, with an emphasis on the latter.[2]

Manuscripts

The manuscript collections are important for those interested in nineteenth-century scientific and literary history; there are three main groups. The first, now deposited in the Bodleian, is concerned with Mary Somerville (1780–1872), a distinguished scientist after whom the College was named, and her physician husband William (1771–1860), which includes their scientific manuscripts and autobiographies together with public correspondence and family letters, as well as some family papers.[3] The Somervilles moved in the leading intellectual circles of the first half of the nineteenth century, so there are many letters from, for instance, Charles Babbage, Michael Faraday, Charles Darwin, George Washington[4] and Sir John Herschel. Typescript calendars are kept in Duke Humfrey.

The second group, the papers of Amelia B. Edwards (1831–1892), the first secretary of the Egypt Exploration Fund, was received in 1907. Miss Edwards wrote novels before becoming an Egyptologist and, moreover, knew many prominent literary contemporaries; among her correspondence can be found letters from Matthew Arnold, Robert Browning, Charles Dickens, Edward Lear, H. W. Longfellow, D. G. Rossetti, and John Ruskin. A group of seventy-two letters written to her between 1878 and 1891 by G. Maspero, the archaeologist, have been published in summary form by Warren R. Dawson.[5] A typescript list is available in Duke Humfrey.

[1]   Muriel St. Clare Byrne and Mrs. Catherine Hope Mansfield, *Somerville College, 1879–1921* [1922], record the chief benefactions up to 1920 on pp. 25–9.

[2]   Dame Janet Vaughan, 'Somerville College', *American Oxonian*, xlvii (1960), 122–5.

[3]   Mrs. E. C. Patterson, 'Mary Somerville', *British Journal for the History of Science*, iv (1969), 311–39.

[4]   Mrs. E. C. Patterson, 'A Washington letter', *BLR*, viii (1970), 201–5.

[5]   'Letters from Maspero to Amelia Edwards', *Journal of Egyptian Archaeology*, xxxiii (1947), 66–89.

The third, and smallest, group is also of literary interest, comprising the manuscripts of Violet Paget (1856–1935), better known as 'Vernon Lee', together with about a hundred letters addressed to her, but these cannot be examined until 1985.

Printed Books

The library at first acquired such books as were necessary for undergraduate studies, spending about £100 a year on acquisitions, so that by 1898 it contained about 6,000 volumes. The earliest important benefaction was received in 1884, when about 500 items from the library of Mark Pattison (1813–1884), selected by his widow and Ingram Bywater, were added. The next, and most significant from the research angle, was the library of John Stuart Mill (1806–1873), presented to the College in 1905 by his step-daughter, Miss Helen Taylor, at the suggestion of John Morley. This is a collection of about 2,000 volumes ranging from Aldines to the writings of Mill himself and is especially interesting for Mill's annotations and notes in his own works.

A large portion of Amelia B. Edwards' printed books came to the College with her papers, already mentioned, in 1907; the subjects are many and wide-ranging, including older volumes such as a second folio Shakespeare and a Chaucer of 1550 besides numerous standard works.

Since these benefactions, there have been many smaller ones, not so spectacular, but helping to build up a strong, working academic library. In 1920 part of the library of Sir William Bousfield (1842–1910), who had been much concerned with education and social work in London, was received, and shortly afterwards about 300 books formerly owned by A. V. Dicey (1834–1922), the legal historian.

The historical library of Maude Violet Clarke, including a number of books printed in the early seventeenth century and scarce monographs, came to the College on her death in 1935. In 1955 a selection of works on diplomatic history from the collections of David Morier (1784–1807) and Sir Robert Morier (1826–1893) was presented by the Hon. Mrs. Cunnack.

In recent years the library has benefited from bequests of books from distinguished *alumnae*; philological works, for instance, came from Mildred Pope (1872–1956); English literature of all periods from Helen Darbishire (1881–1961); and history from Rose Graham (1875–1963), besides more general books from Margery Fry (1874–1958). Recently, a collection of works on the Crusades, medieval French and Italian history has been presented in memory of Lady Woodward, sometime history tutor.

Catalogues

There is an author catalogue on cards. Compilation of a register of donors covering the whole period of the College's existence was begun in 1952.

TRINITY COLLEGE
1555

Incunabula: 21 *STC*: 289

The site and derelict buildings of Durham College, the Oxford house of the Benedictine monks of Durham, were acquired by the Founder, Sir Thomas Pope, a successful civil servant, in February 1555, a few months before Sir Thomas White obtained the nearby St. Bernard's College for St. John's. The idea of both a chained and a circulating collection, as at St. John's, was embodied in the statutes,[1] signed on 30 May 1556, but whereas Trinity had a library immediately, there was none at St. John's till 1583.[2] The Founder gave about a hundred volumes himself and there was a steady stream of gifts during the sixteenth century, but where these books were housed is uncertain.

During the Presidency of Ralph Kettell between 1599 and 1643, the library increased in size through purchases, mostly classics, and gifts of money. A bequest from Edward Hyndmer (d. 1618) enabled the present Old (or Fellows') Library to be fitted with bookcases; the first librarian was appointed in 1629.[3] It is in a part of the east wing of the Chapel (or Durham) Quadrangle, used by the former Durham College as a library in the fifteenth century.[4] Richard Rands, a former fellow who died in 1640, left land to provide £20 annually for Trinity library, and this fund considerably augmented in 1869, continues today.[5] Money for books was also given by William Craven, 1st Earl of Craven (1608–1697), during Kettell's Presidency. After the Restoration, fixed sums were taken for the library from graduation fees.[6] As in several other colleges at this period, a Benefactors' Book, not used exclusively for the library, was started in 1665, the year after Ralph Bathurst became President, and gifts from 1615 were entered in it.[7] Three volumes were used to bring the record intermittently to the mid-nineteenth century and the fourth, in current use, was begun in 1890.[8]

Chaining continued well into the eighteenth century; separate library accounts begin in 1726 and the purchase of chains is recorded in 1727,

1 Cap. 24.
2 Ker, 486.
3 H. E. D. Blakiston, *Trinity College* (College Histories, 1898), 104.
4 *VCH Oxon.* iii, 251.
5 Blakiston, 110–1.
6 Ibid., 166.
7 Ker, 462.
8 Blakiston, 232; Trinity College, Oxford, *Report 1956–8*, 12–3.

1730 and 1733. A note of a College order on the inner cover of an account book dated 1765, in the hand of Thomas Warton, reads: 'The Librarian to review ye unchained Books once a quarter at least'. The word 'unchained' was later scratched out, implying that *all* books should be reviewed, presumably since all chains had been taken off, possibly during 1765 when the room was re-glazed and painted. Some books were being kept in chests as late as 1697 when a separate catalogue of them was made.[9]

During the later eighteenth and early nineteenth centuries, the library went through a period of stagnation; some of the older volumes were sold or thrown away,[10] while in 1807 a quantity of books, presented in the sixteenth and seventeenth centuries, were discovered, greatly decayed, walled up in part of the old Vestry under the southern end of the library, and only some fragments survive.[11]

James Ingram, President from 1824 to his death in 1850, bequeathed his large collection of books, doubling the library's size, so the old book-cases dating from the 1620s were raised to the roof and thereby spoiled.[12] The library has since been little altered physically, and subsequent large benefactions have chiefly been more suitable for the undergraduates' Reading Room. The current Borrowers' Register for the Old Library begins in 1859; another, covering 1783–1801, has survived.

Trinity has the distinction of having provided the earliest recorded separate undergraduates' library; John Harris (1666?–1719), scholar 1684–88, wrote in his manuscript autobiography, owned by the College: 'There was also in this Coll. a very Good Collection of Philos: & Mathem: Books of all kinds as also of the Classicks placed in a room wch we called the *Lower Library*, where every undergraduate had the liberty to go & study as long as he pleased, wch was a mighty advantage to the House, and ought to be imitated by other Colleges'.[13] Queen's had a separate library for taberdars about this time, while Christ Church undergraduates had set up their own coffee house library in 1668, but no other college is known to have made similar provision. This arrangement, however, does not seem to have survived the eighteenth century, and Cardinal Newman when an undergraduate had to have permission to use the Fellows' Library in 1819–20.[14] Another separate undergraduate library was provided in 1877 in a new lecture room,[15] but it was soon too small and an undergraduates' Reading Room was provided during the Presidency of John Percival, 1883–87.[16] A new building was erected between 1925 and 1928 as a memorial to

[9] Streeter, 231–2.
[10] Ker, 486.
[11] Blakiston, 76.
[12] Streeter, 225–30; Blakiston, 216.
[13] Quoted in Blakiston, 173–4, and M. Maclagan, *Trinity College: a Short Guide and History*. Revised ed. (1963), 22.
[14] Blakiston, 220.
[15] Ibid., 238, n. 1.
[16] *VCH Oxon.* iii, 249.

members of the College who fell in the 1914–18 War, and this was enlarged with a gallery in 1954.[17] In recent years also, Trinity and its neighbour Balliol have collaborated in the provision of a joint library for natural sciences for their undergraduates.

Manuscripts

Coxe lists eighty-nine manuscripts; five more have since been added.[18] It is a miscellaneous collection, chiefly of theological texts such as an eleventh-century Augustine, a Boethius of the twelfth and another of the fifteenth century, and the winter part of the Abingdon Missal, dated 1461. There are a few English items like a Brute's *Chronicle*, and three versions of Richard Rolle of Hampole's *Pricke of Conscience*, all of the fifteenth century. The principal donor was the Founder, Sir Thomas Pope, with twenty-six items surviving today; others were Francis Baber, Ralph Bathurst and Thomas Unton (d. 1693), while there were many single gifts from individuals. All the manuscripts listed by Coxe have been deposited in the Bodleian, where they can be consulted.

Besides these, there are some manuscripts housed in the College. A missal once owned by the Founder was purchased in 1945 and is in the custody of the President. There are also groups of letters and papers connected with Ralph Bathurst (1620–1704) and Thomas Warton (1728–1790). Some modern groups are kept in the College library; a list has been prepared for the National Register of Archives and a copy placed in Duke Humfrey. They include some poems by William Collins (1721–1759); antiquarian notes concerned with the history of the College made by James Ford (1779–1850), who also presented in 1808 Arthur Charlett's catalogue of Trinity scholars, 1555–1692; correspondence of James Ingram (1774–1850) and H. E. D. Blakiston (1862–1942), both former Presidents; (the last-named collection is restricted); letters of W. A. Greenhill (1814–1894), sometime exhibitioner, an Oxford physician interested in medical history and a friend of A. P. Stanley. There are also a few letters of Cardinal J. H. Newman, sometime an undergraduate at Trinity.

Archives

There is no full calendar of the archives of Trinity College available for consultation. There is a nineteenth-century handlist and another has recently been compiled for the History of the University Project. H. T. Riley described a few records for the Historical Manuscripts Commission in 1871.[19] President Blakiston's *College History* of 1898 is based on a full survey of the archives, to which references are given in footnotes, while

---

17 Maclagan, 15.
18 Pt. 2, sect. 5; additions made in typescript to the Duke Humfrey copy. A separate copy of Coxe, kept in the College library (shelfmark: B. 14. 14), also has these additions, together with notes made by M. R. James in 1924.
19 *2nd Report* (1871), 142–3.

there is a summary in the more recent *VCH Oxon*.[20] A list of the archives in a 'Cupboard in the Tower', mainly deeds relating to College property, has been prepared for the National Register of Archives, and is available in Duke Humfrey.[21]

Bursars' accounts are nearly complete from 1556, though defective 1639–51. A list of fellows and scholars covering 1556–1640 was drawn up by Ralph Kettell, President 1599–1643, who has left numerous memoranda and copies in the archives.[22] Caution money books and a regular admission register were begun in 1648, but commoners do not appear in the latter regularly before 1664. A register of leases, indentures and other documents concerned with College property begins with the foundation in 1556. Court rolls and documents relating to College land in Oxfordshire, Buckinghamshire, Leicestershire, Gloucestershire, and Essex are listed in the National Register of Archives calendar mentioned above. Two maps of Wroxton, Oxfordshire, of 1756 and 1768, have been reproduced in facsimile.[23]

The minute book of the College Debating Society for 1873–95 is kept in the College library; the honorary secretaries include Sir Arthur Quiller-Couch, A. E. W. Mason and Sir Michael Sadler.

The archives are in the custody of the librarian.

Printed Books

Printed books were among Sir Thomas Pope's gift in 1555, mainly sixteenth-century theological works; some had been acquired from Greenwich and Hampton Court palaces when Pope was winding up the estate of Henry VIII, and are finely bound.[24] A good proportion have survived, unlike those presented shortly afterwards by Thomas Slythurst, John Arden and Thomas Allen, which were among the decayed fragments found in 1807. There has been a steady stream of gifts and bequests from members of small numbers of books, or single volumes, from the earliest days, such as Thomas Rawes, and indeed these were the principal sources of supply till the bequest of Richard Rands in 1640. For example, Henry Cuffe, a former fellow and benefactor of Queen's College library, gave some books in 1594; Arthur Yeldard, President, a set of the Magdeburg *Centuriae* in 1599; Sir Edward Hoby a set of the Eton Chrysostom in 1613, finely bound by Williamson[25]; and Alexander Gil junior, several theological works in 1623. Ralph Kettell, President 1599–1643, gave more theology, and Ralph Bathurst, President 1664–1704, bequeathed £20

20 *VCH Oxon*. iii, 238; a calendar of the sixteenth-century records is being prepared for use by the History of the University Project.
21 Shelfmark: R. 7. 62/3, ff. 225–48.
22 Blakiston, 100.
23 J. L. Mowat, *Sixteen Old Maps of Properties in Oxfordshire in the Possession of Some of the Colleges* (1888), Maps 15–6.
24 *Fine Bindings 1500–1700*, nos. 62, 67, and the references there cited.
25 Hoby's letter to Kettell about this gift, pasted in the first volume, is printed in *Gentleman's Magazine*, lvi (1786), 6.

worth of books, having given many during his lifetime, according to Thomas Warton.[26]

During the eighteenth century there were not so many acquisitions of note, though Samuel Johnson gave a Baskerville Virgil in 1769 through his friend Thomas Warton, suitably inscribed to mark a stay in Trinity. The following century, however, saw a great increase in the library's size. John Skinner of Camerton (1750–1823) bequeathed his books, strong in British topography and a foretaste of the extensive library left by James Ingram, President, in 1850. This immediately doubled the number of printed books; but there are many subjects represented among Ingram's books besides topography, revealing wide-ranging interests, such as linguistics, history and theology. A small collection of seventeenth- and eighteenth-century works by or about non-jurors was given by W. J. Copeland (1804–1885) in 1847 when he was a fellow. Former members also presented their own writings occasionally such as Cardinal Newman with a set of his works in 1885, but there has been no attempt to build up an *alumnus* collection.

During the present century the generosity of members has continued, mainly of working libraries suitable for the undergraduates' reading room, such as Latin texts bequeathed by Robinson Ellis in 1913,[27] books on natural science left by D. H. Nagel in 1920,[28] historical works from William Hunt and English literature from W. G. Peterson, both in 1930–31.[29] Canon W. Sanday left theological works to Trinity in 1920, as well as to Mansfield, Queen's and St. Edmund Hall.

Other groups are more relevant for research purposes. A collection of over eighty editions of the Bible, mostly English of the sixteenth and seventeenth centuries, but ranging from the Coverdale of 1535 to the Doves Press edition of 1903, made by Sir Charles E. H. Chadwyck-Healey, bart. (1845–1919), was presented to the College during the last War; it is now deposited in the Bodleian. Eric Rücker Eddison bequeathed a collection of Icelandic books in 1945.[30] Not long afterwards C. H. Wilkinson (1888–1960), of Worcester College, presented an interesting group of early editions of Walter Savage Landor (1775–1864), a Trinity man.[31] A recent gift has been a Bible with Newman family entries.

Catalogues

The separate list of books kept in chests, of 1697, was mentioned above. There is an interesting series of ledgers containing library catalogues. The earliest, made about 1700, has a list of the College manuscripts at the end. The next was made about 1730, followed by another in the

26   *Life of Ralph Bathurst* (1761), 197.
27   [*Report*] *1913–1914*, 5.
28   [*Report*] *1920–1921*, 5.
29   [*Report*] *1930–1931*, 5.
30   [*Report*] *1944–1946*, 15.
31   [*Report*] *1947–1949*, 16.

early nineteenth century, in turn superseded by the current catalogue, in use since 1860. A separate catalogue of James Ingram's books was made after his death in 1850.

For the undergraduates, a *Catalogue of Books in the New Library* was printed about 1891, with supplements covering additions 1891–92, 1898–99 and 1899–1901. There is now a card catalogue covering the contents of the New Library.

UNIVERSITY COLLEGE
1249?

Stock: 20,000   Incunabula: 3   *STC*: 427

Although it is the senior Oxford college, the present holdings of manuscripts and printed books of University College do not reflect its antiquity, probably since it has never been particularly large or wealthy, while shortage of space has prevented accumulations of little-used material. The earliest statutes of 1292 included regulations about books, indicating that the best copy of each text, or the only copy if not more than one, was to be kept chained and the rest circulated among members, a system followed by all the early colleges, except that provision was made here for loans to non-members.[1] There was apparently at first no special room set aside for books and they may have been stored in chests; the first mention of a library occurs in 1391.[2] Bequests were received during the fourteenth and fifteenth centuries, and there was still a collection of manuscripts in the College in the mid-sixteenth century, but all had disappeared before Thomas James published his *Ecloga Oxonio-Cantabrigiensis* in 1600.[3]

Like Balliol and Exeter, University College only had a small income in its early days, and no printed books seem to have been bought before the 1580s, when some essential patristic texts were purchased.[4] George Abbot, Master 1597–1610, later Archbishop of Canterbury, gave £100 for the purchase of books in 1632.[5] After the Restoration, a new building combining a library over the kitchen (as at Wadham) was erected between 1668 and 1670. As in other Oxford colleges, benefactions were much sought after at this period, and subscriptions were invited to complete the library, not entirely fitted out until 1675.[6] A separate Benefactors' Register for gifts to the library was acquired,[7] but only used between 1674 and 1678; it contains a charming drawing of this room.[8]

---

[1] William Carr, *University College* (College Histories, 1902), 24–5; R. W. Hunt, 'The Manuscript Collection of University College, Oxford: Origins and Growth', *BLR* iii (1950), 14; *VCH Oxon.* iii, 70.

[2] Carr, 51; Hunt, 15.

[3] Hunt, 17–8.

[4] Ker, 484.

[5] *VCH Oxon.* iii, 70.

[6] Ibid., 77.

[7] University College is exceptional in having three separate Benefactors' Registers, for recording general gifts, gifts of plate, and gifts of books. See *University College Record*, iii (1956), 33–4; *VCH Oxon.* iii, 61, n. 1; *Fine Bindings 1500–1700*, nos. 225, 233.

[8] Reproduced in Hunt, plate 1 facing p. 12; *VCH Oxon.* iii, plate facing p. 76; *University College Record* iii (1956), plate facing p. 34.

A separate undergraduates' library was in existence from at least the second half of the eighteenth century,[9] but a new library building completed in 1862 was opened to all members of the College.[10] The old library, first converted into sets of rooms, was restored to its original proportions in 1953.[11]

By 1915 space was a problem, so a considerable number of seventeenth-century pamphlets were presented to the Bodleian, and classical texts put on one side to give to the pillaged University of Louvain after the War.[12] During the mastership of Sir Michael Sadler between 1923 and 1935, the book-stock was again reconsidered; many early printed and scarce works were presented to the Bodleian, where they are kept as a separate collection named after the College, while the older theological section[13] was removed, stored at first in the College and later deposited in the Bodleian, where it remains. At the same time, there was a certain amount of weeding out and discarding from all parts. An upper floor was inserted and taken into use in 1937,[14] in spite of which space is still a problem although rooms elsewhere in the College have been converted into special subject libraries. Some shelf-space was gained in recent years by placing about 150 volumes of early scientific works on permanent loan in the Museum of the History of Science.

Until the nineteenth century, books were acquired mainly by gift and bequest,[15] as in the majority of college libraries, rather than by purchase; library fines were used to increase the funds available at least after the new library was opened in 1862.[16] The absence of gift inscriptions on the books themselves, found in most other colleges, makes provenances difficult to establish.

Although the pre-Reformation manuscripts have been lost and so much material, both manuscript and printed, has either been presented to or deposited in the Bodleian and elsewhere, there is still a fair number of early printed books in the College library, though perhaps little that is outstanding or not available in other Oxford libraries. There is a small *alumnus* collection, but it is not actively maintained.

Manuscripts

The manuscripts owned by University College in medieval times, though apparently there in the mid-sixteenth century, had disappeared by 1600, but Bernard's *Catalogi Librorum Manuscriptorum Angliæ et Hiberniæ* (1697)

9   *Gentleman's Magazine* lvi (1780), 7; Carr, 215; inscriptions in surviving books.
10   Carr, 215; *VCH Oxon.* iii, 81.
11   *University College Record 1953–1954*, 11–2, and iii (1956), 32–40.
12   *University College [Report] 1915–1916*, 3.
13   *University College Record 1935*, 10; *i.e.* all books with press-marks beginning II.
14   Ibid., *1937*, 8–9; *1938*, 10–1; *VCH Oxon.* iii, 81.
15   Since 1933, gifts have been listed in *University College Record*.
16   Stated in the printed library regulations in 1862, 1864, 1871, 1873, preserved in the University College scrapbook in the Bodleian Library (G. A. Oxon. c. 288, ff. 1–2).

describes 165 items.[17] In the mid-nineteenth century, Coxe listed 188,[18] and there are now 208.[19] The absence of gift inscriptions, already mentioned, makes provenances obscure, but the chief donors were John Bancroft (d. 1640), Master 1609–32; Thomas Walker (d. 1665), Master 1632–46; and William Rogers of Painswick, admitted a commoner in 1670, with the possible addition of Mary Langbaine, widow of Gerard Langbaine (d. 1692). It seems probable that Obadiah Walker, Master 1676–89, a convert to Roman Catholicism, gave about half of the existing collection.[20]

Besides the patristic, theological, and liturgical manuscripts to be expected, a notable feature is the number of fifteenth-century English ones, such as *The Cloud of Unknowing* and Hilton's *Ladder of Perfection.* Several are illuminated, such as a fourteenth-century Dominican *Horae* and a fine copy of Bede's *Life of St. Cuthbert*, possibly written in Durham in the twelfth century, part of William Rogers' gift. There are also some legal manuscripts and documents concerned with English ecclesiastical foundations, like two cartularies of Fountains Abbey and rentals of Beverley Minster and Chichester and Salisbury Cathedrals.[21] Only one or two testify to the College's long connection with scientific studies: a description of John of Montpellier's quadrant, with astronomical tables, and a fragment of the second part of Robert Morison's *Plantarum Historia Universalis Oxoniensis* (1680).[22]

Among those from later periods are some theological manuscripts of Abraham Woodhead (d. 1678), marked up for printing on the press installed by Obadiah Walker in the Master's Lodgings during the reign of James II; a copy of one of William Somner's transcripts made in 1710–11 by William Elstob, a fellow, and his sister, Elizabeth, both Anglo-Saxon scholars; and some official documents in Turkish once owned by Sir William Jones (1746–1794), the pioneer oriental scholar, an *alumnus.* Lecture notes taken by William Scott, Lord Stowell (1745–1836), as an undergraduate in the Old Ashmolean have been deposited in the Museum of the History of Science.[23]

In the Robert Ross Memorial Collection, made by Walter Edwin Ledger and presented by Donald C. L. Cree in 1930, which consists mainly of printed material, there are some letters and notes concerned with Oscar Wilde, as well as the manuscript of Stuart Mason's *Bibliography* of that writer.[24]

17   For the growth of the manuscript collection, see Hunt, 17–20.
18   Pt. I, section 1.
19   Nos. 189–208 are described in Hunt, 31–4.
20   Donors of manuscripts now lost are listed in *VCH Oxon.* iii, 70 and Hunt, 16.
21   See the anonymous summary description in *University College [Report] 1928–1929*, 4; and S. Gibson, 'The Manuscripts of University College', *University College Record 1934*, 13–6; Carr, 216–7 describes a few.
22   E. J. Bowen, 'The Study of Science in University College, Oxford', *University College [Report] 1929–1930*, 49–51.
23   *Tenth Annual Report for 1933 of the Committee of Management of the Lewis Evans Collection*, [1934], 9.
24   *University College Record 1930–1931*, 22–3.

Of political interest are the papers of C. R. Attlee, 1st Earl Attlee, (1883–1967), accumulated while Prime Minister, who deposited them in his old College in 1961.[25] A typescript calendar has been prepared. The College was unable to accept the papers of William Beveridge, 1st Baron Beveridge, Master 1937–45, so they have been sent to the London School of Economics.[26]

All University College manuscripts were deposited in the Bodleian in 1882;[27] the Attlee papers, however, are kept in the College library.

Archives

The historian of University College has written that 'the documents in the possession of the College, considering its great antiquity, are somewhat disappointing'.[28] Deeds relating to College property within Oxford date from the twelfth century, but for property outside Oxford, including Essex and Yorkshire, only from the fifteenth. Ledger books of leases have been kept since 1588. Bursars' rolls are nearly complete for the years 1381 to 1597, though their successors, the Bursars' journals, are incomplete from 1616. The College Register starts in 1509, but the admission register only dates from about 1600. General accounts go back to 1632 and buttery accounts to 1639, while building accounts for the seventeenth century, when much of the College was reconstructed, have also survived. The three registers of benefactions, acquired in the late seventeenth century, were only maintained for short periods.[29]

William Smith (1651–1735), fellow 1675–1705, made eleven volumes of transcripts from the muniments for his *Annals of University College*, published in 1728; these are now in possession of the College, but twenty-seven other volumes of transcripts of related material are now owned by the Society of Antiquaries of London.

The archives are in the custody of the College Archivist.

Printed Books

It has already been indicated above that considerable portions of the older printed books have been given to or deposited in other libraries, but a certain number remain within the College. Taken as a whole, the collection is fairly strong in theology and classics, and also scientific works, not surprisingly in view of the strong connections with Oxford scientists over the centuries.[30]

---

[25] Ibid., iv, no. 1 (1961), 7; v, no. 3 (1968), 161.

[26] Ibid., iii, no. 3 (1958), 162; v, no. 1 (1966), 25–6.

[27] Sir E. Craster, *History of the Bodleian Library, 1845–1945* (1952), 189, where they are briefly described.

[28] Carr, viii.

[29] A good summary of the archives is in *University College Record 1952–1953*, 11–2; shorter ones are in Carr, viii, and *VCH Oxon.* iii, 61, n. 1. The account by H. T. Riley for the Historical Manuscripts Commission, *5th Report*, (1876), 477–9 is not satisfactory.

[30] Gunther, 39; and also his 'Contributions to Science by Early Members of University College', *University College Record 1935*, 11–14; see also Bowen, 49–55.

As elsewhere, benefactions from members have accounted for a large proportion before the 1580s;[31] William Holcott (d. 1575), a Berkshire squire who retired to the College from time to time to pursue his theological studies, had bequeathed part of his library shortly before this period.[32] John Reiner (d. 1613), a fellow, also left some sixteenth-century theological works. More general collections were bequeathed by William Pindar (d. 1678) and John Ledgard (d. 1683), a fellow. Obadiah Walker (d. 1699), Master 1679–89, a professed Roman Catholic convert, not unsurprisingly gave a folio Sarum Missal of 1534 (now in the Bodleian) as well as some liturgical and controversial books, while his interleaved and annotated Bodleian catalogue of 1674 is still in the College library. Another Catholic fellow, Timothy Nourse (d. 1699), also made a bequest of books, including some relative to his writings on husbandry.

Possibly the largest bequest numerically came from Thomas Hudson (d. 1719), a former fellow, later Bodley's Librarian; it consists chiefly of seventeenth-century theology.

Another good general collection strong in theology was left by John Browne, Master 1745–64, intended to be used as the personal library of his successors in that office; although some rarities have been presented to the Bodleian and a considerable number discarded as duplicates, the residue is still shelved in the Master's Lodgings. A century later another Master, F. C. Plumptre, who held office from 1836 to his death in 1870, presented a quantity of theological books in 1861. Smaller gifts, of course, were numerous; Goldwin Smith, for instance, sometime a fellow, gave books in 1864, and the anthropological and historical books of E. J. Payne came to the College after his death in 1904.[33]

Some distinctive and unusual benefactions have been received during the present century. In 1930, Donald C. L. Cree, an *alumnus*, presented a remarkable collection of editions and translations of Oscar Wilde, together with works illustrating literary movements in England in the 1890s; it had been gathered together by Walter Edwin Ledger, but is designed to commemorate the work of Robert Ross, Wilde's friend, and so is called after him.[34] This collection was immediately deposited in the Bodleian, where it remains; the printed books are incorporated in the Bodleian general catalogue. Another unusual group is a collection of works published by the Nonesuch Press, mostly bequeathed by Sir Alleyne Percival Boxall, bart., in 1945, with a few left by Oliver Bell in 1952; the latter's bequest included some Aldines as well.

---

[31]   Ker, 484.

[32]   He also left books to Queen's College, Carr, 83–4; J. R. Magrath, *The Queen's College*, ii (1921), 272; cf. *VCH Berks.* iv, 457.

[33]   *University College [Report] 1904–1905*, 1; the items of Unitarian interest among Goldwin Smith's books were given in 1925 to the Unitarian College Library in Manchester; see H. McLachlan, *The Unitarian College Library* (1939), 95.

[34]   *University College Record 1930–1931*, 22–3; masterly circumlocution avoids mentioning Oscar Wilde by name.

Catalogues

The published lists of the College manuscripts, first Bernard and later Coxe, mentioned above, have been used as guides to the main group. There is a separate typescript calendar of the Attlee papers. There is no guide to the manuscript material in the Robert Ross Memorial Collection.

The College is unusual in not annotating the published Bodleian catalogues to show its holdings of printed books, though the 1738 and 1843 editions are on the shelves. The first volume, covering A–Lu, of an early eighteenth-century manuscript catalogue has survived. This was succeeded by another handwritten catalogue in 1805, to which acquisitions were added up to about 1924, when the library was re-arranged and catalogued on cards,[35] now in the process of revision.

A separate catalogue of the Browne Library in the Master's Lodgings was made in 1930;[36] as mentioned, the printed books in the Robert Ross Memorial Collection are included in the Bodleian general catalogue.

Obadiah Walker does not seem to have entered College books into his 1674 Bodleian catalogue.

---

[35] *University College [Report] 1925*, 2.
[36] *University College Record 1930–1931*, 24.

WADHAM COLLEGE
1612

Incunabula: 53   STC: 690

The library of Wadham, much of which is still housed in the original room erected for the purpose between 1609 and 1613, has a very distinctive character. There is the usual emphasis on theology and classics, but several large gifts from the eighteenth century onwards make it important in the fields of English literature and Spanish studies, while the close connections of the College with the early days of the Royal Society make it strong in the history of the natural sciences.[1]

Regulations for the custody and care of books were laid down in the first statutes[2] and chains were used until the late eighteenth century on the larger volumes. Money from admission fees was spent on books in the late seventeenth century. Duplicates were sold in 1736 and again in 1788; the proceeds were used for buying other books.[3] Library accounts exist from 1800.

There is no adequate description of the collections in print, though E. Gordon Duff, the distinguished bibliographer and an *alumnus*, supplied a very deprecatory account appended to Sir Thomas Jackson's history,[4] concentrating on the incunabula and sixteenth-century books. Jackson himself gave a very general survey describing the architectural history and the principal donors.[5]

Manuscripts

Coxe[6] describes fifty-three manuscripts, while a list of omitted items and later accessions has been compiled for the National Register of Archives. They are of a most miscellaneous character, ranging from a few medieval texts given by Philip Stubbs (1665–1738) to plays and poems by Humbert Wolfe (1886–1940); some are mentioned by E. Gordon Duff.[7] Of special interest are various Spanish manuscripts, from two sources; the earlier are those formerly owned by Sir William Godolphin (1634?–1696), ambassador in Madrid 1671–78, mainly decrees of an historical nature,

---

[1]   Gunther, 48–9.
[2]   Sir Thomas G. Jackson, *Wadham College, Oxford* (1893), 192.
[3]   Jackson, 196.
[4]   Ibid., 198–200. A summary account, naming some donors and a few books is in *VCH Oxon.* iii, 282.
[5]   Jackson, 191–97.
[6]   Pt. II, section 8.
[7]   Jackson, 200.

which were presented by a relative, Charles Godolphin (c. 1650–1720). H. Butler Clarke gave a report on this group to the Real Academia de la Historia of Madrid about 1890, but it is not published.[8] A direct result of the presence of this collection was the gift of the manuscripts of Benjamin Barron Wiffen (1794–1867), a Quaker much interested both in the history and contemporary struggles of church reform in Spain. Hence, besides various sixteenth- and seventeenth-century documents chiefly concerned with Juan de Valdes (c. 1500–1541), there is also much correspondence between Wiffen and the Spanish reformers of his time, such as Luis de Usoz y Rio, besides English people interested in the same subject, such as John and Maria Betts.

There are also various individual manuscripts presented by or connected with former members of the College; for example, the diary of John Swinton (1703–1777) while travelling in the Mediterranean in 1730 and 1731, or a *Hortus siccus* of West Country flowers made by William Paine in 1729 and bequeathed with Richard Warner's library in 1775. Worthy of mention also are some interleaved and annotated volumes from Warner reflecting his two interests of English literature and botany. Philip Stubbs also left copies of interleaved and copiously annotated copies of *Bibliotheca Classica* (1625) and *Bibliotheca Exotica* (1625), both by Georg Draud.

Archives

Calendars of the muniments made in 1856 and 1930 were incorporated into one made by Lawrence Stone, completed in 1962, available through the National Register of Archives; this covers all documents except for some deeds of Mountpilliers Farm at Writtle, Essex. These, however, were described by H. T. Riley for the Historical Manuscripts Commission[9] and also transcribed by C. R. Cheney in 1933; a typescript copy is kept in Duke Humfrey.

Numerous records concerned with the foundation and building of the College have survived. Admission registers begin in 1613 and 'Convention Books', dealing with College government, in 1616. Bursarial accounts start in 1649, though buttery books only date from 1729 and battels from 1740. There is an exceptionally good series of minute books or papers of College and undergraduate societies, including some early nineteenth-century ones, such as a Book Club, running from 1824 to 1871. The Boat Club minutes and accounts begin in 1837, the Debating Society in 1843 and a Beefsteak Club in 1849. Various literary and theological societies have left records from the 1880s onwards.

Printed Books

At its foundation, Wadham was given what must have been one of the largest privately-owned libraries then in existence—that of Philip Bisse

[8] J. Wells, *Wadham College* (College Histories, 1898), 101, note.
[9] *5th Report* (1876), 479–81.

(*c.* 1541–1613), formerly of Brasenose and Magdalen and latterly Arch-deacon of Taunton. The College's Somerset origins and connections are thus reflected in its first books and Bisse is specifically mentioned in the statutes relating to the library.[10] Bisse's library consists chiefly of Continental printed theology, with a few other subjects, like classics, sparsely represented, and well reflects the academic interests of his time.

During the next hundred years there was no large benefaction, though there were many individual gifts and small bequests, such as the theological books left by Gilbert Drake, a fellow, in 1629, and the medical library of John Goodrich, another fellow, which came in 1651. The fact that scientific books of the period are well represented is not surprising since John Wilkins (1614–1672), the centre of the group who founded the Royal Society and its first secretary, was Warden 1648–59, while Thomas Sprat (1635–1713), the Society's historian, was also a Wadham man.

Several large bequests received during the eighteenth century have given this library a permanent, distinctive character. The earliest came from Charles Godolphin, an *alumnus* and prominent Civil Servant already mentioned; it includes many books from all periods printed in Spain, chiefly of historical and theological interest, which, like the manuscripts, formerly belonged to his relative, Sir William Godolphin, a convert to Roman Catholicism, whose will was declared invalid by Parliament. A varied collection, strong in European literature and reflecting an interest in many subjects, was bequeathed by Alexander Thistlethwayte (1718–1771), an *alumnus*. Shortly afterwards, another former member also left a library of over 4,000 volumes; this was Richard Warner, already mentioned, of Woodford Row, Essex, whose twin interests were English literature and botany. A friend of David Garrick, he had planned an edition of Shakespeare but abandoned the project after George Steevens issued his advertisement in 1766. Warner acquired all four folios of Shakespeare besides many other plays and works by Elizabethan and Jacobean writers; he did not apparently pay much attention to condition, or perfection, so the state of his books leaves much to be desired and is well below the standard of the similar and larger group in Worcester College. Warner also owned much European literature and botany, as befitted the author of *Plantae Woodfordienses*; furthermore, he had the admirable habit of noting the precise provenance in each volume.

Another large accession soon followed Warner's library, in the form of the bequest of the books of Samuel Bush, Vicar of Wadhurst, Sussex, who died in 1783. This collection is strong in eighteenth-century works, especially theology. These two bequests naturally caused shelving problems which led to the refitting of the library, financed by subscriptions and compulsory levies from all members of the College. It is noticeable that none of Warner's or Bush's books bear traces of chains, which were presumably discontinued about this time.

[10] Jackson, 192.

Apart from the usual small gifts, no notable collection was acquired for nearly a century; then, in 1867, was received the library of B. B. Wiffen, whose manuscripts have been mentioned. The printed books relate to the same subject of church reform in Spain and include many editions of Juan de Valdes besides nineteenth-century works. Wiffen annotated his books extensively and was also a skilful facsimilist when making up imperfect copies.[11]

Two smaller, but important, groups also call for comment; those of John Griffiths (1806–1885), Warden 1871–81 and editor of the Elizabethan *Homilies*, who left his collection of original editions of that work; and of Henry King (1817–1888), fellow 1844–88, who left a small group of mid-nineteenth-century novels and poetry which though not extensive is unusual in its context.

In the early years of this century, R. B. Gardiner, the compiler of the College *Register*, left his collection of works by *alumni*; unfortunately, it has not been maintained as strenuously as in some other colleges. Since then there have been two further bequests. Benjamin Bickley Rogers (1828–1919), fellow 1852–61 and translator of Aristophanes, left his large collection of editions of that author, together with the rest of his library, strong in classical subjects although it included a few early editions of Jane Austen, Benjamin Disraeli, and other early nineteenth-century novelists. More recently F. W. Hirst (1873–1953) bequeathed his books, which are very miscellaneous in character with an emphasis on the classics. A special section devoted to Iran is currently being planned.

Catalogues

No early catalogue has survived; presumably it was the 'Public Catalogue of the Library' into which Warner's books were incorporated in 1780.[12] The next was made by Parker, the Oxford bookseller, in 1809, while there are two copies of the Bodleian Catalogue of 1843 annotated with the College holdings. These were in use until recently when a new card catalogue was begun and the older books reclassified, but the work is not yet complete and individual books are frequently difficult to trace.

A short Catalogue of Books Printed in England and English Books Printed abroad before 1641 in the Library of Wadham College, Oxford, was compiled in 1918 by an *alumnus*, H. A. Wheeler, and published in 1929, but it unfortunately omits all relevant items in the Wiffen collection and some other groups; neither was it collated with the Bibliographical Society's *Short-title Catalogue* of the same period which appeared three years before the Wadham list. However, it gives full descriptions with details of imperfections.

---

[11]   S. R. Pattison, *The Brothers Wiffen* (1880) has on pp. 106–142 an autobiographical account of B. B. Wiffen's book-hunting experiences. Edward Boehmer, *Bibliotheca Wiffeniana; Spanish Reformers of two Centuries*, 3 vols. (1874, 1883, 1904) indicates books now in Wadham.

[12]   Jackson, 196.

There is a separate manuscript catalogue of the Wiffen collection of printed books, together with a chronological list of items printed between 1526 and 1699; both were completed in 1881 and were the work of George Parker, of the Bodleian, where there is a copy of the former, and also a manuscript catalogue of incunabula, made in 1884.

WORCESTER COLLEGE

1714

Stock: 175,000   Incunabula: 29   *STC*: 1,659

Worcester College stands on the site of Gloucester College, the Oxford home of the Benedictines from 1283 to the Reformation, but from this period only part of one manuscript is in the library today.[1] However, several hundred volumes from Gloucester Hall, the post-Reformation foundation, are still present. During the eighteenth century the College received several notable bequests, detailed below, while in the late nineteenth-century scheme of subject specialization by college, classical archaeology and ancient history were allotted to Worcester.[2]

The College has been fortunate in having had at least two great book-lovers acting as librarians over long periods: in the last century, H. A. Pottinger (1824–1911), an inveterate and omnivorous collector; in the present, C. H. Wilkinson (1888–1960), chiefly interested in English literature, especially of the seventeenth century. Colonel Wilkinson published a detailed history and description of his College's library, a pioneer of its type, which has made its riches widely known, and to which this present summary is greatly indebted.[3]

Worcester College library should not be neglected by the student of the history of architecture, of English literature and history of the seventeenth and eighteenth centuries, while its holdings of Spanish printed books of the sixteenth and seventeenth centuries, book-sale catalogues, and of privately printed English books are unusual for a college library. The number of interesting bindings is similarly higher. The College wisely adds to its special collections as opportunity occurs, and since 1960 has issued an annual list of principal accessions.

Manuscripts

Coxe[4] lists sixty manuscripts, all of which belong to the seventeenth century except a fourteenth-century life of the Black Prince by the Chandos Herald.[5] They were also briefly outlined, with a few additions,

---

[1] Transferred from Merton College in 1938; N. R. Ker, *Medieval Libraries of Great Britain*, 2nd ed. (1964), 146.

[2] *A Catalogue of the Books Relating to Classical Archaeology and Ancient History in the Library of Worcester College* (1878). [And] *Additions* 1878/79–1879/80.

[3] C. H. Wilkinson, 'Worcester College Library', *Oxford Bibliographical Society Proceedings and Papers*, i (1927), 263–320.

[4] Pt. 2, section 9.

[5] See Wilkinson, 278, for details of published versions.

by H. T. Riley for the Historical Manuscripts Commission.[6] The seven-teenth-century manuscripts came in a bequest from George Clarke (1661–1736), politician and fellow of All Souls, the bulk consisting of the papers of his father, Sir William Clarke (1623?–1666), sometime Secretary at War. Of great importance historically, several volumes of selections, edited by Sir Charles Firth and others,[7] have been published. Not all the Clarke papers came to Worcester, and it is unfortunate that those formerly at Littlecote House, Hungerford, can no longer be traced.[8] Besides his father's papers, George Clarke's bequest included a quantity of archi-tectural drawings, many by Inigo Jones (1573–1652) and John Webb (1611–1672),[9] and also a good number of designs by eighteenth-century architects.[10]

There are a few manuscripts from sources other than Clarke worthy of mention: some papers and letters of Thomas de Quincey (1785–1859), an *alumnus*; several manuscripts of A. C. Swinburne (1837–1909), collected by C. H. Wilkinson; five seventeenth-century plays,[11] and some mathe-matical notes by John Aubrey (1626–1697), the antiquary.[12] Besides these, there are two interesting family collections: papers and letters of C. H. O. Daniel (1836–1919), sometime Provost, and his family; and of the family of Sir W. H. Hadow (1859–1937), sometime a fellow, including an account of the experiences of Gilbert Hadow in India in the 1840s and 1850s. Access is restricted to the extremely full diaries kept by John Amphlett of Clent (1845–1918), an *alumnus* who became a prominent Worcestershire figure.

Archives

There is no comprehensive list available, though there is a schedule made in 1839 giving a list of title-deeds to manors and livings. A handlist has been prepared for the History of the University Project.

A few documents of the late sixteenth and the seventeenth centuries relating to Gloucester Hall have survived,[13] but the bulk of the records only go back to the College's foundation in 1714. The Register with details of College government begins in that year, as do the bursar's journals and the battels books, though there are gaps. Provost's accounts for 1736–75 have survived. Annual lists of members also start in 1714, but there is in addition a modern slip index of members of both Gloucester Hall and

[6]   *2nd Report* (1871), 143.
[7]   See Wilkinson, 279, for details.
[8]   Historical Manuscripts Commission, *16th Report* (1904), 79–86; *Leybourne-Popham Manuscripts* (1899).
[9]   Wilkinson, 303–4; J. A. Gotch, 'Catalogue of Drawings attributed to Inigo Jones . . . at Worcester College. . .', *Journal of the Royal Institute of British Architects*, series 3, xx (1913), 349–57.
[10] H. M. Colvin, *A Catalogue of the Architectural Drawings of the Eighteenth- and Nineteenth-Centuries in . . . Worcester College. . . .* (1964).
[11] Wilkinson, 277–8.
[12] Ibid., 266; *Times Literary Supplement*, 20 Jan. 1950, p. 48.
[13] Historical Manuscripts Commission, *2nd Report* (1871), xv.

Worcester College from about 1580 to 1900, giving particulars of both academic and subsequent careers.

There is a good collection of student notebooks of the seventeenth and eighteenth centuries, on philosophy, theology and the classics. Senior Common Room accounts from 1761 to 1885 are in the archives, and comparable records for the Junior Common Room from 1892 to 1935. Cricket Club minutes start in 1861, while the De Quincey Society, a literary club, has only left papers for 1878 to 1880. The library account book for 1845 to 1909 is also here.

The archives are in the custody of the College Archivist.

Printed Books

Of the volumes surviving from the Gloucester Hall period, an interesting group of about forty, mostly mathematical, was formerly owned and annotated by John Aubrey, the antiquary.[14] A small general library of over 400 items was given by Samuel Cooke in 1714, when the Hall became Worcester College.

The bequest of George Clarke in 1736 was as important for printed material as for the manuscripts already mentioned. It is fully outlined in C. H. Wilkinson's paper, and in the present context it will have to suffice to say that the printed section is remarkable for its seventeenth-century pamphlets and Civil War tracts, many of Scottish interest, collected by his father, Sir William; old English plays and literature; and books on architecture, some once belonging to Inigo Jones; it is truly a remarkable collection, unparalleled in any other Oxford college.

In 1742, the library of John Loder, an *alumnus* of Gloucester Hall, was received, and in 1761 that of Daniel Godwyn, but neither was very large, unlike the next accession, the bequest of William Gower, Provost from 1736 to his death in 1777. This remarkable library, apart from the standard works and texts of the time, was notable for its Spanish printed books, some of which were formerly owned by Sir William Godolphin (1634?–1696), sometime ambassador in Spain, much of whose library is now at Wadham, and for its collection of old English plays. The latter was possibly more numerous than that of Clarke, but duplicates were sold, probably at the time of receipt and certainly in 1920.[15] There are now about 1,200 plays printed before 1750.[16]

During the last century there were no really important benefactions, though L. M. Eastly presented his library, including some printed missals and Aldines; in the last fifty years, three interesting groups have been

14 Wilkinson, 266; *Times Literary Supplement*, 13 Jan. 1950, p. 32; 20 Jan. 1950, p. 48.
15 Wilkinson, 274.
16 *A Handlist of English Plays and Masques printed before 1750 in . . . Worcester College, Oxford* (1929). [And] *Plays added up to March 1948, including some previously omitted* [1948]. Relevant holdings are also shown in Sir W. W. Greg, *A Bibliography of the English printed Drama to the Restoration*. 4 vols. (Bibliographical Society Illustrated Monographs, no. xxiv, 1939–59.)

received: H. A. Pottinger, already mentioned as a devoted College librarian, bequeathed his own collections in 1911, which cover every possible subject and period. Though described with some reserve by Colonel Wilkinson,[17] they will possibly be more appreciated and used in the future, for the immense number of nineteenth-century tracts and ephemera include much scarce material. Obviously, Pottinger never discarded any printed matter that came his way and possibly scavenged other people's waste-paper baskets as well.

Sir Charles Firth (1857–1936), the historian who first appreciated and edited the Clarke papers, bequeathed much of his working library together with a small group of late-seventeenth- and early-eighteenth-century tracts and books to Worcester, although he was not a member of the College.[18]

The most recent important benefaction came from C. H. Wilkinson (1888–1960) who had looked after the library for so many years and knew every book in it from close examination. His special interest was English literature of the later seventeenth century, and in his will he bequeathed a select collection; to this have been added some judicious purchases from the sale of his books.[19] It was Colonel Wilkinson who built up a collection of modern private press books, as well as a comprehensive set of the products of the press maintained, for many years within the College, by C. H. O. Daniel, Provost from 1903 until his death in 1919.

The parochial library of Denchworth, Berks., founded in 1693, has been deposited in the College library for safe-keeping; this contains about 150 volumes, mainly of theological interest;[20] its greatest treasure, a Caxton Golden Legend, was sold by an 'erring Vicar' in 1832, but it has fortunately come to rest in the Bodleian.[21]

Catalogues

The main author catalogue is in the form of ledgers with manuscript entries, dating from the late nineteenth century, but it does not cover the Pottinger collection, for which the donor's own, rather rough, manuscript catalogue has to be used. The published lists of books on classical archaeology, and of plays, have been mentioned above. An unusual and extremely valuable tool is a chronological list of pamphlets up to 1900 in the library, arranged by year of publication and including Pottinger's tracts. Two further aids are unfortunately incomplete: a list of book-plates, and one of signatures of former owners, both made by E. M. Girling, 1900–01.

[17]   P. 313.
[18]   Other parts of Sir Charles' library are now in the Bodleian and Sheffield University Library.
[19]   Sotheby's, 24–5 Oct. 1960, 27–9 Mar., 26–7 June 1961.
[20]   Central Council for the Care of Churches, *The Parochial Libraries of the Church of England* (1959), 76–7.
[21]   Arch. G b. 2.

II

*FACULTY, DEPARTMENTAL, INSTITUTE
AND OTHER LIBRARIES*

INSTITUTE OF AGRICULTURAL ECONOMICS

LITTLE CLARENDON STREET

1913

Stock: 8,500   Current periodicals: 700   *Union List*

This Institute was established in 1913 with C. S. Orwin (1876–1955) as Director, a post he retained until 1945. It has since absorbed the Institute of Agrarian Affairs. As its present name implies, the library specialises in both agriculture and economics. There is a large stock of periodicals and government publications, with reports and bulletins from U.S.A. institutions well represented.[1]

There is a small collection of manuscripts and older books, including some eighteenth- and nineteenth-century items. The manuscripts (deposited in the Bodleian) relate chiefly to the history of agriculture; among the correspondents in a mixed batch of letters are Thomas Percy, Bishop of Dromore, the third Earl of Hardwicke, Charles Kingsley, Philip Pusey (1799–1855), and Arthur Young. Some Fortescue papers are concerned with the reclamation of Exmoor. There are also documents connected with enclosures at Dunkeswell, Devon; Cossington, Leicestershire; and Flaxton, Yorkshire. The printed books, nearly all of British interest, include long runs of older periodicals, such as the *Farmers' Magazine*, *Annals of Agriculture*, and the Board of Agriculture's *Reports*, while there are also some rare pamphlets.

The library issues lists of new accessions monthly, as well as a regular survey of articles in periodicals not specifically dealing with agriculture or economics. Holdings of works concerned with the modern Middle East are included in the Union Catalogue maintained by the Middle East Centre of St. Antony's College.

---

[1] A *List of Annual Publications currently Received* was published in 1964, and a *List of Periodicals (other than Annuals)* in April 1965, to which *Supplementary Lists* were added in December 1965 and 1966.

DEPARTMENT OF AGRICULTURAL SCIENCE

PARKS ROAD

1907

Stock: 8,300   Current periodicals: 130   *Union List*   Pre-1641 foreign
books: 11   Wing: 9

Although a chair of Rural Economy was founded and endowed by a bequest from John Sibthorp in 1798, the present Department really originated in 1907 when St. John's College supplemented the endowment sufficiently for a full-time professor to be appointed. The name was changed to Agriculture in 1945 and to Agricultural Science in 1968.

The library includes a small collection of older books and sets of nineteenth-century periodicals and official reports on agriculture, mainly of British interest. Some came from a bequest of Sir William Somerville, Sibthorpian Professor 1906–25; others were placed here on permanent loan in 1940 by the Department of Botany. The only manuscript is the tithe book of Deddington, Oxfordshire, for 1743–46, formerly owned by Sir William Osler and presented by his widow.

ASHMOLEAN MUSEUM OF ART AND ARCHAEOLOGY

BEAUMONT STREET

1677

Stock: 150,000   Incunabulum: 1   *STC*: 20

There have been books and manuscripts in the Ashmolean Museum[1] from its earliest days; the original building in Broad Street (now the Museum of the History of Science) once housed those formerly owned by the founder, Elias Ashmole (1617–1692), his fellow antiquaries Anthony Wood (1632–1695), and John Aubrey (1626–1697), and some of the papers of Sir William Dugdale (1605–1686), besides the medical library of Martin Lister (1638–1712). All these, however, were transferred to the Bodleian in 1860.[2] The library as it now exists is very much a nineteenth-century creation that has grown proportionately with the development of classical, archaeological, and fine art studies within the University, and with the evolving character of the Museum itself. Since 1970, the Library has been a separate unit, administratively independent of the Museum.

Its nucleus was formed by bringing together in 1901 the Eldon (Art) Library (established 1873), the Archaeological (Classical) Library (established 1888), and the books of the Museum not transferred to the Bodleian.[3] Since this date it has developed into a complex library with several components, but which 'in combination . . . form an indivisible whole'.[4] The parts give a good picture of the subjects covered; they are: the departments of Western Art, Eastern Art, and Antiquities; the libraries of Classical Archaeology, of Classical Literature, of European Archaeology and of the Heberden Coin Room; the Haverfield Library of Ancient History; the Griffith Institute Library (for Near Eastern Archaeology and Egyptology); the Grenfell and Hunt Papyrological Library.

The library's original function was to provide a working collection for the Museum staff in looking after and studying the objects in their charge, but over the years it has come also to serve as a faculty library in the fields of Near Eastern, classical and European archaeology, classical history and literature. Nevertheless, as has been recently written, 'The presence of so comprehensive a Library, for some purposes the best in the world, in

---

[1]   For a summary of its history and a general description, see University of Oxford, *Report of the Committee on the Ashmolean Museum* (Supplement no. 1 to the *Gazette* xcviii, Nov. 1967), especially 8–17, 45–52.

[2]   Sir E. Craster, *History of the Bodleian Library, 1845–1945* (1952), 64–5.

[3]   R. F. Ovenell, *Ashmolean Museum; Library Development* (1963), 1.

[4]   Ibid.

close proximity to the collections, is one of the most important advantages our Museum possesses'.[5] It is also an important library for research in the fields mentioned, while the Print Room in the Department of Western Art contains much material relevant to a wide range of subjects.

Notable accessions were listed in the librarian's report included in the annual reports of the Visitors, published separately and in the *Gazette* between 1899 and 1970 (from 1964 as a supplement).

Archives and Manuscripts

An irregular series of documents connected with the working of the Museum from its earlier days has survived, including the original Accessions Register begun in 1683 (similar to a college Benefactors' Book), and manuscript catalogues of objects prepared by Robert Plot, the first keeper from 1683 to 1690, and his assistant and successor, Edward Lhwyd, keeper 1690–97.[6] These archives are in the custody of the librarian. There are also a few documents connected with the Museum in the University Archives.

Manuscript material, with one exception, consists chiefly of the archaeological notes and drawings of a number of distinguished scholars with Oxford connections, ranging from Sir Arthur Evans (1851–1941), of Cretan fame, to F. J. Haverfield (1860–1919) and Sir Ian Richmond (1902–1965), the historians of Roman Britain. Papers of Near Eastern archaeologists and Egyptologists are possibly more numerous and are kept in the Griffith Institute; a list has been published;[7] this group includes the notes and diaries of Howard Carter (1873–1939) relating to the Tomb of Tut'ankhamûn; those of Nina (d. 1965) and Norman de Garis Davies (d. 1941), F. L. Griffith (1862–1934), A. H. Sayce (1845–1933), and Sir Alan Gardiner (1879–1963). Egyptological and Near Eastern drawings and paintings by W. J. Bankes (d. 1855) are on deposit. St. John's College has placed on loan the notebooks of Sir W. M. Ramsay (1851–1939) and the Department of Zoology the archaeological sections of the papers of George Rolleston (1829–1881). The exception to this type of material is the Pissaro correspondence kept in the Print Room, mainly letters of Camille (1830–1903) and Lucien (1863–1944), presented by the latter's widow.

Prints

The Department of Western Art has the custody of several remarkable collections of prints, mostly transferred from the Bodleian. The most extensive is that of F. W. Hope (1797–1862), founder of the Hope Department of Entomology, presented to the University in 1859, first housed in the Radcliffe Camera and transferred to the Ashmolean in 1924.[8] This

---

[5]   *Report of the Committee on the Ashmolean Museum*, 15.
[6]   Ashmolean Museum, Dept. of Antiquities, *A Summary Guide to the Collections* (1951), 7.
[7]   *List of Records in the Griffith Institute, Ashmolean Museum, Oxford* (1947).
[8]   Craster, 314.

collection is classified by subject and partially indexed; it contains approximately 140,000 portraits, 70,000 topographical and more than 20,000 natural history engravings.[9] Another collection is that of Francis Douce (1757–1834), originally bequeathed to the Bodleian and transferred over a period of years, which is especially rich in the German Renaissance period.[10] In 1943 it was decided to transfer from the Bodleian the magnificent collection of more than 19,000 prints presented by Mrs. Alexander Sutherland in 1837, which had first been formed as extra illustrations to Clarendon's *History of the Rebellion* and Bishop Burnet's *History of his Own Time*, with other works added later.[11]

In connection with the subject of prints, it should perhaps be mentioned that the original water-colour drawings made as designs for Oxford *Almanacks* by such artists as Turner, Cotman, De Wint, have been deposited in the Ashmolean by the Delegates of the University Press.

Printed Books

The comprehensive coverage of archaeology, classical history and literature maintained by the library has already been mentioned. As in so many Oxford libraries, the Ashmolean has been fortunate in its benefactors, mainly distinguished scholars whose personal, specialized libraries have greatly helped to establish this intensive coverage. There has been a continuous flow of such bequests and gifts, of varying size and importance, during the last half century and it is invidious to name only a few, such as F. J. Haverfield and Sir Alan Gardiner whose papers have already been mentioned; art books from C. D. E. Fortnum (1820–1899), the 'second founder'; numismatics from F. P. Barnard (1854–1931); British archaeology from E. T. Leeds (1877–1955); classical literature from Gilbert Murray (1866–1957); Roman archaeology from Margaret Venables Taylor (d. 1965). The books of F. L. Griffith (1862–1934) and Nora C. C. Griffith (d. 1965) are the basis of the Egyptological library in the Griffith Institute, opened in 1939, and those of B. P. Grenfell (1869–1926) and A. S. Hunt similarly are the foundation of the Papyrological Library named after them.

Brasenose College has deposited about 150 books from its Pelham Library, and Oriel College works on comparative philology from its Monro Collection. The library of the Oxford Architectural and Historical Society has been kept in the Museum since 1888, and is useful for English topographical works. As in several other Oxford libraries, about a score of books formerly in Magdalen Hall, acquired in 1931, are on the shelves here.

[9] Craster, 114, quoting *Gentleman's Magazine* 1862, pt. 1, 787. *Cf.* a broadsheet by J. O. Westwood, January 1860, giving variant estimates (Bodleian Library, G. A. Oxon. c. 76 (36)).
[10] J. G. Mann, 'Francis Douce as a collector', *BQR* vii, no. 81 (1934), 360–65; Craster, 16, 113, 314.
[11] Craster, 113, 314; Mrs. Sutherland's *Catalogue of the Sutherland Collection* was published in 1837.

There are also two specialist collections of interest. First, that part of the library of F. W. Hope not concerned with entomology, which is kept as a separate entity. It contains a fair collection of eighteenth- and nineteenth-century travel and European guide books, many of an ephemeral nature and not found elsewhere in Oxford. There are also volumes from which Hope had removed the illustrations for his print collection, a strong group of biographical works from the seventeenth to the nineteenth centuries, for use with his portraits, and good runs of Sotheby and Christie sale catalogues.

The other group is a collection of editions of, and works relating to, John Ruskin (1819–1900), gathered by Sir Edward Tyas Cook and Alexander Wedderburn when preparing their great edition of his *Works*, published between 1903 and 1912. Received in 1919 and housed originally in the Ruskin School of Drawing, it was transferred to the library of the Department of Western Art in 1950.

Catalogues

A catalogue of the various separate parts of the library was made by George Parker in 1895. There is now a central card catalogue covering all holdings except for a portion of the Department of Eastern Art, and including the library of the Department of the History of Art. Subsidiary catalogues are maintained for the Grenfell and Hunt Papyrological Library, the Griffith Institute, the Heberden Coin Room and the Department of Western Art.

Holdings of Arabic works in the Department of Eastern Art are entered also in the union catalogue at the Middle East Centre.

BOTANY DEPARTMENT

SOUTH PARKS ROAD

1621

Stock: 16,000   Current periodicals: 270   *Union List*   *STC*: 24

The history of the Botany Department is closely linked to that of the Botanic Garden, the oldest in England, founded by Henry Danvers, first Earl of Danby, in 1621.[1] Over the years a large collection of manuscripts and printed books has been built up, but on account of shortage of space in the Department, about 750 books, mostly printed before 1750, and 480 manuscript volumes were deposited in the Bodleian Library in 1960.[2] Designated the Sherard Collection after William Sherard (1659–1728), an early benefactor, the printed books have been entered in the Bodleian catalogues and a separate calendar of the manuscripts compiled, kept in Duke Humfrey. A few older books and some of the papers of G. C. Druce (1850–1932) are kept in the Department.

The foundations of this collection are the libraries of Jacob Bobart (1641–1719), son of the first *Horti Praefectus*, and William Sherard; later bequests or gifts came from John Sibthorp (1758–1796) and George Williams (1762–1834), both holders of the chair of botany, and more recently S. H. Vines (1849–1934), G. C. Druce and J. B. Davy (1870–1940). The books were essentially the working libraries of the donors, reflecting both their interests and, in the presentation copies, their connections with international botanical circles. There was a sale of duplicates from the Bobart collection to Bohn in 1835.

The manuscripts consist mainly of papers connected with the benefactors already mentioned, with the addition of those of Robert Morison (1620–1683), the first Oxford Professor of Botany, J. J. Dillenius (1687–1747) and C. G. B. Daubeny (1795–1867). The Daubeny papers include voluminous correspondence, much of it about rural economy rather than botany.

The Botany Department Library also collaborates closely with its neighbour, the Forestry Library, and certain sections have been amalgamated for joint use.

[1] Mrs. Hermia Newman Clokie, *An Account of the Herbaria of the Department of Botany in the University of Oxford* (1964), 1–9.

[2] *BLR*, vi (1960), 581–83.

EDWARD GREY INSTITUTE OF FIELD ORNITHOLOGY
ALEXANDER LIBRARY
DEPARTMENT OF ZOOLOGY, ST. CROSS ROAD

1938

Stock: 7,000 Current periodicals: 300 *Union List* *STC*: 1
Pre-1641 foreign books: 6 Wing: 17

Ornithology as an academic subject in Oxford began in 1930 with the appointment of W. B. Alexander (1885–1965) as Director of the Bird Census, though, of course, as an amateur study it goes back much further.[1] Edward Grey, first Viscount Grey of Falloden (1862–1933), a noted bird lover, lent his support as Chancellor of the University when a public appeal for money for an Institute of Field Ornithology was launched shortly before his death in 1933, and in 1938, when the Institute was formally set up, his name was included in its title.

Over the years W. B. Alexander built up an extensive collection of books, journals and offprints, many presented by himself, so that it is very appropriate that the Institute's library now bears his name; after his retirement as Director in 1945, he continued to look after the library for another ten years.[2] It has been described as 'the most complete ornithological library in Europe for works published during the twentieth century'.[3] Over half the books and the majority of the journals have come as gifts from individuals or societies. But it contains not only modern publications, but also a large number of the nineteenth century as well as a fair collection of earlier books, including editions of Aldrovandus, Gesner, Pliny, Buffon and others. The library is strong in studies of birds found in particular localities. An author and subject card catalogue covers both books and offprints.

Besides printed material, this library also houses an important group of manuscripts, mainly ornithological field notes, many by distinguished people. It is difficult to single out individuals, but the papers of W. B. Alexander, T. A. Coward, G. Eliot Howard and Edmund Selous should be mentioned. There is a separate card index to the manuscripts, arranged by author and subject.

Taken as a whole, this is a most interesting small library, covering in depth a subject whose fringe is only touched elsewhere in Oxford.

[1] B. W. T[ucker], 'Ornithology in Oxford', *Oxford Magazine*, lvii (1939), 304–6, 341–3. See also Sir C. Norwood, 'The Edward Grey Institute of Field Ornithology', *Oxford*, vi (1939), 52–7.
[2] *Nature*, 209 (1966), 759–60.
[3] D. Lack, 'The Position of the Edward Grey Institute, Oxford', *Bird Notes*, xxiii (1949), 232.

ENGLISH FACULTY LIBRARY

ST. CROSS BUILDING, MANOR ROAD

1914

Stock: 50,000   Current periodicals: 100   *STC*: 251

The English Faculty Library is unusual among the faculty and departmental libraries of Oxford in so far that research material has been steadily acquired since its foundation, and it is, therefore, much more than a collection of text-books and journals for the use of undergraduates.

The first Merton Professor of English Language and Literature, A. S. Napier (1853–1916), held the post from 1885 until 1904, and the final honour school of English began in 1893,[1] but the library was not founded until 1914. For the next twenty years it was built up sagaciously and energetically by Percy Simpson (1865–1962), the editor of Ben Jonson. After a brief spell in Acland House,[2] it was housed in the upper part of the Examination Schools until 1965, and now shares the St. Cross Building with the Bodleian Law Library and the Institute of Economics and Statistics.

Archives and Manuscripts

Apart from the library's own records, dating from 1914, there are few manuscripts besides those in the E. H. W. Meyerstein and Vigfússon-York Powell collections mentioned below. There are some papers and letters concerned with his own studies and the English School from A. S. Napier's library,[3] and similar material in the Percy Simpson donation, besides a few miscellaneous ones. A brief calendar has been prepared for the National Register of Archives and a copy can be consulted in Duke Humfrey.

Printed Books

The library's first books were 213 volumes given by the Delegates of the Press and 129 volumes presented by Joseph Wright (1855–1930), the philologist, while in 1916 a subscription was started for the purchase of A. S. Napier's library. This was mainly concerned with Anglo-Saxon and

---

[1] Sir C. H. Firth, *The School of English Language and Literature; a Contribution to the History of Oxford Studies* (1909), 30; D. J. Palmer, *The Rise of English Studies* (1965), 104–17, 143.

[2] MS. history of the library by Percy Simpson, among his papers, on which much of this account is based.

[3] N. R. Ker, 'A. S. Napier, 1853–1916', in *Philological Essays . . . in Honour of Herbert Dean Merritt* (Janua Linguarum, series maior 37, 1970), 152.

Germanic philology, including many pamphlets and German disserta-
tions. After the death of the second Merton Professor, Sir Walter Raleigh,
in 1922, 800 volumes were purchased by subscription from his library. To
this foundation, early editions of English literary authors have since been
steadily added, helped by numerous gifts from members of the English
Faculty and friends, such as R. W. Chapman (1881–1960), Sir W. A.
Craigie (1867–1957), Canon F. E. Hutchinson (1871–1947), and F. P.
Wilson (1889–1963); the last two reflect the donors' interests in sixteenth-
and seventeenth-century authors. Mrs. Gerrans, widow of H. T. Gerrans,
fellow of Worcester College, in 1931 presented a small collection relating
to Samuel Johnson in memory of her husband, incidentally himself a
benefactor of several Oxford libraries. Percy Simpson, during his lifetime,
donated his collection of editions of the works of Ben Jonson, as well as
many other books connected with his varied interests. More recently,
H. J. Davis (1893–1967), Professor of Textual Criticism, gave a number of
editions of Jonathan Swift, while Miss Herma E. Fiedler has presented
philological works once owned by her father, H. G. Fiedler (1862–
1945), sometime Professor of German. In 1970 Professor Nevill Coghill
gave a nearly complete set of first editions, many with interesting associations, of the writings of Edith Œnone Somerville (d. 1949) and 'Martin
Ross', the pseudonym of Violet Florence Martin (1862–1915).

There are also two important groups placed here on deposit. Christ
Church has loaned the Icelandic and Scandinavian collections bequeathed
by Frederick York Powell (1850–1904), many of which were given to him
by Guðbrandr Vigfússon (1828–1889); this is strong in printed editions
of Icelandic sagas, Bibles and history, with a fair number of early imprints
and a few manuscripts.

The literary executors have placed on deposit the printed books and
manuscripts of E. H. W. Meyerstein (1889–1952) the poet and novelist,
a former member of Magdalen College. Besides Meyerstein's own manuscripts, typescripts, proofs and letters, there is a comprehensive assemblage
of his own published writings; not all have yet been fully arranged and
calendared.

Taken as a whole, this library is strong in editions of seventeenth- and
eighteenth-century authors like Jonson, Milton, Dryden, Pope and Swift;
Anglo-Saxon and Germanic philology; and runs of eighteenth-century
English journals; it is also one of the minority of Oxford libraries to have
preserved some runs of booksellers' catalogues. It should be remembered
that since English was a comparatively late addition to the Oxford curriculum, the subject is only spasmodically represented in most college
libraries, and therefore this Faculty Library cannot be neglected by those
interested in the relevant fields.

Catalogues

There are separate card catalogues for the philological pamphlets in the
A. S. Napier collection (now being incorporated into the main catalogue),

and for the Icelandic and Scandinavian collection deposited by Christ Church. An author catalogue on cards covers the rest of the library, while a sheaf catalogue, listing the books in classified order as they are arranged on the shelves, is also available to readers. A leaflet entitled *Notes for Readers and Library Regulations*, frequently reprinted and revised, gives a list of over 130 periodicals taken currently in 1971.

FORESTRY DEPARTMENT

SOUTH PARKS ROAD

1905

Stock: 19,200   Current periodicals: 240   *Union List*   Wing: 5

The School of Forestry was founded in 1905 and in 1938 was merged with the Commonwealth Forestry Institute (a Government department started in 1924 as the Imperial Forestry Institute) to form the Department of Forestry.

This specialist library, which has an international reputation, is rich in rare monographs and periodicals,[1] together with a small number of seventeenth- and eighteenth-century printed books.

A remarkable feature is the Oxford Card Catalogue of World Forestry Literature, compiled and maintained from classified cards received from the Commonwealth Forestry Bureau and based on the library. This classified catalogue of nearly a million cards references and abstracts anything published about forestry from 1934 to the present day. Literature between 1822 and 1933 has also been referenced and recorded in folder files; a small part, covering general forestry, silviculture and forest protection has been published.[2] The Microfilm Unit attached to the Department ensures that copies can be speedily despatched to enquirers.

The Forestry library also collaborates closely with the neighbouring Botany Department library and certain sections have been amalgamated for joint use. Since 1951 a *Library Bulletin* listing new accessions has been regularly published.

[1] *List of Periodicals and Serials in the Forestry Library.* 3rd edition (1968).
[2] *Forest Bibliography to 31st December 1933.* Pts. 1–3, 1936–38.

SCHOOL OF GEOGRAPHY & RADCLIFFE METEOROLOGICAL COLLECTION

MANSFIELD ROAD

1899

Stock: 55,000   Current periodicals: 150   *Union List*   Maps: 47,000
STC: 4   Pre-1641 foreign books: 3   Wing: 3

Geographical studies in Oxford go back to at least the sixteenth century; Baldwin Norton lectured on the subject at Magdalen College in 1541, and Richard Hakluyt did the same at Christ Church sometime after 1574, for example,[1] so most of the older college libraries have some relevant books and atlases. A Readership in Geography was established in 1887 and the present School in 1899.[2] A library of some 55,000 items (including atlases, pamphlets and government publications) has since been built up, which has been described as the largest and best collection of geographical books in Great Britain except for those in the national libraries and the Royal Geographical Society.[3] There is a small number of older books and atlases of the seventeenth and eighteenth centuries, some rare, while voyages and travels of the eighteenth and nineteenth centuries are well represented. Some duplicates from the Bodleian and Queen's College can be found here. A collection of books about the Isle of Wight was presented in memory of L. H. D. Buxton in 1971.[4]

The map collection attempts to cover the world at a scale of at least 1:1,000,000; it now numbers 47,000, not including over 1,000 wall maps.

When the Radcliffe Observatory was transferred to South Africa in 1935, the extensive collection of weather reports maintained by the associated Radcliffe Meteorological Station was taken over by the School of Geography.[5] Some of this is kept up to date; it is currently being reorganized and part has been transferred to the National Library for Science and Technology.

There are author and map catalogues on cards, and a subject catalogue of periodical articles.

---

[1] The *Information Pamphlet*, issued by the School in 1969, summarises the history of the subject in Oxford.
[2] Sir C. H. Firth, *The Oxford School of Geography* (1918), 4, 9.
[3] University of Oxford, *Report of the Committee on University Libraries* (Supplement no. 1 to the *Gazette*, xcvii, 1966), 113, para. 226.
[4] *Gazette*, ci (1971), 1114.
[5] *Report of the Committee on University Libraries*, para. 225; see also *The Times*, 7 August 1936, 11–2.

DEPARTMENT OF GEOLOGY AND MINERALOGY, & THE ARKELL JURASSIC COLLECTION

PARKS ROAD

1813

Stock: 7,000   Current periodicals: 90   *Union List*   Pre-1641 foreign
books: 2   Wing: 2

There has been a readership or professorship in mineralogy at Oxford since 1813, and a professorship of geology since 1819. The libraries of these two departments are now in adjacent rooms and virtually one unit physically and administratively.

There is one collection of manuscripts: the papers, geological maps and drawings of William Smith (1769–1839), known as the 'Father of English Geology';[1] a detailed calendar is in the course of preparation.

The printed books include many rare specialist items formerly owned by William Buckland (1784–1856), the first Professor, and later by F. W. Hope (1797–1862), which came to the University with the latter's benefactions. Subsequent holders of the chairs in these departments who bequeathed their books include William John Sollas (1849–1936), notable for its extensive collection of offprints and pamphlets, and Herbert Lister Bowman (1874–1942), strong in mineralogical topics, as are the bequests of Charles Judson in 1946 and L. R. Wager in 1966. There are also numerous discarded duplicates from the Radcliffe Science Library, the Ashmolean Museum and the Hope Department of Entomology.

The Jurassic Collection bequeathed by W. J. Arkell (1904–1956) is housed separately; it is a remarkably complete assemblage, including some older books.

---

[1]   A description of these papers, with references to articles based on them, is given by Joan M. Eyles in 'William Smith; some Aspects of his Life and Work', in *Towards a History of Geology*. C. J. Schneer, editor. (Proceedings of the New Hampshire Inter-Disciplinary Conference on the History of Geology, 1967. 1969), 145–7.

DEPARTMENT OF THE HISTORY OF ART

35 BEAUMONT STREET

1955

Stock: 2,000   *STC*: 1   Pre-1641 foreign books: 121   Wing: 8

A chair in the history of art was founded in 1955 with the late Dr. Edgar Wind as the first Professor. A library has been established which includes a large collection of slides and photographic material and a smaller number of books, mainly of iconographic interest. Apart from modern works, this library includes about two hundred books printed before 1700, chiefly in Italy, as well as smaller numbers of eighteenth- and nineteenth-century items, many not to be found elsewhere in Oxford. Several of these earlier volumes were formerly owned by Sir William Stirling Maxwell (1818–1878). There is also a section devoted to French art exhibitions and criticism of the eighteenth and nineteenth centuries.

There is a card catalogue, of which a duplicate copy is kept in the Western Art Library, while entries are also incorporated into the main library catalogue of the Ashmolean Museum.

HISTORY FACULTY LIBRARY

MERTON STREET

1908

Stock: 27,000   Current periodicals: 50   *STC*: 3

The History Faculty Library is the descendant of the Maitland Library of Legal and Social History founded in memory of the Cambridge historian, F. W. Maitland (1850–1906) in 1908,[1] mainly through the efforts of A. L. Smith of Balliol and Sir Paul Vinogradoff.[2] About 300 volumes from Maitland's own library were presented by his widow, and later, works acquired during his researches on economic history were given by the family of Frederic Seebohm (1833–1912); some 2,000 books on legal and social studies from Sir Paul Vinogradoff's bequest to the University in 1926, not selected by the Bodleian, were also added.

It was first administered as a seminar library for legal and social historical studies by Sir Paul Vinogradoff until 1925 and then looked after by the Bodleian until 1933, when it was re-organized. About 1,200 volumes were put into the History Library as the Maitland section, about 200 were incorporated into the Bodleian and the remainder dispersed among other libraries, many in Oxford.[3] Since 1933 it has been administered by the History Faculty; in 1956 it acquired its own building and a full-time librarian and is now primarily a lending library for undergraduates. It contains a collection of standard works, reference books and journals planned to assist the teaching of history and palaeography, together with a few seventeenth- and eighteenth-century books. It is particularly strong in works on medieval and on British history.

---

[1] Sir Maurice Powicke, *The Maitland Library* [1957] gives a detailed account of the library's early days, with his own reminiscences. (Duplicated typescript, University Archives, 89/292.) See also *Oxford Magazine*, xxvi, 12 (17 Oct. 1907) and 134–5 (5 Dec. 1907); Sir C. H. Firth, *Modern History in Oxford, 1841–1918* (1920), 46; Sir E. Craster, *The History of All Souls College Library* (1971), 98–9.

[2] H. A. L. Fisher, *Sir Paul Vinogradoff; a Memoir* (1927), 40.

[3] *BQR*, vii (1933), 294; *Gazette*, lxv, (1934), 245.

HOPE DEPARTMENT OF ENTOMOLOGY

UNIVERSITY MUSEUM, PARKS ROAD

1849

Stock: 9,500   *Union List   STC*: 6   Pre-1641 foreign books: 2
Wing: 17

The wealth of the collections of entomological specimens housed in this Department is well known; the library serving it is equally remarkable. The Department was founded in 1849 with the gift of the collections of insects and *Crustacea* of the Rev. F. W. Hope (1797–1862), who later gave a large quantity of engraved portraits to the University, now in the Ashmolean Museum. A library was apparently begun about the same time, based on gifts from Mr. and Mrs. Hope, and the first Hope Professor, J. O. Westwood (1805–1893). In due course all their entomological books came to the Department. Later the specialist libraries of Francis P. Pascoe (1813–1893) and Octavius Pickard-Cambridge (1828–1917) were received. Relevant books were also presented by the Ashmolean Society in the nineteenth century and the Botanic Garden in 1953. As a consequence, this library is rich in older publications and foreign journals hard to find elsewhere in Oxford, while an extensive collection of offprints has been built up over the years and separately catalogued.

Besides the printed material, the Hope Department also houses the papers, journals and correspondence of nearly fifty entomologists, ranging from Dru Drury (1725–1803), and William Jones of Chelsea (d. 1818), through F. W. Hope, J. O. Westwood, J. C. and C. W. Dale, to the present day. Some complement the collections of specimens, such as the Greek *Hymenoptera* of Sir S. S. Saunders and a portion of the large collections of W. W. Saunders, the sawflies of F. D. Morice and the beetles of J. J. Walker and H. St. J. K. Donisthorpe. There are also entomological papers of persons perhaps more widely known in other fields, such as W. M. Geldart (1870–1922), the lawyer, and Sir A. W. Pickard-Cambridge (1873–1952), the classical scholar.

173

MAISON FRANÇAISE
1946
Stock: 24,000

The Maison Française, founded in 1946, is a unique institution belonging both to the University of Oxford and to the University of Paris.[1] It acts as the Oxford centre for French culture. The maintenance of a library has been one of its objects from the start and a considerable collection of modern French books has been built up. Though mainly of interest to undergraduates, it does include works not to be found elsewhere in Oxford.

Nearly half the library is devoted to French literature and the remainder to the fine arts, philosophy, history, geography, political and economic science, linguistics, archaeology, sociology and education. The cinema, theatre, music and sport are well represented also. No attempt is made to cover the natural sciences and medicine. There are also good runs of French newspapers and periodicals.

Since 1956, accession lists arranged by subject have been issued, but the main catalogue is not published.

[1]   Mme. R. J. P. Bedarida, 'The Maison Française', *Oxford Magazine*, lxxxv, (1967), 15-7.

MUSEUM OF THE HISTORY OF SCIENCE

BROAD STREET

1924

Stock: 10,000   Current periodicals: 12   Incunabula: 2   *STC*: 58

The Museum of the History of Science, housed in the Old Ashmolean Building in Broad Street, erected by the University between 1679 and 1683, owes its existence to the enthusiasm of R. T. Gunther (1869–1940), sometime fellow of Magdalen College, and the generosity of Lewis Evans (1853–1930), a member of the family of distinguished archaeologists and historians. With the remarkable collection of mathematical instruments (astrolabes, calculating and surveying instruments, and sundials) presented by Lewis Evans in 1924 was included his library of relevant early and modern books. Since then, an active policy has been adopted of acquiring by purchase, loan or gift early and modern works on scientific instruments, as well as standard works and journals on the history of science generally, or anything concerned with the Museum's possessions. Consequently this specialist library is of particular importance for research connected with all aspects of the history of science, containing much that cannot be found elsewhere in Oxford.[1]

Manuscripts

The miscellaneous group of manuscripts kept in the Museum comes from varied sources. That there are some papers of Lewis Evans, F.R.S. (1755–1827), great-grandfather of the first benefactor is not surprising, while there are also some bequeathed by another benefactor, George H. Gabb (d. 1948). Forty volumes of manuscripts and letters of R. T. Gunther, the first Curator, dealing with the establishment of the Museum, have been deposited,[2] as have the records of the Radcliffe Trustees relating to the University Observatory. A collection, made by H. E. Stapleton, of manuscripts and copies of manuscripts all to do with Islamic alchemy is

---

[1]   Full details of the establishment of the Museum (called the Lewis Evans Collection between 1924 and 1935) can be found in A. E. Gunther, *Robert T. Gunther: a Pioneer in the History of Science, 1869–1940* (Early Science in Oxford, xv, 1967). Books and manuscripts have been included in the lists of accessions in the annual *Reports*, published in the *Gazette*, and also separately; those covering 1924–39 are reprinted in A. E. Gunther's book cited above; selected items only appear between 1924–5 and 1950–1 and full lists from 1951–2 to 1965–6; since then, full lists have been published separately with the reprints of the *Report*, which can be obtained from the Museum.

[2]   Other documents are in the Bodleian and Magdalen College (A. E. Gunther, 312).

particularly valuable.[3] Some papers of L. H. Dudley Buxton (1889–1939), deposited in the Museum, are especially interesting since they include some written by Charles Babbage (1792–1871), the pioneer of calculating machines, and others connected with Sir John Ross (1777–1856), the Arctic explorer.[4] Besides these larger groups there are also numerous smaller ones, such as the papers of Frederick Jervis-Smith (1848–1911), the founder of the Engineering Laboratory in Oxford, or the lecture notes taken by Jacob Wragg at Cambridge in 1725,[5] to give two examples.

The Museum also houses the index cards, compiled by Dr. H. J. J. Winter, describing Arabic optical manuscripts in British libraries,[6] as well as an expanding collection of photographs of and notes on scientific instruments to be seen elsewhere.

Printed Books

The foundation of the library is the collection presented by Lewis Evans to accompany his instruments. Many more relevant items have since been acquired, though it is to be regretted that some Oxford libraries in the past have dispersed unwanted stock in the sale-room instead of letting this Museum have an opportunity to improve its holdings. A particularly flagrant example occurred in 1935 when the Radcliffe Trustees, having given the Bodleian a chance of making a selection, sold the rest of the library of Stephen Peter Rigaud (1774–1839), formed to illustrate the history of British science. The Museum, however, after vigorous protests by R. T. Gunther, was allowed to acquire some eighty volumes about scientific instruments and 115 bound volumes containing about 1,500 tracts.[7] This last group consists mainly of eighteenth- and nineteenth-century offprints, with some earlier items, from all over Europe, many of some rarity. More recently, other libraries have deposited appropriate works: about 150 items, chiefly early printings, from University College; the residue of the Observatory library from the Department of Astrophysics; early pamphlets from the Hope Department of Entomology and the Institute of Agricultural Economics,[8] while about 1,500 books from R. T. Gunther's library were deposited after his death in 1940.[9]

The H. E. Stapleton collection on Islamic alchemy, already mentioned for its manuscripts, includes printed matter as well. A classified collection of offprints, which has been built up and is currently added to, is useful to those approaching a problem from the subject angle.

About 300 volumes and forty journals concerned with optics as well as general science were purchased from the Royal Microscopical Society in

[3]   *Report for 1947–8*, 1–2.
[4]   *Report for 1939*, 4; A. E. Gunther, 512.
[5]   *Report for 1933*, 2, 8; A. E. Gunther, 446, 462.
[6]   *Report for 1955–6*, 3.
[7]   A. E. Gunther, 259–66, 470, 475, 481–2.
[8]   *Report for 1941–2*, 2.
[9]   *Report for 1940*, 2.

1970; there are some late seventeenth-century works, many from the eighteenth- and early nineteenth-centuries,[10] from all European countries.

Catalogues

A typescript calendar of the manuscripts, with an index, is kept in the Museum; it is being prepared for publication. There is a separate card catalogue for the Radcliffe tracts while the remainder of the printed material is catalogued on cards. Those for the Lewis Evans collection were written by the donor himself and contain details of other editions known to him, occasionally with additional notes.

[10]   *Gazette,* ci (1970), 409.

MUSIC FACULTY LIBRARY

33 HOLYWELL STREET

1944

Stock: 15,000   Current periodicals: 53   *STC*: 3   Pre-1641
foreign books: 3   Wing: 6

A library was established in the Music School in 1627 with the bequest of
William Heather, the founder of the music lecturership, comprising some
printed volumes of madrigals and songs, together with a manuscript
collection of masses by Tudor composers made by William Forrest *(fl.*
1581), a protégé of Queen Mary I. All these, with the substantial later
additions, were transferred to the Bodleian in 1885.[1]

A Faculty of Music was instituted in 1944, since when a Faculty Library
has been created, absorbing the Music Students' Library. It contains a
small number of eighteenth-century scores and books on music,[2] together
with a few earlier works, mainly acquired through gifts, including items
formerly owned by C. B. Heberden (1849–1921), sometime Principal of
Brasenose, Sir William Henry Hadow (1859–1937), and J. H. Mee
(1852–1919).

[1] Sir E. Craster, *History of the Bodleian Library, 1845–1945* (1952), 186–7; the manuscripts are described in *BQR*, iv (1924), 147–8, and the Heather Collection in
the same journal, v (1926), 23–4; see also Margaret Crum, 'Early Lists of the Oxford
Music School Collection', *Music and Letters*, xlviii (1967), 23–24.

[2] Holdings are shown in [M. Moon], *A Checklist of Books on Music Published before 1800
in the Bodleian* [1965].

OXFORD CITY LIBRARIES

ST. ALDATE'S

1854

A Public Library was established by Oxford City Council in 1854.[1] Today, besides the lending, reference, and children's departments and the branch libraries customarily associated with this type of municipal institution, the Central Library also houses material of research interest. There is a notable local history collection specializing in printed matter relating to Oxford and its surrounding area, the northern parts of Oxfordshire and Berkshire. There are very nearly complete runs of local newspapers, from *Jackson's Oxford Journal*, which started in 1753, onwards. Supplementing these are the photographs[2] of H. W. Taunt (1840–1922), especially valuable for views of Oxford around the turn of the century, and microfilms of the 1841, 1851, and 1861 Census returns for Oxfordshire, not available elsewhere in Oxford. The surviving log-books of Oxford City schools, going well back into the nineteenth century, are also housed here.

The long series of the City's archives, kept in the Town Hall, going back to the thirteenth century, are in the custody of the Town Clerk (to whom application must be made for permission to consult), but can be read in the Central Library, where it is hoped to transfer them when its new building is finished. A catalogue, made in 1958, is available through the National Register of Archives; a copy is kept in Duke Humfrey.

Besides local history, the City Library has been allocated several subjects under the regional specialization scheme set up in 1960, for which all current and some older British books are acquired. These are: motor cars (including workshop manuals); general genealogy and heraldry; and Brahmanism.

[1] See *Oxford City Libraries, 1854–1954* [1954], for a detailed history.
[2] The negatives are now with the National Monuments Record, in London.

OXFORD UNION SOCIETY

FREWIN COURT, CORNMARKET STREET

1823

Stock: 60,000   Current periodicals: 75   *STC*: 4   Pre-1641 foreign
books: 2

*Note: The use of this library is restricted to subscribing members of the Society;
membership is normally only open to matriculated members of the University of
Oxford.*

The Union Society, a club open to all members of the University on
payment of a subscription, though perhaps more widely known for its
debates, has maintained a general lending library from its earliest days,
and indeed took over books from its short-lived predecessor, the United
Debating Society.[1] A member was first appointed librarian in 1830.[2] The
objects have been to maintain a general library on the lines of a London
club together with a collection of the main works required by undergraduates in their studies. It should be remembered that most college
libraries did not allow access to undergraduates until fairly late in the
nineteenth century,[3] so presumably a much wanted need was supplied.

Not a great deal has been discarded over the years, so this library of some
60,000 volumes now contains much interesting material, especially from
the nineteenth century, for English literature, history, theology and
philosophy; there are a few earlier items. Collections of nineteenth-
century pamphlets and scientific text-books can be found here, as well as
good runs of many British periodicals, both of a scholarly and a general
nature.

A catalogue was first printed in 1836, followed by revisions until the
sixth edition was reached in 1875. After this, supplementary annual lists
of additions were printed until 1924. A subject catalogue was published
in two parts, 1883–85. There are now author and dictionary subject
catalogues on cards.

[1]   C. Hollis, *The Oxford Union* (1965), 28.
[2]   T. D. Acland (1809–1898), of Christ Church; ibid., 235.
[3]   See the Introduction above, xii–xiii.

PHILOSOPHY LIBRARY

12 MERTON STREET

1904

Stock 3,700   Current periodicals: 56   Incunabula: 3   *STC*: 26

Although mainly a working collection of text-books and journals for Faculty use, the Philosophy Library also houses an important group of about 600 volumes on logic, psychology, Aristotle and Bacon bequeathed by Thomas Fowler (1832–1904), sometime President of Corpus Christi College, to the holders of the Wykeham Professorship of Logic. His other books and papers went to Corpus. Among the books here are several early editions and a few manuscripts, but the Aristotelian section is not as extensive as the Chandler Collection at Pembroke College. A catalogue was published in 1906.[1] A list of the few manuscripts, which include some early students' notes, is available through the National Register of Archives. Harriet Shaw Weaver (1876–1961), the patroness of James Joyce, deposited the typescript of her symposium on *Time*, completed in 1955, in this library.[2]

---

[1]   *Catalogue of Books Bequeathed by Thomas Fowler, D.D., President of Corpus Christi College, Oxford,and sometime Wykeham Professor of Logic, to the Wykeham Chair of Logic* (1906).
[2]   Jane Lidderdale *and* Mary Nicholson, *Dear Miss Weaver* (1970), 435.

PITT RIVERS MUSEUM: BALFOUR LIBRARY

UNIVERSITY MUSEUM, PARKS ROAD

1882

Stock: 33,000   *Union List   STC*: 1   Wing: 5

General A. H. L. F. Pitt Rivers (1827–1900) presented his ethnological and archaeological collections to the University in 1882, and this library has since been built up to illustrate and explain the subjects in the Museum. It is named after Henry Balfour (1863–1939), Curator from 1891 until his death, whose own books and pamphlets, numbering about 10,000 in all, form the backbone of the library.

There are several groups of manuscripts, in the main consisting of the ethnological papers and correspondence of scholars such as Henry Balfour himself; Sir Edward Burnett Tylor (1832–1917, whose printed books are divided between the Museum and the Institute of Social Anthropology); Sir Francis Knowles (1886–1951); Sir Baldwin Spencer (1860–1929). In addition there are a few papers and journals connected with European exploration of Asia and Africa in the nineteenth century. A list can be consulted in Duke Humfrey.

The library has fortunately had a series of benefactors,[1] though none so munificent as Henry Balfour, including Sir John Evans (1823–1908), Sir John Myres (1869–1954), Dr. L. H. Dudley Buxton (1889–1939), H. G. Beasley (d. 1911), Captain Robert Powley Wild, and the family of R. R. Marett (1866–1943). Taken as a whole, the Balfour Library is a most interesting and fascinating collection of books on a wide range of topics, many of them outside the more usual fields of academic work. There are a few seventeenth- and eighteenth-century printed books, as well as many scarce monographs produced in out of the way places.

---

[1] *Gazette*, lxxx (1949/50), 529.

PUSEY HOUSE

61 ST. GILES

1884

Stock: 54,000   Current periodicals: 53   Incunabula: 5   *STC*: 98

Pusey House, opened in 1884, was founded to commemorate the work of Edward Bouverie Pusey (1800–1882), especially his contribution to the revival of English Church life in the nineteenth century. An independent Anglican institution, it is a place of worship and study for men and women of the University.

The library of the Faculty of Theology is also accommodated here, divided into an undergraduate lending library (instituted 1961) and a senior library, begun in 1953;[1] the books of the latter are classified and shelved with those of Pusey House. This library is important for its large holdings of both manuscript and printed material dealing with the Oxford Movement and the Tractarians in the nineteenth and twentieth centuries, while the generosity of benefactors has also enriched it with many rare works besides those to be expected in a working theological library.

Manuscripts

The manuscripts[2] at Pusey House are all modern, consisting chiefly of the papers and correspondence of various important figures and organisations in the High Church movement. The basis is about 120 volumes of E. B. Pusey's own correspondence, original and transcribed, put together by H. P. Liddon for his *Life*, first published between 1893 and 1897, and a collection by or relating to Cardinal Newman. Some of Liddon's own theological papers are also here, as well as some of Philip Pusey (1830–1880), son of E. B. Pusey. Besides these, there are extensive collections of papers and letters of several lesser figures in the Tractarian movement, such as Edward (1800–1874) and William Ralph Churton (1838–1897); W. K. Hamilton (1808–1869), sometime Bishop of Salisbury; Charles Marriott (1811–1858); Robert Scott (1811–1887),[3] and H. A. Woodgate (1800–1874), to name a few. The records of several organisations are also to be found here, such as the Association for the Promotion of the Unity of Christendom (1857 to 1921); the Church Union; the Brotherhood of

---

[1] University of Oxford, *Report of the Committee on University Libraries* (Supplement no. 1 to the *Gazette*, xcvii, 1966), para. 166.

[2] J. A. Fenwick, *Nineteenth century pamphlets at Pusey House; an Introduction for the Prospective User by Father Hugh* (1961), 10, gives a brief account of some of the groups.

[3] J. M. Prest, *Robert Scott and Benjamin Jowett* (*Balliol College Record*, supplement 1966), includes some of Scott's letters.

the Holy Trinity (1844 to 1923); the Oxford branch of the Christian Social Union, later the Industrial Christian Fellowship (1889 to 1923); Friends in Council (1911 to 1940); and the Oxford Society for Reunion (1924 to 1927).

The papers and correspondence of three men important in Anglo-Catholic circles in the twentieth century are deposited at Pusey House: Cuthbert Hamilton Turner (1860–1930); Darwell Stone (1859–1941), closely connected with the House from 1903 until his death, and Principal 1909 to 1934;[4] and Sidney Leslie Ollard (1875–1949).

A slip index, compiled by various hands, partially covers these collections, except those of C. H. Turner, Darwell Stone, the Church Union and the Association for the Promotion of the Unity of Christendom, which are not yet fully arranged.

Printed Books

The strength of Pusey House library, so far as printed matter is concerned, lies in the large number of theological pamphlets, described below, and nineteenth-century books. It does, however, contain a fair number of earlier items. As in college libraries, there have been many benefactions or bequests, frequently of individual items.

As with the manuscripts, the foundation of the library was laid with the acquisition of E. B. Pusey's own books, which reflect not only his own part in nineteenth-century Anglican church history, but also his interest in Hebrew studies. After Pusey's books, the next to arrive were those of Charles Jacomb in 1887, a general collection. H. P. Liddon (1829–1890), most of whose library is (or was) in Keble with portions at St. Edmund Hall and Cuddesdon, gave a choice group of early Sarum missals and early liturgies. In 1892 a bequest, containing many early printed theological books, came from C. C. Balston, a member of Corpus Christi College.

The first large bequest after the turn of the century, from Robert Barrett (1859–1915), was of a general character; the second came in 1930, with his papers, from Cuthbert Hamilton Turner, sometime Ireland's Professor of Exegesis of Holy Scripture at Oxford, and was particularly notable for works on canon law. Another collection, which, though general has a bias towards ecclesiastical law, was the library of Walter George Frank Phillimore, the first Baron of that name (1845–1929), presented by relatives in 1935. Besides his papers, Darwell Stone, Principal 1909–34, bequeathed his large and wide-ranging library of printed books, reflecting his interests not only in all the theological questions of the day, but also in the Fathers of the Church and the lexicon of Patristic Greek later published in 1961–68 under the editorship of G. W. H. Lampe. In more recent years Pusey House has received a selection of books formerly owned by Claude Jenkins (1877–1959), the ecclesiastical historian, much of whose library is now in St. Anne's College.

---

[4] F. L. Cross, *Darwell Stone* (1943), vii.

Father Hugh's introduction to the Pusey House pamphlets, already mentioned, gives some indication of the wealth of topics to be found covering all aspects of theological discussion in the nineteenth century, as well as the main donors. Those from Archdeacon G. A. Denison (1805–1896) number about 2,500 and are especially strong in Oxford subjects. Another 2,000 derive from R. W. Church (1804–1871), largely concerned with controversial matters, but also including many on subjects other than theology, such as art, literature, Darwinism and science in general, philology and seven volumes of booksellers' catalogues. Three smaller groups came from William Ince (1825–1910), sometime Regius Professor of Divinity at Oxford, Thomas Keble (1826–1903), nephew of John, and Richard Jenkyns (1782–1854), a Master of Balliol. All those so far mentioned were in sympathy with the Tractarian movement, so two large groups, formed by men with the opposite viewpoint, help to keep a balance; these were W. Hayward Cox (1804–1871), sometime Vice-Principal of St. Mary Hall, who had 1,400 pamphlets mostly connected with Oxford controversies, and A. J. Stephens, Q.C. (1808–1880), who had 1,900 which by-pass the main theological questions to concentrate on rather more fringe matters such as religious controversies in the British colonies.

The collection of the second Viscount Halifax (1839–1934), a High Churchman, is here, numbering about 1,000 pamphlets containing much on the English Church Union, reunion and European affairs. About 800 formerly owned by Edward Mason Ingram are chiefly concerned with social work and church life in London parishes between 1855 and 1909. Another 3,500 came from the Church House and the English Church Union libraries, dealing with most theological topics and strong in tracts of parochial interest.

In addition, those pamphlets formerly owned by the donors of printed books already named, such as E. B. Pusey and Darwell Stone, help to swell the numbers. It should perhaps be indicated here that nearly all college libraries, especially the older ones and those with denominational affiliations, have pamphlet collections concerned with the many nineteenth-century theological controversies, but that at Pusey House is undoubtedly the largest and, therefore, the most valuable to those working in the appropriate fields.

Catalogues

The slip index of manuscripts is mentioned above. A card catalogue of printed books supersedes a manuscript ledger one, but neither is complete and a considerable number of books is awaiting re-organisation. The pamphlets, housed in a separate room, are catalogued on cards by author, title and subject, while a handlist contains very useful notes of replies, sequels and other editions.[5]

[5]   Fenwick, 17–22.

INSTITUTE OF SOCIAL ANTHROPOLOGY
TYLOR LIBRARY

51 BANBURY ROAD

1910

Stock: 10,000   Current periodicals: 58   *STC*: 1   Pre-1641 foreign
books: 1

Social anthropology, as distinct from the general aspect of the subject,
has been taught at Oxford since 1910 when R. R. Marett was appointed
Reader.[1] In the following year, Sir Edward Burnett Tylor, the first holder
of the Oxford chair of anthropology, presented his own books, later added
to by his widow after Sir Edward's death in 1917. The second important
gift came from Walter W. Skeat (1866–1953), which included works on
Indonesia and Malaya together with a few Oriental manuscripts and the
typescript of his *Reminiscences* as a civil servant in Malaya, written between
1947 and 1951.

Basically a working, specialist library, it has also some books on voyages
and travels, including a few older ones, and a good collection of pamphlets and offprints. Important acquisitions are mentioned in the annual
Report, which since that for 1948–49 has been published in the *Gazette*.
Relevant holdings are incorporated in the union catalogue maintained by
the Middle East Centre of St. Antony's College.

[1]   E. E. E[vans]-P[ritchard], 'The Institute of Social Anthropology', *Oxford Magazine*,
lxix (1951), 354.

TAYLOR INSTITUTION

ST. GILES

1845

Stock: 250,000   Current periodicals: 600   Incunabula: 56   *STC*: 52

The Taylor Institution is the centre for the teaching and study of the modern European languages (except English) in Oxford. Since 1914 its library has had claims to be the largest separate collection of its kind in Britain.[1] The Institution owes its name to Sir Robert Taylor (1714–1788), the sculptor and architect, who left the residue of his estate for establishing a foundation for the teaching of modern languages in the University of Oxford. The bequest did not become effective until 1834.[2] A building was erected between 1841 and 1844; the regulations of the Institution were approved in part in 1845 and finally in 1847; and its first librarian was appointed in the same year.[3]

After the establishment of the Taylor Institution library, the Bodleian considered itself relieved of any obligation to acquire modern European literature,[4] but its large collections of older manuscripts and books mean that research in most literary subjects (especially if it is concerned with the period before 1800) involves the use of both libraries.[5] Although there was much activity in the field of comparative philology during the nineteenth century, the study of modern languages did not really flourish in Oxford until an Honour School was established in 1903, but the library grew steadily during the nineteenth, and its growth has rapidly accelerated during the present, century. Policy was originally 'not so much to collect rare books, as to keep abreast with the progress of philological and literary learning in the various nations, so that a scholar might find on the shelves . . . all the tools needed for his equipment'[6] but at the same time works on

---

[1] Cf. *Gazette*, xliv (1914), 487; D. M. Sutherland, *Modern Language Libraries: a Rapid Survey of their Resources* (*French Studies* Supplementary Publication no. 1, 1963), 11; *Handbook to the University of Oxford* (1967), 189.

[2] For the history of the library, see Sir C. H. Firth, *Modern Languages at Oxford, 1724–1929* (1929) *passim*; for the controversies surrounding the early history of the Institution see J. S. G. Simmons, 'Slavonic Studies at Oxford, I', *Oxford Slavonic Papers*, iii (1952), 125–52.

[3] John Macray, librarian 1847 to 1871, and father of W. D. Macray, the annalist of the Bodleian (Firth, 50–1).

[4] Sir E. Craster, *History of the Bodleian Library, 1845–1945* (1952), 70.

[5] The problem of co-ordinating the two libraries is discussed in University of Oxford, *Report of the Committee on University Libraries* (Supplement no. 1 to the *Gazette*, xcvii, 1966), paras. 205–15.

[6] Firth, 62.

the history and culture of the countries concerned have been steadily acquired. Up to 1914 about £250 was spent each year on book purchase but after the First World War this sum increased steadily[7] and in the past twenty-five years it has attempted to keep up with both rising prices and a considerably increased demand. Although a few older books and manuscripts were bought or given before 1914 (such as the Finch and Martin Collections), from the 1920s onwards there has been a steady growth in the acquisition of this type of material, and at the same time there have been a number of transfers of relevant collections from other Oxford libraries. The Taylorian, as the library of the Taylor Institution is often called, is consequently rich in research material connected with all the languages and literatures of Continental Europe, especially those studied in Oxford: French, German, Italian, Spanish and Portuguese (including the modern literature of Latin America), Slavonic, and Modern Greek. It also possesses smaller but valuable Afrikaans, Albanian, Basque, Celtic, Dutch, Icelandic and Scandinavian collections.

The Annual Reports of the Curators have been published in the *Gazette* since 1889; these contain information about the library, including details of notable accessions and expenditure.

The Modern Languages Faculty library, associated with this Institution, is designed to serve undergraduate needs, and is not a research library. It is the descendant of the separate seminar or departmental libraries set up first for the Honour School generally in 1906 and later split into German (1908), French (1909), Italian (1927) and Spanish (1928). These were amalgamated in 1961.[8]

Manuscripts

Although there are two volumes of designs (for rococo chimney pieces and for funeral monuments) which came from the founder, Sir Robert Taylor, manuscripts seem not to have been acquired before the appointment of Heinrich Krebs as the second librarian in 1871. Between that year and 1950 a small number of a miscellaneous character were obtained. Since then, the volume has greatly increased. Miss Herma Fiedler has presented many letters and documents formerly owned by her father, H. G. Fiedler (1862–1945), Professor of German 1907–37, particularly valuable for German studies; a catalogue of the pre-1850 items has been published.[9] In addition, Miss Fiedler has given letters of a more general nature, principally derived from the Peyton and Harding families of Birmingham, and concerned with musical, literary and political life during the later

---

[7] Possibly also not unconnected with the retirement of Heinrich Krebs, the second librarian, in 1921, and the advent of the third, Dr. L. F. Powell, the Johnsonian scholar.

[8] Sutherland, 5; *Report of the Committee on University Libraries*, para. 211.

[9] *Catalogue of the Fiedler Collection: Manuscript Material and Books up to and including 1850* (1962).

nineteenth and earlier twentieth centuries.[10] Funds arising from Sir Basil Zaharoff's gift for promoting the teaching and study of French have been used since 1950 for the purchase of modern French autograph material. A catalogue, describing with lengthy quotations over eight hundred items acquired between 1950 and 1970, was issued in duplicated form in 1970 to mark the retirement of the fourth librarian, Mr. D. M. Sutherland,[11] while a list of the miscellaneous manuscripts has been prepared for the National Register of Archives. Copies of this list are available in the Taylorian and in Duke Humfrey.

There are a few manuscripts of the fourteenth and fifteenth centuries, including a Cecco d'Ascoli, *Acerba*;[12] Leonardo Bruni Aretino, *Historia Fiorentina*; early Dante[13] and Petrarch codices; a *Recueil de rondeaux singuliers*, once ascribed to Gringoire and Marot,[14] and an English *Magna Carta*. Among the seventeenth-century items are manuscripts concerned with Antonio Perez (1539–1611) (formerly in the possession of Fernando de Arteaga y Pereira, Professor of Spanish, 1921–28); and a Mediterranean portolan drawn up by a Monegasque surgeon, G. F. Monno, in 1620. Manuscripts of eighteenth-century interest include a *Recueil de Chansons*; modern typescript copies of plays by C. A. Coypel (1694–1752) (presented by Miss Irene Jamieson in 1931); and botanical notes by Richard Richardson on English and Welsh plants made in connection with a journey by J. J. Dillenius into North Wales in 1726.

The Taylor Institution has a remarkable collection of nineteenth- and twentieth-century autograph material, especially letters from European literary figures, as well as a few connected with English writers such as Thomas Carlyle, S. T. Coleridge, George Eliot, G. H. Lewes, Amelia Opie and C. M. Yonge; or other prominent figures, such as the first Duke of Wellington, J. S. Mill, and Bishop Colenso. There is a large group of letters from Sir Robert Peel (1788–1850) and his son, Sir Robert (1822–1895). But it is invidious to select names when letters from so many well-known authors and artists could be mentioned. Similarly, a very large number of French and German writers and artists are represented by one or more items, such as Augier, Banville, Barrès, Béranger, Leon Bloy, Champfleury, Claudel, Constant, Dumas père et fils, Comtesse de Genlis, Gide, Gounod, Melchior Grimm, Guizot, Hugo, Lamartine, Leconte de Lisle, Michelet, Proust, Ravel, J. B. Rousseau, Scribe and Thiers; Goethe, Freiligrath, Kainz, Laube, J. P. Richter and Tieck. There are representa-

10 Complementary material, including letters from the same correspondents but concerned more with Birmingham affairs, has been presented by Miss Fiedler to the library of the University of Birmingham.

11 *Catalogue of Autograph Material Acquired by the Library during the Years 1950–70* (1970).

12 J. P. Rice, 'Notes on the Oxford Manuscripts of Cecco d'Ascoli *Acerba*', *Italica*, xii (1935), 137; H. Pflaum, 'L'*Acerba* di Cecco d'Ascoli', *Archivum Romanicum*, xxiii (1939), 195–6.

13 E. Moore, *Contributions to . . . Criticism of the* 'Divine Comedy' (1889), 549–50.

14 K. Chesney, '*More Poèmes de Transition*' (*Medium Aevum* Monographs viii, 1965).

tive groups of letters concerned with the Princesse de Belgiojoso, Louise Colet, Renan, Saint-Beuve, Sully Prudhomme and Laurent Tailhade; and Hauptmann, Rilke and Richard Strauss. There is also a growing collection of autographed editions of modern French authors. There are a few items in other languages as well; the largest group is probably that of the copies of 128 letters between 1827 and 1849 from Gabriele Rossetti to Charles Lyell, which were transcribed by Vitale de Tivoli, the Taylorian Teacher of Italian from 1861 to 1883.

Miss Fiedler's gifts have not been confined to literary material, since her family was also interested in the late nineteenth-century artistic and especially the musical world; letters from *e.g.* Sir Edward Elgar, Granville Bantock and Sir Lawrence Alma Tadema are present.

The Taylor Institution, therefore, contains important sources for research into the literary history of Europe, particularly for nineteenth-century authors, although few are as yet represented in depth.

Printed Books

General. The printed books in the Taylor Institution Library form the most important collection in Oxford for research into European languages and literatures, more particularly perhaps for the post-1800 period, although in its own field it has unrivalled collections of critical works and texts for the medieval period and the sixteenth and seventeenth centuries. It is, however, essential to bear in mind that, as has been mentioned above, the Taylorian and the Bodleian complement one another, and both must be used.

Not unexpectedly, Sir Robert Taylor, the founder, is represented by his own copies of the great architectural publications of his own day, such as the *Vitruvius Britannicus* (*c.* 1725–1771), Robert Adam, *Ruins of Spalatro* (1764), James Gibbs, *Book of Architecture* (2nd ed., 1739), G. B. Piranesi, *Opere* (1746–1778), P. Decker, *Fürstlicher Baumeister* (1711) and similar works. Robert Finch (1783–1830), a Balliol man, bequeathed his large general library to the University and the collection was at first deposited in a special room at the Taylorian; a catalogue was published in 1874,[15] and in 1921 the books were divided among the Bodleian, the Taylor Institution and the Ashmolean Museum.[16] The Taylorian therefore now contains many literary and linguistic works of the sixteenth to the eighteenth centuries from this source. Finch also made a monetary bequest used, from 1929,[17] for the purchase of rare works, the principal items being listed each year since 1931 in the Annual Report published in the *Gazette*.

During the last half-century, several Oxford colleges have either given or placed on revocable loan bequests or collections specializing in European

---

[15]   *A Catalogue of the Books in the Finch Collection, Oxford* (1874).

[16]   Craster, 283; the Bodleian copy of the *Catalogue* is marked to indicate items not in that library and those which were ultimately brought to it; it is noticeable that many of the first class, especially if of linguistic interest, were not acquired.

[17]   *Gazette*, lix (1929), 282.

literature or philology, and these are mentioned below in the appropriate section. The earliest college to adopt this practice was Oriel, which presented a general collection of 'valuable books' in 1922.[18]

Amongst smaller collections from private owners can be mentioned an interesting gift cutting across language boundaries received in 1955 when the Misses Esther Catherine, Susan Mary, and Josephine Fry gave many books from their family library including in particular just under a hundred volumes of herbals and botanical works, over seventy of which were printed before 1800.[19]

Apart from its major collections of literary interest, the library's principal strength and richness lie in its large holdings of general philological and linguistic works, represented by its many early dictionaries and grammars vital for comparative philological work, its very extensive philological collections in the general Romance, Germanic, and Slavonic fields, its wide holdings of runs of periodicals and reference books, its rich collection of linguistic atlases, its facsimiles and microfilms of manuscripts, and, to meet more recent needs, its collection of works for the study of foreign bibliography and its comparative literature section. A bequest by Professor Benjamin Slack has made it possible to keep abreast of the large output of learned works on linguistics, and the library's collection of these works is particularly valuable. The specialization in these subjects for over a century and a quarter has made the library unrivalled in its field, and the possessor of many works not available in the national libraries of many European countries.

The following languages, listed alphabetically, are those represented in the library, though not all of them are studied in the Honour School of Modern Languages; if no comment is made, the implication is that though there may not be any special collection of research interest, an active policy of purchases in these fields is in force.

Afrikaans. The only non-European language studied; the Rhodes Trustees have helped with special grants.[20]

Albanian. A collection of Albanian material, the property of Mrs. M. M. Hasluck, was presented in 1950,[21] together with a fund named after Lef Nosi of Elbasan. This language is also represented in the library of R. M. Dawkins (1871–1955), placed on loan by Exeter College (see *Greek, Modern*, below).

Basque. An interesting collection was presented by the Basque scholar, Edward S. Dodgson.

Bulgarian—see *Slavonic*.

Catalan.

[18]   *Gazette*, lii (1922), 541.
[19]   I am indebted to Mr. David Gilson both for permitting me to see the typescript of his forthcoming article on the Fry Collection in *The Book Collector*, and for help with other information about the library.
[20]   *Gazette*, lxix (1938), 197.
[21]   Ibid., lxxx (1950), 980.

Celtic (Cornish, Welsh, Gaelic, Manx, Erse, Breton). Built up by the generosity of the Sir John Rhys Trustees.

Czech—see *Slavonic*.

Danish—see *Scandinavian languages*.

Dutch. Besides a small representative collection other works were presented by R. W. Lee in 1940.

Finnish.

French. The Taylor Institution Library is especially strong in works of French literature, history and culture, notably of the seventeenth and eighteenth centuries, and there is also a good number of earlier printed books. One hundred and sixty eighteenth-century plays were bought in 1930[22] and five hundred volumes of periodicals from the same period were placed on deposit by St. John's College in 1936.[23] A number of seventeenth- and eighteenth-century plays were bought from the Mendelssohn–Bartholdy family in the late 1940s. In 1960, over a thousand volumes formerly owned by Gustave Rudler (1872–1957), Professor of French Literature 1920–49, were received; they include rare eighteenth-century editions and Benjamin Constant material as well as numerous autographed works by twentieth-century authors.[24] More eighteenth-century French literature as well as some sixteenth- and seventeenth-century editions[25] were included in the bequest of Margaret Grey Skipworth in 1965.

German. The collection of German literature, philology and history is no less remarkable than the French. During the last century, F. Max Müller was influential in obtaining funds to fill gaps.[26] The first specialist group, consisting of about 350 tracts and *Flugschriften* by Luther, Melanchthon, Zwingli, Erasmus and others, was acquired between 1875 and 1879.[27] More recently the Institution has benefited from the generosity of H. G. Fiedler, Professor of German, 1907–37, and his daughter, who have been previously mentioned in connection with their gifts of manuscripts. Besides literature and philology, Professor Fiedler was interested in the influence of German literature on English literature, and in accounts of travels by Germans in Britain, and Britons in Germany. A catalogue of the pre-1850 items has been published.[28] The library of Marshall Montgomery, sometime Reader in German, was given in 1938 and contains over five hundred works largely on German literature and history, while the gift from Queen's College in the same year included German history. Other donations in this field are associated with the names of H. T. Gerrans

[22] Ibid., lxi (1931), 275.
[23] Ibid., lxvii (1936), 263.
[24] Ibid., xci (1960), 369; Sutherland, 12.
[25] *Gazette*, xcvi, Supplement 7 (1966), 13.
[26] Firth, 63.
[27] Ibid.; about 300 were duplicates from the Universitätsbibliothek in Heidelberg, of which there is a list in the Bodleian.
[28] See Note 9 above.

(1923),[29] R. W. Lee (1940),[30] J. Knight Bostock (1964), and H. G. Barnes (1969).

Greek, Modern. R. M. Dawkins (1871–1955), sometime Bywater and Sotheby Professor of Byzantine and Modern Greek, left his remarkable collection of works on the literature and history of Modern Greece, Albania and Byzantium to Exeter College for the use of subsequent holders of his Chair. The College has deposited this in the Taylor Institution in order to make it more widely accessible.

Hungarian.

Icelandic—see *Scandinavian languages.*

Irish—see *Celtic.*

Italian. The Finch library was rich in works of Italian literature and the library has always maintained a strong interest in this field. Paget Toynbee, the Dante scholar, gave a collection in 1909.[31] A good Mazzini collection was presented by Mrs. E. F. Richards in 1926,[32] and L. R. Farnell, sometime Rector of Exeter College, gave works on Italian history in 1928.[33] An important group of nearly 900 volumes of editions of Dante, and related works made by Edward Moore (1864–1916) and bequeathed by him to Queen's College, was placed on revocable loan in 1939.[34] Professor J. N. Mavrogordato gave early editions of writers such as Andrew Calso, Luigi Grotto and Luigi Tansillo in 1940–41[35] and about 300 volumes came in a bequest from Cesare Foligno in 1950.[36] There is also a small collection of Futurist manifestos and other writings.

Norwegian—see *Scandinavian languages.*

Polish—see *Slavonic.*

Portuguese. A number of sixteenth-century Portuguese books was purchased from the library of the late Fernando de Arteaga y Pereira in 1935[37] and in recent years much progress has been made in developing both the Portuguese and Brazilian holdings.

Provençal.

Rumanian.

Russian—see *Slavonic.*

Scandinavian languages. The collection of saga literature and works concerning Danish and Anglo-Saxon bequeathed by W. P. Ker (1855–1923) to All Souls has been deposited on loan in the Taylorian.[38] The presence of Guðbrandr Vigfússon in Oxford (from 1866 to his death in 1889)[39] and

29 *Gazette*, liii (1923), 337.
30 Ibid., lxxi (1940), 164.
31 Ibid., xl (1910), 398; Firth, 63.
32 *Gazette*, lvii (1927), 282.
33 Ibid., lix (1929), 282.
34 Ibid., lxx (1939), 199.
35 Ibid., lxxi (1940), 164; lxxii (1941), 133.
36 Ibid., lxxx (1950), 980.
37 Ibid., lxi (1935), 233.
38 Ibid., lxxiv (1953), 273.
39 Firth, 55–7.

later that of Sir William Craigie, mean that Icelandic is well represented, helped by a small collection placed on loan by Trinity College in 1952.[40] Except for some interesting editions of Kierkegaard and a representative collection of Ibsen and Strindberg modern Scandinavian literature is not represented in depth.

Serbo-Croat—see *Slavonic*.

Slavonic. The interest in Slavonic culture in Oxford is of long standing,[41] but its academic study began as the result of a bequest for the encouragement of Polish and other Slavonic languages, literatures and history from the fourth Earl of Ilchester (1795–1865), to be used to provide an occasional course of lectures. The first was delivered by W. R. Morfill in 1870.[42] Morfill (1834–1909), the first Professor of Russian, left his Slavonic library to Queen's College (where some additions were made to it) and it was deposited in the Taylorian by the College in 1936; it numbers nearly 4,000 volumes;[43] a widely ranging collection, it is strong in folk-literature and the Russian Symbolist movement. Nevill Forbes (1883–1929), Morfill's successor, left about 2,600 volumes of Slavonic interest to the Taylor Institution, as well as others to the Bodleian and Queen's College.[44] Since the end of the Second World War there has been a great expansion of Slavonic studies in Oxford, as elsewhere; the main Oxford centres are: the Taylorian (for language and literature); St. Antony's College (for post-1917 politics, literature and history, and current and recent newspapers); the Bodleian (for other disciplines and material received under copyright deposit). Acquisition policies have been co-ordinated since 1947, but there is a certain amount of conscious duplication.[45] The Taylorian is strong in nineteenth- and twentieth-century Russian journals and subscribes to most of the important current literary and linguistic serials in the Slavonic field.

Spanish. About a thousand Spanish and Portuguese books were bequeathed by Miss W. M. Martin in 1895, including early editions of Cervantes, Calderon and Lope de Vega, besides historical works.[46] In 1933 St. John's College placed on revocable loan the 'greater part' of the collection left by H. Butler Clarke (1863–1904)[47] together with Spanish books from

40 *Gazette*, lxxii (1952), 246.
41 For a description of the library's Slavonic holdings with historical details and a select list of periodicals, see J. S. G. Simmons, 'Slavica Tayloriana Oxoniensia', *Cahiers du Monde Russe et Soviétique*, x (1969), 536–45. This is a revised version of the same writer's *Notes on the Slavonic Collections in the Library of the Taylor Institution*, issued in duplicated form earlier in 1969. See also his article on Slavonic studies in Oxford quoted in Note 2 above.
42 Firth, 57–8.
43 *Gazette*, lxvii (1936), 263.
44 Ibid., lx (1930), 271.
45 Simmons, 537.
46 Firth, 63; *Gazette*, xxvi (1896), 449.
47 [F. de Arteaga y Pereira], *A Catalogue of a Portion of the Library of H. Butler Clarke, Now in the Library of St. John's College* (1916); the Arabic manuscripts have been retained by the College.

the College library.[48] This bequest specialized in Latin, Arabic and Spanish books concerned with the study of Spanish civilization before the re-discovery of America, and is also rich in literary works. The Parry Committee on Latin American studies made Oxford a centre of research in this field and the modern language and literature of this area have consequently been the subject of much development recently. The Latin-American Centre at St. Antony's College includes relevant Taylorian holdings in its card catalogue.

Welsh—see *Celtic*.
Yugoslav—see *Slavonic*.

Catalogues

A catalogue of the library was printed in 1861 and additions were published almost annually from 1878 to 1920. The actual original catalogue was abandoned by Henrich Krebs, the second librarian, who re-classified and re-catalogued the library during his long tenure of office from 1871 to 1921.[49] This handwritten catalogue in large folio ledgers was, in its turn, revised and superseded by a system of printed entries pasted into guard-books, similar to that used by the Bodleian, under the third librarian, Dr. L. F. Powell, between 1921 and 1949. There are separate card catalogues for items in the W. P. Ker and R. M. Dawkins collections for which no entries appear in the General Catalogue.

There are some useful specialist catalogues:

(i) An author and subject card index to articles in about thirty of the main periodicals.

(ii) A card-index of offprints not entered in the main catalogue.

(iii) Card-indexes of selected fine bindings and of some provenances.

(iv) Card-index arranged chronologically of fifty-six incunabula.

(v) Manuscript catalogue of the Morfill Collection.[50]

(vi) Catalogue of the H. Butler Clarke Collection.

(vii) Catalogue of the pre-1850 items in the Fiedler Collection.

(viii) A card-index of the Cesare Foligno collection of pamphlets on Italian literature.

(ix) Duplicated list of selected autograph accessions 1950–70.

---

[48]   *Gazette*, lxiv (1934), 347.
[49]   Firth, 64.
[50]   Simmons, 539.

UNIVERSITY ARCHIVES

by T. H. Aston, *Keeper,* and D. G. Vaisey, *Deputy Keeper*

The main archives of the University are kept in the two upper rooms of the tower in the Bodleian Schools quadrangle. The tower was built between 1613 and *c.* 1618–19 and, although there is no direct evidence that it was designed for the purpose, it appears that there were some muniments in the lower room by 1629 when the delegacy appointed to codify the University statutes began work.[1] Most of the archives were in the old Congregation House adjoining St. Mary's church, but were removed from there to the lower room of the Bodleian tower by the first two Keepers of the Archives, Brian Twyne (Keeper 1634–44) and Gerard Langbaine (Keeper 1644–58), in the 1630s and 1640s.[2] Three of the presses in which the archives are still kept date from this period. The upper room of the tower was not assigned to the Keeper of the Archives until 1854, having previously served various purposes including that of a storehouse for 'powder and shott' during the Civil War, and later having been employed for 'astronomical uses'.[3] Both rooms were reconstructed during the restoration of the Bodleian Library between 1953 and 1955, when the wooden floors were replaced with reinforced concrete and the levels altered. This unfortunately necessitated the removal of the ingenious beamed ceiling of the lower room about which John Wallis, third Keeper of the Archives (1658–1703) and Savilian Professor of Geometry, had delivered lectures in the 1650s; but the appearance of the upper room was enhanced by the placing there of the eighteenth-century plaster ceiling removed from the tower section of the Bodleian upper reading room.[4] In recent years additional strong-room facilities for modern records have been provided in the Indian Institute building, and in the Examination Schools.

The Keeper of the Archives and the Deputy Keeper are appointed by the Delegates of Privileges.[5] The Deputy Keeper must always be a member of the staff of the Bodleian Library, and archives are produced for consultation in Duke Humfrey's reading room there. Application to

---

[1]   Anthony Wood, *History . . . of the University of Oxford* [edited] by J. Gutch, ii (1796), 386.

[2]   R. L. Poole, *A Lecture on the History of the University Archives* (1912), 20–25.

[3]   Ibid., 21.

[4]   See the reports of the Keeper of the Archives (printed with the annual reports of the Delegates of Privileges) for 1954–55 (*Gazette,* lxxxvi, 855) and for 1968–69 (*Gazette,* c, supplement no. 6).

[5]   *Statutes, Decrees and Regulations of the University of Oxford* (1971), Chap. iv, sect. vi (pp. 266–7). Prior to 1969 the Keeper was elected by Convocation.

consult the archives should be made to the Keeper at the Bodleian Library.

The University archives were, until quite recently, kept primarily for evidential purposes in safeguarding the University's property, jurisdiction, rights and privileges. Historically, therefore, only the more formal documents were consigned to the archives. Many other papers generated by those bodies which produced the formal records were either not offered as archives or, if offered, were rejected by the Keeper. Although in 1909 the University statutes were changed and the duty of receiving administrative and other records was laid on the Keeper,[6] it was still possible for R. L. Poole (Keeper 1909–27) to speak in 1912 of 'the fatal inability of Keepers to destroy things when they are done with and to refuse to accept papers which do not concern them.'[7] Not until 1941 were 'historical or other grounds' statutorily recognized as reasons for a document's retention in the archives.[8] Prior to that date the proximity of the great manuscript collections of the Bodleian Library meant that some University records whose proper home was with the archives were placed with the Library's collections. In this way, for example, 101 volumes of nineteenth-century administrative papers, mainly financial, were transferred by the University Registrar in 1907 and 1908 not to the archives but to the Bodleian, and, again, in 1929, fifty-two volumes of the papers of the University Police covering the period from 1829 to 1869 were placed in the Bodleian by the Proctors.[9]

The main series of records kept in the University archives are:

1. The University's charters, its grants of privilege, the foundation deeds of professorships, prizes, etc., and the title deeds of property purchased by the University or forming part of its endowment, dating from the thirteenth century.[10]

2. Statutes of the University from the fourteenth century onwards[11] and orders in Council from 1857.

---

[6] Stat. Tit. xvii, s. vii, cl. 3.

[7] Poole, 5. Poole regarded Philip Bliss (Keeper 1826–57) as the worst offender of this kind: ibid., 27. Most of Bliss's papers were subsequently transferred to the Bodleian Library.

[8] Stat. Tit. xx, s. i, cl. 1.

[9] Bodleian Library, MSS. Top. Oxon. b. 49–70, d. 73–79, e. 106–107, f. 27, described in *Summary Catalogue of Western Manuscripts in the Bodleian Library*, vi, nos. 33792–813, 33816–94; MSS. Top. Oxon. b. 129–63, c. 320–22, e. 242–55, described in P. S. Spokes, *Summary Catalogue of Manuscripts in the Bodleian Library relating to the City, County and University of Oxford, Accessions from 1916 to 1962* (1964), 14–16.

[10] The medieval deeds and charters were printed with some other documents from the archives in two volumes by H. E. Salter in *Medieval Archives of the University of Oxford* (Oxford Historical Society, lxx, 1917 and lxxiii 1919.)

[11] The statutes up to 1634 are edited from those in the archives and elsewhere in S. Gibson, *Statvta Antiqva Vniversitatis* (1931).

3. Registers of Congregation and Convocation from 1448,[12] and one letter book of the University, 1421–1503.[13]

4. Registers of matriculations from 1572[14] and of subscriptions from 1581 until the end of subscription in 1891—the latter volumes containing very large numbers of individual signatures.

5. Registers of examinations: candidates from 1638 and class lists from 1801.[15]

6. Minutes of the Hebdomadal Board and Council from 1788[16] and of Boards of Studies from 1871. Minutes and reports of Faculty Boards from 1883 and of the General Board of the Faculties from 1913. Minutes and various papers of some individual delegacies and committees.

7. Financial records: the Proctors' accounts beginning in 1464 but running continuously only from 1564, and the Vice-Chancellor's accounts from 1547 incorporating the accounts of the Schools from 1621 and of the Theatre from 1663. The ledgers, accounts, and minutes of the University Chest from 1868, but including some account books of individual funds from earlier dates.[17]

8. The records of the Chancellor's jurisdiction. These begin as registers which run from 1434 to c. 1561;[18] after this date the series are: act books of the court, depositions of witnesses, testamentary records (wills, inventories, administration bonds and accounts),[19] and miscellaneous papers (citations, mandates, interrogatories, exhibits, costs, licences, etc.).[20]

---

12   With a few entries for 1440. The earliest register (Reg. Aa) is shortly to be published by the Oxford Historical Society.

13   Reg. F. A later letter book covering the period 1508–97 (Reg. FF) has been in the Bodleian Library since 1603–5 where its press-mark is MS. Bodley 282. For other University letters see, e.g., A. Clark, The Life and Times of Anthony Wood . . . , iv, 132-3. (Oxford Historical Society, xxx, 1895). Reg. F. was edited in two volumes by H. Anstey as Epistolae Academicae Oxon. (Oxford Historical Society, xxxv and xxxvi, 1898.) Reg. FF is scheduled for publication by the Oxford Historical Society.

14   With a list of members in 1565. See A. Clark, Register of the University of Oxford, II (pts. i-ii) (Oxford Historical Society, x and xi, 1887.)

15   The class lists are printed in The Historical Register of the University of Oxford, 1220–1900 (1900), and its three subsequent supplements, 1901–30, 1931–50 and 1951–65.

16   An earlier register contains entries for the years 1738 and 1759 only. The first minute book of the Delegates of the Press, at the Clarendon Press, contains records of meetings for some years between 1701 and 1728; see p. 203, note 8.

17   The records of the Chest also include a series of 'General Ledgers' which contain copies of university leases, licences etc. (many with plans) from 1791.

18   The earliest register of the court (Reg. Aaa) was edited in 1932 in two volumes by H. E. Salter as Registrum Cancellarii Oxoniensis, 1434–1469 (Oxford Historical Society, xciii and xciv, 1932.)

19   The testamentary documents are listed in J. Griffiths, An Index to Wills Proved in the Court of the Chancellor of the University of Oxford. . . . (1862).

20   The records of the Chancellor's Court are summarily listed as Appendix II in Poole.

Very few of the administrative papers which lay behind these formal series survive in the archives. There is, for instance, virtually no corrrespondence of any kind (apart from chance survivals)[21] save for that of Keepers of the Archives and even this is very small in quantity; only a small group from the seventeenth and eighteenth centuries of vouchers or other supporting documents for the long series of accounts;[22] very little material in the way of accounts, surveys, or other papers of agents in the administration of the University's estates;[23] and very few maps or plans of its property at any period.

Apart from the material in the series noted above, such administrative papers as remain from the period before the late nineteenth and early twentieth centuries have normally survived because of their connection with one of the University's privileges. This, for instance, accounts for the survival of the three volumes of tax assessments of privileged persons[24] from 1589 to 1625, or the papers concerned with the licensing by the University of common carriers and coach proprietors from the sixteenth to the nineteenth centuries. Occasionally the possession of a privilege has led to some unexpected survivals. For instance, the enforcement of the assize of bread and ale and the control of the market has left not only a broken series of Court Leet rolls for the period between 1546 and 1733,[25] but also a series of books recording the prices of corn (and sometimes of other provisions) for the years 1617–45, 1653–58, 1673–1722, 1754–60 and 1779–1836.

Because of the continual disputes between the University and the City over trading privileges, a considerable body of material survives on these matters from the sixteenth century onwards in the form of copies of documents, extracts from registers, notes of precedents, and documents produced in the course of lawsuits. Much of this material was gathered and annotated by John Wallis, who was himself very active in protecting the University's privileges; but the rather more antiquarian collections of Brian Twyne relative to the University and the City were transferred from the archives to the Bodleian Library during the Keepership of Strickland Gibson (1927–45).[26]

[21] Such as, for example, a letter book of the delegacy for building the new schools, 1875–88, or the letters received by the Craven committee from its scholars from 1910 onwards.

[22] Though for some from the nineteenth century see above note 9.

[23] An exception to this is the collection of 28 volumes of deeds and papers relating to the Savilian estates which had passed to the Bodleian Library with the Savile books and manuscripts in 1844 and which were restored to the archives in 1936. They are described in *Summary Catalogue of Western Manuscripts in the Bodleian Library*, v, nos. 26174–82, 26184–202.

[24] I.e. those college servants, traders and others who were admitted to the privileges of the University.

[25] See I. G. Philip, 'The Court Leet of the University of Oxford', *Oxoniensia*, xv (1950), 81–91.

[26] MSS. Twyne 1–19, 21–24 and MSS. Twyne–Langbaine 1–6, listed in A. Clark, *The Life and Times of Anthony Wood. . .*, iv, 204–18. MS. Twyne 20 and the two 'extra numerum' (*ibid.*, 218–9) remain in the archives.

Apart from the formal records of some central departments such as the Faculties Office of the Registry and the University Chest, and of some institutions such as the Indian Institute which have now ceased to function, few departmental records have yet been transferred to the archives. The Delegacy of Lodgings[27] and the University Museum, however, have recently deposited their records from their foundation until the middle of the present century, and there is every reason to believe that more departments will be depositing their material in the archives in future.

Lists of the archives exist from the time before their removal from the old Congregation House. Brian Twyne's list of 1631 was printed as Appendix I to R. L. Poole's *Lecture on the History of the University Archives* (1912). As a working tool this was superseded in 1664 when John Wallis[28] drew up his 'Repertorie of Charters, Muniments, and other Writings, belonging to the University of Oxford, under the care of the Custos Archivorum'.[29] Although a fuller list of the University registers and many of the other series of muniments to the beginning of the eighteenth century was printed by Andrew Clark in 1895,[30] it was Wallis's list of 1664 which continued as the standard catalogue of the archives for two and a half centuries until Strickland Gibson became Keeper. Between 1929 and 1945 Gibson compiled a detailed handlist of the contents of the lower archive room, and a xerox copy of this handlist, annotated with subsequent additions, is available on the open shelves in Duke Humfrey's reading room at the Bodleian; there are manuscript indices to the handlist in the lower archive room. Papers now in the upper room and the strong-rooms are not covered by Gibson's handlist, and these (besides the Chancellor's Court papers and the Savilian documents) comprise the 'Long Boxes' containing the larger charters,[31] Registry and faculty papers of the late nineteenth and twentieth centuries, some individual committee and delegacy minute books, modern departmental records, and certain collections of modern deeds.

Despite its limitations the University's archive comprises, particularly from the fifteenth century onwards, a large and very rich accumulation of papers relating to the University and its history. Much very useful information on the scope and historical value of these records is to be found in the annual reports of W. A. Pantin during his Keepership (1945–69). These reports are printed with those of the Delegates of

---

[27]   The wide responsibilities of the Delegacy of Lodgings from its foundation in 1868 generated a very rich collection of papers bearing on the social and economic history of Oxford.

[28]   Aided by Anthony Wood, see A. Clark, *The Life and Times of Anthony Wood. . .* , ii, 11 (Oxford Historical Society, xxi, 1892.)

[29]   Wallis's original list was for many years lost but was rediscovered in 1815 by James Ingram (Keeper 1815–18) 'among heaps of loose papers and rubbish'. There are three contemporary copies in the archives: one originally for the Vice-Chancellor, one for the Senior Proctor and the working copy.

[30]   *The Life and Times of Anthony Wood. . .* , iv, 122–52.

[31]   Listed in Poole, 93–100.

Privileges,[32] and a selection from them was published in collected form by the Clarendon Press in 1972.[33]

[32] From 1947 to 1962 they were published in the *Gazette*, but since 1963 they have formed part of one of the supplements to it.
[33] W. A. Pantin, *Oxford Life in Oxford Archives* (1972).

UNIVERSITY PRESS

WALTON STREET

Incunabula: 2 *STC*: 447

The Press is a University department placed under the charge of the Delegacy to the University Press. Though there is an extensive literature on the history of printing in Oxford and on the University Press in particular,[1] little has been published about the collection of books built up by the Press at various times for internal use. The earliest regular collection seems to be a small library formed during the late eighteenth and early nineteenth centuries in the Delegates' Room at the Clarendon Building, partly by the accumulation of material connected with works published by the Press, and partly by purchase and bequest. It consisted largely of notes and correspondence concerned with classical scholarship and theological studies. This library was deposited in the Bodleian by the Delegates in 1885 and presented outright in 1922.[2]

The Walton Street premises, where the Press moved in 1830, now house two organizations (a) the University Press proper, *i.e.* the printing business under the direction of the Printer to the University, and (b) the Clarendon Press, *i.e.* the University publishing house, of which the Secretary to the Delegates is the executive head. Both organizations possess libraries and archives.

The libraries maintained by the Printer to the University are as follows:[3]

    (i)   the Printer's Library
    (ii) the Bible Library
    (iii) the Typographical Library
    (iv) the Technical Library
    (v) the Management Library

The first three are of potential interest to research workers, and are described below; the other two are self-explanatory and contain modern works. Since 1955, a select list of additions has been published in *The Clarendonian*, the house magazine of the Press.

It should be noted that the Constance Meade Collection of Ephemeral

---

[1]   For a list of works on the Press, see E. H. Cordeaux and D. H. Merry, *A Bibliography of Printed Works Relating to the University of Oxford* (1968), nos. 6270–6400. For a comprehensive survey of the Press, see *Report of the Committee on the University Press, Gazette*, c, Supplement 7*, (1970); its history is summarised on 17–20.

[2]   Sir E. Craster, *History of the Bodleian Library, 1845–1945* (1952), 193–4.

[3]   *The Clarendonian*, n.s. ix (1955), 3–4.

Printing, formed by John Johnson (1882–1956), Printer to the University from 1925 to 1946, was transferred to the Bodleian in 1968.[4]

Private ownership and independence are two features of Oxford libraries which were stressed in our opening Notice. In the case of the Press the need to write in advance (to the Printer to the University or the Secretary to the Delegates at the Clarendon Press as the case may be) if one wishes to consult archive or printed materials at Walton Street is of particular importance in view of the confidential nature of some of its work. Undergraduate members of the University are not admitted to the printing departments.

Archives

Some materials relating to the Press are in the University archives;[5] other documents, including materials collected by Sir William Blackstone (1723–1780)[6] and the Clarendon Press accounts for 1827–1833 and 1847–1852 are now in the Bodleian.[7] The archives at the Press fall into two classes: firstly, the archives of the Delegates, dating from the second half of the seventeenth century,[8] and secondly, the Printer's archives, which begin in the early nineteenth century.

Printed Books

The following libraries maintained by the Printer to the University contain material of historical value:

(i) *The Printer's Library*, which is divided into two sections: (a) file copies of books printed at the Press since 1851, 'forming an "archive" maintained for reference purposes';[9] and (b) an historical collection of books printed in Oxford from the earliest times to 1850, begun by John Johnson while he was Printer to the University. The historical collection contains about 2,000 volumes and includes a number of items formerly owned by Falconer Madan (1851–1935), the historian of Oxford printing. The file-copy collection (which is far from complete for the period 1851–1944) is arranged in accordance with the Dewey classification and has a combined author, title, and series card-index. The books in the historical collection are shelved in chronological order in three size-series; there are slip-catalogues by author, date, and printer.

---

[4]   *Times Literary Supplement.* 3 Oct. 1968, 1136.
[5]   Some are printed in J. Johnson and S. Gibson, *Print and Privilege at Oxford to the Year 1700* (1946).
[6]   For these, with documents from the Delegates' Archives, see I. G. Philip, *Sir William Blackstone and the Reform of the Oxford University Press in the Eighteenth Century* (Oxford Bibliographical Society publications, n.s. vii, 1957).
[7]   MSS. Top. Oxon. b. 33; d. 33; d. 37.
[8]   Cf. *The First Minute Book of the Delegates of the Oxford University Press, 1668–1756.* Edited by S. Gibson and J. Johnson (Oxford Bibliographical Society. Extra publication, 1943.).
[9]   *The Clarendonian,* loc. cit.

(ii) *The Bible Library* contains a miscellaneous collection of early printed English Bibles as well as current Oxford-printed Bibles and Prayer Books. A flysheet of 1834 answering complaints about Oxford Bibles, shows that the Delegates then owned seventeen English editions printed between 1611 and 1686[10] and there are now considerably more. There are title and chronological card-indexes.

(iii) *The Typographical and Reference Library* is intended to serve the needs of research into the history of the University Press as well as general printing history. There is a good and well-organised selection of modern works on the history of printing, type-design and typefounding, Biblical scholarship and Bible printing, and the history of the University of Oxford and the University Press. The library also contains a collection of about 400 type specimens (including a number of unique items)[11] and a selection of early printed and more recent books which are of interest as examples of printing.

The library is arranged in accordance with a specially devised classification scheme; there are author and classified catalogues and card-indexes of provenances and of printers and other craftsmen.

Besides those libraries maintained by the Printer to the University, there are two others connected with the Press:

The *Clarendon Press*, which shares the Walton Street premises with the University Press, keeps a library of file copies of modern books published by them (much on the lines of the Printer's library file-copy collection). They also have a collection of about 1,000 Clarendon Press books published from the eighteenth until the middle of the nineteenth-century, most of them in the original publishers' paperboards or cases.

Wolvercote Mill, controlled by the Delegates of the Press, maintains a technical library concerned with paper and paper-making; it also has the custody of title-deeds going back to the seventeenth century, and records of the management and working of the Mill from the mid-nineteenth century.[12]

10 Bodleian Library, G. A. Oxon. b. 175, f. 2.
11 Some of the specimens are listed (with an introduction which includes information about the Printer's libraries) in J. S. G. Simmons, 'Specimens of Printing Types before 1850 in the Typographical Library at the University Press, Oxford', *Book Collector*, viii (1959), 397–410.
12 Harry Carter, *Wolvercote Mill: a Study in Paper-Making at Oxford* (Oxford Bibliographical Society. New series, extra publication, 1957) makes references to these, as well as to documents in the custody of the Clarendon Press and the Printer to the University.

APPENDIX I

List of Libraries not described in Parts I–II

Note: Where the information is available, the approximate size of the bookstock (including pamphlets) and number of current periodicals taken has been added in brackets after the address.

*   a few older books present.

\+   included in *Union List of Serials in the Science Area, Oxford: Stage II*, 1970.

(NU)   not connected with the Libraries Board of the University of Oxford, the United Oxford Hospitals, or the Regional Hospital Board.

(PG)   requests for access must be made through the Librarian, Postgraduate Medical Library, Radcliffe Infirmary.

(PG)   Accident Service, Radcliffe Infirmary

* +   Anaesthetics, Nuffield Department of, Radcliffe Infirmary

\+   Anatomy, Department of Human, University Museum, Parks Road (3,900; 86 current pers.)

\+   Animal Ecology Research Group (Elton Library), Department of Zoology, St. Cross Road (5,800; 207 current pers.)

Archaeology and the History of Art Research Laboratory, 42 Banbury Road (380; 20 current pers.)

Archaeology, Institute of, 35 Beaumont Street (750; 26 current pers.)

\+   Astrophysics, Department of (University Observatory), South Parks Road (5,260; 400 current pers.)

Atmospheric Physics, Department of, Clarendon Laboratory, Parks Road (100)

(PG)   Bacteriology and Virology Laboratory, Radcliffe Infirmary

Bacteriology, Nuffield Department of Pathology and, *see* Pathology and Bacteriology

Balliol and Trinity Science Library

Barnett House, *see* Social and Administrative Studies, Department of

\+   Biochemistry, Department of, South Parks Road (5,170; 142 current pers.)

\+ (PG) Biochemistry, Nuffield Department of Clinical (Gibson Laboratories), Radcliffe Infirmary

Biomathematics, Department of, Pusey Street (800; 19 current pers.)

Chemical Crystallography Laboratory, South Parks Road (150)

205

+ (PG) Chest Clinic, Churchill Hospital
+   Children, Park Hospital for, Old Road, Headington [Psychiatry; apply to the Librarian, Warneford Hospital in the first instance]
+   Churchill Hospital Medical Library, Churchill Hospital, Headington (200; 21 current pers.)
+   Clarendon Laboratory [physics], Parks Road (4,920; 100 current pers. including Atmospheric and Theoretical Physics)
   Commonwealth Studies, Institute of, 20–1 St. Giles (7,950; 295 current pers.)
+   Computing Laboratory, 19 Parks Road and 45 Banbury Road (1,020; 40 current pers.)
   Dental Department, *see* Oral and Maxillo-Facial Dental Department
(PG) Dermatology, Department of, Slade Hospital
+   Dyson Perrins Laboratory [organic chemistry], South Parks Road (2,800; 34 current pers.)
*   Economics and Statistics, Institute of, St. Cross Building, Manor Road (48,600; 2, 280 current pers.)
   Educational Studies, Department of, 15 Norham Gardens (27,300; 260 current pers.)
   Elton Library, *see* Animal Ecology Research Group
+   Engineering Science, Department of, Parks Road (14,000; 156 current pers.)
+   Experimental Psychology, Institute of, South Parks Road (5,000; 93 current pers.)
   External Studies, Department of, Rewley House, Wellington Square (64,900; 43 current pers.)
   Geodesy, *see* Surveying and Geodesy, Department of
(PG) Geriatrics, Department of, Cowley Road Hospital
   Gibson Laboratories, *see* Biochemistry, Nuffield Department of Clinical; and, Pathology and Bacteriology, Nuffield Department of
(PG) Infectious Diseases, Department of, Slade Hospital
+   Inorganic Chemistry Laboratory, South Parks Road (2,150; 23 current pers.)
   Kilner Library, *see* Plastic Surgery, Department of
(NU) Latimer House [theological], 131 Banbury Road (4,500; 25 current pers.)
   Linacre College
+   Littlemore Hospital Medical Library [psychiatry], Littlemore
(NU) Management Studies, Oxford Centre for, Kennington Road, Kennington (2,000; 57 current pers.)
+   Mathematical Institute (Whitehead Library), 24–9 St. Giles (6,050; 98 current pers.)
+   Medical Research, Nuffield Institute for, Headington (780; 22 current pers.)

+(PG) Medicine, Nuffield Department of Clinical, Radcliffe Infirmary
+   Metallurgy, Department of, Parks Road (1,800; 50 current pers.)
          Modern Languages Faculty Library, Taylor Institution, St. Giles
             (47,500; 60 current pers.)
+   Molecular Biophysics Laboratory, Department of Zoology, St.
             Cross Road
(NU)   Music and Scores, Library of Recorded, 39A St. Giles (5,000
             records)
(NU)   Musical Club and Union, Holywell Street (2,000)
+   Nuclear Physics Laboratory, Keble Road (3,000; 79 current pers.)
          Observatory, *see* Astrophysics, Department of.
+   Obstetrics and Gynaecology, Division of, Research Institute,
             Churchill Hospital
+   Ophthalmology, Nuffield Laboratory of, Radcliffe Infirmary,
             Walton Street (1,900; 25 current pers.)
(PG)   Oral and Maxillo-Facial and Dental Departments, Churchill
             Hospital
          Organic Chemistry, *see* Dyson Perrins Laboratory
*   Oriental Institute, Pusey Lane (45,000; 173 current pers.; 12
             pre-1641 foreign books)
+   Orthopaedic Centre, Nuffield, Headington (1,000)
+(PG) Osler House [medical school], Radcliffe Infirmary (1,100)
(PG)   Otolaryngology, Department of, Radcliffe Infirmary
(NU)   Oxford Mail and Times Ltd., Osney Mead (2,000; 750,000
             clippings and 80,000 photographs)
(NU)   Oxford Polytechnic, Gipsy Lane, Headington (45,000; 650 current
             pers.)
(PG)   Paediatrics, Department of, Radcliffe Infirmary
* +   Pathology and Bacteriology, Nuffield Department of (Gibson
             Laboratories), Radcliffe Infirmary (*STC*: 1; pre-1641 foreign
             books: 2; Wing: 3)
* +   Pathology, Sir William Dunn School of, South Parks Road
             (2,900; 58 current pers.)
* +   Pharmacology, Department of, South Parks Road (2,600; 40
             current pers.)
+   Physical Chemistry Laboratory, Parks Road (2,900; 24 current
             pers.)
(NU)   Oxford Polytechnic, Gipsy Lane, Headington (45,000; 650 current
             pers.)
(PG)   Physical Medicine, Department of, Radcliffe Infirmary
+   Physiology, Department of, South Parks Road (2,600; 52 current
             pers.)
+   Plastic Surgery, Department of (Kilner Library), Churchill
             Hospital (1,000; 22 current pers.)
          Population Genetics Unit (Medical Research Council), Old
             Road, Headington (2,500; 50 current pers.)

+   Postgraduate Medical Library [formerly Nuffield III], Radcliffe
         Infirmary (880; 184 current pers.)
(NU)   Pressed Steel Fisher Ltd. (Information Section, Research and
         Development Division, Cowley (8,000; 300 current pers.)
(PG)   Radiology, Nuffield Department of, Radcliffe Infirmary
(PG)   Radiotherapy, Department of, Churchill Hospital
+   Regional Hospital Board, Oxford, Old Road, Headington [medi-
         cal administration] (2,500; 110 current pers.)
+(PG)   Regius Professor of Medicine, Radcliffe Infirmary
(NU)   Ripon Hall [theological], Boars Hill (15,000; 26 current pers.)
(NU)   Ruskin College, Walton Street (7,000; 20 current pers.)
(NU)   Ruskin School of Drawing, Ashmolean Museum
*   St. Benet's Hall, 38 St. Giles (pre-1941 foreign books: 5; Wing: 2)
*   St. Catherine's College (10,000; *STC*: 2; pre-1641 foreign books:
         2; Wing: 6)
         St. Cross College
*   St. Peter's College (*STC*: 2; pre-1641 foreign books: 1; Wing: 5)
(NU)   St. Stephen's House [theological], 17 Norham Gardens (6,000)
         Social and Administrative Studies, Department of, New Barnett
         House, Little Clarendon Street (2,860; 12 current pers.)
+   Social Medicine, Department of, 8 Keble Road (830; 15 current
         pers.)
         Social Studies Library, 45 Wellington Square (26,000; 50 current
         pers.)
         Statistics, *see* Economics and Statistics, Institute of
+(PG)   Surgery, Nuffield Department of, Radcliffe Infirmary
+   Surveying and Geodesy, Department of, 62 Banbury Road
         (1,530; 36 current pers.)
         Theology Faculty Library, Pusey House, 61 St. Giles (7,720;
         22 current pers.)
         Theoretical Physics, Department of, Clarendon Laboratory,
         Parks Road
(PG)   Thoracic Surgery, Department of, Churchill Hospital
         Virology, *see* Bacteriology and Virology Laboratory
+   Warneford Hospital Medical Library [psychiatry], Headington
(NU)   Westminster College of Education, North Hinksey
         Whitehead Library, *see* Mathematical Institute
         Wolfson College
(NU)   Wycliffe Hall [theological], 54 Banbury Road
         X-ray Department, *see* Radiology, Nuffield Department of
* +   Zoology, Department of, St. Cross Road (7,150; 146 current pers.;
         pre-1641 foreign books: 1)

ALEXANDER, Wilfred Backhouse (1885–1965)	Ornithological papers and correspondence	Edward Grey Inst.
ALLEN AND HANBURY Ltd.	Papers, c. 1802–1894	Museum of the History of Science [deposited]
ALLEN, Sir Hugh (1869–1946)	Notebooks	New College
ALLEN, Percy Stafford (1896–1933)	(1) Papers (2) „	Corpus Merton
ALLSOP, Thomas (1795–1880)	28 letters from Charles Voysey 1869–71	Manchester
AMPHLETT, John (1845–1918)	Diaries [restricted access]	Worcester
ANDERSON, J. G. C. (1870–1952)	Archaeological notebooks	Ashmolean Museum
ANGUS, Joseph (1816–1902)	Papers	Regent's Park
APLIN, Oliver Vernon (1879–1918)	Ornithological notebooks, diaries, correspondence	Edward Grey Inst.
ARMSTRONG, George (d. 1945)	Papers	Manchester
ARNOLD, Matthew (1822–1888)	Papers	Balliol
ASPLAND, Robert Brook (d. 1869)	Notes on dissenting history	Manchester
ASSOCIATION FOR THE PROMOTION OF THE UNITY OF CHRISTENDOM	Papers and correspondence 1857–1921	Pusey House
ATTLEE, Clement Richard, 1st Earl (1883–1967)	Papers	University College
AUBREY, John (1626–1697)	Papers	Worcester
AUGSBURGER ALLGEMEINE ZEITUNG	MS. extracts from 1843–66 by F. von Ellrodt	Taylor Inst.
AUSTEN, Jane (1775–1817)	Seven letters	St. John's
BABBAGE, Charles (1792–1871)	Papers	Museum of the History of Science [deposited]
BALFOUR, Henry (1863–1939)	Ethnological papers	Balfour Library, Pitt Rivers Museum
BANKES, William John (d. 1855)	Archaeological drawings	Griffith Inst., Ashmolean Museum [deposited]
BARHAM, R. H. (1788–1845)	Two poems	Brasenose
BARLOW, Thomas, bp. (1607–1691)	Papers	Queen's

BARRY, *Sir* Charles (1795–1860)   Architectural and archae-   Griffith Inst.,
                                    ological drawings   Ashmolean Mus-
                                                                eum

BATHURST, Ralph (1620–1704)   Papers   Trinity
BEAZLEY, *Sir* John Davidson   Notes on Greek vases   Ashmolean
  (1885–1970)   Museum
BELSHAM, Thomas (1750–1829)   Lecture notes   Manchester
BENSLEY, Edward   Working papers on English   English Faculty
                                    literature, *c.* 1903–07   Library
BETTS, John *and* Maria   Correspondence with B. B.   Wadham
                                    Wiffen (1794–1867)
BLACKSTONE, *Sir* William   Law lecture notes   All Souls
  (1728–1780)
BLAKISTON, Herbert E. D.   Correspondence [restricted Trinity
  (1862–1942)   access]
BLOOD, B. N. (d. 1948)   Entomological papers   Hope Dept. of
                                                                Entomology
BLOXAM, J. R. (1807–1891)   Papers; diary   Magdalen
BLUNDEN, Edmund Charles   (1) MS. poems   Queen's
  (1896–)   (2)   ,,   Worcester
BOBART, Jacob (1641–1719)   Botanical papers   Botany [deposited
                                                                in Bodleian]
BODLEY, John Edward Courtenay   Journal; memoir by wife   Balliol
  (1854–1925)
BONOMI, Joseph (1796–1878)   Diary, 1824–32 (Typescript Griffith Inst.,
                                    copy)   Ashmolean
                                                                Museum
BOURNE, W. R. P.   Ornithological notes,   Edward Grey Inst.
                                    1952–63
BOWRING, *Sir* John (1792–1872)   Papers and letters   Manchester
BRADLEY, Andrew Cecil   Papers   Balliol
  (1851–1935)
BRADLEY, Francis Herbert   Papers and letters   Merton
  (1846–1924)
BRAY, Thomas (1706–1785)   Letters   Exeter
BRICKNELL, William Simcox   Theological papers   Pusey House
  (1806–1888)
BRITTEN, Benjamin (1913–)   A few letters   Merton
BRODRICK, George C. (1831–1903)   Papers   Merton
BROOKE, Ralph (1553–1625)   Heraldic MSS.   Queen's
BROTHERHOOD OF THE HOLY   Minutes, 1844–1923   Pusey House
  TRINITY
BROWN, Henry Rowland   Entomological diaries,   Hope Dept. of
  (1865–1922)   1840–1908   Entomology
BROWNE, John   Papers, 1589–1613   University College

BROWNE, *Sir* Richard (1605–1683) — Correspondence with 1st Earl of Clarendon — Christ Church (Evelyn deposit)

BROWNING, Robert (1812–1889) — MSS. of later poems and papers about — Balliol

BUCKLAND, William (1784–1856) — Geological notes — Geology Dept.

BURCHELL, William John (1782–1863) — Entomological papers — Hope Dept. of Entomology

BURROWS, Christine M. E. (1872–1959) — Papers — St. Hilda's

BUTLER, Edward (d. 1745) — Correspondence, 1740–43 — Magdalen

BUXTON, *Sir* John Jacob, *bart.*, (d. 1842) — Pocket diary, 1809 — Christ Church

BUXTON, L. H. Dudley (1889–1939) — Papers — Museum of the History of Science [deposited]

BYRON, G. G. N. *6th baron* (1788–1824) — MS. of *Monk of Athos* — Balliol

CAPPER FAMILY — Papers, 1887–1943 — St. Hilda's

CARPENTER, Geoffrey Douglas Hale (1882–1953) — Entomological papers — Hope Dept. of Entomology

CARPENTER, Joseph Estlin (1844–1927) — Papers — Manchester

CARPENTER, Lant (1780–1840) — Papers and correspondence — Manchester

CARTER, Howard (1873–1939) — Archaeological papers — Griffith Inst., Ashmolean Museum

CAVENDISH, George (*c.* 1500– *c.* 1561) — Life of Wolsey — Christ Church

CERNY, Jaroslav (1898–1970) — Egyptological notes — Griffith Inst., Ashmolean Museum

CHERWELL, F. A. Lindemann, *1st viscount* (1886–1957) — Papers — Nuffield

CHESTER, Greville (1831–1892) — Oriental MSS. — Balliol

CHINNERY, George Robert — Correspondence, Jan. 1808– Nov. 1811 — Christ Church

CHRISTIAN SOCIAL UNION — Minutes, 1889–1920 — Pusey House

CHURCH, Richard William (1815–1890) — Theological papers — Pusey House

CHURCH UNION — Papers, nineteenth-twentieth centuries — Pusey House

CHURTON, Edward (1800–1874) — Theological papers — Pusey House

CHURTON, H. B. W. — Sermons, *c.* 1842–76 — Pusey House

CHURTON, William Ralph (1838–1897) — Theological papers — Pusey House

CLARENDON, Edward Hyde, *1st earl of* (1609–1674) — Correspondence with Sir Richard Browne — Christ Church (Evelyn deposit)

CLARKE, George (1661–1736) — MS. drawings — Worcester

CLAY, *Sir* Henry (1883–1954)	Papers	Nuffield
CLOUGH, Arthur Hugh (1819–1861)	Papers	Balliol
CLOVER, J. T. (*fl.* 1880)	Medical notebooks	Nuffield Dept. of Anaesthetics
COBBETT, William (1762–1835)	Papers	Nuffield
COLE, George Douglas Howard (1889–1959)	Papers	Nuffield
COLE, Sydney J. B. (b. 1870)	Autobiography	Balliol
COLLIN, James Edward (1876–1968)	Entomological papers	Hope Dept. of Entomology
COLLINS, William (1721–1759)	Poems	Trinity
COPELAND, William John (1804–1885)	Theological papers	Pusey House
CORNISH, Charles Lewis (1810–1870)	Correspondence	Pusey House
COWARD, Thomas Alfred (1867–1933)	Ornithological papers, field notes	Edward Grey Inst.
COYPEL, C. A. (1694–1752)	Typescript copies of 21 plays	Taylor Inst.
CRAWFORD, Osbert Guy Stanhope (1886–1957)	Archaeological photographs	Ashmolean Museum
CRAWLEY, Walter Cecil (*fl.* 1898–1930)	Entomological papers	Hope Dept. of Entomology
CRIPPS, *Sir* Stafford (1889–1952)	Papers	Nuffield [deposited]
CRUM, Walter Ewing (1865–1944)	Archaeological papers	Griffith Inst., Ashmolean Museum
CURTIS, John (1792–1862)	Notes and drawings	Hope Dept. of Entomology
CYNWAL, William (*d.* 1587–8)	Welsh poems to Salusbury family	Christ Church
DALE, Charles William (1852–1906)	Entomological notebooks and correspondence	Hope Dept. of Entomology
DALE, James Charles (1792–1872)	Entomological notebooks and correspondence	Hope Dept. of Entomology
DALGAIRNS, John Dobrée [Bernard] (1818–1876)	Letters	Keble
DANIEL FAMILY	Letters, including correspondence of C. H. O. Daniel, nineteenth to twentieth centuries	Worcester
DAUBENY, Charles Giles Bridle (1795–1867)	(1) Botanical papers, correspondence on library and rural economy	Botany [deposited in Bodleian]
	(2) Lectures on chemistry, etc., 1822–57	Magdalen

DAVIES, *Mrs.* Nina de Garis (d. 1965)	Archaeological papers	Griffith Inst., Ashmolean Museum
DAVIES, Norman de Garis (1864–1941)	Archaeological papers	Griffith Inst., Ashmolean Museum
DAWKINS, Richard McGillivray (1871–1955)	Modern Greek, etc., MSS. owned by him	Exeter [deposited Taylor Inst.]
DENNISTON, John Dewar (d. 1949)	Notebooks	Ashmolean Museum
DENNY, Henry (1803–1871)	Anoplura MSS.	Hope Dept. of Entomology
DE QUINCEY, Thomas (1785–1859)	Papers	Worcester
DEVAUD, Eugene (*fl.* 1915–1930)	Etymological papers	Griffith Inst., Ashmolean Museum
DEWAR, *Capt.* Alfred Charles, *R.N.* (d. *c.* 1968)	Naval history notes	New College
DICKENSON, Thomas	Shorthand notes of sermons, 1694–96	Manchester
DILLENIUS, Johann Jacob (1687–1747)	Botanical papers	Botany [deposited in Bodleian]
DIXEY, Frederick Augustus (1855–1935)	(1) Entomological correspondence	Hope Dept. of Entomology
	(2) Theological correspondence	Pusey House
	(3) Papers and correspondence about Roger Bacon and Greyfriars in Oxford, 1913	Greyfriars Priory
DIXON, W. D.	Papers on Federation of Catholic Priests, *c.* 1917–50	Pusey House
DODDRIDGE, Philip (1702–1751)	Lecture notes	Manchester
DODGSON, Charles Lutwidge (1832–1898)	S.C.R. papers	Christ Church
DODWELL, Henry (1641–1711)	Papers	St. Edmund Hall [deposited in Bodleian]
DODWELL, William (1709–1785)	Papers	St. Edmund Hall [deposited in Bodleian]
DONISTHORPE, Horace St. John Kelly (1870–1951)	Entomological papers	Hope Dept. of Entomology
DONOVAN, Edward (1768–1837)	Entomological papers	Hope Dept. of Entomology
DONOVAN COMMITTEE ON TRADE UNIONS AND EMPLOYERS' ASSOCIATIONS, 1965–68	Papers	Nuffield
DORSETT, Robert	11 letters to Sir Philip Sidney (1554–86)	Christ Church

DOWNING, *Sir* George (1623? – 1684) Letters to Sir William Temple 1664–67 All Souls

DOWSON, William (d. 1800) Commonplace book St. Edmund Hall [deposited in Bodleian]

DRAKE, Francis (1721–1795) Italian tour journal, 1755–56 Magdalen

DRUCE, George Claridge (1850–1932) Botanical papers Botany

DRURY, Dru (1725–1803) Entomological papers Hope Dept. of Entomology

DYSON, Humphrey (d. 1632) Six notebooks All Souls

EAST INDIA COMPANY Papers, 1619–85 All Souls

EDWARDS, Amelia Ann Blandford (1831–1892) (1) Letter and papers; Somerville

(2) Archaeological notebook Griffith Inst., Ashmolean Museum

ELIOT, George (1819–1880) Letters to, on death of G. H. Lewes, 1878 Balliol

ELIOT, Thomas Stearns (1888–1965) A few letters Merton

ELLRODT, Ferdinand von Extracts from *Augsburger Allgemeine Zeitung*, 1843–66 Taylor Inst.

ELTRINGHAM, Harry (1873–1941) Entomological papers Hope Dept. of Entomology

EMMOTT, Alfred, *1st baron* (1858–1926) Papers Nuffield

ENSOR, *Sir* Robert Charles Kirkwood (1877–1958) Letters and papers Corpus

EVANS, *Sir* John (1823–1908) Archaeological papers Ashmolean Museum

EVANS, Lewis (1755–1827) Papers Museum of the History of Science

EVANS, Mary Ann, *see* ELIOT, George

EVELYN, JOHN (1620–1706) (1) Papers, diaries, etc.; Christ Church [deposited]

(2) Almanacs (1636–37) Balliol

FABIAN SOCIETY Papers Nuffield

FEDERATION OF CATHOLIC PRIESTS Papers, *c.* 1917–50 Pusey House

FIEDLER, Hermann Georg (1862–1945) Correspondence and miscellaneous MSS. owned by him Taylor Inst.

FISHER, Henry (1685–1773) Sermons; verses Balliol

FISHER, Herbert Albert Laurens (1865–1940) (1) Personal and household accounts Inst. Econ. & Statistics

(2) Correspondence during General Strike, 1926 New College

FORBES, Nevill (1883–1929) Papers Taylor Inst.

FORD, James (1779–1850) Antiquarian papers Trinity

Fowke, *Sir* Frederick (1782–1856)	Cattle and sheep sale invoices 1831–33	Inst. Agric. Econ. [deposited in Bodleian]
Fowler, Thomas (1832–1904)	Papers and letters	Corpus
Fowler, William Warde (1847–1921)	Papers and diaries	Lincoln
Fox, Arthur William (1863–1945)	Papers, verse	Manchester
Fox, Charles James (1749–1806)	Nine letters	Hertford
Foxe, Richard (1448?–1528)	Correspondence with John Claimond, 1514–28	Corpus
Fraenkel, Eduard (1888–1963)	Papers and letters	Corpus
Friends in Council	Minutes, 1911–40	Pusey House
Gabb, George H. (d. 1948)	Papers	Museum of the History of Science
Gainford, J. A. Pease, *1st baron* (1860–1943)	Papers	Nuffield
Gaisford, Thomas (1779–1855)	Papers	Christ Church
Gardiner, *Sir* Alan Henderson (1879–1963)	Egyptological papers	Griffith Inst., Ashmolean Museum
Garrod, Heathcote William (1878–1960)	Papers	Merton
Geldart, William Martin (1870–1922)	Entomological papers	Hope Dept. of Entomology
Germany, *Auswärtiges Amt*	Archives, *c.* 1870–1920 (microfilm)	St. Antony's
Gibbon, Edward (1737–1794)	Accounts, 1749–94	Magdalen
Glover, Joseph (*c.* 1580)	Heraldic MS.	Queen's
Godolphin, *Sir* William (1634?–1696)	Spanish MSS. owned by him	Wadham
Goethe, J. W. von (1749–1832)	Letters	Taylor Inst.
Goffe, Edward Rivenhall (1887–1952)	Entomological papers	Hope Dept. of Entomology
Gorham, Henry Stephen	Entomological papers, *c.* 1870–80	Hope Dept. of Entomology
Grandorge, John (1669–1729)	Commonplace book, *c.* 1690	St. Edmund Hall [deposited in Bodleian]
Great Britain, *Foreign Office*	Italian diplomatic files, *c.* 1920–40 (photocopies)	St. Antony's
Green, John Richard (1837–1883)	Papers; transcripts of correspondence, 1859–72	Jesus
Green, Thomas Hill (1836–1882)	Papers	Balliol
Greenhill, William Alexander (1814–1894)	Correspondence	Trinity
Gregory, David (1661–1708)	Mathematical papers	Christ Church

GRESLEY, William (1801–1876) | Theological papers | Pusey House
GREY, Edward Grey, *1st viscount* (1862–1933) | Some ornithological letters | Edward Grey Inst.
GRIER, Mary Lynda Dorothea (1880–1967) | Papers | Lady Margaret Hall
GRIFFITH, Francis Llewellyn (1862–1934) | Archaeological papers | Griffith Inst., Ashmolean Museum
GRUB STREET JOURNAL | Minutes and accounts of proprietors, 1730–38 | Queen's
GUILLEBAUD COMMITTEE ON RAILWAY PAY | Papers, 1958–59 | Nuffield
GUNTHER, Robert T. (1869–1940) | (1) Scientific papers (2) History of science papers | Magdalen Museum of the History of Science
HADOW, Grace Eleanor (1875–1940) | Papers on education | Worcester
HADOW, *Sir* William Henry (1859–1937) | Diaries and correspondence, 1874–1900; MS. music | Worcester
HAINES, A. G. (*c.* 1885–*c.* 1960) | Notes on J. H. Newman | Pusey House
HALCOMB, Thomas R. (1799–1840) | Papers | Lincoln
HALDANE, *Mrs.* Louisa Kathleen, *née* Trotter (1863–1961) | Literary MSS. | Lady Margaret Hall
HALL, George William (1770–1843) | Correspondence | Pembroke
HALLAM, Henry (1777–1859) | Papers | Christ Church
HAMILTON, Walter Ker, *bp.*, (1808–1869) | Theological papers | Pusey House
HANITSCH, Robert (1860–1940) | Entomological papers | Hope Dept. of Entomology
HARDEN, John Mason, *bp.* (d. 1931) | Ethiopic notes | Griffith Inst., Ashmolean Museum
HARDING FAMILY, *of Birmingham* | Correspondence, late nineteenth century | Taylor Inst.
HARDWICKE, Philip Yorke, *3rd earl of* (1757–1834) | 50 letters to | Inst. Agric. Econ. [deposited in Bodleian]
HARRIS, Charles, *of Oxford* | Account book, 1686–1705 | Queen's
HARRIS, John (*c.* 1666–1719) | MS. autobiography | Trinity
HARRISON, Thomas | Mathematical and geometrical notes, *c.* 1690 | Magdalen
HARWOOD, Philip (1882–1957) | Entomological papers | Hope Dept. of Entomology
HAVERFIELD, Francis John (1860–1919) | Archaeological papers | Ashmolean Museum

HAWKER, Edward James	Drawings, etc.	Griffith Inst., Ashmolean Museum
HAWKINS, Edward (1789–1882)	Correspondence	Oriel
HEATH, Edward	Correspondence with Thomas Tullie, 1657–8	St. Edmund Hall [deposited in Bodleian]
HEBER, Reginald, *bp.* (1783–1826)	Papers	Brasenose
HEINEKIN, Thomas (1763–1840)	Notes taken at lectures	Manchester
HEMMING, Arthur Francis (1893–1964)	Papers, mainly concerned with the International Committee for Non-Intervention in Spain, 1936–39	Corpus
HENRY, Philip (1631–1696)	Sermons	Manchester
HESS, Jean Jacques (*fl.* 1890–1938)	Demotic notes	Griffith Inst., Ashmolean Museum
HIGGS, Griffin (1589–1659)	Diaries, 1637, 1638	Merton
HILL, John (1787–1855)	Diaries, 1803–5, 1820–55	St. Edmund Hall [deposited in Bodleian]
HINCKS, Edward (1792–1866)	Correspondence	Griffith Inst., Ashmolean Museum
HINGSTON, R. W. G.	Entomological papers, Greenland, 1928	Hope Dept. of Entomology
HOARE, Joseph (d. 1802)	Commonplace book	Jesus
HOLLICK, A. T.	Paintings, 1867–70	Hope Dept. of Entomology
HOPE, Frederick William (1797–1862)	Entomological papers	Hope Dept. of Entomology
HOPKINS, Gerard Manley (1844–1889)	Papers	Campion Hall
HORNEMAN, Frederick	African journal, 1797–98,	Inst. of Social Anthropology
HORNSBY, Thomas (1733–1810)	Astronomical papers	Corpus
HOSKINS, George Alexander (*fl.* 1835–1863)	Drawings, etc.	Griffith Inst., Ashmolean Museum
HOUSMAN, Alfred Edward (1859–1936)	Few letters	St. John's [loan]
HOWARD, Henry Eliot (b. 1873)	Ornithological notebooks, 1918–31	Edward Grey Inst.
HULTON, William	Papers and letters, Venice, *c.* 1880–1910	Ashmolean Museum
HUTTON, William Holden (1860–1930)	Papers and letters	St. John's
HYDE, *see* CLARENDON		
IMAGE, Selwyn (1849–1930)	Entomological notebooks	Hope Dept. of Entomology

INDUSTRIAL CHRISTIAN FELLOWSHIP	Minutes, 1920–23	Pusey House
INGRAM, James (1774–1850)	Correspondence	Trinity
IONIAN BANK Ltd.,	Papers	St. Antony's
ITALY, *see* GREAT BRITAIN, *Foreign Office*		
JEDDERE-FISHER, *Mrs.* J.	Travel diaries, 1837, 1844	Balfour Library, Pitt Rivers Museum
JEFFERY, R. W. (1877–1956)	Letters and MSS. [not available until A.D. 2006]	Brasenose
JENKINS, Claude (1877–1959)	Papers	Christ Church
JENKINS, *Sir* Leoline (1623–1685)	(1) Papers (2) Latin letters to	All Souls Jesus
JENKYNS, Richard (1782–1854)	Papers	Balliol
JERVIS-SMITH, Frederick (1848–1911)	Papers	Museum of the History of Science
JOHNSON, Samuel (1709–1784)	Papers and letters relating to	Pembroke
JONES, Inigo (1573–1652)	Drawings	Worcester
JONES, *Sir* William (1746–1794)	Oriental MSS.	University College
JONES, William, *of Chelsea* (d. 1818)	Entomological notes and drawings	Hope Dept. of Entomology
JOSEPH, Horace William Brindley (1867–1943)	Travel diaries, 1895–1902	New College
JOURDAIN, Francis Charles Robert (1865–1940)	(1) Entomological papers;	Hope Dept. of Entomology
	(2) Ornithological papers, 1876–1937	Edward Grey Inst.
JOWETT, Benjamin (1817–1893)	Papers	Balliol
KEBLE, John (1792–1866)	(1) Papers and letters; (2) Letters	Keble Oriel
KENRICK, John (1788–1877)	Papers and letters	Manchester
KILNER, Thomas Pomfret (1890–1964)	Professional papers and correspondence	Kilner Library of Plastic Surgery
KINGSLEY, Charles (1819–1875)	*c.* 20 letters to J. M. F. Ludlow, 1850–51	Inst. Agric. Econ. [deposited in Bodleian]
KNOWLES, *Sir* Francis H. S., *bart.* (1886–1953)	Ethnological papers	Balfour Library, Pitt Rivers Museum
KODALY, Zoltán (1882–1967)	1 score, 4 letters	Merton
LAMBOURNE, William Alfred Stedwell (1877–1959)	Entomological papers	Hope Dept. of Entomology
LAMPLUGH, Thomas, *abp.* (1615–1691)	Papers	Queen's
LANE, Edward William (1801–1876)	Archaeological papers	Griffith Inst., Ashmolean Museum

LANGDON, Stephen Herbert (1876–1937) — Sumerian notes — Griffith Inst., Ashmolean Museum

LEE, Vernon, *pseud.*, *see* PAGET, Violet

LEIGH, Theophilus (1694–1785) — Letters to 1st Duchess of Chandos — Balliol

LEPSIUS, Karl Richard (1810–1884) — Diary about Thebes — Griffith Inst., Ashmolean Museum [loan]

LIDDON, Henry Parry (1829–1890) — (1) Papers and letters; (2) ,, ,, ; (3) ,, ,, — Keble; St. Edmund Hall; Pusey House

LINANT DE BELLE FONDS, Louis Maurice Adolphe — Diary in Egypt, 1821–22 — Griffith Inst., Ashmolean Museum [loan]

LINDEMANN, *see* CHERWELL

LINDSEY, Theophilus (1723–1808) — Sermons — Manchester

LINGARD, John (1771–1851) — Papers — Manchester

LITTLE, Andrew George (1863–1945) — Correspondence and notes about Franciscans in Britain — Greyfriars Priory

LITTLE, William (1848–1922) — Letters to *c.* 1867–1871 — Corpus

LONDON, *Glasshouse Yard Baptist Church* — Minute book, 1680–1740 — Regent's Park

LONDON SOCIETY OF ENTOMOLOGISTS — Minute book, 1780–82 — Hope Dept. of Entomology

LONGSTAFF, George Blundell (1849–1921) — Entomological papers and diaries, 1895–1915 — Hope Dept. of Entomology

LOSH, William — Exercises, 1730 — St. Edmund Hall [deposited in Bodleian]

LOVEDAY, Alexander (1888–1962) — Papers — Nuffield

LUCAS, William John (1858–1932) — Entomological drawings — Hope Dept. of Entomology

LUTTRELL, Narcissus (1657–1732) — Papers — All Souls

LYNES, Hubert — Ornithological diaries, 1928–30 — Edward Grey Inst.

MCINDOE, *Sir* Archibald Hector (1900–1960) — Professional papers and correspondence — Kilner Library of Plastic Surgery

MACKAIL, John William (1859–1945) — Two notebooks — Balliol

MCLACHLAN, Robert (1837–1904) — Entomological correspondence — Hope Dept. of Entomology

MADRAS PRESIDENCY — Papers, *c.* 1767–1807 — All Souls

MAGRATH, John Richard (1839–1930) — Genealogical and topographical MSS. — Queen's

MAKERETI, *Chieftainess of the Arawa Tribe, New Zealand*	Ethnological papers	Balfour Library, Pitt Rivers
MALCHAIR, Johan Baptist [John] (*c.* 1727–1812)	Sketch books	Corpus
MALCOLM, *Sir* Neill (1869–1953)	Papers on Rhineland occupation, *c.* 1919–29	St. Antony's
MALMESBURY, James Harris, *1st earl of* (1746–1820)	Papers	Merton
MARCHANT, S.	Ornithological notebooks, 1938–62	Edward Grey Inst.
MARETT, R. R. (1866–1943)	Ethnological papers	Balfour Library, Pitt Rivers Museum
MARGOLIOUTH, David Samuel (1858–1940)	Papers and diaries, 1916–35	Griffith Inst., Ashmolean Museum
MARRIOTT, Charles (1811–1858)	Correspondence	Pusey House
MARSHALL, Thomas (1621–1685)	Papers	Lincoln
MARSHALL, Thomas Ansell (1827–1906)	Poemata, 1845	New College
MARTINEAU, Harriet (1802–1876)	Papers	Manchester
MARTINEAU, James (1805–1900)	Papers	Manchester
MATHESON, Percy Ewing (1859–1946)	Memories (written 1929); and travel diaries, 1887–88	New College
MATTHEWS, Arnold Gwynne (1881–1962)	Papers	Mansfield
MAX MUELLER, Friedrich (1823–1900)	Papers	Taylor Inst.
METCALFE, Theophilus (*fl.* 1757)	Verses	Queen's
MEYERSTEIN, E. H. W. (1889–1952)	MSS, typescripts, proofs and letters	English Faculty Library [deposited]
MIERS, John (1789–1879)	Entomological papers	Hope Dept. of Entomology
MILL, John (1645–1707)	Theological papers	St. Edmund Hall [deposited in Bodleian]
MILNER, Alfred, *1st viscount* (1854–1925)	Papers and letters	New College [deposited in Bodleian]
MILNER, Maud M. (*fl.* 1870–1920)	Papers and letters	Pusey House
MITCHINSON, John, *bp.* (1833–1918)	Papers	Pembroke
MOND, *Sir* Robert Ludwig (1867–1938)	Archaeological papers	Griffith Inst., Ashmolean Museum

MOORE, Edward (1835–1916)	Papers	St. Edmund Hall [deposited in Bodleian]
MOREAU, Reginald Ernest	Ornithological notes, correspondence, 1954–66	Edward Grey Inst.
MORICE, Francis David (1849–1926)	Entomological papers	Hope Dept. of Entomology
MORIER, *Sir* Robert Burnett (1826–1893)	Papers	Balliol
MORISON, Robert (1620–1683)	Botanical papers	Botany [deposited in Bodleian]
MORRIS, Reginald Owen (1886–1948)	Musical scores	New College
MORRISON, Herbert Morrison, *1st baron* (1888–1965)	Papers	Nuffield
MOTTISTONE, J. E. B. Seely, *1st baron* (1868–1947)	Papers	Nuffield
MUNRO, Alexander (1697–1767)	Lecture notes	Manchester
MURE, Geoffrey R. G. (1893–)	Papers	Merton
NAPIER, A. S. (1853–1916)	Papers and letters on his studies	English Faculty Library
NARES, Edward (1762–1841)	Autobiography	Merton
NEATE, Charles (1806–1879)	Notebook on Roman history, 1828	Oriel
NEWCOME, Peter (*fl.* 1703–1738)	Papers	St. Edmund Hall [deposited in Bodleian]
NEWMAN, John Henry, *cardinal*, (1801–1890)	(1) Correspondence; (2) A few letters	Oriel Trinity
NEWMAN, William (1723–1855)	Diary	Regent's Park
NEWTON, Alfred (1829–1907)	28 letters to Sir Philip Manson Bahr (1903–07)	Edward Grey Inst.
NEWTON, *Sir* Isaac (1642–1727)	Chronological and theological papers	New College
NICOLSON, *Sir* Harold George (1886–1968)	Diaries [typescript; restricted access]	Balliol
NOWELL, Alexander (1507?–1602)	Papers	Brasenose
NUFFIELD SOCIAL RECONSTRUCTION SURVEY, 1940–44	Papers	Nuffield
NUFFIELD TRUST FOR SPECIAL AREAS, *c.* 1935–39	Papers	Nuffield
OAKELEY, Frederick (1802–80)	Papers	Balliol
ODGERS, James Edwin (1843–1925)	Papers	Manchester
ODGERS, John Collins (d. 1928)	Historical notebooks	Manchester

OLDHAM, Charles	Ornithological notebooks, 1883–1939	Edward Grey Inst.
OLLARD, Sidney Leslie (1875–1949)	Theological papers	Pusey House
O'MAHONY, Eugene (1899–1951)	Entomological notes	Hope Dept. of Entomology
ONSLOW, R. W. A. Onslow, 5th earl of (1876–1945)	Letters to Sir Dougal Malcolm [restricted]	All Souls
ORIELTON DECOY BOOKS	1877–1919, 1935–46	Edward Grey Inst.
OWEN, J. H.	Ornithological notes, 1944–50	Edward Grey Inst.
OWEN, Sidney George (1858–1940)	Latin lecture notes	Christ Church
OXFORD DANTE SOCIETY	Abstracts of papers, 1876–98	Taylor Inst.
OXFORD MOVEMENT (Newman, Keble, Hawkins, Hampden, etc.	(1) Correspondence; (2) ,,	Oriel Pusey House
OXFORD SOCIETY FOR REUNION	Papers 1924–1927	Pusey House
PAGET, Violet [Vernon Lee, pseud.] (1856–1935)	Papers [not to be consulted before 1985]	Somerville
PALMER, Thomas Fyshe (1747–1802)	Letters	Manchester
PASCOE, Francis Polkinghorne (1813–1893)	Entomological papers	Hope Dept. of Entomology
PEAKE, Edward	Ornithological diaries and notebooks, 1894–1941	Edward Grey Inst.
PEASE, see GAINFORD		
PEEL, Sir Robert (1822–1895)	51 letters to Harding family	Taylor Inst.
PERCY, Thomas, bp. (1729–1811)	4 letters to	Inst. Agric. Econ. [deposited in Bodleian]
PEREZ, Antonio (1539–1611)	2 MSS.	Taylor Inst.
PEREZ, Juan	Lineage de España [c. 1750]	Taylor Inst.
PETRIE, Sir W. M. Flinders (1853–1942)	Journals and letters, 1880–1933	Griffith Inst., Ashmolean Museum
PEYTON FAMILY, of Birmingham	Correspondence, late nineteenth century	Taylor Inst.
PHELPS, Lancelot Ridley (1853–1936)	Correspondence	Oriel
PICKARD-CAMBRIDGE, Sir Arthur William (1873–1952)	Entomological diaries, c. 1888–1910	Hope Dept. of Entomology
PICKARD-CAMBRIDGE, Octavius (1828–1917)	Entomological papers	Hope Dept. of Etnomology
PISSARRO, Camille (1830–1903)	Correspondence	Ashmolean Museum (Print Room)
PISSARRO, Lucien (1863–1944)	Correspondence	Ashmolean Museum (Print Room)

Pogson-Smith, W. G.	Entomological papers, c. 1896–1906	Hope Dept. of Entomology
Pointer, John (1668–1754)	Misc. MSS. owned by him	St. John's
Poulton, Sir Edward Bagnall (1856–1943)	Entomological papers	Hope Dept. of Entomology
Powell, S. P.	Ethnological papers	Balfour Library, Pitt Rivers Museum
Priestley, Joseph (1733–1804)	Notebooks and letters	Manchester
Pudsey, George	Bills, accounts, memoranda, c. 1680–90	Magdalen
Pusey, Edward Bouverie (1800–1882)	Papers and correspondence	Pusey House
Pusey, Philip (1799–1855)	35 letters to	Inst. Agric Econ. [deposited in Bodleian]
Pusey, Philip Edward (1830–1880)	Theological papers	Pusey House
Radcliffe Trustees	Papers, nineteenth-century	Museum of the History of Science
Ramanujan, Srinivasa	Notebooks. Transcribed by G. N. Watson, 1928	Mathematics Inst.
Ramsay, Sir William Mitchell (1851–1939)	Notebooks	St. John's [deposited in Ashmolean Museum]
Randall, Richard William (1824–1906)	Diaries, 1851–61	Pusey House
Rashdall, Hastings (1858–1924)	(1) Correspondence with mother and family, 1866–1923;	New College
	(2) Sermon notes [ascribed to Rashdall]	Pusey House
Rawlinson, Richard (1690–1755)	Papers	St. John's
Reeve, Henry (1813–1895)	Correspondence with A. de Tocqueville, 1835–59	All Souls
Renouf, Sir Peter Le Page (1822–1897)	Correspondence	Pembroke
Richmond, Sir Ian (1902–1965)	Archaeological papers	Ashmolean Museum
Rigaud, Stephen Peter (1774–1839)	Papers	Museum of the History of Science
Rilke, Rainer Maria (1875–1926)	Six letters	Taylor Inst.

RIPPON, John (1751–1836)	Papers	Regent's Park
ROBINSON, Alfred (1841–1895)	Papers	New College
ROBLEY, Horatio Gordon (c. 1865–1900)	Ethnological papers	Balfour Library, Pitt Rivers Museum
ROLLESTON, George (1829–1881)	(1) Scientific papers;	Zoology Dept.
	(2) Archaeological papers	Zoology Dept. [deposited in Ashmolean Museum]
Ross, Sir John (1777–1856)	Papers	Museum of the History of Science
ROSSETTI, Dante Gabriel (1828–1882)	Letters	Worcester
ROSSETTI, Gabriele (1733–1854)	Copies of 128 letters to Charles Lyell, 1827–40	Taylor Inst.
ROTHNEY, G. A. J. (1849–1922)	Entomological papers	Hope Dept of Entomology
ROUTH, Martin (1755–1854)	Papers; letters to	Magdalen
RUDLER, Gustave (1872–1957)	Papers	Taylor Inst.
RUSKIN, John (1819–1900)	18 letters	Balliol
RUSSELL, John Fuller (1814–1884)	Papers and correspondence	Pusey House
ST. GEORGE, Sir Henry (1625–1715)	Heraldic MSS.	All Souls
ST. GEORGE, Sir Richard (d. 1635)	Heraldic MSS.	Queen's
SAINTSBURY, George (1845–1933)	Papers	Merton
SALISBURY, Robert Arthur Talbot Gascoyne-Cecil, 3rd marquess of (1830–1903)	Papers	Christ Church [deposited]
SAUNDERS, Sir Sidney Smith (1809–1884)	Entomological papers	Hope Dept. of Entomology
SAUNDERS, William Wilson (1809–1879)	Entomological papers	Hope Dept. of Entomology
SAYCE, Archibald Henry (1845–1933)	Archaeological papers	Griffith Inst., Ashmolean Museum
SCOTT, Charles Prestwich (1846–1932)	Correspondence	Balliol
SCOTT, Robert (1811–1887)	(1) Correspondence;	Pusey House
	(2) ,,	Balliol
SECKER, Thomas, abp. (1692–1768)	Account book, 1746–58	St. Edmund Hall [deposited in Bodleian]
SEDDON, John (1725–1770)	Correspondence	Manchester

SELOUS, Edmund (1858–1934)	Ornithological notebooks and papers	Edward Grey Inst.
SEWELL, James Edwards (1810–1903)	Papers	New College
SEWELL, William (1804–1874)	Journal, 1844–45	Pusey House
SHELDON, Ralph (1623–1684)	MSS. owned by him	Jesus
SHELLEY, Percy Bysshe (1792–1822)	One letter	Balliol
SHEPHERD, William (1768–1847)	Papers	Manchester
SHERARD, William (1659–1728)	Botanical papers	Botany [deposited in Bodleian]
SHIRLEY, *Sir* Thomas (*c.* 1590–1654)	Heraldic MSS.	Queen's
SIBTHORP, John (1758–1796)	Botanical papers	Botany [deposited in Bodleian]
SIDGWICK, Arthur (1840–1920)	Entomological papers and diaries	Hope Dept. of Entomology
SIDGWICK, Nevil Vincent (1873–1952)	Papers and diaries	Lincoln
SIMPSON, Percy (1865–1962)	Papers and letters	English Faculty Library
SIXESMITH, Thomas (d. 1650)	Commonplace book	Brasenose
SKEAT, Walter William (1866–1953)	(1) Ethnological papers;	Balfour Library, Pitt Rivers Museum
	(2) ,, ,,	Tylor Library, Inst. of Social Anthropology
SMITH, A. F.	Ornithological notebooks, 1922–41	Edward Grey Inst.
SMITH, Arthur Lionel (1850–1924)	Papers	Balliol
SMITH, John Alexander (1864–1939)	Papers	Balliol
SMITH, Joseph (1670–1756)	Genealogical and topographical MSS.	Queen's
SMITH, Sydney (1771–1845)	Letters	New College
SMITH, William (1711–1787)	Commonplace book, 1729	New College
SMITH, William (1769–1839)	Geological papers, maps, drawings	Dept. of Geology
SMYTH, John (1744–1809)	Diaries of travels, 1793–*c.* 1802	Pembroke
SOMERVILLE, *Mrs.* Mary (1780–1872)	Letters	Somerville [deposited in Bodleian]

SPENCER, *Sir* Baldwin (1860–1929)	Ethnological papers	Balfour Library, Pitt Rivers Museum
SPOONER, William Archibald (1844–1930)	Correspondence; papers	New College
STALLYBRASS, W. T. S. (1883–1948)	Notebooks	Brasenose
STANLEY, Arthur Penrhyn (1815–1881)	Letters to B. Jowett; papers	Balliol
STAPLETON, H. E. (d. *c.* 1948)	Papers, mainly on alchemy	Museum of the History of Science
STEVENSON, William Henry (1858–1924)	Papers	St. John's
STONE, Darwell (1859–1941)	Papers, diaries, correspondence	Pusey House
STRACHAN-DAVIDSON, James Leigh (1843–1916)	Papers	Balliol
STREETER, Burnett Hillman (1874–1937)	Theological notes	Queen's
STUBBS, Philip (1665–1738)	MSS. and printed books owned and annotated by him	Wadham
SWINBURNE, Algernon Charles (1837–1909)	(1) MSS.; (2) ,,	Balliol Worcester
SWINTON, John (1703–1777)	Diary on Mediterranean tour, 1730–31.	Wadham
SWYNNERTON, Charles Francis Massey (1877–1938)	Entomological papers	Hope Dept. of Entomology
TAYLER, John James (1797–1869)	Papers	Manchester
TAYLOR, John (1750–1826)	Papers	Manchester
TAYLOR, Margaret Venables (d. 1963)	Archaeological papers	Ashmolean
TEMPLE, *Sir* William (1628–1699)	Letters from Sir George Downing, 1664–67	All Souls
THOMAS, Edward (1878–1917)	MSS. of *Roads* (poem); and *Oxford* (essays)	Lincoln
THOMPSON, Gladys Scott (d. 1966)	Papers	Somerville
THORN-DRURY, George (1860–1931)	Papers	Worcester
TOCQUEVILLE, Alexis de (1805–1859)	Correspondence with Henry Reeve, 1835–59	All Souls
TODD, Hugh (1658?–1728)	Papers on history of Diocese of Carlisle	St. Edmund Hall [deposited in Bodleian]
TRIMEN, Roland (1840–1916)	Entomological papers and correspondence	Hope Dept. of Entomology

TRIPPETT, Charles (d. 1707)	Notes on theology	Magdalen
TULLY, H.	Ornithological notes and diaries, 1932–51	Edward Grey Inst.
TURNER, Cuthbert Hamilton (1860–1930)	Theological papers	Pusey House
TURNER, Daniel (1710–1798)	Papers	Regent's Park
TYLOR, Sir Edward Burnett (1832–1917)	Ethnological papers	Balfour Library, Pitt Rivers Museum
URQUHART, David (1805–1877)	Papers	Balliol
VALDES, Juan de (c. 1500–1541)	Documents concerned with	Wadham
VAUGHAN, Sir Charles Richard (1774–1849)	Papers	All Souls
VOYSEY, Charles (1828–1912)	28 letters to Thomas Allsop, 1869–71	Manchester
WADE, Nugent (1809–1893)	Correspondence	Pusey House
WAKE, William, abp. (1657–1737)	Papers	Christ Church
WALKER, Ernest (1870–1949)	Diary 1888–94; papers	Balliol
WALKER, George (1734?–1807)	Lecture notes	Manchester
WALKER, John James (1851–1939)	Coleoptera notebooks	Hope Dept. of Entomology
WALLACE, Alfred Russell (1823–1913)	Correspondence	Hope Dept., of Entomology
WARD, Mrs. Mary Augusta [Mrs. Humphry Ward] (1851–1920)	Some correspondence	Pusey House
WARNER, Richard (1713?–1775)	Botanical annotations	Wadham
WARRINGTON ACADEMY	Trustees' minutes, 1757–86	Manchester
WARTON, Thomas (1728–1790)	Papers	Trinity
WASE, Christopher (1625?–1690)	Papers	Corpus
WATERS, Edwin George Ross (1890–1930)	Entomological notes and diaries, 1909–28	Hope Dept. of Entomology
WEAVER, Harriet (1876–1961)	On Time	Philosophy Library
WELLBELOVED, Charles (1769–1858)	Papers	Manchester
WESLEY, John (1703–1791)	Letters	Lincoln
WESSELING, Peter (1692–1764)	Classical notes	Hertford
WESTWOOD, John Obadiah (1805–1893)	Entomological papers and correspondence	Hope Dept. of Entomology
WHITE, Gilbert (1720–1793)	Account book, 1793	Oriel
WHITE, Henry Julian (1859–1934)	Papers on Biblical texts	Christ Church
WHITE, Joseph Blanco (1775–1841)	Papers	Manchester

WHITLEY, William Thomas (1861–1947)	Papers and letters	Regent's Park
WICKSTEED, Charles (d. 1885)	Papers	Manchester
WIFFEN, Benjamin Barron (1794–1867)	Letters and papers on church reform in Spain	Wadham
WILBERFORCE, Henry William (1807–1873)	Theological notebook, 1831	Oriel
WILD, James William	Archaeological papers	Griffith Inst., Ashmolean Museum
WILKES, Arthur Hamilton Paget	Field notes on birds of Kenya and Uganda, c. 1935	Edward Grey Inst.
WILKINSON, Henry (1610–1675)	Theological notes and sermons	Hertford
WILKINSON, Sir John Gardner (1797–1875)	Archaeological papers	Griffith Inst., Ashmolean Museum [loan]
WILLIAMS, George (1762–1834)	Botanical papers	Botany [deposited in Bodleian]
WILLIAMS, Henry Williamson	Latin dictionary, 1853	New College
WILLIAMSON, Sir Joseph (1633–1701)	Papers	Queen's
WITHERBY, Harry Forbes	Ornithological letters, maps and notes on Spain [c. 1930–5?]	Edward Grey Inst.
WODHULL, Michael (1740–1816)	Papers	Brasenose
WOLFE, Humbert (1886–1940)	Literary MSS.	Wadham
WOODGATE, Henry Arthur (1800–1874)	Papers and correspondence, 1835–70	Pusey House
WOODTHORPE, Robert Gossett (1844–1898)	Far Eastern diaries	Balfour Library, Pitt Rivers Museum
WREN, Sir Christopher (1632–1723)	Architectural plans	All Souls
WRIGHT, Joseph (1855–1930)	Papers	Taylor Inst.
WRIGHT, Richard (1764–1836)	Papers	Manchester
WRIGHT, Thomas (fl. 1800)	Notes on Flaxton, Yorks.	Inst. Agric. Econ. [deposited in Bodleian]
WYNNE, Owen (fl. 1670)	Papers	All Souls
YERBURY, John William (1804–1858)	Entomological papers	Hope Dept. of Entomology
YOUNG, Arthur (1741–1820)	Letter to Sidney, his printer, 1814	Inst. Agric. Econ. [deposited in Bodleian]
ZUNTZ, Leonie	Hittite notes	Griffith Inst., Ashmolean Museum

INDEX

Institutions outside Oxford have been entred under place-names; those in Oxford under their own names. Subject entries have not been attempted for the manuscript collections listed in Appendix II, but all proper names have been included.

229